AF248920

Applied Mathematical Sciences
Volume 140

Editors
J.E. Marsden L. Sirovich

Advisors
S. Antman J.K. Hale
P. Holmes T. Kambe J. Keller
B.J. Matkowsky C.S. Peskin

Springer
New York
Berlin
Heidelberg
Barcelona
Hong Kong
London
Milan
Paris
Singapore
Tokyo

Applied Mathematical Sciences

(continued following index)

Andrej Cherkaev

Variational Methods for Structural Optimization

With 94 Figures

Springer

Andrej Cherkaev
Department of Mathematics
The University of Utah
Salt Lake City, UT 84112
USA
cherk@math.utah.edu

Editors

J.E. Marsden
Control and Dynamical Systems, 107-81
California Institute of Technology
Pasadena, CA 91125
USA

L. Sirovich
Division of Applied Mathematics
Brown University
Providence, RI 02912
USA

Mathematics Subject Classification (1991): 73-02, 73Kxx, 35B27, 49-xx

Library of Congress Cataloging-in-Publication Data
Cherkaev, Andrej, 1950–
    Variational methods for structural optimization / Andrej Cherkaev.
      p. cm. — (Applied mathematical sciences ; 140)
    Includes bibliographical references and index.
    ISBN 0-387-98462-3 (hardcover : alk. paper)
    1. Structural optimization.   2. Calculus of variations. I. Title. II. Applied mathematical
sciences (Springer-Verlag New York, Inc.) ; v. 140.
    QA1 .A647 vol. 140
    [TA658.8]
    510 s—dc21
    [624.1′7713]   99-052755

Printed on acid-free paper.

Production managed by Lesley Poliner; manufacturing supervised by Erica Bressler.
Camera-ready copy prepared from the author's LaTeX files.
Printed and bound by Edwards Brothers, Inc., Ann Arbor, MI.
Printed in the United States of America.

9 8 7 6 5 4 3 2 1

ISBN 0-387-98462-3 Springer-Verlag New York Berlin Heidelberg SPIN 10659110

Contents

List of Figures

Preface

Optimal Design, Structures, and Composites

This book discusses problems of structural optimization. The problem is to lay out several materials throughout a given domain to maximize or minimize an integral functional associated with the conductive or elastic state of an assembled medium. We assumed that several materials are available, and one is asked to arrange them on the volume of the body of a given shape. It turns out that the materials in the optimal body are mixed on an infinitely fine scale: The finer the scale, the better the construction. From an engineering point of view, optimization problems require the use of composites of given materials rather than materials singly.

As a rule, an optimal design is made of composites. Physically speaking, we use composites in designs because we prefer materials with properties that are not immediately available but can be obtained by mixing available materials; such a mixture can be more suitable than any of the individual ingredients. For example, composites assembled of isotropic materials can be anisotropic. Moreover, they can possess such exotic features as a negative thermal expansion coefficient, or a negative Poisson ratio. These and similar unusual features could be useful for solving optimization problems.

Optimal composites correspond to rapidly oscillating state variables, such as stresses and strains in elasticity or currents and fields in conductivity. The oscillation of optimal solutions is well understood in the theory of one-dimensional control problems. In some problems, the solution has to zigzag to satisfy the optimality requirement. The functional decreases as the zigzags become more finely scaled. It is not surprising that such generalized

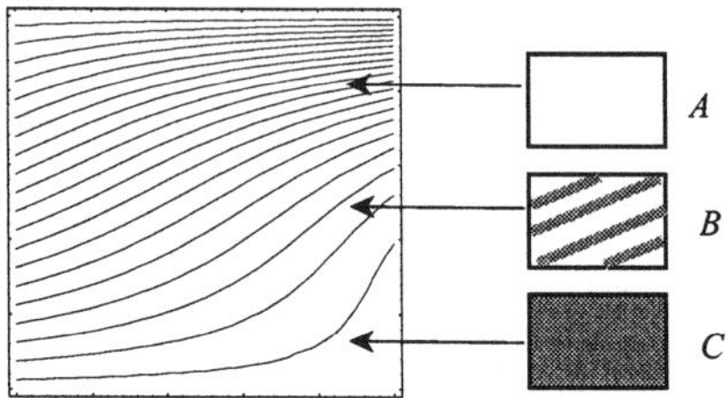

FIGURE P.1. The scheme of a composite structure that transforms the homogeneous boundary potential into an inhomogeneous boundary current. The horizontal sides are insulated, and the potential on each vertical side is constant. The current lines are shown. The inhomogeneity of the current is caused by the inhomogeneity of the material layout. The good conductor A attracts the current, the bad conductor C pushes the current away, and the anisotropic composite B turns the current in a desired direction.

controls also appear in the multidimensional problems of optimal layout of materials; here they correspond to microstructures of composites. Investigation of multidimensional optimization problems requires determination of the geometry of optimal composite structures. The one-dimensional analogue of the problem of the best microstructure is relatively simple because the only way to form a mixture in one dimension is to alternate materials along the line.

Example P.1 Let us consider the problem of an optimal inhomogeneous conducting structure that transforms the given boundary potentials to the desired boundary currents, as shown in Figure P.1P. Suppose that one has a set of materials of different isotropic conductivity and the layout of materials in the designed domain must be optimized. Clearly, one can control the boundary currents by varying the materials' layout, because the variation in conductivity forces the current out of regions of low conductivity and attracts it into regions of high conductivity. Careful consideration shows an additional mechanism of control through the use of anisotropic materials. The current is controlled and sent in the desired direction by refraction in an anisotropic composite. The last mechanism is specific to multidimensional problems and has no one-dimensional analogues. It shows the usefulness of the anisotropic composite media assembled of initially isotropic materials.

The use of anisotropy to control a process in a medium is well known. Observe a skier on a slope. The skier can control the direction of his motion because the resistance to sliding along the ski is much less than the resistance to sliding in the orthogonal direction. This mechanism allows the skier to traverse across the slope and make turns. Anisotropy is also used to steer a sailboat in a direction different from the direction of the wind. When a current of passive particles moves in a medium due to an applied

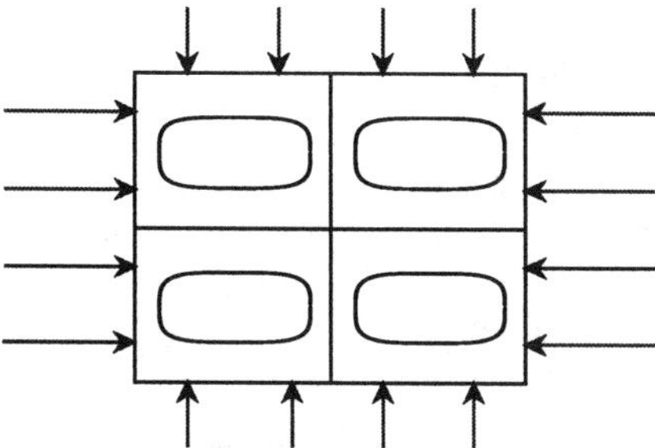

FIGURE P.2. The scheme of an elastic structure with cavities showing the maximum stiffness under a given loading. The intensity of the loading is anisotropic, and so is the corresponding optimal structure. Greater intensity of the loading corresponds to the direction of greater stiffness.

force field, the anisotropy of the medium plays a similar control role: It generates a current in a direction different from the direction of the force.

In the optimization of elastic designs, we also find intuitive reasons for using an anisotropic composite rather than isotropic materials.

Example P.2 Let us consider an elastic material that shows maximal stiffness under some anisotropic external loading; see Figure P.2P. One would assume that the larger the stress, the more stiffness is needed to resist. Therefore, we anticipate that the structure tends to be stiffer in the direction of a larger stress, even at the expense of being weakened in the direction of a smaller stress. Hence, we expect that anisotropic composites with controllable degrees of anisotropy are more suitable than isotropic materials for maximization of the stiffness.

Generally, structural optimization determines the structure, that is best adapted to the object of the design and the loading conditions. The adaptation implies uniform exploitation of the material. For example, the stiffness optimization is achieved by a structure that evenly stresses the material inside the structure. To keep the stress level constant, the fine-scale geometrical parameters of an optimal structure vary from one point to another. Sometimes, one needs to organize the layout in several length scales to optimize a structure, as seen in the structure of bones, leaves, airplane wings, or domes.

Structural Optimization in Engineering and Mathematics

In practice, the process of design always includes a mysterious element: The designer chooses the shape and materials for the construction using intuition and experience. Since ancient times this technique has proved effective, and for centuries engineering landmarks such as aqueducts, cathedrals, and ships were all built without mathematical or mechanical theories.

However, from the time of Galileo and Hooke, engineers and mathematicians have developed theories to determine stresses, deflections, currents and temperature inside structures. This information helps in the selection of a rational choice of structural elements. Certain principles of optimality are rooted in common sense. For example, one wants to equalize the stresses in a designed elastic construction by a proper choice of the layout of materials. The overstressed parts need more reinforcement, and the understressed parts can be lightened. These simple principles form a basis for rational construction of amazingly complicated mechanical structures, like bridges, skyscrapers, and cars. Still, knowledge of the stresses in a body is mostly used as a checking tool, parallel with the design proper, which remains the responsibility of the design engineer.

In the past few decades, it has become possible to turn the design process into algorithms thanks to advances in computer technology. Large contemporary projects require the use of computer-aided design systems. These systems often incorporate algorithms that gradually improve the initial design by a suitable variation of design variables, namely, the materials' cost and layout. Optimization techniques are used to effect changes in a design to make it stronger, lighter, or more reliable.

This progress has stimulated an interest in the mathematical foundations of structural optimization. These foundations are the main topic of this book. The theory of extremal problems is used to address problems of design. A design problem asks for the best geometry of layouts of different materials in a given domain. Of course, this approach simplifies (or, as a mathematician would say, idealizes) the real engineering problem, because questions such as convenience or cost of manufacturing are not considered. Analysis of optimal structures allows us to formulate general principles of an optimally designed construction. In particular, we can extend the intuitive principle of equally stressed construction to a multidimensional situation and find optimal structures that are, in a sense, hybrids of simple mechanisms.

The Purpose of the Book

A gap exists between mathematical approaches to variational problems and the practical use of results in structural optimization, theory of composites, and other engineering applications. On the one hand, we shall see how mathematicians develop advanced theories such as quasiconvexity and G-convergence for this purpose. On the other hand, the engineering and numerical community develops software for numerical optimization of complicated structures and successfully optimizes constructions of airplanes, bridges, and so on.

Progress in the area of numerical approaches is often ahead of mathematical methods required for an adequate formulation and rigorous solution to corresponding optimization problems. Mathematics deals with its own objectives: Standards of rigor are higher and models are simpler. This tends to make mathematical papers not too exciting for engineers. As usual, mathematicians use advanced methods to solve simple equations, and engineers use simple methods but work with complicated models. As a result, many practically oriented researchers are skeptical about the usefulness of refined mathematical theories. An opposite tendency, to interpret abstract mathematical results as prophecy, is no less risky.

These two approaches should be used in concert, each highlighting supplementary ideas of optimal design. I hope to present the foundations of structural optimization in a sufficiently simple form to make them available for practical use and to allow their critical appraisal for improving and adapting these results to specific models. I also hope that the reader will enjoy the beauty and elegance of the presented mathematical methods.

Often, mathematical analysis of an optimization problem leads to "unusual" solutions that are characterized by fractal geometries and are hardly suitable for manufacturing. This is acceptable in the framework of the chosen approach: We are looking for a mathematically correct solution, and we accept its features. From a practical point of view, the emergence of "strange" solutions reveals certain hidden features of optimality. These solutions should not be rejected as mathematical extravagance, but rather should be understood and interpreted in depth; often, they point to better solutions that may be approximated with available resources.

The Contents of the Book

Let us outline the contents of the five parts of the book.

Preliminaries. The exposition starts with an introductory Chapter 1 that discusses instabilities in one-dimensional variational problems. Specifically, we study variational problems with rapidly oscillating solutions and ways to describe these solutions. We also introduce the concept of relaxation of a nonstable variational problem by replacing the Lagrangian with its convex envelope.

Chapter 2 introduces the subject of optimization. We discuss conductivity of inhomogeneous materials and composites. The properties of a composite significantly depend on its microstructure. We introduce homogenization methods to describe the effective behavior of structures and calculate effective properties of special structures. Homogenization theory, in turn, puts forward the so-called G-closure problem (Chapter 3) that asks for bounds of effective properties of composites assembled from given materials. Bounds of G-closures correspond to composites of extreme effective properties that arise in optimal design.

Optimization of Conducting Composites. A large class of optimization problems of conducting composites requires only the simplest laminate structures for solution. These problems are used in the book as the testing ground for methods of structural optimization. We introduce all the control methods, including sufficient and necessary conditions of optimality and minimizing sequences. Chapter 4 deals with the optimization of the total conductivity of a domain. This problem does not have a classical solution; the optimal layout is a fine-scale mixture or a composite. We reformulate (relax) the problem, replacing the layout of available materials with the layout of optimal composites made of them. We also investigate the fields in optimal structures. Chapter 5 treats the problems of minimization of a large class of functionals associated with the solution to the conductivity problem, such as the minimization of the mean temperature in a part of the domain or the maximization of the boundary current.

Quasiconvexity and Relaxation. The second part deals with the relaxation technique of multidimensional variational problems with nonconvex integrands. This part contains most of the new mathematical results. In Chapter 6, we briefly discuss instabilities, the Weierstrass test, and we introduce the main tool for relaxation–the quasiconvex envelope.

In Chapter 7 we obtain upper bounds of the quasiconvex envelope by constructing some special minimizing sequences. The optimal layouts are represented by alternatng materials in laminate microstructures. We introduce special layouts with hierarchical geometries called "laminates of a high rank" and we derive their properties.

In Chapter 8 we derive lower bounds for the relaxed functional that correspond to sufficient conditions of optimality. The lower bound is built by a so-called translation method. We develop this method using the theories of quasiconvexity and compensated compactness.

In Chapter 9 we develop a technique of minimal extensions based on necessary conditions of the Weierstrass type. The extension we obtain gives an upper bound for the functional but avoids the explicit consideration of minimizing sequences.

All of these three approaches are illustrated by the solution of an optimization problem of a conducting structure that minimizes a sum of energies caused by several external sources.

G-Closures. To find the optimal structure of a composite, one first describes the set of effective properties of all possible microstructures. This set is called the *G*-closure of the properties of initially given materials. The fourth part discusses the knotty problems of *G*-closures. Chapter 10 deals with techniques used to describe the boundaries of the closures, i.e., the extreme effective properties of composites. The techniques are based on the variational methods introduced in Part III.

In Chapter 11 several examples of *G*-closures are constructed. These include the *G*-closures of conducting materials, the exact coupled bounds for conducting properties of composites, and bounds for properties of polycrystals.

Chapter 12 discusses multimaterial composites. The methods for these problems are less developed and more diverse. In particular, the technique of necessary conditions allows us to address the problem of bounds for a three-material composite.

Chapter 13 deals with the problem of complex conductivity. We suggest a variational principle for this problem, and we apply the variational technique to find coupled bounds on the real and imaginary parts of conductivity tensor.

Optimization of Elastic Structures. The last part of the book deals with optimal design of elastic structures. We begin with a discussion of the equations and variational principles for elasticity of inhomogeneous media and the algebra of fourth-rank tensors of elastic moduli (Chapter 14). In this chapter we also derive effective properties of elastic composites.

In Chapter 15 we consider the problem of minimization of the compliance of an elastic body, exploiting its similarity to the problems discussed in earlier chapters; we obtain elastic structures of extreme stiffness. We also discuss optimization of the shapes of cavities.

In Chapter 16 we survey the results regarding bounds for elastic moduli. Specifically, we consider an isotropic composite of two isotropic materials (plane problem), and we describe the bounds on its shear and bulk moduli. These bounds are coupled. We also consider the problem of isotropic polycrystals with extreme properties and describe the fractal geometry of optimal polycrystals. These examples demonstrate advanced applications of the variational technique described in Part III.

Chapter 17 discusses new formulations of a number of problems of structural optimization. We consider the minimization of the sum of elastic energies of different processes, the optimization of a periodic composite, the optimization of a nonenergetic functional, and the optimization in an unknown class of loadings. This last problem is formulated as a min-max game between the applied loadings and the responding structure.

Mathematical Methods

Mathematically, the book considers one type of problems in different settings. We describe optimal solutions to unstable variational problems. The goal is to define a solution that is reasonably smooth; particularly, it should not depend on the mesh in a discretization scheme. However, it often turns out that the optimal solution is characterized instead by infinitely fine oscillations. Special tests are developed to distinguish variational problems with smooth and nonsmooth solutions, and suitable frameworks for describing the solution with fine oscillations are worked out.

Both aspects deal with a special property of Lagrangians of the variational problem called quasiconvexity. Variational problems with quasiconvex Lagrangians possess stable solutions and problems with nonquasiconvex Lagrangians may not. Therefore, the test for oscillatory solutions requires consideration of the quasiconvexity of the Lagrangian. For one-dimensional variational problems and for some multidimensional problems, quasiconvexity degenerates to convexity, which makes the determination easy. Generally, however, the property of quasiconvexity is not geometric, and we need more refined tools to determine that a Lagrangian is quasiconvex.

If the Lagrangian lacks quasiconvexity, the minimizers generally are replaced with oscillating minimizing sequences. We perform the relaxation of the problem, also called the minimal extension, by averaging the solution over an infinitesimal volume. This corresponds to replacing the original nonquasiconvex Lagrangian with its quasiconvex envelope. In this way we obtain a new variational problem that possesses the same cost as the original one, but its solution is smooth and equal to the mean value of the fast oscillatory solution.

If quasiconvexity degenerates to convexity, the convex envelope can be built by systematic geometrical methods. There is no systematic universal method for constructing quasiconvex envelopes, so we instead build two extensions of the original Lagrangian, one above and one below the quasiconvex envelope (Chapters 7–9). Sometimes, these extensions coincide, in which case the quasiconvex envelope is determined.

The technique of bounds is addressed three times: first, in the context of one-dimensional variational problems (Chapter 1), then for the simplest multidimensional problems with a scalar potential (Chapter 3), and then in the general case (Chapters 6–9) of multidimensional problems with several state variables. This technique is used many times to solve various problems of G-closure (Part IV) and optimal elastic structures (Part V).

Related Topics

The theory of structural optimization lies at a busy intersection of several mathematical disciplines—optimal control, calculus of variations, homogenization, convex analysis—and is strongly influenced by materials science. Its applications include traditional optimal design, theory of composites, phase transition in solids, "smart" materials, nondestructive testing, self-organization in physics, biomaterials, and so on. Each of these fields has its own philosophy, its history, and a huge literature. Here we mention several of the related fields in mathematics and engineering. Each field could probably be identified by a representative, but not complete, list of the contributors. Specific references are placed in the body of the text.

The variational problems and problems of optimal control require methods of selecting and describing solutions with infinitely fast oscillations. It is known in control theory that minimization is generally achieved by an infinitely rapid oscillating control function, called the chattering control. This theory was originated by Pontryagin and Young and developed by Gamkrelidze, Krotov, Rozonoer, Varga, and others. The variational methods for nonconvex problems were introduced in works by Carathéodory, Morrey, and Young and developed in the works by Dacorogna, Ekeland,

Kohn, Lions, Lurie, Müller, Murat, Raitum, Rockafellar, Strang, Tartar, Temam, and many others.

An average description of the layout for the highly oscillatory materials is the subject of the theory of homogenization. It was originated in the works by Babuška, Bakhvalov, Bensoussan, Hashin, Keller, Khruslov, Lions, Olejnik, Papanicolaou, Sanchez-Palencia, and Shtrikman, and developed in many respects in the works by Benveniste, Bergman, Bruno, Golden, Kohn, Kozlov, Markov, Milton, Norris, Panasenko, Telega, Torquato, Vigdergauz, Vogelius, Zhikov, and others. The advanced theories of solution to differential equations with rapidly oscillating coefficients can be found in the papers by Berlyand, Buttazzo, Cioranescu, Dal Maso, de Georgi, Fonseca, Francfort, Kinderlehrer, Kohn, Müller, Pedregal, Sukey, and Tartar, among others.

Approaches for bounds on the effective properties of composites are especially useful for our goals. This area, initiated around the beginning of the twentieth century by Rayleigh, Reuss, Voigt, and Wiener, was developed by Bruggeman, Hill, Hashin, Shtrikman, and Walpole and recently updated by Avellaneda, Benveniste, Beran, Francfort, Gibiansky, Kohn, Lurie, Markov, Milton, Murat, Nesi, Ponte Castañeda, Schulgasser, Talbot, Tartar, Torquato, Willis, and Zhikov, among others.

The physical side of the picture was highlighted by the mechanicians and applied mathematicians who formulated and solved structural optimization problems for several decades, starting from the works by Prager. We mention here the works of Armand, Arora, Banichuk, Bendsøe, Diaz, Eshenauer, Fuchs, Haber, Haftka, Kikuchi, Kirsch, Litvinov, Lipton, Mota Soares, Mroz, Olhoff, Pedersen, Rasmussen, Rozvany, Sigmund, Taylor, Tortorelli, and Zowe.

Computational techniques of structural optimization deserve special considerations, yet we feel that it does not fit the scope of this book, which is devoted exclusively to mathematical foundations of structural optimization. A detailed discussion of the computational techniques can be found, for example, in the books by Bendsøe, Haftka and Gürdal, Rozvany, and Papalambros and Wilde.

Natural Phenomena. Natural phase transitions, shape memory alloys, and naturally optimal biomaterials form a novel area of application of the discussed mathematical techniques. These problems, involving complicated materials, are in many respects similar to structural optimization. In both cases one deals with several materials or solid phases that are distributed in a domain in a specific way. The optimality requirement posed by a designer is parallel to a natural variational principle of minimization of the total energy of the system (the Gibbs principle). The transformation from one phase to another is parallel to the use of different materials in a design. In minimizing its energy, a natural system exhibits phase separation and forms a sort of natural composite that possesses optimal microstructure.

These similarities suggest that corresponding approaches could be applied to describe natural mixtures with minimal energy. This concept was put forward in the works of Ericksen, Khachaturyan, and Kinderlehrer and developed in the works of Ball, Bhattarcharya, Kohn, Fonseca, Grinfield, James, Luskin, Roitburd, Rosakis, Truskinovsky, and others. The methods of quasiconvexity are successively implemented for an explanation of structures arriving at some natural phase transitions; we refer to the works of the above-mentioned authors.

However, natural phenomena are much deeper than the problems of structural optimization. Indeed, the best engineering system should reach the global minimum of the minimizing functional that represents the quality of the system. On the contrary, an equilibrium state of a natural system corresponds to any local minimum of the energy. The energy of complicated natural systems is typically characterized by a large class of metastable local minima.

There are other differences, too. Contrary to an optimal engineering construction, a realizable equilibrium state of a natural system corresponds to a dynamical process that has led to it. Finally, natural composites usually are a random mixture of the states that correspond to local minima. The search for a distribution of local minima requires different techniques from those discussed here; we do not touch on this subject in the book.

Biomaterials. The amazing rationality of biological "constructions" also calls for the use of mathematical methods of structural optimization to model them. Consider, for example, the problem of the structure of a bone. A bone is a mechanical structure made of composites with variable parameters that adapts itself to its working conditions. It performs the clear mechanical task of supporting the organism. These features are similar to such man-made composite structures as masts, bridges, and towers. Therefore, it would be natural to apply optimization methods developed for engineering constructions to bone structures.

However, the two problems are not the same. In addition to the problems of local minima, stable evolutionary dynamics, and randomness already mentioned, it is not clear what quantity is minimized in natural evolutionary biomaterials (we mean the explicit optimality criterion of a natural structure, not a general reference to the evolution that perfects organisms). In engineering problems, the goal is the minimization of a given functional that is not the subject of a search or even a discussion. The problem is to find the structure that minimizes a functional prescribed by a designer. On the other hand, the structure of a bone is known, but it is not clear in what sense (if any) the bone structure is optimal.

The corresponding problem is the search for the cost functional of an optimization problem with a known solution. This problem has not been sufficiently investigated, to our knowledge.

Indexes, References, etc.

The electronic version of the manuscript for the book was prepared with the help of Professor Nelson Beebe using special BibTeX and LaTeX macros that he developed. In addition to the detailed table of contents, it contains the list of figures, references, the author/editor index, and the index of topics. Each item in the references points to the pages on which the source was referred to. The references section is ordered alphabetically by the name of primary author.

The author/editor index refers to the pages that contain the reference. Boldface author names indicate primary authors, while names in Roman text are nonprimary authors.

The book's Web site, `http://www.math.utah.edu/books/vmso`, contains an expanded bibliography in BibTeX form, an errata list, and other related resources. Please email your comments to `cherk@math.utah.edu`.

Use in the Classroom

The book is an extended and edited version of the author's lecture notes for courses delivered at the University of Utah. The contents of the book may be used for a year-long graduate course for students in applied mathematics, science, and engineering. We do our best to keep the exposition simple and do not hesitate to sacrifice rigor in favor of vividness, and generality in favor of vigorous illustrations. The references point to more rigorous formulations. The problems for discussion are in the end of chapters. Some of them are simple exercises; the others require more serious analysis and can be used for course projects.

A course in calculus of variations may be based on the classical material (Gelfand and Fomin, 1963; Ewing, 1969; Weinstock, 1974), supplemented by Chapter 1 (nonconvex one-dimensional problems), Chapter 4 (an example of a variational problem for a distributed system), and Chapters 6–9 (relaxation of nonconvex multivariable problems), with examples from Chapters 10–12 (G-closures).

A course in homogenization may use chapters from the "homogenization" books (Bensoussan, Lions, and Papanicolaou, 1978; Jikov, Kozlov, and Oleĭnik, 1994) and Chapters 2 and 3 (conductivity, homogenization, G-closure), Chapter 7 (laminates, various structures of laminates of high rank), Chapter 12 (multiphase structures), and Chapter 14 (elasticity, homogenization and matrix laminates). Chapters 4 and 5 (optimization by laminates) may be used as examples of the use of composites.

A course in structural optimization may use Chapters 4 and 5 (optimization of conducting bodies), Chapters 6–9 (relaxation of nonconvex multivariable problems), and Chapters 14–17 (elasticity, optimization of elastic structures).

Credits

The author has been tempted to present his view of the history of structural optimization and nonconvex variational methods, but he has given that up because the topic is too awesome, controversial, and sensitive. The very length of the preceding list of names and topics testifies to this. We give credit here to authors in a specific context of themes discussed. It is hardly possible to give a complete survey of even the recent development of related topics: New branches of the theory are constantly appearing. Instead, we concentrate our attention on underlying ideas and methods that should help to find solutions to new problems.

Most of the results and opinions presented are based on or related to the author's research, conducted for the most part in collaboration with Tim Burns, Elena Cherkaeva, Andrey Fedorov, Leonid Gibiansky, Lars Krog, Konstantin Lurie, Graeme Milton, Robert Palais, and Shmul Vigdergauz.

Many new results obtained by Grégoire Allaire, Marco Avellaneda, Martin Bendsøe, John Ball, Gilles Francfort, Leonid Gibiansky, Zvi Hashin, Robert Kohn, Robert Lipton, Konstantin Lurie, Graeme Milton, François Murat, Vincenzo Nesi, Niels Olhoff, Ole Sigmund, Gil Strang, Vladimír Šverák, Luc Tartar, Salvatore Torquato, Smul Vigdergauz, Vasily Zhikov, and others are explicitly used in the text.

Acknowledgments

The author gratefully acknowledges support and encouragement from colleagues and students in the Department of Mathematics, University of Utah, as well from friends and family. My special thanks are to Elena Cherkaev, Leonid Gibiansky, Konstantin Lurie, and Graeme Milton, who discussed different topics of the book and made valuable suggestions. Other useful suggestions were made by anonymous referees, whom the author wants to thank. Many suggestions were made by the members of my classes, especially O. Hamed, I. Kucuk, and T. Robbins. The author is indepted to colleagues: Martin Bendsøe, Robert Kohn, Robert Lipton, Konstantin Lurie, Graeme Milton, Robert Palais, Ole Sigmund, and John Taylor who looked through the manuscript, suggested improvements of the text, and provided additional references. The electronic manuscript was produced with the help of Nelson Beebe, whose expertise in LaTeX problems and various suggestions greatly helped improve the text.

My special thanks go to the staff at Springer-Verlag. The book was copyedited by David Kramer and Shari Chappell, who greatly amended the English as well as the presentation style. Achi Dosanjh, senior editor, Lesley Poliner, senior production editor, and Christina Torster, production associate, were extremely effective and encouraging.

Salt Lake City, Utah   Andrej Cherkaev
                                                          1 March 2000

Part I

Preliminaries

1
Relaxation of One-Dimensional Variational Problems

This introductory chapter gives a brief review of nonconvex variational problems. We examine stability of solutions to one-dimensional extremal problems associated with ordinary differential equations. The reader familiar with nonconvex variational problems can skip this chapter. However, this material is necessary to understand the one-dimensional analogue of the multidimensional ill-posed problems that are in the focus of this book.

Optimizing design is a variational problem. Such problems ask for the minimization of a functional that measures the quality of a structure choosing by a proper control function (the materials' layout). In this chapter we detect and describe unstable solutions of these *extremal problems* in a one-dimensional setting. The solutions to these problems are characterized by fine-scale oscillations. To deal with these oscillations, we introduce a relaxation procedure. *Relaxation* essentially means the replacement of an unstable optimization problem with another that has a classical (differentiable) solution.

1.1   An Optimal Design by Means of Composites

Let us start with an example that demonstrates why composites appear in optimal design. Here we find an optimal solution using only commonsense arguments.

The Elastic Beam

Consider an elastic beam with variable stiffness $d(x)$. The beam is loaded by the load $f(x)$, and its ends $x = 0$ and $x = 1$ are simply supported. The deflection $w = w(x)$ of the points of the beam satisfies the classical boundary value problem (see, for example (Timoshenko, 1970))

$$\theta = dw'', \quad \theta'' = f,$$
$$w(0) = w(1) = 0, \quad \theta(0) = \theta(1) = 0, \tag{1.1.1}$$

where $\theta = \theta(x)$ is the bending moment and $d = d(x)$ is the material's stiffness at point x. Suppose that the beam can be made of two materials with effective stiffness d_1 and d_2, so that the stiffness takes one of these two values, $d(x) = d_1$ or $d(x) = d_2$ at each point $x \in [0, 1]$. The deflection w depends on the layout $d = d(x)$ and the loading $f = f(x)$: $w = w(f, d)$.

Optimization Problem

Let us state the following optimal design problem: Lay out the given materials with the stiffness d_1 and d_2 along the beam to approximate in the best way some desired function w_* with the deflection $w(f, d)$. Specifically, we want to minimize the square of the L_2-norm of the difference between the actual displacement $w(d, f)$, which depends on the layout $d = d(x)$ and the loading f, and the desired function[1] $w_*(x)$:

$$I = \min_{d} \int_0^1 (w(d, f) - w_*)^2. \tag{1.1.2}$$

Let us assume that the desired function w_* is the deflection of a homogeneous beam of an intermediate stiffness d_*,

$$d_* = \frac{d_1 + d_2}{2}, \tag{1.1.3}$$

which is subject to the same boundary conditions and the same loading f: $w_* = w_*(d_*, f)$. The deflection w_* satisfies the equation

$$\theta = d_*(x)w_*'', \quad \theta_*'' = f,$$
$$w_*(0) = w_*(1) = 0, \quad \theta_*(0) = \theta_*(1) = 0, \tag{1.1.4}$$

similar to (1.1.1). The optimization problem becomes

$$I = \min_{d(x)} \int_0^1 (w(d, f) - w(d_*, f))^2. \tag{1.1.5}$$

---

[1]The symbol "dx" of the differential is omitted in the integrals over the explicitly defined interval of the independent variable x.

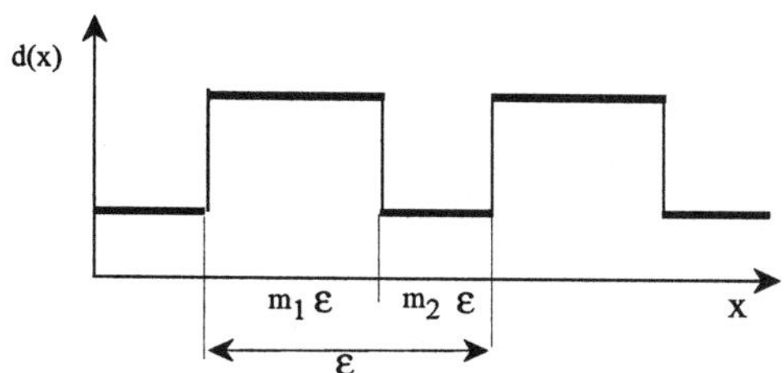

FIGURE 1.1. Oscillation of the pointwise stiffness of an optimal inhomogeneous beam.

The Minimizing Sequence

The solution to the optimization problem (1.1.4) is intuitively obvious: One should mix the given materials d_1 and d_2 in special proportions m_1 and $m_2 = 1 - m_1$ to imitate the intermediate stiffness d_* of the beam and therefore to make the nonnegative cost I (see (1.1.2)) arbitrarily close to zero. The stiffness becomes an oscillatory function of x that alternates between the values d_1 and d_2. The approximation improves when the frequency of the oscillations increases. Therefore, an optimal design does not exist: the higher the frequency, the better (see Figure 1.1). Formally, the minimizing layout of the material corresponds to the limit $\lim_{\varepsilon \to 0} d_\varepsilon(x)$, where

$$
d_\varepsilon(x) = \begin{cases} d_1 & \text{if } x \in [n\varepsilon, (n + m_1)\varepsilon], \\ d_2 & \text{if } x \in [(n + m_1)\varepsilon, (n + 1)\varepsilon], \end{cases} \quad n = 1, \ldots N,
$$

$\varepsilon \ll 1$ is a small parameter, and $N = \left[\frac{1}{\varepsilon}\right]$. The remaining problem is the computation of the needed proportion m_1. We will demonstrate that $m_1 \neq \frac{1}{2}$, contrary to the intuitive expectation.

Homogenization

This consideration poses the question of an adequate description of rapidly oscillating sequences of control. To describe these sequences we use the method of homogenization, which simplifies the problem: Details of the behavior of minimizing sequences become intractable, and the equations depend only on average characteristics of them.

Let us derive equations for an average deflection $\langle w \rangle$ of the beam. The averaging operator $\langle \ \rangle$ is introduced by the formula

$$
\langle z(x) \rangle = \frac{1}{2\varepsilon'} \int_{x+\varepsilon'}^{x-\varepsilon'} z(\xi) d\xi, \tag{1.1.6}
$$

where $[x - \varepsilon', \ x + \varepsilon']$ is the interval of the averaging and $z = z(x)$ is the averaged variable.

We suppose that the interval ε' is much larger than the period ε of oscillation of the control but much smaller than the length of the beam:

$$
0 < \varepsilon \ll \varepsilon' \ll 1. \tag{1.1.7}
$$

Homogenized Equation

To derive the homogenized equation for the average deflection $\langle w \rangle$, we mention that the variable $\theta(x)$ is twice differentiable (see (1.1.1)); therefore it is continuous even if $d(x)$ is discontinuous:

$$\langle \theta \rangle (x) = \theta(x) + O(\varepsilon').$$

This implies that discontinuities in $d(x)$ are matched by discontinuities in $w''(x)$, leaving the product $\theta = d(x)w''(x)$ continuous. Therefore $\langle w''(x) \rangle$ is computed as

$$\langle w''(x) \rangle = \left\langle \frac{\theta(x)}{d(x)} \right\rangle = \theta(x) \left\langle \frac{1}{d(x)} \right\rangle + O(\varepsilon'). \qquad (1.1.8)$$

Notice that we take the smooth function $\theta(x)$ out of the averaging because its derivation is arbitrarily small in the small interval of averaging.

Note also that the function $\frac{1}{d(x)}$ takes only two values, and it alternates faster than the averaging (1.1.7). Therefore its average is found (up to terms of the order of ε') as

$$\left\langle \frac{1}{d(x)} \right\rangle = \frac{m_1}{d_1} + \frac{m_2}{d_2}, \quad m_2 = 1 - m_1. \qquad (1.1.9)$$

The homogenized equation for the average value $\langle w \rangle''$ of the deflection of the beam can easily be found from (1.1.1), (1.1.8), and (1.1.9):

$$\begin{aligned} \theta &= \left\langle \tfrac{1}{d} \right\rangle^{-1} \langle w \rangle'', & \theta'' &= f, \\ \langle w(0) \rangle &= \langle w(1) \rangle = 0, & \theta(0) &= \theta(1) = 0 \end{aligned} \qquad (1.1.10)$$

(these equations are satisfied up to ε'). They are called the *homogenized equations* for the composite beam.

Homogenized Solution

We are able to approximate the desirable deflection w_* by the deflection of an inhomogeneous beam. Comparing (1.1.4) and (1.1.10), we conclude that the solutions to these two equations are arbitrarily close to each other if $\varepsilon' \to 0$ and if the fraction m_1 corresponds to the equality

$$\frac{1}{d_*} = \frac{m_1}{d_1} + \frac{m_2}{d_2}. \qquad (1.1.11)$$

Then the cost of (1.1.5) goes to zero together with ε'. Strictly speaking, the minimizing layout does not exist: the smaller the period, the better. The actual minimum of the functional I is not achievable. Nothing bounds the period ε of oscillation of $d(x)$ from zero.

The minimizing sequence corresponds to the optimal volume fraction m_1 that can be found from (1.1.3), (1.1.11):

$$m_1 = \frac{d_1}{d_1 + d_2}.$$

Notice that $m_1 \neq \frac{1}{2}$ and $\langle d(x) \rangle = \frac{d_1^2 + d_2^2}{d_1 + d_1} \neq d_*$.

Remark 1.1.1 *In a general situation the fraction m_1 may vary from point to point. Then the outlined homogenization procedure introduces a smoothly varying quantity $m(x)$ (the volume fraction of the material in the composite) that describes the fine-scale oscillating sequence of control.*

Remark 1.1.2 *The described solution with fine-scale oscillations is an example of so-called chattering controls, which are well known in the theory of one-dimensional control problems (Gamkrelidze, 1962; Young, 1969). Chattering regime of control occurs when the interval of the independent variable x is split into infinitely many subintervals, and each of them is characterized by alternation of the value of control. The value of the minimizing functional decreases as the scale of alternation of intervals becomes finer.*

1.2 Stability of Minimizers and the Weierstrass Test

1.2.1 Necessary and Sufficient Conditions

Extremal Problems

Consider an extremal problem:

$$I(u) = \min_{u(x)} \int_0^1 F(x, u(x), u'(x)), \quad u(0) = u_0, \ u(1) = u_1, \qquad (1.2.1)$$

where x is a point of the interval $[0, 1]$, $u(x)$ is a function that is differentiable almost everywhere in $[0, 1]$, and F is a function of three arguments called the Lagrangian. We assume that the Lagrangian F is a continuous and almost everywhere differentiable function of its arguments. Problem (1.2.1) asks for a function $u_0(x)$ that delivers the minimum of $I(u)$:

$$I(u_0) \leq I(u) \quad \forall u(x).$$

This function is called the *minimizer*.

There are several approaches to the solution of this extremal problem (see, for example (Ioffe and Tihomirov, 1979)). The simplest approach is based on sufficient conditions. One guesses solutions using special algebraic properties of the Lagrangian F; typically, the convexity of F is used.

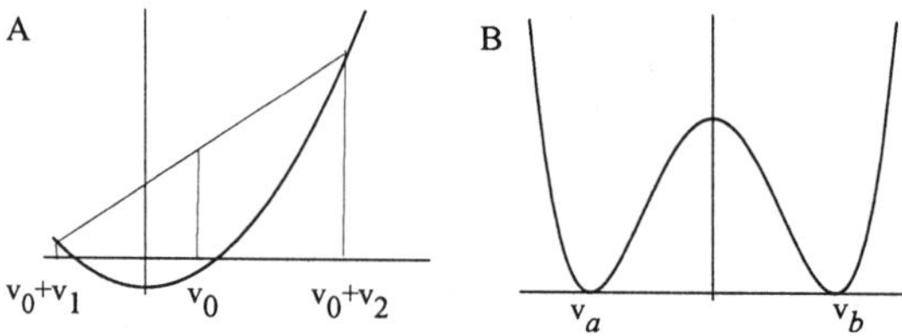

FIGURE 1.2. (A) The definition of convexity; (B) graph of a nonconvex function; $(\mathbf{v}_a, \mathbf{v}_b)$–interval of nonconvexity.

Convexity

Let us briefly discuss convexity. For a detailed exposition of properties of convex functions and convex functionals the reader is referred to (Krasnosel'skiĭ and Rutickiĭ, 1961; Rockafellar, 1997; Ekeland and Temam, 1976; Hardy, Littlewood, and Pólya, 1988).

Here we define convexity through Jensen's inequality. We consider a continuous function $F(\mathbf{v})$ of an n-dimensional vector argument $\mathbf{v} = [v_1, \ldots, v_n]$. Suppose that $\mathbf{v}$ varies in the whole space R^n.

Definition 1.2.1 The function $F(\mathbf{v})$ is convex at the point $\mathbf{v}_0$ if the following inequality (called Jensen's inequality) holds:

$$F(\mathbf{v}_0) \leq \frac{1}{r} \sum_{k=1}^{r} F(\mathbf{v}_0 + \mathbf{v}_k) \quad \forall \mathbf{v}_k : \sum_k \mathbf{v}_k = 0. \qquad (1.2.3)$$

The function $F(\mathbf{v})$ is strongly convex at the point $\mathbf{v}_0$ if (1.2.3) becomes a strong inequality.

This inequality expresses the geometrical fact that the graph of a convex function F lies below any secant hyperplane to that graph. The secant is supported by the graph of F at points $\mathbf{v}_0+\mathbf{v}_k$ (see Figure 1.2). For example, the function $F_1(v) = v^2$ of a scalar argument[2] v is convex everywhere, and $F_2(v) = (v^2 - 1)^2$ is convex at the points of the intervals $[-\infty, -1]$ and $[1, \infty]$.

We list here several properties of the convex function that will be used (for a detailed exposition, we refer to the mentioned books):

- A strongly convex function has only one minimum.

- For convexity of F, it is necessary and sufficient that for any point $\mathbf{v}$ there exists an affine function (supporting hyperplane)

$$l(\mathbf{v}) = a_1 v_1 + \ldots + a_n v_n + a_0$$

---

[2] As a rule, we use Roman letters for scalar and boldface letters for vectors.

such that the graph $F(\mathbf{v})$ lies above the graph of $l(\mathbf{v})$ or coincide with it,

$$F(\mathbf{v}) = l(\mathbf{v}), \quad F(\mathbf{v}') \geq l(\mathbf{v}') \quad \forall \mathbf{v}'.$$

- If F is twice differentiable, then the eigenvalues of its Hessian

$$H = \{H_{ij}\}, \quad H_{ij} = \frac{\partial^2 F}{\partial v_i \partial v_j}, \quad i, j = 1, \ldots, n$$

are nonnegative: $H \geq 0$. The limiting case when F belongs to the boundary of the domain of the convexity corresponds to the vanishing of an eigenvalue of H or to the condition $\det H = 0$.

Integral Form. We also use an integral form of the definition of convexity. In the limit $r \to \infty$ Jensen's inequality takes the form of an integral inequality. In this case, the vectors $\{\mathbf{v}_1, \ldots, \mathbf{v}_r\}$ are replaced by a vector function $\boldsymbol{\xi}(x) = [\xi_1(x), \ldots, \xi_n(x)]$ of a scalar argument $x \in [0, \, l]$. In this notation, (1.2.3) yields the following inequality.

The function $F(\mathbf{v})$ is convex at point $\mathbf{v}_0$ if and only if the following inequality holds:

$$F(\mathbf{v}_0) \leq \frac{1}{l} \int_0^l F(\mathbf{v}_0 + \boldsymbol{\xi}), \quad \forall \boldsymbol{\xi} : \int_0^l \boldsymbol{\xi} = 0. \tag{1.2.4}$$

Of course, we assume existence of the integrals in (1.2.4).

An equivalent form of Jensen's inequality is obtained by setting $\mathbf{v}(x) = \mathbf{v}_0 + \boldsymbol{\xi}(x)$ and using the identity $l\,F(\mathbf{v}_0) = \int_0^l F(\mathbf{v}_0)$. The inequality is

$$\int_0^l F(\mathbf{v}_0) \leq \int_0^l F(\mathbf{v}(x)) \quad \text{if} \quad \frac{1}{l} \int_0^l \mathbf{v}(x) = \mathbf{v}_0. \tag{1.2.5}$$

This inequality states that the integral of a convex Lagrangian $F(\mathbf{v})$ takes its minimal value if the minimizer $\mathbf{v}$ is constant.

Inequality (1.2.5) introduces a convex functional of $\mathbf{v}$. The properties of convex functionals are discussed in many classical books, such as (Hardy et al., 1988). Particularly, if $f_1(\mathbf{v})$ and $f_2(\mathbf{v})$ are convex functionals and α_1 and α_2 are positive numbers, then $\alpha_1 f_1(\mathbf{v}) + \alpha_2 f_2(\mathbf{v})$ is also a convex functional.

Convexity and the Extremal Problems

For some problems, the convexity of the Lagrangian can be immediately used to find the solution.

Example 1.2.1 Consider the problem of the shortest path between two points in a plane. Suppose that the coordinates of these points are $A =$

$(0,0)$ and $B = (c,d)$ and that the path between them is given by a curve $y = u(x)$. The minimal length of the path is the solution to the problem

$$I = \min_{u(x)} \int_0^c \sqrt{1 + (u'(x))^2}, \quad u(c) - u(0) = \int_0^c u'(x) = d. \tag{1.2.6}$$

The function $f(v) = \sqrt{1 + v^2}$ is convex. The integral $\int_0^c v$ is fixed. Therefore (see (1.2.5)) the minimal value of I corresponds to the constant minimizer $v(x)$, $v(x) = \text{constant}(x)$. Applying inequality (1.2.5) to (1.2.6) and using the constraint in (1.2.6), we find that the solution to (1.2.6) is a straight line with slope $u'(x) = \frac{d}{c}$ that passes through the prescribed starting point. We have $u(x) = \frac{d}{c}x$. The cost is $I = \sqrt{c^2 + d^2}$.

More advanced sufficient conditions yield to isoperimetric inequalities (Pólya and Szegö, 1951), symmetrization, Lyapunov functions, etc. If applicable, these conditions immediately detect the true minimizer. However, they are applicable to a very limited number of problems.

Generally, there is no guarantee that sufficient conditions result in strict inequalities that are realizable by a function $u(x)$. If the inequalities are not strict, they can still serve as a lower bound of the cost, but in this case they do not lead to the minimizer.

1.2.2 Variational Methods: Weierstrass Test

More general methods are based on an analysis of infinitesimal variations of a minimizer. We suppose that the function $u_0 = u_0(x)$ is a minimizer and replace u_0 with a varied function $u_0 + \delta u$, assuming that the norm of δu is infinitesimal. The varied function $u_0 + \delta u$ satisfies the same boundary conditions as u_0. If indeed u_0 is a minimizer, the increment of the cost $\delta I(u_0) = I(u_0 + \delta u) - I(u_0)$ is nonnegative:

$$\delta I(u_0) = \int_0^1 \left(F(x, u_0 + \delta u, (u_0 + \delta u)') - F(x, u_0, u_0') \right) \geq 0. \tag{1.2.7}$$

To effectively compute $\delta I(u_0)$, we also assume the smallness of δu. Variational methods yield to the necessary conditions of optimality because it is assumed that the compared trajectories are close to each other. On the other hand, variational methods are applicable to a wide variety of extremal problems of the type (1.2.1), called *variational problems*. Necessary conditions are the true workhorses of extremal problem theory, while exact sufficient conditions are rare and remarkable exceptions.

There are many books that expound the calculus of variations, including (Bliss, 1946; Courant and Hilbert, 1962; Gelfand and Fomin, 1963; Lavrent'ev, 1989; Weinstock, 1974; Mikhlin, 1964; Leitmann, 1981; Fox, 1987; Dacorogna, 1989).

Euler–Lagrange Equations

The calculus of variations suggests a set of tests that must be satisfied by a minimizer. These conditions express realization of (1.2.7) by various variations δu. To perform the test one must specify the type of perturbations δu. The simplest variational condition (the Euler–Lagrange equation) is derived by linearizing the inequality (1.2.7) with respect to an infinitesimal small and localized variation

$$\delta u = \begin{cases} \rho(x) & \text{if } x \in [x_0, x_0 + \varepsilon], \\ 0 & \text{otherwise.} \end{cases} \tag{1.2.8}$$

Here $\rho(x)$ is a smooth function that vanishes at points x_0 and $x_0 + \varepsilon$ and is constrained as follows:

$$|\rho(x)| < \varepsilon, \quad |\rho'(x)| < \varepsilon \quad \forall x \in [x_0, x_0 + \varepsilon].$$

Linearizing with respect to ε and collecting main terms, we rewrite (1.2.7) as

$$\delta I(u_0) = \varepsilon \left(\int_0^1 \left(\frac{\partial F}{\partial u}(\delta u) + \frac{\partial F}{\partial u'}(\delta u)' \right) \right) + o(\varepsilon) \geq 0. \tag{1.2.9}$$

Integration by parts of the last term on the right-hand side of (1.2.9) gives

$$\delta I(u_0) = \varepsilon \int_0^1 S(u, u', x)\delta u + \left. \frac{\partial F}{\partial u'}\delta u \right|_{x=0}^{x=1} + o(\varepsilon) \geq 0, \tag{1.2.10}$$

where

$$S(u, u', x) = -\frac{d}{dx}\frac{\partial F}{\partial u'} + \frac{\partial F}{\partial u}. \tag{1.2.11}$$

The second term on the right-hand side of (1.2.10) is zero, because the boundary values of u are prescribed

$$u(0) = u_0, \quad u(1) = u_1$$

and their variations $\delta u|_{x=0}$ and $\delta u|_{x=1}$ are zero.

Due to the arbitrariness of δu we conclude that a minimizer u_0 must satisfy the differential equation

$$S(u, u', x) = 0, \quad u(0) = u_0, \quad u(1) = u_1, \tag{1.2.12}$$

called the *Euler–Lagrange equation* and the corresponding boundary conditions. The Euler–Lagrange equation is also called the *stationary condition*. Indirectly, we assume in this derivation that u_0 is a twice differentiable function of x. We do not discuss here the properties of the Euler–Lagrange equations for different types of Lagrangians; we refer readers to mentioned books on the calculus of variations.

It is important to mention that the stationarity test alone does not allow us to conclude whether u_0 is a true minimizer or even to conclude that a solution to (1.2.12) exists. For example, the function u that *maximizes* $I(u)$ satisfies the same Euler–Lagrange equation.

The Weierstrass Test

In addition to being a solution to the Euler equation, the true minimizer satisfies necessary conditions in the form of inequalities. These conditions distinguish the trajectories that correspond to the minimum of the functional from trajectories that correspond either to its maximum or to a saddle point stationary solution. One of these conditions is the Weierstrass test; it detects stability of a solution to a variational problem against strong local perturbations.

Suppose that u_0 is the minimizer of variational problem (1.2.1) that satisfies the Euler equation (1.2.11). Additionally, u_0 should satisfy another test that uses a type of variation δu different from (1.2.8). The variation used in the Weierstrass test is an infinitesimal triangle supported on the interval $[x_0, x_0 + \varepsilon]$ in a neighborhood of a point $x_0 \in (0, 1)$ (see Figure 1.3):

$$\Delta u(x) = \begin{cases} 0 & \text{if } x \notin [x_0, x_0 + \varepsilon], \\ v_1(x - x_0) & \text{if } x \in [x_0, x_0 + \alpha\varepsilon], \\ v_1\alpha\varepsilon + v_2(x - x_0 - \alpha\varepsilon) & \text{if } x \in [x_0 + \alpha\varepsilon, x_0 + \varepsilon], \end{cases}$$

where the parameters α, v_1, v_2 are related by

$$\alpha v_1 + (1 - \alpha)v_2 = 0.$$

This relation provides the continuity of $u_0 + \Delta u$ at the point $x_0 + \varepsilon$, because it yields to the equality $\Delta u(x_0 + \varepsilon - 0) = 0$.

Note that this variation (the Weierstrass variation) is localized and has an infinitesimal absolute value (if $\varepsilon \to 0$), but its derivative $(\Delta u)'$ is finite, unlike the variation in (1.2.8) (see Figure 1.3):

$$(\Delta u)' = \begin{cases} 0 & \text{if } x \notin [x_0, x_0 + \varepsilon], \\ v_1 & \text{if } x \in [x_0, x_0 + \alpha\varepsilon], \\ v_2 & \text{if } x \in [x_0 + \alpha\varepsilon, x_0 + \varepsilon]. \end{cases}$$

Computing δI from (1.2.7) and rounding up to ε, we find that

$$\delta I = \varepsilon[\alpha F(x_0, u_0, u_0' + v_1) + (1 - \alpha)F(x_0, u_0, u_0' + v_2) \\ - F(x_0, u_0, u_0')] + o(\varepsilon) \geq 0$$

if u_0 is a minimizer.

The last expression yields to the Weierstrass test and the necessary Weierstrass condition. Any minimizer $u(x)$ of (1.2.1) satisfies the inequality

$$\alpha F(x_0, u_0, u_0' + v_1) + (1 - \alpha)F(x_0, u_0, u_0' + v_2) - F(x_0, u_0, u_0') \geq 0.$$

Comparing this with the definition of convexity (1.2.2), we observe that the Weierstrass condition requires convexity of the Lagrangian $F(x, y, z)$ with respect to its third argument $z = u'$. The first two arguments x, $y = u$ here are the coordinates x, $u(x)$ of the testing minimizer $u(x)$. Recall that minimizer $u(x)$ is a solution to the Euler equation.

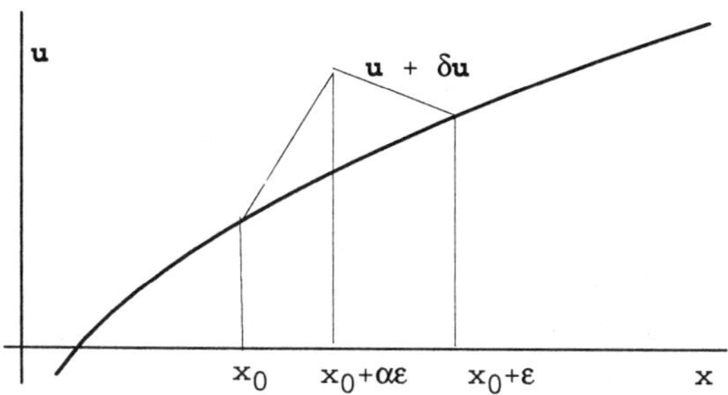

FIGURE 1.3. Weierstrass variation.

Vector-Valued Minimizer. The Euler equation and the Weierstrass condition can be naturally generalized to the problem with the vector-valued minimizer

$$I(u) = \min_{\mathbf{u}} \int_0^1 F(x, \mathbf{u}, \mathbf{u}'),$$

where x is a point in the interval $[0, 1]$ and $\mathbf{u} = [u_1(x), \ldots, u_n(x)]$ is a vector function. We suppose that F is a twice differentiable function of its arguments. The classical (twice differentiable) local minimizer $\mathbf{u}_0$ of the problem (1.2.1) is given by a solution to the vector-valued Euler equations,

$$\frac{d}{dx}\frac{\partial F}{\partial \mathbf{u}_0'} - \frac{\partial F}{\partial \mathbf{u}_0} = 0,$$

which expresses the stationarity requirement of a minimizer to small variations of the variable $\mathbf{u}$.

The Weierstrass test requires convexity of $F(x, \mathbf{y}, \mathbf{z})$ with respect to the last vector argument. Here again $\mathbf{y} = \mathbf{u}_0(x)$ represents a minimizer.

Remark 1.2.1 *Convexity of the Lagrangian does not guarantee the existence of a solution to a variational problem. It states only that the minimizer (if it exists) is stable against fine-scale perturbations. However, the minimum may not exist at all, see, for example (Ioffe and Tihomirov, 1979; Zhikov, 1993).*

If the solution of a variational problem fails the Weierstrass test, then its cost can be decreased by adding infinitesimal wiggles to the solution. The wiggles are the Weierstrass trial functions, which decrease the cost. In this case, we call the variational problem ill-posed, and we say that the solution is unstable against fine-scale perturbations.

1.3 Relaxation

1.3.1 Nonconvex Variational Problems

Typical problems of structural optimization correspond to a Lagrangian $F(x, \mathbf{y}, \mathbf{z})$ that is nonconvex with respect to $\mathbf{z}$. In this case, the Weierstrass test fails, and the problem is ill-posed.

Let us consider a problem of this type. Suppose that the Lagrangian $F(x, \mathbf{y}, \mathbf{z})$ is a nonconvex function of its third argument; is bounded from below (say, by zero),

$$F(x, \mathbf{y}, \mathbf{z}) \geq 0 \quad \forall x, \mathbf{y}, \mathbf{z};$$

and satisfies the condition

$$\lim_{|z| \to \infty} \frac{F(x, \mathbf{y}, \mathbf{z})}{|z|} = \infty.$$

Then the infimum I_0

$$I_0 = \inf_{\mathbf{u}} I(\mathbf{u}), \quad I(\mathbf{u}) = \int_0^1 F(x, \mathbf{u}, \mathbf{u}')$$

is nonnegative, $I_0 \geq 0$.

One can construct a minimizing sequence $\{\mathbf{u}^s\}$ such that $I(\mathbf{u}^s) \to I_0$. Due to the preceding condition, the minimizing sequence $\{\mathbf{u}^s\}$ consists of continuous functions with L_1-bounded derivatives; see (Dacorogna, 1989).

Because $F(., ., \mathbf{z})$ is not convex, this minimizing sequence cannot tend to a differentiable curve in the limit. Otherwise it would satisfy the Euler equation and the Weierstrass test, but the last requires the convexity of $F(., ., \mathbf{z})$.

We will demonstrate that a minimizing sequence tends to a "generalized curve." It consists of infinitesimal zigzags. The limiting curve has a dense set of points of discontinuity of the derivative. A detailed explanation of this phenomenon can be found, for example, in (Young, 1942a; Young, 1942b; Gamkrelidze, 1962; Young, 1969; Warga, 1972; Gamkrelidze, 1985). Here we give a brief description of it, mainly by working on several examples.

Example 1.3.1 Consider a simple variational problem that yields to the generalized solution (Young, 1969):

$$\inf_u \ I(u) = \inf_u \int_0^1 G(u, u'), \tag{1.3.1}$$

where

$$G(u, v) = u^2 + \min\{(v - 1)^2, \ (v + 1)^2\}, \quad u(0) = u(1) = 0. \tag{1.3.2}$$

The graph of the function $G(., v)$ is presented in Figure 1.2B.

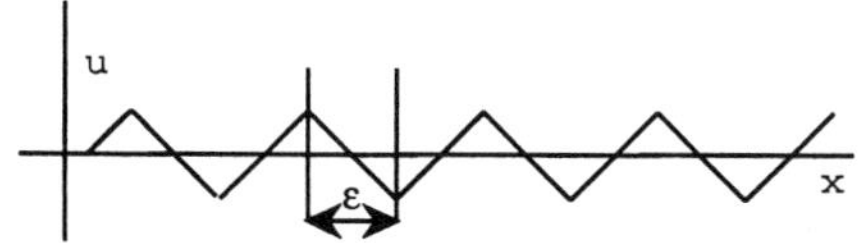

FIGURE 1.4. Oscillating minimizing sequence.

The Lagrangian G penalizes the trajectory u for having the speed $|u'|$ different from ± 1 and penalizes the deflection of the trajectory u from zero. These contradictory requirements cannot be resolved in the class of classical trajectories.

Indeed, a differentiable minimizer satisfies the Euler equation (1.2.12) that takes the form

$$u'' - u = 0 \quad \text{if} \ \ u' \neq 0. \tag{1.3.3}$$

Next, the Weierstrass test additionally requires convexity of $G(u, v)$ with respect to v; the Lagrangian $G(u, v)$ is nonconvex in the interval $v \in (-1, 1)$ (see Figure 1.2). The Weierstrass test requires that the extremal (1.3.3) is supplemented by the inequality (recall that $v = u'$)

$$u' \notin (-1, 1) \quad \text{at the optimal trajectory.} \tag{1.3.4}$$

and it is not clear how to satisfy it

On the other hand, the minimizing sequence for problem (1.3.1) can be immediately constructed. Indeed, the infimum of (1.3.1) obviously is nonnegative, $\inf_u I(u) \geq 0$. Therefore, a sequence u^s with the property

$$\lim_{s \to \infty} I(u^s) = 0$$

is a minimizing sequence.

Consider a set of functions $\tilde{u}^s(x)$ that belong to the boundary of the *forbidden interval* $\tilde{u}'(x) = -1$ or $\tilde{u}'(x) = 1$ of the nonconvexity of $G(., v)$. These functions make the second term in the Lagrangian (1.3.2) vanish, $\min\{(\tilde{u}' - 1)^2, (\tilde{u}' + 1)^2\} = 0$, and the problem becomes

$$I(\tilde{u}^s, (\tilde{u}^s)') = \int_0^1 (\tilde{u}^s)^2.$$

The term $\tilde{u}^s$ oscillates near zero if the derivative $(\tilde{u}^s)'$ changes its sign on intervals of equal length. The cost $I(\tilde{u}^s)$ depends on the density of switching points and tends to zero when the number of these points increases (see Figure 1.4). Therefore, the minimizing sequence consists of the saw-tooth functions $\tilde{u}^s$; the heights of the teeth tend to zero and their number tends to infinity as $s \to \infty$.

Note that the minimizing sequence $\{\tilde{u}^s\}$ does not converge to any classical function but rather to a distribution. This minimizer $\tilde{u}^s(x)$ satisfies the contradictory requirements, namely, the derivative must keep the absolute value equal to one, but the function itself must be arbitrarily close to zero:

$$|(\tilde{u}^s)'| = 1 \quad \forall x \in [0, 1], \quad \max_{x \in [0,1]} \tilde{u}^s \to 0 \quad \text{as } s \to \infty.$$

The limiting curve u_0 has zero norm in $C_0[0,\ 1]$ but a finite norm in $C_1[0,\ 1]$.

Remark 1.3.1 *If boundary values were different, the solution could correspond partly to the classical extremal (1.3.3), (1.3.4), and partly to the saw-tooth curve; in the last case u' belongs on the boundary of the forbidden interval $|u'| = 1$.*

This considered nonconvex problem is an example of an ill-posed variational problem. For these problems, the classical variational technique based on the Euler equation fails to work. Other methods are needed to deal with such problems. Namely, we replace an ill-posed problem with a *relaxed* one.

1.3.2 Convex Envelope

Consider a variational problem with a nonconvex Lagrangian F. We want to replace this problem with a new one that describes infinitely rapidly oscillating minimizers in terms of averages. This will be done by the construction of the *convex envelope* of a nonconvex Lagrangian. Let us give the definitions (see (Rockafellar, 1997)).

Definition 1.3.1 The *convex envelope* CF is a solution to the following minimal problem:

$$CF(\mathbf{v}) = \inf_{\boldsymbol{\xi}} \frac{1}{l} \int_0^l F(\mathbf{v} + \boldsymbol{\xi}) \quad \forall\, \boldsymbol{\xi} : \int_0^l \boldsymbol{\xi} = 0. \tag{1.3.5}$$

This definition determines the convex envelope as the minimum of all parallel secant hyperplanes that intersect the graph of F; it is based on Jensen's inequality (1.2.4).

To compute the convex envelope CF one can use the Carathéodory theorem (see (Carathéodory, 1967; Rockafellar, 1997)). It states that the argument $\boldsymbol{\xi}(x) = [\xi_1(x), \dots, \xi_n(x)]$ that minimizes the right-hand side of (1.3.5) takes no more than $n+1$ different values. This theorem refers to the obvious geometrical fact that the convex envelope consists of the supporting hyperplanes to the graph $F(\xi_1, \dots, \xi_n)$. Each of these hyperplanes is supported by no more than $(n + 1)$ arbitrary points.

The Carathéodory theorem allows us to replace the integral in the right-hand side of the definition of CF by the sum of $n + 1$ terms; the definition

(1.3.5) becomes:

$$CF(\mathbf{v}) = \min_{m_i \in M} \min_{\boldsymbol{\xi}_i \in \Xi} \left(\sum_{i=1}^{n+1} m_i F(\mathbf{v} + \boldsymbol{\xi}_i) \right), \qquad (1.3.6)$$

where

$$M = \left\{ m_i : \quad m_i \geq 0, \quad \sum_{i=1}^{n+1} m_i = 1 \right\} \qquad (1.3.7)$$

and

$$\Xi = \left\{ \boldsymbol{\xi}_i : \quad \sum_{i=1}^{n+1} m_i \boldsymbol{\xi}_i = 0 \right\}. \qquad (1.3.8)$$

The convex envelope $CF(\mathbf{v})$ of a function $F(\mathbf{v})$ at a point $\mathbf{v}$ coincides with either the function $F(\mathbf{v})$ or the hyperplane that touches the graph of the function F. The hyperplane remains below the graph of F except at the tangent points where they coincide.

The position of the supporting hyperplane generally varies with the point $\mathbf{v}$. A convex envelope of F can be supported by fewer than $n+1$ points; in this case several of the parameters m_i are zero.

On the other hand, the convex envelope is the greatest convex function that does not exceed $F(\mathbf{v})$ in any point $\mathbf{v}$ (Rockafellar, 1997):

$$CF(\mathbf{v}) = \max \phi(\mathbf{v}) : \quad \phi(\mathbf{v}) \leq F(\mathbf{v}) \; \forall \mathbf{v} \quad \text{and} \quad \phi(\mathbf{v}) \text{ is convex.}$$

Example 1.3.2 Obviously, the convex envelope of a convex function coincides with the function itself, so all m_i but m_1 are zero in (1.3.6) and $m_1 = 1$; the parameter $\boldsymbol{\xi}_1$ is zero because of the restriction (1.3.8).

The convex envelope of a "two-well" function,

$$\Phi(\mathbf{v}) = \min \left\{ F_1(\mathbf{v}), \; F_2(\mathbf{v}) \right\},$$

where F_1, F_2 are convex functions of $\mathbf{v}$, either coincides with one of the functions F_1, F_2 or is supported by no more than two points for every $\mathbf{v}$; supporting points belong to different wells. In this case, formulas (1.3.6)–(1.3.8) for the convex envelope are reduced to

$$C\Phi(\mathbf{v}) = \min_{m,\boldsymbol{\xi}} \left\{ m F_1(\mathbf{v} - (1 - m)\boldsymbol{\xi}) + (1 - m) F_2(\mathbf{v} + m\boldsymbol{\xi}) \right\}.$$

Indeed, the convex envelope touches the graphs of the convex functions F_1 and F_2 in no more than one point. Call the coordinates of the touching points $\mathbf{v} + \boldsymbol{\xi}_1$ and $\mathbf{v} + \boldsymbol{\xi}_2$, respectively. The restrictions (1.3.8) become $m_1 \boldsymbol{\xi}_1 + m_2 \boldsymbol{\xi}_2 = 0$, $m_1 + m_2 = 1$. It implies the representations $\boldsymbol{\xi}_1 = -(1 - m)\boldsymbol{\xi}$ and $\boldsymbol{\xi}_2 = m\boldsymbol{\xi}$.

Example 1.3.3 Consider the special case of the two-well function,

$$F(v_1, v_2) = \begin{cases} 0 & \text{if } v_1^2 + v_2^2 = 0, \\ 1 + v_1^2 + v_2^2 & \text{if } v_1^2 + v_2^2 \neq 0. \end{cases} \tag{1.3.9}$$

The convex envelope of F is equal to

$$CF(v_1, v_2) = \begin{cases} 2\sqrt{v_1^2 + v_2^2} & \text{if } v_1^2 + v_2^2 \leq 1, \\ 1 + v_1^2 + v_2^2 & \text{if } v_1^2 + v_2^2 > 1. \end{cases} \tag{1.3.10}$$

Here the envelope is a cone if it does not coincide with F and a paraboloid if it coincides with F.

Indeed, the graph of the function $F(v_1, v_2)$ is axisymmetric in the plane v_1, v_2; therefore, the convex envelope is axisymmetric as well: $CF(v_1, v_2) = f(\sqrt{v_1^2 + v_2^2})$. The convex envelope $CF(\mathbf{v})$ is supported by the point $\mathbf{v} - (1 - m)\boldsymbol{\xi} = \mathbf{0}$ and by a point $\mathbf{v} + m\boldsymbol{\xi} = \mathbf{v}^0$ on the paraboloid $\phi(\mathbf{v}) = 1 + v_1^2 + v_2^2$. We have

$$\mathbf{v}^0 = \frac{1}{1 - m}\mathbf{v}$$

and

$$CF(\mathbf{v}) = \min_m \left\{ (1 - m)\phi\left(\frac{1}{1 - m}\mathbf{v}\right) \right\}.$$

The calculation of the minimum gives (1.3.10).

Example 1.3.4 Consider the nonconvex function $F(v)$ used in Example 1.3.1:

$$F(v) = \min\{(v - 1)^2, (v + 1)^2\}.$$

It is easy to see that the convex envelope CF is

$$CF(v) = \begin{cases} (v + 1)^2 & \text{if } v \leq -1, \\ 0 & \text{if } v \in (-1, 1), \\ (v - 1)^2 & \text{if } v \geq 1. \end{cases}$$

Hessian of Convex Envelope. We mention here a property of the convex envelope that we will use later. If the convex envelope $CF(\mathbf{v})$ does not coincide with $F(\mathbf{v})$ for some $\mathbf{v} = \mathbf{v}_n$, then $CF(\mathbf{v}_n)$ is convex, but not strongly convex. At these points the Hessian $H(F) = \frac{\partial^2}{\partial v_i \partial v_j}F(\mathbf{v})$ is semipositive; it satisfies the relations

$$H(CF(\mathbf{v})) \geq 0, \quad \det H(CF(\mathbf{v})) = 0 \quad \text{if } CF < F,$$

which say that $H(CF)$ is a nonnegative degenerate matrix. These relations can be used to compute $CF(\mathbf{v})$.

1.3.3 Minimal Extension and Minimizing Sequences

The construction of the convex envelope is used to reformulate (relax) a nonconvex variational problem. Consider again the variational problem

$$I(\mathbf{u}) = \min_{\mathbf{u}} \int_0^1 F(x, \mathbf{u}, \mathbf{u}') \qquad (1.3.11)$$

where F is a continuous function that is not convex with respect to its last argument. Recall that this problem does not satisfy the Weierstrass test on the intervals of nonconvexity of F.

Definition 1.3.2 We call the *forbidden region* $\mathcal{Z}_\mathrm{f}$ the set of $\mathbf{z}$ for which $F(x, \mathbf{y}, \mathbf{z})$ is not convex with respect to $\mathbf{z}$,

$$Z_\mathrm{f} = \{z : \ \mathcal{C}_z F(x, \mathbf{y}, \mathbf{z}) < F(x, \mathbf{y}, \mathbf{z})\}.$$

The notation $\mathcal{C}_z F(x, \mathbf{y}, \mathbf{z})$ is used to show the argument $\mathbf{z}$ of the convexification: The other two arguments are considered to be parameters when the convex envelope is calculated. (Later, we omit the subindex $(\ )_z$ when this does not lead to misunderstanding.)

Note that the derivative $\mathbf{u}'$ of a minimizer $\mathbf{u}$ of (1.3.11) never belongs to the region Z_f:

$$\mathbf{u}' \notin Z_\mathrm{f}.$$

This additional constraint on the minimizer is satisfied in the construction of a minimizer of a nonconvex problem.

To deal with a nonconvex problem, we "relax" it. *Relaxation* means that we replace the problem with another one that has the same cost but whose solution is stable against fine-scale perturbations; particularly, it cannot be improved by the Weierstrass variation. The relaxed problem has the following two basic properties:

- The relaxed problem has a classical solution.

- The infimum of the functional (the cost of the problem) in the initial problem coincides with the cost of the relaxed problem.

Here we will demonstrate two approaches to relaxation. Each of them yields to the same construction but uses different arguments to achieve it. In the next chapters we will see similar procedures applied to variational problems with multiple integrals; sometimes they also yield the same construction, but generally they result in different relaxations.

Minimizing Sequences

The first construction is based on local minimization. Consider the extremal problem (1.3.11) and the corresponding solution $\mathbf{u}_0(x)$. Let us fix two neighboring points $A = (x_0, \mathbf{u}_0(x_0))$ and $B = ((x_0 + \varepsilon), \mathbf{u}_0(x_0 + \varepsilon))$ on this solution. Using the smallness of ε, we represent B as

$$B = ((x_0 + \varepsilon), \mathbf{u}_0(x_0) + \varepsilon \mathbf{u}'(x_0) + o(\varepsilon)).$$

The impact to the cost of problem (1.2.1) due to this interval is

$$I_\varepsilon(\mathbf{u}_0) = \int_{x_0}^{x_0+\varepsilon} F(x, \mathbf{u}_0, \mathbf{u}_0') = \varepsilon F(x_0, \mathbf{u}_0(x_0), \mathbf{u}_0'(x_0)) + o(\varepsilon).$$

Let us examine a local variation of the solution $\mathbf{u}_0(x)$: We replace it with a zigzag piecewise linear curve that passes through the points A and B.

Consider a continuous curve $\mathbf{u}_\varepsilon$ that contains $p - 1$ subintervals of the constancy of the derivative $\mathbf{v} = \mathbf{u}_\varepsilon'$. The variable $\mathbf{v}(x)$ takes several values $\mathbf{v}_1 + \mathbf{u}_0'(x_0), \ldots, \mathbf{v}_p + \mathbf{u}_0'(x_0)$; each value is kept on the subinterval of length εm_i, where $\sum m_i = 1$. The derivative $\mathbf{u}_\varepsilon'(x)$ is

$$\mathbf{u}_\varepsilon'(x) = \mathbf{u}_0'(x_0) + \mathbf{v}_k \quad \text{if} \quad x \in \left[x_0 + \varepsilon \sum_{i=1}^{k} m_i,\ x_0 + \varepsilon \sum_{i=1}^{k+1} m_i \right],$$

where $k = 1, \ldots, p - 1$. The saw-tooth curve $\mathbf{u}_\varepsilon$ is

$$\mathbf{u}_\varepsilon(x) = \mathbf{u}(x_0) + \int_{x_0}^{x} (\mathbf{u}'(x_0) + \mathbf{v}(x))dx. \tag{1.3.12}$$

We require that any admissible solution $\mathbf{u}_\varepsilon$ passes through point B. More exactly, we require that its value at the point $x_0 + \varepsilon$ is equal to $\mathbf{u}_0(x_0 + \varepsilon)$ up to the terms of the order of $o(\varepsilon)$,

$$\mathbf{u}_\varepsilon(x_0 + \varepsilon) - \mathbf{u}_0(x_0 + \varepsilon) = \sum_{i=1}^{p} m_i \mathbf{v}_i = o(\varepsilon). \tag{1.3.13}$$

Let us compute the effect of replacing the differentiable curve $\mathbf{u}_0$ with the zigzag curve $\mathbf{u}_\varepsilon$. We estimate the integral of $F(x, \mathbf{u}_\varepsilon, \mathbf{u}_\varepsilon')$ over this interval, up to terms of order of $o(\varepsilon)$. To estimate, we use the smallness of the interval of variation. Replace $\mathbf{u}_\varepsilon(x)$ with $\mathbf{u}_0(x_0)$

$$\mathbf{u}_\varepsilon(x) = \mathbf{u}_0(x_0) + O(\varepsilon)$$

and compute

$$F(x, \mathbf{u}_\varepsilon(x), \mathbf{v}_i + \mathbf{u}_0'(x_0)) = F(x_0, \mathbf{u}_0(x_0), \mathbf{u}_0'(x_0) + \mathbf{u}_\varepsilon'(x)) + O(\varepsilon)$$

for any $x \in [x_0, x_0 + \varepsilon]$. The Lagrangian (rounded up to $O(\varepsilon)$) is piecewise constant in the interval $[x_0, x_0 + \varepsilon]$. The impact $I_\varepsilon(\mathbf{u}_\varepsilon)$ becomes

$$I_\varepsilon(\mathbf{u}_\varepsilon) = \varepsilon \sum_{i=1}^{p} m_i F(x_0, \mathbf{u}_\varepsilon(x_0), \mathbf{u}_0'(x_0) + \mathbf{v}_i) + o(\varepsilon). \tag{1.3.14}$$

Calculate the minimum of (1.3.14) with respect to the arguments $\mathbf{v}_1, \ldots, \mathbf{v}_p$ and $m_1, \ldots, m_p$, which are subject to the constraints (see (1.3.13))

$$m_i(x) \geq 0, \quad \sum_{i=1}^{p} m_i = 1, \quad \sum_{i=1}^{p} m_i \mathbf{v}_i = 0. \qquad (1.3.15)$$

This minimum coincides with the convex envelope of the original Lagrangian with respect to its last argument (see (1.3.7)):

$$\min_{m_i, \mathbf{v}_i \in (1.3.15)} \sum_{i=1}^{p} m_i F(x, \mathbf{u}, \mathbf{v}_i) = C F_{\mathbf{v}}(x, \mathbf{u}_0, \mathbf{v}). \qquad (1.3.16)$$

By referring to the Carathéodory theorem (1.3.7) we conclude that it is enough to split the interval into

$$p = n + 1 \qquad (1.3.17)$$

parts so that $\mathbf{v} = \mathbf{u}'$ takes $k + 1$ values.

Note that the constraint (1.3.15) leaves the freedom to choose inner parameters m_i and $\mathbf{v}_i$ to minimize the function $\sum_{i=1}^{p} m_i F(u, \mathbf{v}_i)$ and thus to minimize the cost of the variational problem (see (1.3.16)).

Compare the costs $I_\varepsilon(\mathbf{u}_0)$ and $I_\varepsilon(\mathbf{u}_\varepsilon)$ of (1.3.11) corresponding to the smooth solution $\mathbf{u}_0$ and to the zigzag solution $(\mathbf{u}_\varepsilon)$. Using the definition of the convex envelope we obtain the inequality:

$$\frac{1}{\varepsilon} \left(I_\varepsilon(\mathbf{u}_0) - I_\varepsilon(\mathbf{u}_\varepsilon) \right) = F(x_0, \mathbf{u}_0(x_0), \mathbf{u}_0'(x_0))$$

$$- C F(x_0, \mathbf{u}_\varepsilon(x_0), \mathbf{u}'(x_0)) + O(\varepsilon) \geq 0.$$

We see that the zigzag solution $\mathbf{u}_\varepsilon$ corresponds to lower cost if $F > C_z F$, that is, in the regions of nonconvexity of F.

Passing to the variational problem (1.3.11) in the whole interval $[0, 1]$ we perform the preceding extension in each interval of length ε. This extension replaces the Lagrangian $F(x, y, z)$ with the convex envelope $C_z F(x, y, z)$ so that the relaxed problem becomes

$$I = \min_{\mathbf{u}} \int_0^1 C_{\mathbf{u}'} F(x, \mathbf{u}(x), \mathbf{u}'(x)). \qquad (1.3.18)$$

The curve $\mathbf{u}_\varepsilon$ strongly converges to the curve $\mathbf{u}_0$:

$$\|\mathbf{u}_\varepsilon - \mathbf{u}_0\|_{L_\infty[0,1]} \to 0, \quad \text{as } \varepsilon \to 0,$$

but its derivative converges to $\mathbf{u}_0'$ only weakly in L_p,

$$\int_0^1 \phi(\mathbf{u}_\varepsilon' - \mathbf{u}_0') \to 0 \quad \forall \, \phi \in L_q(0, 1), \quad \frac{1}{p} + \frac{1}{q} = 1.$$

For the definition and discussion of the weak convergence we refer the reader to books on analysis, such as (Shilov, 1996).

Remark 1.3.2 *The choice of the proper space L_p depends on the Lagrangian F because $F(., ., \mathbf{u}')$ must be integrable.*

The cost of the reformulated (relaxed) problem (1.3.18) corresponds to the cost of the problem (1.3.11) on the minimizing sequence (1.3.12). Therefore, the cost of the relaxed problem is equal to the cost of the original problem (1.3.11). The extension of the Lagrangian that preserves the cost of the problem is called the *minimal extension*. The minimal extension enlarges the set of classical minimizers by including generalized curves in it.

Generally speaking, this extension leads to an attainable upper bound of the cost of an unstable problem because we cannot guarantee that the extension cannot be further improved. However, the Lagrangian of the relaxed problem is convex, which guarantees that its minimizers satisfy the Weierstrass test and is stable against fine-scale perturbations.

Minimal Extension, Based on the Weierstrass Test

We introduce an alternative method of relaxation that leads to the same results but does not require consideration of the structure of minimizing sequences.

Consider the class of Lagrangians $\mathcal{N}F(x, y, z)$ that are smaller than $F(x, y, z)$ and satisfy the Weierstrass test $\mathcal{W}(\mathcal{N}F(x, y, z)) \geq 0$:

$$\begin{cases} \mathcal{N}F(x, y, z) - F(x, y, z) \leq 0, \\ \mathcal{W}(\mathcal{N}F(x, y, z)) \geq 0 \end{cases} \quad \forall\, x, y, z.$$

Let us take the maximum on $\mathcal{N}F(x, y, z)$ and call it $\mathcal{S}F$. Clearly, $\mathcal{S}F$ corresponds to turning one of these inequalities into an equality:

$$\begin{array}{llll} \mathcal{S}F(x, y, z) = F(x, y, z), & \mathcal{W}(\mathcal{S}F(x, y, z)) \geq 0 & \text{if} & z \notin \mathcal{Z}_{\mathsf{f}}, \\ \mathcal{S}F(x, y, z) \leq F(x, y, z), & \mathcal{W}(\mathcal{S}F(x, y, z)) = 0 & \text{if} & z \in \mathcal{Z}_{\mathsf{f}}. \end{array}$$

This variational inequality describes the extension of the Lagrangian of an unstable variational problem. Notice that

1. The first equality holds in the region of convexity of F and the extension coincides with F in that region.

2. In the region where F is not convex, the Weierstrass test of the extended Lagrangian is satisfied as an equality; this equality serves to determine the extension.

These conditions imply that $\mathcal{S}F$ is convex everywhere. Also, $\mathcal{S}F$ is the maximum over all convex functions that do not exceed F. Again, $\mathcal{S}F$ is equal to the convex envelope of F:

$$\mathcal{S}F(x, y, z) = \mathcal{C}_z F(x, y, z).$$

The cost of the problem remains the same because the convex envelope corresponds to a minimizing sequence of the original problem.

Remark 1.3.3 *Note that the geometrical property of convexity never explicitly appears here. We simply satisfy the Weierstrass necessary condition everywhere. Hence, this relaxation procedure can be extended to more complicated multidimensional problems for which the Weierstrass condition and convexity do not coincide.*

Properties of the Relaxed Problem

- Recall that the derivative of the minimizer never takes values in the region $\mathcal{Z}_{\mathrm{f}}$ of nonconvexity of F. Therefore, a solution to a nonconvex problem stays the same if its Lagrangian $F(x, \mathbf{y}, \mathbf{z})$ is replaced by any Lagrangian $\mathcal{N}F(x, \mathbf{y}, \mathbf{z})$ that satisfies the restrictions

$$\mathcal{N}F(x, \mathbf{y}, \mathbf{z}) = F(x, \mathbf{y}, \mathbf{z}) \quad \forall \ z \notin \mathcal{Z}_{\mathrm{f}},$$
$$\mathcal{N}F(x, \mathbf{y}, \mathbf{z}) > \mathcal{C}F(x, \mathbf{y}, \mathbf{z}) \ \forall \ z \in \mathcal{Z}_{\mathrm{f}}.$$

Indeed, the two Lagrangians $F(x, \mathbf{y}, \mathbf{z})$ and $\mathcal{N}F(x, \mathbf{y}, \mathbf{z})$ coincide in the region of convexity of F. Therefore, the solutions to the variational problem also coincide in this region. Neither Lagrangian satisfies the Weierstrass test in the forbidden region of nonconvexity. Therefore, no minimizer can distinguish between these two problems: It never takes values in Z_{f}. The behavior of the Lagrangian in the forbidden region is simply of no importance. In this interval, the Lagrangian cannot be computed from the minimizer.

- The infimum of the functional for the initial problem coincides with the minimum of the functional in the relaxed problem. The relaxed problem has a convex Lagrangian. The Weierstrass test is satisfied, and the minimal solution (if it exists) is stable against fine-scale perturbations. To be sure that the solution of the relaxed problem exists, one should also examine other sources of possible nonexistence (see, for example (Ioffe and Tihomirov, 1979)).

- The number of minimizers in the relaxed problem is increased. Instead of one n-dimensional vector minimizer $\mathbf{u}(x)$ in the original problem, they now include $n + 1$ vector minimizers $\mathbf{v}_i(x)$ and $n + 1$ minimizers $m_i(x)$ (see (1.3.17)) connected by two equalities (1.3.15) and the inequalities $m_i(x) \geq 0$. The relaxed problem is controlled by the larger number of independent parameters that are used to compute the relaxed Lagrangian $\mathcal{C}F(x, \mathbf{u}, \mathbf{u}')$.

In the forbidden region, the Euler equations degenerate. For example, suppose that u is a scalar; the convex envelope has the form

$$\mathcal{C}F = au' + b(x, u)$$

if it does not coincide with G. This representation implies that the first term in the left-hand side (1.2.11) of the Euler equation (1.2.12) vanishes

Average derivative	Pointwise derivatives	Optimal concentrations	Convex envelope $CG(u,v)$
$v < -1$	$v_1^0 = v_2^0 = v$	$m_1^0 = 1,\ m_2^0 = 0$	$u^2 + (v-1)^2$
$\|v\| < 1$	$v_1^0 = 1,\ v_2^0 = -1$	$m_1^0 = m_2^0 = \frac{1}{2}$	u^2
$v > 1$	$v_1^0 = v_2^0 = v$	$m_1^0 = 0,\ m_2^0 = 1$	$u^2 + (v+1)^2$

TABLE 1.1. Characteristics of an optimal solution in Example 1.3.1.

identically: $\frac{d}{dx}\frac{\partial}{\partial u'}CF \equiv 0$. The Euler equation degenerates into an algebraic equation $\frac{\partial}{\partial u}CF = 0$.

In the general case, the order of the system of Euler equations decreases (for details, see (Gamkrelidze, 1962; Gabasov and Kirillova, 1973; Clements and Anderson, 1978)).

1.3.4 Examples: Solutions to Nonconvex Problems

Example 1.3.5 We revisit Example 1.3.1. Let us solve this problem by building the convex envelope of the Lagrangian $G(u,v)$:

$$\mathcal{C}_v G(u,v) = \min_{m_1,m_2} \min_{v_1,v_2} \left\{ u^2 + m_1(v_1 - 1)^2 + m_2(v_2 + 1)^2 \right\},$$
$$v = m_1 v_1 + m_2 v_2, \quad m_1 + m_2 = 1, \quad m_i \geq 0.$$

The form of the minimum depends on the value of $v = u'$. The convex envelope $CG(u,v)$ coincides with either $G(u,v)$ if $v \notin [0,\ 1]$ or $CG(u,v) = u^2$ if $v \in [0,\ 1]$; see Example 1.3.4. Optimal values v_1^0, v_2^0, $m_1^0\ m_2^0$ of the minimizers and the convex envelope CG are shown in Table 1.1. The relaxed form of the problem with zero boundary conditions

$$\min_u \int_0^1 CG(u,\ u'), \quad u(0) = u(1) = 0,$$

has an obvious solution,

$$u(x) = u'(x) = 0,$$

that yields the minimal (zero) value of the functional. It corresponds to the constant optimal value m_{opt} of $m(x)$: $m_{\mathrm{opt}}(x) = \frac{1}{2}\ \forall x \in [0,1]$.

The relaxed Lagrangian is minimized over four functions $u, m_1, v_1,\ v_2$ bounded by one equality, $u' = m_1 v_1 + (1 - m_1)v_2$ and the inequalities $0 \leq m \leq 1$, while the original Lagrangian is minimized over one function u. In contrast to the initial problem, the relaxed one has a differentiable solution in terms of these four controls.

A Two-Well Lagrangian

We turn to a more advanced example of the relaxation of an ill-posed nonconvex variational problem. This example highlights more properties

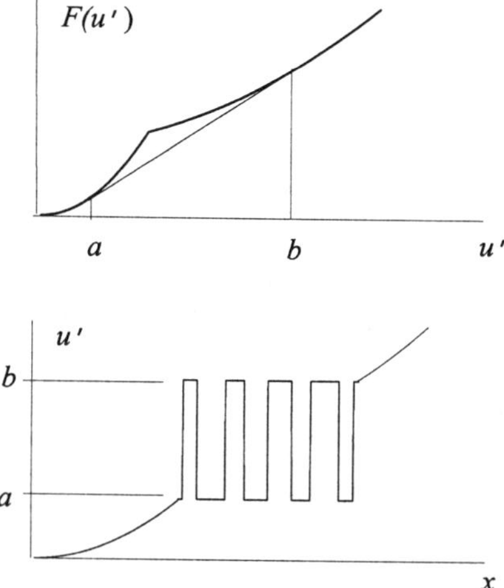

FIGURE 1.5. Convexification of the Lagrangian (top) and the minimizer (bottom); points a and b are equal to v_1 and v_2, respectively.

of relaxation and introduces piecewise quadratic Lagrangians that are the main tool in the investigation of optimal composites.

Example 1.3.6 Consider the minimization problem

$$\min_{u(x)} \int_0^z F_p(x, u, u'), \quad u(0) = 0, \ u'(z) = 0 \qquad (1.3.19)$$

with a Lagrangian

$$F_p = (u - \alpha x^2)^2 + F_n(u'), \qquad (1.3.20)$$

where

$$F_n(v) = \min\{a\,v^2, \ b\,v^2 + 1\}, \quad 0 < a < b, \ \alpha > 0.$$

Note that the second term F_n of the Lagrangian F_p is a nonconvex function of u'.

The first term $(u - \alpha x^2)^2$ of the Lagrangian forces the minimizer u and its derivative u' to increase with x, until u' at some point reaches the interval of nonconvexity of $F_n(u')$. The derivative u' must vary outside of the forbidden interval of nonconvexity of the function F_n at all times.. Formally, this problem is ill-posed because the Lagrangian is not convex with respect to u' (Figure 1.5); therefore, it needs relaxation.

To find the convex envelope CF we must transform $F_n(u')$ (in this example, the first term of F_p (see (1.3.20)) is independent of u' and it does not change after the convexification). The convex envelope CF_p is equal to

$$CF_p = (u - \alpha x^2)^2 + CF_n(u').$$

Let us compute $CF_n(v)$ (again we use the notation $v = u'$). The envelope $CF_n(v)$ coincides with either the graph of the original function or the linear function $l(v) = Av + B$ that touches the original graph in two points (as it is predicted by the Carathéodory theorem; in this example $n = 1$). This function can be found as the common tangent $l(v)$ to both convex branches (wells) of $F_n(v)$:

$$\begin{cases} l(v) = av_1^2 + 2av_1(v - v_1), \\ l(v) = (bv_2^2 + 1) + 2bv_2(v - v_2), \end{cases}$$

where v_1 and v_2 belong to the corresponding branches of F_p:

$$\begin{cases} l(v_1) = av_1^2, \\ l(v_2) = bv_2^2 + 1. \end{cases}$$

Solving this system for v, v_1, v_2 we find the coordinates of the supporting points

$$v_1 = \sqrt{\frac{b}{a(a - b)}}, \quad v_2 = \sqrt{\frac{a}{b(a - b)}},$$

and we calculate the relaxed Lagrangian:

$$CF_n(v) = \begin{cases} av^2 & \text{if } |v| < v_1, \\ 2v\sqrt{\frac{ab}{a-b}} - \frac{b}{a-b} & \text{if } v \in [v_1,\, v_2], \\ 1 + bv^2 & \text{if } |v| < v_2 \end{cases}$$

that linearly depends on $v = u'$ in the region of nonconvexity of F.

The relaxed problem has the form

$$\min_u \int CF_p(x, u, u'),$$

where

$$CF_p(x, u, u') = \begin{cases} (u - \alpha x^2)^2 + a(u')^2 & \text{if } |u'| \le v_1, \\ (u - \alpha x^2)^2 + 2u'\sqrt{\frac{ab}{a-b}} - \frac{b}{a-b} & \text{if } v_1 \le |u'| \le v_2, \\ (u - \alpha x^2)^2 + b(u')^2 + 1 & \text{if } |u'| \ge v_2. \end{cases}$$

Note that the variables u, v in the relaxed problem are the averages of the original variables; they coincide with those variables everywhere when $CF = F$. The Euler equation of the relaxed problem is

$$\begin{cases} au'' - (u - \alpha x^2) = 0 & \text{if } |u'| \le v_1, \\ (u - \alpha x^2) = 0 & \text{if } v_1 \le |u'| \le v_2, \\ bu'' - (u - \alpha x^2) = 0 & \text{if } |u'| \ge v_2. \end{cases}$$

The Euler equation is integrated with the boundary conditions shown in (1.3.19). Notice that the Euler equation degenerates into an algebraic equation in the interval of convexification. The solution u and the variable $\frac{\partial}{\partial u'}CF$ of the relaxed problem are both continuous everywhere.

Integrating the Euler equations, we sequentially meet the three regimes when both the minimizer and its derivative monotonically increase with x (see Figure 1.5). If the length z of the interval of integration is chosen sufficiently large, one can be sure that the optimal solution contains all three regimes; otherwise, the solution may degenerate into a two-zone solution if $u'(x) \leq v_2 \ \forall x$ or into a one-zone solution if $u'(x) \leq v_1 \ \forall x$ (in the last case the relaxation is not needed; the solution is a classical one).

Let us describe minimizing sequences that form the solution to the relaxed problem. Recall that the actual optimal solution is a generalized curve in the region of nonconvexity; this curve consists of infinitely often alternating parts with the derivatives v_1 and v_2 and the relative fractions $m(x)$ and $(1 - m(x))$:

$$v = \langle u'(x) \rangle = m(x)v_1 + (1 - m(x))v_2, \quad u' \in [v_1, v_2], \qquad (1.3.21)$$

where $\langle \ \rangle$ denotes the average, u is the solution to the original problem, and $\langle u \rangle$ is the solution to the homogenized (relaxed) problem.

The Euler equation degenerates in the second region into an algebraic one $\langle u \rangle = \alpha x^2$ because of the linear dependence of the Lagrangian on $\langle u \rangle'$ in this region. The first term of the Euler equation,

$$\frac{d}{dx} \frac{\partial F}{\partial \langle u \rangle'} \equiv 0 \quad \text{if} \ \ v_1 \leq |\langle u \rangle'| \leq v_2,$$

vanishes at the optimal solution.

The variable m of the generalized curve is nonzero in the second regime. This variable can be found by differentiation of the optimal solution:

$$(\langle u \rangle - \alpha x^2)' = 0 \quad \Longrightarrow \quad \langle u \rangle' = 2\alpha x.$$

This equality, together with (1.3.21), implies that

$$m = \begin{cases} 0 & \text{if} \ \ |u'| \leq v_1, \\ \frac{2\alpha}{v_1 - v_2} x - \frac{v_2}{v_1 - v_2} & \text{if} \ \ v_1 \leq |u'| \leq v_2, \\ 1 & \text{if} \ \ |u'| \geq v_2. \end{cases} \qquad (1.3.22)$$

Variable m linearly increases within the second region (see Figure 1.5). Note that the derivative u' of the minimizing generalized curve at each point x lies on the boundaries v_1 or v_2 of the forbidden interval of nonconvexity of F; the average derivative varies only due to varying of the fraction $m(x)$ (see Figure 1.5).

1.3.5 Null-Lagrangians and Convexity

The convexity requirements of the Lagrangian F that follow from the Weierstrass test are in agreement with the concept of null-Lagrangians (see, for example (Strang, 1986)).

Definition 1.3.3 The Lagrangians $\phi(x, \mathbf{u}, \mathbf{u}')$ for which the Euler equation (1.2.12), (1.2.11) identically vanishes are called *Null-Lagrangians*.

It is easy to check that null-Lagrangians in one-dimensional variational problems are linear functions of $\mathbf{u}'$. Indeed, the Euler equation is a second-order differential equation with respect to $\mathbf{u}$:

$$\frac{d}{dx}\left(\frac{\partial}{\partial \mathbf{u}'}\phi\right) - \frac{\partial}{\partial \mathbf{u}}\phi = \frac{\partial^2 \phi}{\partial(\mathbf{u}')^2}\cdot\mathbf{u}'' + \frac{\partial^2 \phi}{\partial \mathbf{u}'\partial \mathbf{u}}\cdot\mathbf{u}' + \frac{\partial^2 \phi}{\partial \mathbf{u}\partial x} - \frac{\partial \phi}{\partial \mathbf{u}} \equiv 0.$$

The coefficient of $\mathbf{u}''$ is equal to $\frac{\partial^2 \phi}{\partial(\mathbf{u}')^2}$. If the Euler equation holds identically, this coefficient is zero, and therefore $\frac{\partial G}{\partial \mathbf{u}'}$ does not depend on $\mathbf{u}'$. Hence, ϕ linearly depends on $\mathbf{u}'$:

$$\phi(x, \mathbf{u}, \mathbf{u}') = \mathbf{u}' \cdot A(\mathbf{u}, x) + B(\mathbf{u}, x);$$
$$A = \frac{\partial^2 \phi}{\partial \mathbf{u}'\partial \mathbf{u}}, \quad B = \frac{\partial^2 \phi}{\partial \mathbf{u}\partial x} - \frac{\partial \phi}{\partial \mathbf{u}}.$$

In addition, if the equality

$$\frac{\partial A}{\partial x} = \frac{\partial B}{\partial \mathbf{u}}$$

holds, then the Euler equation vanishes identically. In this case, ϕ is a null-Lagrangian.

Example 1.3.7 Function $\phi = u\,u'$ is the null-Lagrangian. We have

$$\frac{d}{dx}\left(\frac{\partial}{\partial u'}\phi\right) - \frac{\partial}{\partial u}\phi = u' - u' \equiv 0.$$

Consider a variational problem with the Lagrangian F,

$$\min_{\mathbf{u}} \int_0^1 F(x, \mathbf{u}, \mathbf{u}').$$

Adding a null-Lagrangian to the given Lagrangian does not affect the Euler equation of the problem. The family of problems

$$\min_{\mathbf{u}} \int_0^1 \left(F(x, \mathbf{u}, \mathbf{u}') + t\phi(x, \mathbf{u}, \mathbf{u}')\right),$$

where t is an arbitrary number, corresponds to the same Euler equation. Therefore, each solution to the Euler equation corresponds to a family of Lagrangians $F(x, \mathbf{u}, \mathbf{z}) + t\phi(x, \mathbf{u}, \mathbf{z})$, where t is an arbitrary real number. This says, in particular, that a Lagrangian cannot be uniquely defined by the solution to the Euler equation.

The stability of the minimizer against the Weierstrass variations should be a property of the Lagrangian that is independent of t. It should be a common property of the family of equivalent Lagrangians. On the other hand, if $F(x, \mathbf{u}, \mathbf{z})$ is convex with respect to $\mathbf{z}$, then $F(x, \mathbf{u}, \mathbf{z}) + t\phi(x, \mathbf{u}, \mathbf{z})$ is also convex. Indeed, $\phi(x, \mathbf{u}, \mathbf{z})$ is linear as a function of $\mathbf{z}$, and adding the term $t\phi(x, \mathbf{u}, \mathbf{z})$ does not affect the convexity of the sum. In other words, convexity is a characteristic property of the family. Accordingly, it serves as a test for the stability of an optimal solution.

1.3.6 Duality

Legendre Transform

A useful tool in variational problems is duality. Particularly, duality allows us to effectively compute the convex envelope of a Lagrangian. For detailed exposition, we refer to (Gelfand and Fomin, 1963; Rockafellar, 1967; Rockafellar, 1997; Ekeland and Temam, 1976; Fenchel, 1949; Ioffe and Tihomirov, 1979).

The classical version of the duality relations is based on the Legendre transform of the Lagrangian. Consider the Lagrangian $L(x, u, u')$ that is convex with respect to u'. Consider an extremal problem

$$\max_{u'} \{p\, u' - L(x, u, u')\} \tag{1.3.23}$$

that has a solution satisfying the following equation:

$$p = \frac{\partial L}{\partial u'}. \tag{1.3.24}$$

The variable p is called the *dual* or *conjugate* to the "prime" variable u; p is also called the *impulse*. Equation (1.3.24) is solvable for u', because $L(., ., u')$ is convex. We have

$$u' = \phi(p, u, x).$$

These relations allow us to construct the Hamiltonian H of the system.

Definition 1.3.4 The *Hamiltonian* is the following function of u, p, and x:

$$H(x, u, p) = p\, \phi(p, u, x) - L(x, u, \phi(p, u, x)).$$

The Euler equations and the dual relations yield to exceptionally symmetric representations, called *canonical equations*,

$$u' = -\frac{\partial H}{\partial p}, \quad p' = \frac{\partial H}{\partial u}.$$

Generally, u and p are n-dimensional vectors. The canonical relations are given by $2n$ first-order differential equations for two n-dimensional vectors u and p.

The dual form of the Lagrangian can be obtained from the Hamiltonian when the variable u is expressed as a function of p and p' and excluded from the Hamiltonian. The dual equations for the extremal can be obtained from the canonical system if it is reduced to a system of n second-order differential equations for p.

Example 1.3.8 Find a conjugate to the Lagrangian

$$F(u, u') = \frac{1}{2}\sigma(u')^2 + \frac{\gamma}{2}u^2.$$

The impulse p is

$$p = \frac{\partial F}{\partial u'} = \sigma u'.$$

Derivative u' is expressed through p as

$$u' = \frac{p}{\sigma}.$$

The Hamiltonian H is

$$H = \frac{1}{2}\frac{p^2}{\sigma} - \gamma u^2.$$

The canonical system is

$$u' = \frac{p}{\sigma}, \quad p' = \gamma u,$$

and the dual form F^* of the Lagrangian is obtained from the Hamiltonian using canonical equations to exclude u, as follows:

$$F^*(p, p') = \frac{1}{2}\left(\frac{p^2}{\sigma} - \frac{1}{\gamma}(p')^2\right).$$

The Legendre transform is an involution: The variable dual to the variable $\mathbf{p}$ is equal to $\mathbf{u}$.

Conjugate

The natural generalization of the ideas of the Legendre transform to nonconvex and nondifferentiable Lagrangians yields to conjugate variables. They are obtained by the Young–Fenchel transform (Fenchel, 1949; Rockafellar, 1966; Ekeland and Temam, 1976).

Definition 1.3.5 Let us define $L^*(\mathbf{z}^*)$–the *conjugate* to the $L(z)$–by the relation

$$L^*(\mathbf{z}^*) = \max_{\mathbf{z}}\left\{\mathbf{z}^*\,\mathbf{z} - L(\mathbf{z})\right\}, \tag{1.3.25}$$

which implies that $\mathbf{z}^*$ is an analogue of p (compare with (1.3.23)).

Let us compute the conjugate to the Lagrangian $L(x, \mathbf{y}, \mathbf{z})$ with respect to $\mathbf{z}$, treating $x, \mathbf{y}$ as parameters. If L is a convex and differentiable function of $\mathbf{z}$, then (1.3.25) is satisfied if

$$z^* = \frac{\partial L(\mathbf{z})}{\partial \mathbf{z}},$$

which is similar to (1.3.24). This similarity suggests that the Legendre transform $\mathbf{p}$ and the Young–Fenchel transform $\mathbf{z}^*$ coincide if the Legendre transform is applicable.[3]

However, the Young–Fenchel transform is defined and finite for a larger class of functions, namely, for any Lagrangian that grows not slower than an affine function:

$$L(\mathbf{z}) \geq c_1 + c_2 \|\mathbf{z}\| \quad \forall \mathbf{z},$$

where c_1 and $c_2 > 0$ are constants.

Example 1.3.9 Find a conjugate to the function

$$F(x) = |x|.$$

From (1.3.25) we have

$$F^*(x^*) = \begin{cases} 0 & \text{if} \quad |x^*| < 1, \\ \infty & \text{if} \quad |x^*| > 1. \end{cases}$$

The Use of the Young–Fenchel Transform. We can compute the conjugate to $F^*(\mathbf{z}^*)$, called the *second conjugate F^{**} to F*,

$$F^{**}(\mathbf{z}) = \max_{\mathbf{z}^*} \left\{ \mathbf{z}^* \cdot \mathbf{z} - F^*(\mathbf{z}^*) \right\}.$$

We denote the argument of F^{**} by $\mathbf{z}$.

If $F(\mathbf{z})$ is convex, then the transform is an involution. If $F(\mathbf{z})$ is not convex, the second conjugate is the convex envelope of F (see (Rockafellar, 1997)):

$$F^{**} = \mathcal{C}F.$$

We relax a variational problem with a nonconvex Lagrangian $L(x, \mathbf{u}, \mathbf{u}')$ by replacing it with its second conjugate:

$$\mathcal{C}_{\mathbf{v}} L(x, \mathbf{u}, \mathbf{v}) = L^{**}(x, \mathbf{u}, \mathbf{v}) = \max_{\mathbf{v}^*} \left\{ \mathbf{v}^* \cdot \mathbf{v} - L^*(x, \mathbf{u}, \mathbf{v}^*) \right\}.$$

Note that $x, \mathbf{u}$ are treated as constant parameters during this calculation.

---

[3]Later, we will also use the notation $\mathbf{z}^{\text{dual}}$ for the adjoint variable denoted here as $\mathbf{z}^*$.

1.4 Conclusion and Problems

We have observed the following:

- A one-dimensional variational problem has the fine-scale oscillatory minimizer if its Lagrangian $F(x, u, u')$ is a nonconvex function of its third argument.

- Homogenization leads to the relaxed form of the problem that has a classical solution and preserves the cost of the original problem.

- The relaxed problem is obtained by replacing the Lagrangian of the initial problem by its convex envelope. It can be computed as the second conjugate to F.

- The dependence of the Lagrangian on its third argument in the region of nonconvexity does not effect the relaxed problem.

To relax a one-dimensional variational problem we have used two ideas. First, we replaced the function with its convex envelope and got a stable extension of the problem. Second, we proved that the value of the integral of the convex envelope $CF(\mathbf{v})$ of a given function is equal to the value of the integral of this function $F(\mathbf{v})$ if its argument $\mathbf{v}$ is a zigzag curve. We use the Carathéodory theorem, which tells that the number of subregions of constancy of the argument is less than or equal to $n + 1$, where n is the dimension of the argument of the Lagrangian.

In principle, this construction is also valid for multidimensional variational problems unless the argument of the integral satisfies additional differential restrictions. However, these restrictions necessarily occur in multidimensional problems that deal with the minimization of Lagrangians that depend on gradients of some potentials or vectors of currents. The gradient of a function is not a free vector if the dimension of the space is greater than one; the field $\mathbf{e} = \nabla w$ is curlfree: $\nabla \times \mathbf{e} = 0$. Likewise, the current $\mathbf{j}$ is divergencefree: $\nabla \cdot \mathbf{j} = 0$. These differential restrictions express integrability conditions (the equality of mixed second derivatives) for potentials; they are typical for multidimensional variational problems and they do not have a one-dimensional analogue. Generally, the multidimensional problem cannot be relaxed by convexification of its Lagrangian. In this case, convexity of the Lagrangian $F(\mathbf{x}, w, \nabla w)$ with respect to the last argument is replaced by the more delicate property of *quasiconvexity*, which will be discussed in Chapter 6. Relaxation of multidimensional problems requires replacing the Lagrangian by its *quasiconvex envelope*.

Problems

1. Formulate the Weierstrass test for the extremal problem

$$\min_u \int_0^1 F(x, u, u', u'')$$

that depends on the second derivative u''.

2. Find the relaxed formulation of the problem

$$\min_{u_1, u_2} \int_0^1 \left(u_1^2 + u_2^2 + F(u_1', u_2') \right),$$
$$u_1(0) = u_2(0) = 0, \quad u_1(1) = a, \quad u_2(1) = b,$$

where $F(v_1, v_2)$ is defined by (1.3.9). Formulate the Euler equations for the relaxed problems and find minimizing sequences.

3. Find the relaxed formulation of the problem

$$\min_u \int_0^1 \left(u^2 + \min\{|u' - 1|, \ |u' + 1| + 0.5\} \right),$$
$$u(0) = 0, \quad u(1) = a.$$

Formulate the Euler equation for the relaxed problems and find minimizing sequences.

4. Find the conjugate and second conjugate to the function

$$F(x) = \min\{x^2, \ 1 + ax^2\}, \quad 0 < a < 1.$$

Show that the second conjugate coincides with the convex envelope CF of F.

5. Let $x(t) > 0$, $y(t)$ be two scalar variables and $f(x, y) = x\, y^2$. Demonstrate that

$$f(\langle x \rangle, \langle y \rangle) \geq \langle y \rangle^2 \langle \frac{1}{x} \rangle^{-1}.$$

When is the equality sign achieved in this relation?

Hint: Examine the convexity of a function of two scalar arguments,

$$g(y, z) = \frac{y^2}{z}, \quad z > 0.$$

2
Conducting Composites

We begin the study of structural optimization by optimizing conducting media. This chapter introduces the subject of optimization. We describe the equations for equilibrium of conductivity in inhomogeneous media. We also discuss conducting composites and homogenization–the averaging of fields in micro-inhomogeneous media and the tensor of effective properties of a composite.

2.1   Conductivity of Inhomogeneous Media

2.1.1   Equations for Conductivity

Many physical processes are described by the conductivity or transport equations. The equilibria of electrical and thermal conduction are among them, where the electrical potential and temperature play the role of potentials, and various diffusion equilibria, where the concentration of the diffusive substance is the potential. Transport processes include chemical diffusion, flow in porous media, and the steady-state electrical field in a dielectric. The conductivity equations are derived from a few general conservation laws; they are applicable to various physical situations. In the text we often refer to thermal or electrical conduction when specific problems of structural optimization are discussed. However, the results can be equally well applied to other physical processes.

We consider the steady-state conductivity equilibrium. All variables are independent of time, and they depend only on the space coordinates $\mathbf{x} =$

(x_1, x_2, x_3). A detailed discussion of the conductivity equations can be found in standard textbooks on mathematical physics, such as (Courant and Hilbert, 1962) or physics such as (Landau and Lifshitz, 1984). Here we review the conductivity equations emphasizing the inhomogeneity of media. For definiteness, let us look at the electrical conductivity:

Current

Conductivity assumes that a current of particles passes through a medium. Let us denote the vector of the current by $\mathbf{j} = [j_1, j_2, j_3]$. The current satisfies a differential constraint (called the kinetic equation) that corresponds to the conservation of charge: The total number of particles that cross the boundary of any subdomain from inside and outside equals zero. By Green's theorem, $\mathbf{j}$ is divergencefree in Ω:

$$\nabla \cdot \mathbf{j} = 0 \quad \text{in } \Omega. \tag{2.1.1}$$

If we assume that sources or sinks with intensity $f(\mathbf{x})$ are present, then (2.1.1) takes a more general form:

$$\nabla \cdot \mathbf{j} = f \quad \text{in } \Omega. \tag{2.1.2}$$

It says that the difference between the number of particles that cross the boundary of a domain from inside and outside is equal to the density of the sources in that domain.

Field

The second equation of conductivity specifies the force field $\mathbf{e}$ that causes the motion of particles. We assume that the system is conservative. This implies the existence of a potential $w = w(\mathbf{x})$ for $\mathbf{e}$:

$$\mathbf{e} = \nabla w. \tag{2.1.3}$$

Constitutive Relations

The last equation $\mathbf{j} = \mathbf{j}(\mathbf{e})$ is the constitutive relation. It specifies the material properties mathematically as the dependence of $\mathbf{j}$ on $\mathbf{e}$. This dependence completely defines the conducting material.

Here we assume that this dependence is linear:

$$\mathbf{j} = \boldsymbol{\sigma}\mathbf{e}, \tag{2.1.4}$$

where $\boldsymbol{\sigma}$ is a positive definite symmetric tensor:

$$\boldsymbol{\sigma} = \boldsymbol{\sigma}^T, \quad \boldsymbol{\sigma} > 0.$$

We call $\boldsymbol{\sigma}$ the *conductivity tensor*. Formally, a linear conducting material is specified by its conductivity tensor.

Inhomogeneous Materials

The constitutive relations in isotropic materials express the proportionality between vectors $\mathbf{j}$ and $\mathbf{e}$. Inhomogeneous isotropic materials correspond to conductivity tensors of the form

$$\boldsymbol{\sigma}(\mathbf{x}) = \sigma(\mathbf{x})I,$$

where $\sigma(\mathbf{x})$ is a scalar function and I is the identity matrix.[1]

In an inhomogeneous medium, the value of $\boldsymbol{\sigma}$ differs from one location to another ($\boldsymbol{\sigma} = \boldsymbol{\sigma}(\mathbf{x})$). We are especially interested in a description of a piecewise constant layout $\boldsymbol{\sigma}(\mathbf{x})$ that corresponds to a medium assembled from pieces of materials of different conductivities. Suppose that Ω is parted into several subdomains Ω_i, each of which contains a material with spatially constant properties $\boldsymbol{\sigma}_i$. The conductivity of the assembled medium is represented as

$$\boldsymbol{\sigma}(\mathbf{x}) = \sum_i \chi_i(\mathbf{x})\boldsymbol{\sigma}_i$$

where χ_i is the characteristic function of the ith subdomain:

$$\chi_i = \begin{cases} 1 & \text{if } \mathbf{x} \in \Omega_i, \\ 0 & \text{if } \mathbf{x} \notin \Omega_i. \end{cases} \tag{2.1.5}$$

The Second-Order Conductivity Equation

The system of equations (2.1.2), (2.1.3), and (2.1.4) allows us to determine the potential w from the sources f and the boundary conditions. This system is equivalent to the equation of second order,

$$\nabla \cdot \boldsymbol{\sigma}\nabla w = f \quad \text{in } \Omega, \tag{2.1.6}$$

called the conductivity equation.

Remark 2.1.1 *Notice that $\nabla \cdot \mathbf{A}\nabla w \equiv 0$ if $\mathbf{A}$ is an antisymmetric tensor. This explains the symmetry of the conductivity tensor $\boldsymbol{\sigma}$: The solution to (2.1.6) does not depend on the antisymmetric part of $\boldsymbol{\sigma}$.*

The boundary conditions may have different forms. Generally, we consider the following mixed boundary value problem: The boundary $\partial\Omega$ of Ω consists of two components $\partial\Omega = \partial\Omega_1 \cup \partial\Omega_2$. The potential w is prescribed on $\partial\Omega_1$, and the normal component of the current is prescribed on $\partial\Omega_2$:

$$\begin{aligned} w &= \rho_1 & &\text{on } \partial\Omega_1, \\ \mathbf{n} \cdot \mathbf{j} &= \mathbf{n} \cdot \boldsymbol{\sigma}\nabla w = \rho_2 & &\text{on } \partial\Omega_2, \end{aligned} \tag{2.1.7}$$

---

[1]As a rule, we use the bold letters to denote vectors and tensors and plain letters to denote scalars. For example, $\boldsymbol{\sigma}$ means the conductivity tensor and σ means the scalar isotropic conductivity. However, the unit matrix is denoted by the plain italic I.

where ρ_1 and ρ_2 are given functions of the surface $\partial\Omega_i$, $i = 1, 2$.

If $\partial\Omega = \partial\Omega_1$, then the boundary value problem (2.1.6), (2.1.7) is called the *Dirichlet problem*, and if $\partial\Omega = \partial\Omega_2$, the problem is called the *Neumann problem*.

Note that passing from the system (2.1.2), (2.1.3), and (2.1.4) to the second-order equation (2.1.6) formally requires additional assumptions of smoothness of σ if (2.1.6) is considered in the classical sense. At the same time, the system (2.1.2), (2.1.3), and (2.1.4) does not require even the continuity of σ. Naturally, we want to consider discontinuities in σ no matter what form of equation is used. Therefore, we understand the solution to (2.1.6) in the *weak sense* (Shilov, 1996): The integral equality

$$\int_\Omega (\nabla v \cdot \boldsymbol{\sigma} \nabla w + fv) + \oint_{\partial\Omega_1} \boldsymbol{\sigma}\nabla v \cdot \mathbf{n}(w - \rho_1) + \oint_{\partial\Omega_2} v(\mathbf{n}\cdot\boldsymbol{\sigma}\nabla w - \rho_2) = 0 \quad (2.1.8)$$

holds[2] for any test function $v \in H^1(\Omega)$.

Differential Constraints and Potentials

The system (2.1.2), (2.1.3), and (2.1.4) admits an equivalent representation called the dual form of (2.1.7). To derive this form we notice that the representation $\mathbf{e} = \nabla w$ implies a differential constraint on $\mathbf{e}$, because all components of $\mathbf{e}$ are determined by one scalar field w. The constraints have the form

$$\nabla \times \mathbf{e} \equiv 0. \qquad (2.1.9)$$

Indeed, the vector $\nabla \times \mathbf{e} = \nabla \times \nabla w$ consists of components of the type $\frac{\partial^2 w}{\partial x_i \partial x_j} - \frac{\partial^2 w}{\partial x_j \partial x_i}$ which vanish identically due to integrability conditions.

Similarly, the differential constraint $\nabla \cdot \mathbf{j} = 0$ is identically satisfied if $\mathbf{j}$ corresponds to a vector potential $\mathbf{y}$:

$$\mathbf{j} = \nabla \times \mathbf{y}. \qquad (2.1.10)$$

The vector potential $\mathbf{y} = [y_1, y_2, y_3]$ is determined up to the gradient of a scalar field ψ which can be chosen arbitrarily. Indeed,

$$\nabla \times \mathbf{y} = \nabla \times (\mathbf{y} + \nabla\psi).$$

Therefore, $\mathbf{y}$ depends on two arbitrary potentials: The number of independent functions (two) agrees with the number of components of a current vector $\mathbf{j}$ (three) reduced by one differential constraint $\nabla \cdot \mathbf{j} = 0$.

Table 2.1 summarizes the differential constraints and potentials in conductivity. In Chapter 14 (Table 14.1), we will observe similar duality of the potentials and constraints in elasticity equations.

---

[2]The symbol "dx" of the differential is omitted in the integrals like $\int_\Omega$ over the explicitly defined domain Ω of the independent variable $\mathbf{x}$.

Variable	Constraints	Potential
Field $\mathbf{e}$	$0 = \nabla \times \mathbf{e}$	$\mathbf{e} = \nabla u$
Current $\mathbf{j}$	$0 = \nabla \cdot \mathbf{j}$	$\mathbf{j} = \nabla \times \mathbf{y}$

TABLE 2.1. Differential constraints and potentials in conductivity.

Dual Form of Conductivity Equations

Equation (2.1.10) allows us to introduce the vector potential $\mathbf{y}$ in the conductivity problem:

$$\mathbf{j} = \nabla \times \mathbf{y} + \mathbf{j}_0, \quad \nabla \cdot \mathbf{j}_0 = f, \tag{2.1.11}$$

where $\mathbf{j}_0$ is a particular solution to (2.1.2). Vector field $\mathbf{j}_0$ is not uniquely defined and does not depend on the properties of the medium.

Equations (2.1.9), (2.1.11), and the inverse form of the constitutive relations

$$\mathbf{e} = \sigma^{-1} \mathbf{j} \tag{2.1.12}$$

form a system of equations of conductivity that uses a vector potential $\mathbf{y}$ of currents instead of a scalar potential w of forces. The system (2.1.9), (2.1.11), and (2.1.12) is said to be *dual* to the system (2.1.2), (2.1.3), and (2.1.4) and conversely. These systems are equivalent.

The dual form of equation (2.1.6) is the vector equation

$$\nabla \times \sigma^{-1} (\nabla \times \mathbf{y} + \mathbf{j}_0) = 0. \tag{2.1.13}$$

Its solution should also be understood in the weak sense, similar to (2.1.8).

2.1.2 Continuity Conditions in Inhomogeneous Materials

We have already mentioned that conductivity equations do not require the continuity of σ. What happens to the fields $\mathbf{j}$ and $\mathbf{e}$ on the boundary Γ between the domains where σ takes different constant values σ^+ and σ^-? Denote the normal to Γ by $\mathbf{n}$ and the tangents – by $\mathbf{t}$ and $\mathbf{b}$.

1. The divergencefree nature of the current $\mathbf{j}$ (2.1.1) indicates that the normal component of $\mathbf{j}$ remains continuous (Figure 2.1):

$$[\mathbf{j} \cdot \mathbf{n}] = 0, \tag{2.1.14}$$

where $[z]$ is the jump of a variable z across Γ:

$$[z] = z^+ - z^-.$$

Physically, the normal component of the current is equal to the difference in the number of particles that cross the surface Γ from the

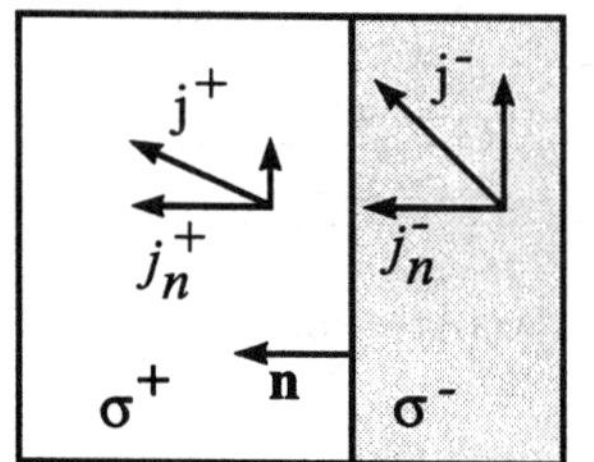

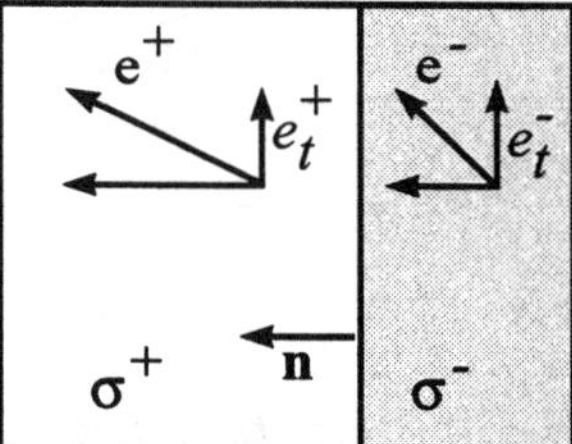

FIGURE 2.1. The refraction of the current and the field on the boundary between two isotropic conductors. The normal component of the current and the tangent component of the field are continuous on the boundary.

left and right (see the kinetic equation (2.1.1)). This number is zero, and therefore $[\mathbf{j} \cdot \mathbf{n}]$ is continuous.

Formally, we also could derive (2.1.14) from equation (2.1.2) written in local coordinates $(\mathbf{n}, \mathbf{t}, \mathbf{b})$:

$$\nabla \cdot \mathbf{j} = \frac{\partial j_n}{\partial n} + \frac{\partial j_t}{\partial t} + \frac{\partial j_b}{\partial b} = f.$$

It implies that the argument $j_n = \mathbf{j} \cdot \mathbf{n}$ of the normal derivative $\frac{\partial}{\partial n}$ is necessarily continuous on the surface Γ. Otherwise, the left-hand side in (2.1.14) would contain a δ-function that lacks its mate on the right-hand side.

Note that the finiteness of the tangent derivatives implies only the continuity of its argument *along* the boundary Γ of both sides, but it does not imply any smoothness of that argument when the boundary is crossed. Generally, we have

$$[\mathbf{j} \cdot \mathbf{t}] \neq 0, \quad [\mathbf{j} \cdot \mathbf{b}] \neq 0. \tag{2.1.15}$$

2. The tangent components of the field $\mathbf{e}$ are continuous due to the continuity of a potential w (Figure 2.1):

$$[\mathbf{e} \cdot \mathbf{t}] = [\mathbf{e} \cdot \mathbf{b}] = 0. \tag{2.1.16}$$

Indeed, the limiting values of w from the left (w^-) and right (w^+) of any point of surface Γ are equal; for two points $\mathbf{x}_1$ and $\mathbf{x}_2$ on Γ we have

$$w_1^+ = w_1^-, \quad w_2^+ = w_2^-;$$

the difference between the potentials at corresponding points is also equal. This implies

$$\frac{w_1^- - w_2^-}{|\mathbf{x}_1 - \mathbf{x}_2|} = \frac{w_1^+ - w_2^+}{|\mathbf{x}_1 - \mathbf{x}_2|},$$

where $w_i = w(\mathbf{x}_i)$. In the limit $|\mathbf{x}_1 - \mathbf{x}_2| \to 0$, the left-hand and right-hand side terms of the last equality represent a tangent derivative on the $(-)$ and the $(+)$ side of Γ. Equation (2.1.16) follows.

Another way to derive this condition is to examine the constraint $\nabla \times \mathbf{e} = 0$. The curl of $\mathbf{e}$ is represented as

$$\nabla \times \mathbf{e} = \left(\frac{\partial e_b}{\partial t} - \frac{\partial e_t}{\partial b}\right)\mathbf{n} - \left(\frac{\partial e_b}{\partial n} - \frac{\partial e_n}{\partial b}\right)\mathbf{t} + \left(\frac{\partial e_t}{\partial n} - \frac{\partial e_n}{\partial t}\right)\mathbf{b} = 0$$

$$(2.1.17)$$

where e_n is the normal and e_t, e_b are tangent components of $\mathbf{e}$. The equality $\nabla \times \mathbf{e} = 0$ requires that the normal derivatives $\frac{\partial e_t}{\partial n}$ and $\frac{\partial e_b}{\partial n}$ be finite, hence e_t and e_b are continuous. Otherwise δ-functions occur in the left-hand side of (2.1.17).

The normal component e_n does not need to be continuous. Generally, we have $[\mathbf{e} \cdot \mathbf{n}] \neq 0$.

3. Let us compute the jumps of the discontinuous components of $\mathbf{e}$ and $\mathbf{j}$. The continuous components $\mathbf{e} \cdot \mathbf{t}$ and $\mathbf{e} \cdot \mathbf{b}$ of the field $\mathbf{e}$ correspond to the discontinuous components $\mathbf{j} \cdot \mathbf{t}$ and $\mathbf{j} \cdot \mathbf{b}$ of the current $\mathbf{j}$, and the discontinuous component $\mathbf{e} \cdot \mathbf{n}$ of the field corresponds to the continuous component $\mathbf{j} \cdot \mathbf{n}$ of the current. Together, the vectors of a current and a field have exactly three continuous and three discontinuous components. Let us denote by $\mathbf{d} = [e_n, j_t, j_b]$ the vector of discontinuous components, and by $\mathbf{c} = [j_n, e_t, e_b]$ – the vector of continuous components.

To compute the jump of the components of $\mathbf{d}$, we solve the state equations (2.1.4) for $\mathbf{d}$:

$$\mathbf{d} = Z(\boldsymbol{\sigma})\mathbf{c},$$

where the matrix Z is

$$Z(\boldsymbol{\sigma}) = \begin{pmatrix} \frac{1}{\sigma_{nn}} & -\frac{\sigma_{nt}}{\sigma_{nn}} & -\frac{\sigma_{nb}}{\sigma_{nn}} \\ -\frac{\sigma_{nt}}{\sigma_{nn}} & \sigma_{tt} - \frac{\sigma_{nt}^2}{\sigma_{nn}} & \sigma_{tb} - \frac{\sigma_{nt}\sigma_{nb}}{\sigma_{nn}} \\ -\frac{\sigma_{nb}}{\sigma_{nn}} & \sigma_{tb} - \frac{\sigma_{nt}\sigma_{nb}}{\sigma_{nn}} & \sigma_{bb} - \frac{\sigma_{nb}^2}{\sigma_{nn}} \end{pmatrix}$$

and $\sigma_{nn}, \sigma_{nt}, \sigma_{nb}, \sigma_{bb}, \sigma_{bt}, \sigma_{tt}$ are the components of the tensor $\boldsymbol{\sigma}$ in the coordinates $\mathbf{n}, \mathbf{t}, \mathbf{b}$,

$$\boldsymbol{\sigma} = \begin{pmatrix} \sigma_{nn} & \sigma_{nt} & \sigma_{nb} \\ \sigma_{nt} & \sigma_{tt} & \sigma_{tb} \\ \sigma_{nb} & \sigma_{tb} & \sigma_{bb} \end{pmatrix}.$$

Now we easily calculate the jump of $\mathbf{d}$ at two neighboring points that lie to the left and right of Γ. Using the continuity of $\mathbf{c}$, we compute

$$[\mathbf{d}] = \left(Z(\boldsymbol{\sigma}^+) - Z(\boldsymbol{\sigma}^-)\right)\mathbf{c}. \qquad (2.1.18)$$

For isotropic materials, these relations become

$$\begin{aligned}
[e_n] &= \left(\tfrac{1}{\sigma^+} - \tfrac{1}{\sigma^-}\right) j_n, \\
[j_t] &= (\sigma^+ - \sigma^-)e_t, \\
[j_b] &= (\sigma^+ - \sigma^-)e_b.
\end{aligned} \qquad (2.1.19)$$

The equations (2.1.18) enable us to determine $\mathbf{e}$ and $\mathbf{j}$ on one side of the boundary Γ if they are known on the other side. These formulas are used for calculation of the average fields of a composite. This technique was described in (Backus, 1962).

2.1.3 Energy, Variational Principles

Multidimensional Variational Problems

We can view the conductivity equation as the Euler equation that corresponds to a minimum of some multidimensional variational functional.

First, let us discuss minimizers of multidimensional variational problems. Consider the problem

$$\min_{w} \int_{\Omega} G(\mathbf{x}, w, \nabla w) - \oint_{\partial \Omega} g(\mathbf{x}, w) \qquad (2.1.20)$$

where G is called the bulk Lagrangian and g is the surface Lagrangian. Suppose that w is a minimizer of (2.1.20). As in the one-dimensional problem, one can derive the necessary condition of optimality for w. The stationary solution to problem (2.1.20) is called the Euler–Lagrange equation. It is a direct multivariable analogue of the one-dimensional Euler equation (1.2.11), (1.2.12). The operator ∇ formally replaces the operator $\frac{d}{dx}$. The Euler–Lagrange equation has the form

$$S(G) = \nabla \cdot G_{\nabla w} - \frac{\partial G}{\partial w} = 0, \qquad (2.1.21)$$

where $G_{\nabla w}$ is the vector

$$G_{\nabla w} = \frac{\partial G}{\partial \nabla w} = \left[\left(\frac{\partial G}{\partial \left(\frac{\partial w}{\partial x_1}\right)} \right), \; \ldots, \; \left(\frac{\partial G}{\partial \left(\frac{\partial w}{\partial x_d}\right)} \right) \right].$$

Any differentiable minimizer w of the problem (2.1.20) satisfies the Euler–Lagrange equation (2.1.21) and the boundary conditions

$$\delta w \left(\sum_{i=1}^{d} (G_{\nabla w})\, n_i - \frac{\partial g}{\partial w} \right) = 0, \quad \text{on } \partial \Omega, \qquad (2.1.22)$$

where n_i is the ith component of the normal to the boundary $\partial \Omega$.

We do not derive these relations here. The derivation is analogous to the one-dimensional case and can be found in any standard course on calculus of variations (see, for example, (Fox, 1987)). However, we derive similar stationary equations in Section 5.2.

The Dirichlet Variational Principle

The steady-state equilibrium of a conducting body corresponds to the minimal solution to a variational problem called the Dirichlet variational principle (Courant and Hilbert, 1962):

$$I_e(\boldsymbol{\sigma}) = \min_{w \in \mathcal{W}} \int_\Omega (W_e(\nabla w, \boldsymbol{\sigma}) + f\, w) + \int_{\partial\Omega_2} w\, \rho_2, \qquad (2.1.23)$$

where

$$\mathcal{W} = \left\{ w: \ w \in H^1(\Omega),\ w|_{\partial\Omega_1} = \rho_1 \right\},$$

ρ_2 is the normal component of applied boundary currents, and w is a potential ($\mathbf{e} = \nabla w$). The quadratic form

$$W_e(\nabla w, \boldsymbol{\sigma}) = \frac{1}{2} \nabla w \cdot \boldsymbol{\sigma} \nabla w$$

is called the *energy* of a conducting body. The Lagrangian $W_e(\nabla w, \boldsymbol{\sigma}) + f\, w$ is composed as a sum of the energy W_e and the work of the sources f in Ω.

The boundary condition

$$w|_{\partial\Omega_1} = \rho_1$$

is called the *main boundary condition*; all minimizers are subject to it. The condition

$$\frac{\partial}{\partial n} \boldsymbol{\sigma} \nabla w \bigg|_{\partial\Omega_2} = \rho_2$$

is called the *natural* or *variational boundary condition*. It is satisfied at the minimum of $I_e(\boldsymbol{\sigma})$.

The Euler–Lagrange equations (2.1.22) for the Dirichlet variational principle coincide with the equilibrium equations (2.1.6) and (2.1.7). One can also check that the minimizer of the energy of an inhomogeneous medium jumps on the dividing surface between the materials, in accord with (2.1.19).

The Thompson Variational Principle

Similarly, the dual system of conductivity equations (2.1.13) correspond to the Euler–Lagrange equations for the variational problem, called the Thomson variational principle:

$$I_\mathbf{y}(\boldsymbol{\sigma}) = \min_{\mathbf{y} \in \mathcal{Y}} \int_\Omega (W_j(\nabla \times \mathbf{y}, \boldsymbol{\sigma}) + \mathbf{j}_0 \cdot \nabla \times \mathbf{y}),$$

where

$$W_j(\nabla \times \mathbf{y}, \boldsymbol{\sigma}) = \frac{1}{2}(\nabla \times \mathbf{y}) \cdot \boldsymbol{\sigma}^{-1}(\nabla \times \mathbf{y}),$$

$\mathbf{j}_0$ is a particular solution to the equation $\nabla \cdot \mathbf{j} = f$, and

$$\mathcal{Y} = \left\{ \mathbf{y}: \ y_i \in H^1(\Omega), \quad (\nabla \times \mathbf{y} + \mathbf{j}_0) \cdot \mathbf{n}|_{\partial\Omega_2} = \rho_2 \right\}.$$

We also assume for simplicity that $\rho_1 = 0$. Recall that $\mathbf{j}_0$ is a particular solution to the equation $\nabla \cdot \mathbf{j} = f$.

Various Expressions for Energy

We have seen that the energy density W in a conducting medium can be written in various forms. It is equal to the scalar product of the current $\mathbf{j}$ and the field $\mathbf{e}$:

$$W(\mathbf{e},\mathbf{j}) = \frac{1}{2}\mathbf{e}\cdot\mathbf{j}, \quad \text{where } \mathbf{e} = \nabla w, \ \mathbf{j} = \nabla \times \mathbf{y}.$$

Using the constitutive relations, the energy can also be represented either as a quadratic form of the field $\mathbf{e}$,

$$W_e(\mathbf{e},\boldsymbol{\sigma}) = W(\mathbf{e},\boldsymbol{\sigma}\mathbf{e}) = \frac{1}{2}\mathbf{e}\cdot\boldsymbol{\sigma}\mathbf{e}, \quad \text{where } \mathbf{e} = \nabla w,$$

or as the quadratic form of the current density $\mathbf{j}$,

$$W_j(\mathbf{j},\boldsymbol{\sigma}) = W(\boldsymbol{\sigma}^{-1}\mathbf{j},\mathbf{j}) = \frac{1}{2}\mathbf{j}\cdot\boldsymbol{\sigma}^{-1}\mathbf{j}, \quad \text{where } \mathbf{j} = \nabla \times \mathbf{y}.$$

Each of these forms corresponds to a variational principle; the Euler–Lagrange equations coincide with the equilibrium equations (2.1.2), (2.1.3), and (2.1.4).

Duality of Variational Principles

The Dirichlet and Thompson variational principles are related. Each of them is dual to the other. The duality of extremal problems was introduced in Chapter 1 for one-dimensional problems. The duality for the variational problems with multiple integrals is defined in the same fashion (see, for example, (Ekeland and Temam, 1976)).

Consider a multivariable Lagrangian $L(\mathbf{x}, u, \nabla u)$. Perform the Legendre transform of L, that is, find the dual vector variable $\mathbf{j}$ from the extremal problem (compare with (1.3.25)

$$L^{\text{dual}}(u,\mathbf{j}) = \min_{\nabla u}\left(\mathbf{j}\cdot\nabla u - L(\mathbf{x},u,\nabla u)\right),$$

which gives $\mathbf{j} = \frac{\partial L}{\partial \nabla u}$. Solving the last equation for ∇u, we express ∇u as a function of $\mathbf{j}$:

$$\nabla u = \phi(\mathbf{j}). \tag{2.1.24}$$

The dual energy is equal to

$$L^{\text{dual}}(u,\mathbf{j}) = \mathbf{j}\cdot\phi(\mathbf{j}) - W(u,\phi(\mathbf{j})).$$

The Euler–Lagrange equation is expressed through $\mathbf{j}$ as:

$$\nabla\cdot\mathbf{j} - \frac{\partial L}{\partial u} = 0. \tag{2.1.25}$$

To satisfy the system (2.1.24), (2.1.25) we introduce a vector $\mathbf{j}_0$, that corresponds to a particular solution to the equation $\nabla \cdot \mathbf{j}_0 - \frac{\partial L}{\partial u} = 0$. The system (2.1.24) and (2.1.25) is equivalent to

$$\mathbf{j} + \mathbf{j}_0 = \nabla \times \mathbf{y}, \quad \nabla \times \phi(\mathbf{j}) = 0. \qquad (2.1.26)$$

The vector $\mathbf{y}$ is the dual to u potential. The system (2.1.26) or the equivalent second-order equation

$$\nabla \times \phi(\mathbf{j}_0 - \nabla \times \mathbf{y}) = 0$$

is the Euler–Lagrange equations in the dual variables $\mathbf{y}$.

Remark 2.1.2 *If the Lagrangian $L(u, \nabla u)$ is convex with respect to ∇u, then the Legendre transform is a convolution: The dual to the Lagrangian $L^* = L^{dual}$ coincides with L, that is, $L^{**}(u, \nabla u) = L(u, \nabla u)$. Otherwise, $L^{**}(u, \nabla u)$ is the convex envelope of $L(u, \nabla u)$ with respect to variable ∇u (see the discussion in Chapter 1 and (Ekeland and Temam, 1976)): $L^{**}(u, \nabla u) = \mathcal{C}L(u, \nabla u).$*

Example 2.1.1 The Dirichlet and Thompson principles correspond to the dual Lagrangians. Indeed, the dual form for the Lagrangian $W = \frac{1}{2}\sigma(\nabla w)^2$ is

$$W^{dual} = \frac{1}{2\sigma}\mathbf{j}^2,$$

where the dual to $\mathbf{e}$ variable $\mathbf{j}$ is defined as

$$\mathbf{j} = (\sigma \nabla w), \quad \mathbf{j} = \nabla \times \mathbf{y}.$$

In this dual form, the conductivity σ is replaced with the resistivity $\frac{1}{\sigma}$.

2.2 Composites

To be prepared to deal with fine-scale oscillating solutions to structural optimization problems, we need to discuss the methods of homogenization. These methods replace oscillating sequences of property layouts with smooth layouts of the effective properties of composite media. The effective properties become controls in the homogenized problem.

We briefly discuss here micro-inhomogeneous media, which are also called media with microstructures or composites. A detailed exposition of micro-inhomogeneous media can be found in many books; we cite (Bensoussan et al., 1978; Sánchez-Palencia, 1980; Jikov et al., 1994; Bakhvalov and Panasenko, 1989).

2.2.1 Homogenization and Effective Tensor

Assumptions

A composite is viewed as a structure assembled from a very large number of fragments of given materials mixed in a prescribed way. Each fragment is assumed to be much smaller than the rate of varying of acting fields and than the size of a considered domain. At the same time, these domains are large enough to assume that the conductivity equation is valid in each fragment of material, which means that fragments are much larger than molecular sizes, the size of a free path, etc. The way of mixing is assumed to be regular in a sense: The microstructure is periodic, quasiperiodic, or statistically homogeneous. The behavior of a piece of the composite is representative of the behavior of neighboring pieces.

It is hopelessly difficult and often useless to describe fields at each point of the composite. For most purposes we do not need to know all the details. Instead, we simplify the problem by introducing an averaged description of a composite. The procedure that replaces the original problem by a simpler averaged problem is called *homogenization.*

In doing homogenization we replace the fine-scale oscillating inhomogeneous material with properties $\sigma(\chi)$ by a homogeneous material with conductivity σ_*. This material imitates some important features of the inhomogeneous system, and σ_* is called the *effective properties tensor* of the composite. In contrast with the rapidly oscillating layout σ, the effective tensor σ_* is a constant or (in the quasiperiodic case) a smoothly varying tensor function of the point $\mathbf{x}$ of the domain.

Clearly, the simplified system cannot preserve all features of the original one, so we should choose the features we would like to preserve by homogenization. The general homogenization concept is to preserve the solution w of a boundary value problem (2.1.6). The requirement that the solution w in the homogenized system stay close to the solution to the initial system leads to an equation for the effective conductivity tensor σ_*. The homogenized system also preserves the mean field $\mathbf{e}$ and the mean current $\mathbf{j}$.

Remark 2.2.1 *Much information about the system is lost. Particularly, the homogenization neglects processes determined by individual behavior of fine pieces of material: field concentration in the corner points of grains, cracks, fine-scale oscillations, percolation, etc. Still, the remaining problems are important: For example, we can obtain the temperature, electrical current density, and so on. In elasticity, homogenization preserves the displacement vector and the averaged tensors of stresses and strains.*

Homogenization is a local procedure: We replace the inhomogeneous medium in a small neighborhood Ω_ε by a uniform medium. The fields are replaced by their mean values, i.e., by averages over Ω_ε. It is assumed that the way of mixing materials in Ω_ε is repeated in the entire composite. This principle makes the homogenization procedure simple enough to be effectively applied.

Homogenization is an asymptotic procedure: Its result becomes better as the size of the representative volume of composite material tends to zero. The size of Ω_ε is compared with the size of the domain and the rate of varying of the external fields. The homogenization assumes that the size of the representative volume is much smaller than the other parameters of the system.

Various methods of homogenization were actively developed in the last decades. They were applied to various physical problems including periodic or random arrays of inclusions, suspensions, nonlinear inhomogeneous materials, diffusion in a stream, percolation problems, checkerboard structures, etc. A review of these problems is beyond the scope of this book; the reader is referred to the recent collections (Dal Maso and Dell'Antonio, 1991; Hornung, 1997; Markov and Inan, 1999; Berdichevsky, Jikov, and Papanicolaou, 1999; Markov and Preziosi, 1999) and the books (Christensen, 1979; Bakhvalov and Panasenko, 1989; Nemat-Nasser and Hori, 1993; Berdichevsky, 1997). The remarkable variety of problems and methods of homogenization are described in many papers that represent different aspects of the approach, such as: (Telega, 1990; Khruslov, 1991; Berlyand and Golden, 1994; Panasenko, 1994; Bourgeat, Kozlov, and Mikelić, 1995; Beliaev and Kozlov, 1996; Ryzhik, Papanicolaou, and Keller, 1996; Levin, 1999; Markov and Preziosi, 1999; Torquato, 1999). The papers (Cioranescu and Murat, 1982; Berdichevsky, Kunin, and Hussain, 1991; Milton, 1992; Cherkaev and Slepyan, 1995; Kozlov and Piatnitski, 1996; Zhikov, 1996; Balk, Cherkaev, and Slepyan, 1999) emphasize unexpected behavior of homogenized systems. The numerical aspects of homogenization were investigated in many papers, such as (Bakhvalov and Knyazev, 1994; Zohdi, Oden, and Rodin, 1996; Helsing, Milton, and Movchan, 1997; Sigmund and Torquato, 1997; Greengard and Rokhlin, 1997; Fu, Klimkowski, Rodin, Berger, Browne, Singer, van de Geijn, and Vemaganti, 1998; Greengard and Helsing, 1998).

Conductivity in an Inhomogeneous Body

To visualize a composite with infinitely small periodic elements we use an iterative process. Consider a periodic two-phase structure. Assume that the domain Ω consists of cubes Ω_i ($\Omega = \cup\Omega_i$) and that the material's layout is the same for all cubes of the size $\frac{1}{2^k}$. Each cube Ω^k is divided into two parts Ω_1^k and Ω_2^k, which are filled with the materials σ_1 and σ_2, respectively. The conductivity $\sigma(\mathbf{x})$ at a point of the cube Ω^k is equal to

$$\sigma(\chi) = \chi^k \sigma_1 + (1 - \chi^k)\sigma_2,$$

where $\chi = \chi^k(\mathbf{x})$ is the space-periodic characteristic function χ^k of the first material in the composite:

$$\chi^k(\mathbf{x}) = \begin{cases} 1 & \text{if} \quad \mathbf{x} \in \Omega_1^k, \\ 0 & \text{if} \quad \mathbf{x} \in \Omega_2^k. \end{cases}$$

Now consider a sequence of $\{\chi^k\}$ layouts. It is built by the following procedure. At each step, the representative cube of periodicity Ω^k is parted into eight cubes Ω^{k+1} half the linear size of each; each cube is filled with a geometrically identical layout of materials but in half the scale of that in the cube Ω^k.

Let us fix the size $\varepsilon = \frac{1}{2^k}$ of the periodicity cell and assume that these cells fill the domain Ω. Consider the conductivity equilibrium (2.1.6) in Ω. The solution w of the conductivity equation (2.1.6) can be represented in the form (Bensoussan et al., 1978)

$$w = w_0(\mathbf{x}) + \varepsilon w_\varepsilon \left(\mathbf{x}, \frac{\mathbf{x}}{\varepsilon} \right) + o(\varepsilon); \tag{2.2.1}$$

it consists of a smooth component $w_0(\mathbf{x}) = O(1)$ that is independent of the size ε and an almost-periodic oscillating component $\varepsilon w_\varepsilon \left(\mathbf{x}, \frac{\mathbf{x}}{\varepsilon} \right)$ that has zero mean value over the periodicity cell:

$$\int_\Omega w_\varepsilon \left(\mathbf{x}, \frac{\mathbf{x}}{\varepsilon} \right) = 0.$$

The magnitude of w_ε is of order one.

The averaging of a process in a composite medium is done by applying the averaging operator $\langle \cdot \rangle(\mathbf{x})$:

$$\langle z \rangle(\mathbf{x}) = \frac{1}{|\Omega_\varepsilon|} \int_{\Omega_\varepsilon(\mathbf{x})} z, \tag{2.2.2}$$

where Ω_ε is a small rectangular domain with the point $\mathbf{x}$ in its center and $|\Omega_\varepsilon|$ is its volume. The operator (2.2.2) is the multidimensional analogue of (1.1.6). The size of the domain of averaging is assumed to be greater than the size ε of the cell of periodicity: $|\Omega_\varepsilon| \gg \varepsilon$. More exactly, we assume that the size of Ω_ε tends to zero, together with the diameter of fragments ω_i, but it remains much greater than the diameter. The one-dimensional averaging introduced in (1.1.9) agrees with the discussed definition.

The field $\mathbf{e} = \nabla w$ can be represented as $\mathbf{e} = \mathbf{e}_0 + \mathbf{e}_\varepsilon$ where

$$\mathbf{e}_0 = \nabla w_0, \quad \mathbf{e}_\varepsilon = \nabla w_\varepsilon, \quad \langle \mathbf{e}_\varepsilon \rangle = 0.$$

However, the magnitude of $\mathbf{e}_\varepsilon$ is at least of order of magnitude $|\sigma_1 - \sigma_2|$ of the jump of $\mathbf{e}$ on the boundary between domains of materials σ_1 and σ_2. Formally, we also observe that ∇w, (2.2.1), is at least of order one because $\nabla \left(\varepsilon w_\varepsilon \left(\mathbf{x}, \frac{\mathbf{x}}{\varepsilon} \right) \right)$ is of order one.

The current $\mathbf{j}_\varepsilon$ in the medium has a similar representation

$$\mathbf{j} = \mathbf{j}_0 + \mathbf{j}_\varepsilon, \quad \langle \mathbf{j}_\varepsilon \rangle = 0;$$

where the magnitude of $\mathbf{j}_\varepsilon$ is at least of order $|\frac{1}{\sigma_1} - \frac{1}{\sigma_2}|$.

The current and field are subject to the constitutive relations

$$\mathbf{j}_0(\mathbf{x}) + \mathbf{j}_\varepsilon(\mathbf{x}) = \boldsymbol{\sigma}(\mathbf{x})(\mathbf{e}_0(\mathbf{x}) + \mathbf{e}_\varepsilon(\mathbf{x})).$$

We are interested in a description of the relation between the smooth components $\mathbf{e}_0$ and $\mathbf{j}_0$ when the size of the periodicity cell is much less than the other parameters of the system: $\varepsilon \to 0$.

Let us average the solution to the conductivity equations (2.1.2), (2.1.3), (2.1.4) by the operator (2.2.2). We find relations between the averaged current $\langle \mathbf{j} \rangle$, averaged field $\langle \mathbf{e} \rangle$, and averaged potential $\langle w \rangle$. The linear operators ∇ and $\langle \cdot \rangle$ commute (up to terms of order $O(\varepsilon)$) (see (Bensoussan et al., 1978; Jikov et al., 1994)); hence we can change their order. We obtain

$$\langle \nabla \cdot \mathbf{j} \rangle = \nabla \cdot \mathbf{j}_0 = f + O(\varepsilon), \quad \langle \mathbf{e} \rangle = \nabla \langle w \rangle + O(\varepsilon). \tag{2.2.3}$$

To complete the procedure we determine the connection between the average current $\langle \mathbf{j} \rangle = \langle \boldsymbol{\sigma} \, \mathbf{e} \rangle$ and the average field $\langle \mathbf{e} \rangle$. We introduce the tensor of *the effective properties of the composite* $\boldsymbol{\sigma}_*$ such that

$$\langle \mathbf{j} \rangle = \langle \boldsymbol{\sigma} \mathbf{e} \rangle = \boldsymbol{\sigma}_* \langle \mathbf{e} \rangle. \tag{2.2.4}$$

The effective tensor links vectors $\langle \mathbf{j} \rangle$ and $\langle \mathbf{e} \rangle$. Using the effective tensor we can formulate the homogenized equations of the medium. The system (2.2.3), (2.2.4) or the equivalent second-order equation

$$\nabla \cdot \boldsymbol{\sigma}_* \nabla \langle w \rangle = f + O(\varepsilon) \tag{2.2.5}$$

describes the conductivity in the medium with the conductivity tensor $\boldsymbol{\sigma}_*$ or the homogenized conductivity. The solution should be understood in the weak sense (Jikov et al., 1994).

Calculation of Effective Tensor

Here we find the ways to compute the effective tensor. We illustrate the idea of the approach with a simple example. Namely, we consider a two-dimensional problem of the conductivity of a composite of two isotropic materials.

Consider a periodic layout of two conducting materials. The element of periodicity Ω is the unit square

$$\Omega = \{x_1, x_2 : \ 0 \le x_1 \le 1, \quad 0 \le x_2 \le 1\}. \tag{2.2.6}$$

Suppose that Ω is parted into two rectangular parts Ω_1 and Ω_2 of the areas m_1 and m_2 respectively. They are occupied by two isotropic materials with conductivities σ_1 and σ_2 respectively.

Applied Fields

The following simple algorithm enables us to compute an effective tensor. Consider again the periodic structure (2.2.6). Suppose that a uniform external field $\mathbf{E} = \mathbf{E}_1$ is applied to the structure that is equal to $\mathbf{E}_1 = \mathbf{i}_1$, where $\mathbf{i}_1$ is a unit vector directed along the x_1-axis:

$$\mathbf{i}_1 = \begin{pmatrix} 1 \\ 0 \end{pmatrix}.$$

We solve the boundary value problem

$$\nabla \cdot \boldsymbol{\sigma}(\mathbf{x})[\nabla w(\mathbf{x}) + \mathbf{E}_1] = 0, \quad w \text{ is periodic in } \Omega. \qquad (2.2.7)$$

and compute the average current $\mathbf{j}_1 = \langle \boldsymbol{\sigma}(\mathbf{x})[\nabla w(\mathbf{x}) + \mathbf{E}_1]\rangle$.

The average current vector $\mathbf{j}_1 = \boldsymbol{\sigma}_* \mathbf{E}_1$ is equal to the first column of the effective tensor $\boldsymbol{\sigma}_*$:

$$\begin{pmatrix} j_1^1 \\ j_2^1 \end{pmatrix} = \begin{pmatrix} (\sigma_*)_{11}, & (\sigma_*)_{12} \\ (\sigma_*)_{21}, & (\sigma_*)_{22} \end{pmatrix} \begin{pmatrix} 1 \\ 0 \end{pmatrix}; \qquad (2.2.8)$$

or, in coordinate form,

$$j_1^1 = (\sigma_*)_{11}, \quad j_2^1 = (\sigma_*)_{21}.$$

This way we determine two elements of $\boldsymbol{\sigma}_*$ by measuring the vector of the averaged current.

The effective tensor cannot be completely determined by one "experiment," that is, by applying one external field. Indeed, the measured current $\mathbf{j}_1$ cannot depend on the conductivity in the direction orthogonal to the direction of the applied field. To determine the second column of the matrix of $\boldsymbol{\sigma}_*$ we solve boundary value problem (2.2.7) for a different external field $\mathbf{E}_2$. We can take as $\mathbf{E}_2$ a unit vector directed along the x_2-axis, $\mathbf{E}_2 = \mathbf{i}_2$, where

$$\mathbf{i}_2 = \begin{pmatrix} 0 \\ 1 \end{pmatrix}.$$

Clearly, the average current $\mathbf{j}_2 = (j_1^2, j_2^2)$ for this problem coincides with the second column of the effective tensor $\mathbf{j}_2 = \boldsymbol{\sigma}_* \cdot \mathbf{E}_2$ or

$$j_1^2 = (\sigma_*)_{12}, \quad j_2^2 = (\sigma_*)_{22}.$$

Remark 2.2.2 *Of course, one must expect symmetry $\sigma_{12} = \sigma_{21}$ in the coefficients of the effective tensor (Jikov et al., 1994). The symmetry can be shown in many ways, for example, by the symmetry of Green's function for the conductivity operator (2.1.5).*

The results are easily represented in matrix notation. Let us form the matrix E of external fields $\mathbf{E}_1, \mathbf{E}_2$ and the matrix J of corresponding currents $\mathbf{j}_1, \mathbf{j}_2$:

$$E = [\mathbf{E}_1, \mathbf{E}_2], \quad J = [\mathbf{j}_1, \mathbf{j}_2].$$

The effective properties can then be determined from the matrix equations

$$J = \boldsymbol{\sigma}_* E \quad \text{or} \quad \boldsymbol{\sigma}_* = JE^{-1}. \tag{2.2.9}$$

The inversion is possible if the external fields $\mathbf{E}_1$, $\mathbf{E}_2$ are linearly independent. Here we are considering the case where $E = I$; therefore, $\boldsymbol{\sigma}_* = J$.

Applied Currents

Alternatively, we may determine $\boldsymbol{\sigma}$ by applying the trial currents $\mathbf{J}_i$ instead of trial fields $\mathbf{E}_i$. This time we assume that the periodic composite is submerged into a uniform currents $\mathbf{J}_i$ instead of the uniform field $\mathbf{E}_i$. The problem for the periodicity cell becomes

$$\nabla \cdot [\boldsymbol{\sigma}(\mathbf{x})\nabla w(\mathbf{x}) + \mathbf{J}_i] = 0 \text{ in } \Omega_\varepsilon, \quad i = 1, 2, \tag{2.2.10}$$

with periodic boundary conditions on w.

The external currents can be chosen as

$$\mathbf{J}_1 = \mathbf{i}_1, \quad \mathbf{J}_2 = \mathbf{i}_2.$$

Solving (2.2.10) for these currents, we obtain average the fields $\mathbf{e}_1$ and $\mathbf{e}_2$, respectively. Measuring the average fields, one measures the coefficients of the inverse tensors $\boldsymbol{\sigma}_*^{-1}$ due to the first equation of (2.2.9);

$$\boldsymbol{\sigma}_*^{-1} = EJ^{-1},$$

where $J = [\mathbf{J}_1, \mathbf{J}_2]$ is the matrix of the applied currents and E is the matrix of the calculated mean fields. For linear media, these procedures lead to the same resulting effective tensor, because they describe the same linear relationship between averaged fields and currents.

2.2.2 Effective Properties of Laminates

As an example, let us compute the effective tensor of a laminate. The laminate geometry allows us to solve the partial differential problem (2.2.7) in closed form.

Effective Tensor for Laminates of Two Conducting Materials

Consider the conductivity problem for a laminate composite in the plane. Let the periodicity cell be the unit square Ω. Assume that the laminates are oriented along the x_2-axis. The rectangles Ω_1 and Ω_2,

$$\Omega_1 : \{0 \le x_1 \le m,\ 0 \le x_2 \le 1\}, \quad \Omega_2 : \{m \le x_1 \le 1,\ 0 \le x_2 \le 1\},$$

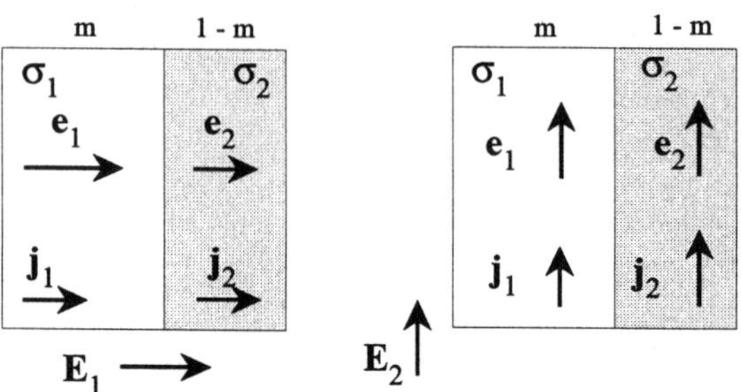

FIGURE 2.2. The fields and currents in a laminate. If the external field is applied across the layers (left), the current stays constant everywhere in the structure. If the external field is applied along the layers (right), the field stays constant everywhere in the structure.

are filled with isotropic conducting materials with conductivities σ_1 and σ_2, respectively; see Figure 2.2.

For physical reasons, we should expect that the laminates are equivalent to an *anisotropic* material, that is, characterized by a tensor of effective properties $\boldsymbol{\sigma}_*$.

Conductivity Across the Layers

First apply the unit external field $\mathbf{E}_1 = \mathbf{i}_1$ perpendicular to the layers (see Figure 2.2, left). The conductivity of the laminates is described by the boundary value problem

$$\nabla \cdot \sigma(\mathbf{x})(\nabla w(\mathbf{x}) + \mathbf{E}_1) = 0, \quad w \text{ is periodic in } \Omega, \tag{2.2.11}$$

where $\sigma(\mathbf{x}) = \sigma_i$ if $\mathbf{x} \in \Omega_i$. It has a simple analytical solution: The potential w is a continuous piecewise linear function of x_1:

$$w = w(x_1) = \begin{cases} \alpha_1 x_1 & \text{in } \Omega_1, \\ m\alpha_1 + \alpha_2 (x_1 - m) & \text{in } \Omega_2, \end{cases} \tag{2.2.12}$$

where α_1, α_2 are constants and we assume that $w(0) = 0$. The gradient ∇w is a piecewise constant vector directed along the x_1-axis:

$$\mathbf{e} = \nabla w = \begin{cases} \mathbf{e}_1 & \text{in } \Omega_1, \\ \mathbf{e}_2 & \text{in } \Omega_2, \end{cases}$$

where

$$\mathbf{e}_1 = \begin{pmatrix} \alpha_1 \\ 0 \end{pmatrix}, \quad \mathbf{e}_2 = \begin{pmatrix} \alpha_2 \\ 0 \end{pmatrix}.$$

The constants α_1, α_2 are determined from two constraints. The periodicity of ∇w states that $\langle \nabla w \rangle = 0$, which yields to $m\alpha_1 + (1 - m)\alpha_2 = 0$.

The second constraint comes from the jump condition $[\mathbf{J} \cdot \mathbf{n}]^{+}_{-} = 0$ on the line $x_1 = m$. It yields to

$$\sigma_1(\mathbf{e}_1 + \mathbf{E}_1) \cdot \mathbf{n} - \sigma_2(\mathbf{e}_2 + \mathbf{E}_1) \cdot \mathbf{n} = 0.$$

From these conditions, we compute α_1, α_2:

$$\alpha_1 = \frac{\sigma_2}{m\sigma_2 + (1-m)\sigma_1} - 1, \quad \alpha_2 = \frac{\sigma_1}{m\sigma_2 + (1-m)\sigma_1} - 1.$$

The solution of (2.2.12) satisfies the equation (2.2.11) and boundary conditions. The mean field in the cell is equal to $\mathbf{i}_1$, and the mean current is

$$\langle \mathbf{j} \rangle = m\sigma_1(\mathbf{e}_1 + \mathbf{E}_1) + (1-m)\sigma_2(\mathbf{e}_2 + \mathbf{E}_1) = \begin{pmatrix} \frac{\sigma_1\sigma_2}{m\sigma_2+(1-m)\sigma_1} \\ 0 \end{pmatrix}.$$

Note that the current $\mathbf{j}$ is constant in the entire cell.

Thus we determine the two coefficients of the effective tensor by (2.2.8). The element σ_{11} is equal to the harmonic mean of the conductivities of the initial materials:

$$(Gs_*)_{11} = \sigma_h = \frac{\sigma_1\sigma_2}{m\sigma_2 + (1-m)\sigma_1} = \left(m\sigma_1^{-1} + (1-m)\sigma_2^{-1}\right)^{-1},$$

where σ_h is the harmonic mean of the conductivities.

We also find that $(Gs_*)_{12} = 0$. The symmetry of $\boldsymbol{\sigma}_*$ implies that $(Gs_*)_{21} = 0$. The normal to the layers is the eigenvector of the effective tensor; the effective conductivity across the layers is equal to one of the eigenvalues of this tensor.

Conductivity Along the Layers

Let us determine the effective conductivity in an orthogonal direction. We direct the applied field $\mathbf{E}_2 = \mathbf{i}_2$ along the layers (see Figure 2.2, right). The boundary value problem,

$$\nabla \cdot \sigma(\mathbf{x})(\nabla w(\mathbf{x}) + \mathbf{E}_2) = 0, \quad w \text{ is periodic in } \Omega,$$

where $\sigma(\mathbf{x}) = \sigma_i$ if $\mathbf{x} \in \Omega_i$, has the uniform solution, $w = x_2$. This solution satisfies the boundary conditions, the jump condition, and the differential equation. This solution implies that the field is constant everywhere, $\mathbf{e} = \nabla w = \mathbf{i}_2$, and the mean current $\langle \mathbf{j} \rangle$ is equal to

$$\langle \mathbf{j} \rangle = \langle \sigma \rangle \mathbf{e} = \begin{pmatrix} 0 \\ m\sigma_1 + (1-m)\sigma_2 \end{pmatrix}.$$

The eigenvalue of the effective conductivity tensor that has been determined is equal to the arithmetic mean σ_a of the conductivities:

$$(Gs_*)_{22} = \sigma_a, \quad \sigma_a = m\sigma_1 + (1-m)\sigma_2;$$

the eigenvector corresponds to the tangent to the layers.

54   2. Conducting Composites

The Effective Tensor

We have found that a laminate in a uniform external field behaves equivalently to a homogeneous but anisotropic medium. We denote the tensor of effective properties $\boldsymbol{\sigma}_*$ of a laminate structure by $\sigma_{\mathrm{lam}} = \boldsymbol{\sigma}_*$. This tensor depends on the structural parameters: the volume fraction m of the materials and the orientation of the structure. The constitutive equation for this medium represents a relationship between the mean value of the current density and the mean value of the field; it depends on the normal $\mathbf{n}$ and the tangent $\mathbf{t}$ to the layers. In the coordinates $\mathbf{n}, \mathbf{t}$, it has the form

$$\begin{pmatrix} j_n \\ j_t \end{pmatrix} = \boldsymbol{\sigma}_{\mathrm{lam}} \begin{pmatrix} e_n \\ e_t \end{pmatrix}; \quad \boldsymbol{\sigma}_{\mathrm{lam}} = \begin{pmatrix} \sigma_h & 0 \\ 0 & \sigma_a \end{pmatrix},$$

where subindices $_n$ and $_t$ denote the normal and tangent components of the corresponding vectors.

Remark 2.2.3 *The asymptotic case of very different conductivities*

$$\sigma_1 \ll \sigma_2$$

corresponds to the asymptotics

$$\sigma_h \approx \frac{\sigma_1}{m_1}, \quad \sigma_a \approx \sigma_2 m_2.$$

This formula demonstrates that the conductivity along the layers is determined mainly by the conductivity of the best conductor σ_2 and the conductivity across the layers by the conductivity of the worst conductor σ_1. This remark emphasizes that composites can emphasize the property of each phase and possess new properties, such as anisotropy.

Generalizations

The obtained formulas permit a straightforward generalization to laminates made from more than two materials. In this case the arithmetic and harmonic means are expressed as

$$\sigma_a = \sum_{i=1}^{p} m_i \sigma_i, \quad \sigma_h = \left(\sum_{i=1}^{p} m_i \sigma_i^{-1} \right)^{-1}, \tag{2.2.13}$$

where p is the number of mixed materials, and σ_i and m_i are the conductivity and volume fraction of the ith material.

The generalization to the three-dimensional case is also straightforward. The effective properties tensor $\boldsymbol{\sigma}_{\mathrm{lam}}$ is equal to

$$\boldsymbol{\sigma}_{\mathrm{lam}} = \begin{pmatrix} \sigma_h & 0 & 0 \\ 0 & \sigma_a & 0 \\ 0 & 0 & \sigma_a \end{pmatrix};$$

the normal direction of laminates corresponds to the eigenvalue σ_h.

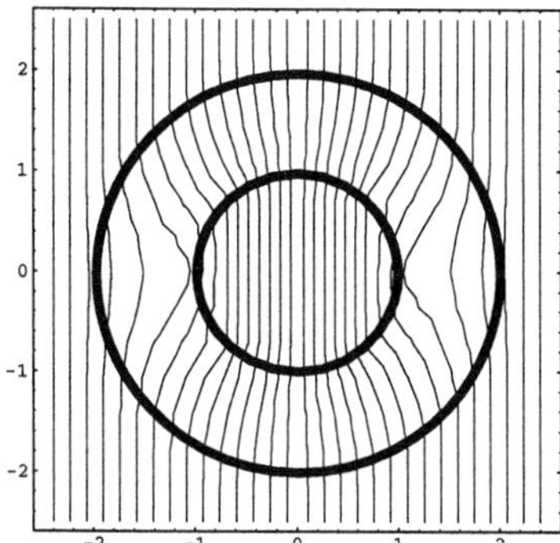

FIGURE 2.3. The geometry of coated circles. The field outside the external disk is homogeneous. The inner disk has higher conductivity than the effective medium, and the exterior annulus has lower conductivity than the effective medium. Observe the complete mutual compensation of the inclusion. The inclusion is "invisible" in a uniform external field.

2.2.3 *Effective Medium Theory: Coated Circles*

Here we describe a way to calculate effective properties for media with located symmetric inclusions. The approach leads to exact formulas for the effective conductivity. It was originated and discussed in (Bruggemann, 1935; Bruggemann, 1937; Hashin and Shtrikman, 1962a; Christensen, 1979) and others. Specifically we discuss the structure of the "coated spheres" suggested in (Hashin and Shtrikman, 1962a).

Consider a homogeneous material with isotropic conductivity σ_*. Suppose we replace the medium in a disk of unit radius with the following two-phase configuration. The inner disk of radius $r_0 < 1$ is filled with material σ_1, and the annulus $r_0 < r < 1$ is filled with material σ_2. This configuration (see Figure 2.3) is called the *coated circles* (or, in three dimensions, the *coated spheres*).

Assume that the configuration is submerged into a uniform external field $\mathbf{e}(r, \theta) \to \cos\theta$, when $r \to \infty$. The corresponding potential w tends to the affine function $w \to r \cos\theta$.

Suppose we manage to define the conductivity σ_* so that the field everywhere outside of the inclusion is a constant vector: $\mathbf{e} = \mathbf{i}_1$. In polar coordinates, this condition takes the form

$$\mathbf{e}(r, \theta) = [\cos\theta, \sin\theta] \quad \forall r > 1. \tag{2.2.14}$$

In this case, we cannot detect the presence of the inclusion by observing the fields anywhere outside of the inclusion. Hence, we cannot distinguish the homogeneous configuration with conductivity σ_* from a configuration with one, or several, or even infinitely many circular inclusions of the described type; see Figure 2.3. In this case we call σ_* the effective conductivity of a composite made of coated circles.

To find σ_* we explicitly calculate the field everywhere in the configuration. The field satisfies the boundary value problem

$$\nabla^2 w = 0 \ \text{ in } R^2, \quad \lim_{r \to \infty} \frac{\partial w(r, \theta)}{\partial r} = \cos \theta,$$

and satisfies the jump conditions on the circles and the effective medium condition (2.2.14).

This problem permits separation of variables; the solution w has the form $w = R(r) \cos \theta$. The function $R(r)$ must satisfy the ordinary differential equation

$$r \frac{d}{dr} \left(r \frac{d}{dr} R \right) - R = 0 \tag{2.2.15}$$

the conditions

$$\begin{aligned}
R(0) &= 0, & R'(0) &= 0, \\
[R(r_0)] &= 0, & [\sigma(r) R'(r_0)] &= 0, \\
[R(1)] &= 0, & [\sigma(r) R'(1)] &= 0, \\
\lim_{r \to \infty} R &= r, &
\end{aligned} \tag{2.2.16}$$

where $[x]$ means the jump of x, and the condition (2.2.14). The conductivity $\sigma(r)$ is

$$\sigma(r) = \begin{cases} \sigma_1 & \text{if } r \in [0, r_0), \\ \sigma_2 & \text{if } r \in [r_0, 1), \\ \sigma_* & \text{if } r \in [1, \infty). \end{cases}$$

We assume that the potential is zero at $r = 0$ (we can always assume this, because the potential is defined up to a constant), and we require the continuity of the field at $r = 0$. The last condition in (2.2.16) says that the field in the system with the inclusion tends to a homogeneous field when $r \to \infty$. The remaining conditions express the continuity of the potential and of the normal current on the circles $r = r_0$ and $r = 1$.

The solution to (2.2.15) that satisfies the conditions (2.2.16) has the form

$$w = \begin{cases} A_0 r & \text{if } 0 < r < r_0, \\ A_1 r + \frac{B_1}{r} & \text{if } r_0 < r < 1, \\ r + \frac{B_2}{r} & \text{if } 1 < r. \end{cases} \tag{2.2.17}$$

To define the four constants A_0, A_1, B_1, and B_2 we use conditions (2.2.16).

The key point of the scheme is the following: We assign the constant σ_* in such a way that $B_2 = 0$ or that the field is homogeneous if $r > 1$. This way, (2.2.14) is satisfied.

Accounting for the constants, we have

$$\begin{aligned}
A_0 &= \frac{2\sigma_2}{m_2 \sigma_1 + (1 + m_1) \sigma_2}, \\
A_1 &= \frac{\sigma_1 + \sigma_2}{m_2 \sigma_1 + (1 + m_1) \sigma_2}, \\
B_1 &= \frac{m_1 (-\sigma_1 + \sigma_2)}{m_2 \sigma_1 + (1 + m_1) \sigma_2},
\end{aligned} \tag{2.2.18}$$

and

$$\sigma_* = \sigma_{HS} = \sigma_1 \frac{(1 + m_1)\,\sigma_1 + m_2\,\sigma_2}{m_2\,\sigma_1 + (1 + m_1)\,\sigma_2}. \qquad (2.2.19)$$

Formula (2.2.19) shows the effective conductivity of the configuration. The conductivity was calculated in (Hashin and Shtrikman, 1962a), where it was also proven that σ_{HS} is the *extreme* isotropic conductivity that one can achieve by arbitrary mixing of two isotropic materials in the prescribed proportion.

Remark 2.2.4 *A generalization of the procedure was suggested in (Milton, 1980), which considered the geometry of "coated ellipses" (one inscribed into another) and found the explicit description of their effective properties. This time, the effective medium is anisotropic. The idea of the calculation is the same: We consider one "coated elliptical inclusion," i.e., two ellipses in an unbounded domain and a homogeneous field applied at infinity.*

2.3   Conclusion and Problems

This chapter introduced the main objects for the structural optimization of conducting composites.

- We described the conductivity of an inhomogeneous medium, the differential constraints and potentials for fields and currents, and the jump conditions on the boundary between different materials. The corresponding pair of dual variational principles was introduced.

- We described the properties of composites and the homogenization procedure. An algorithm has been presented to compute the tensor of effective properties of a composite. We have analytically computed the effective properties of laminates and of coated circles.

Problems

1. Consider the function

$$f(c_1, \ldots, c_n, x) = \sum_{i=1}^{n} \chi_i(x) c_i,$$

   where χ_i are the characteristic functions of nonoverlapping domains of x, and a function $G(z)$. Prove the superposition rule

$$G(f(c_1, \ldots, c_n, x)) = f(G(c_1), \ldots G(c_n), x).$$

2. Consider a conducting composite made of two anisotropic materials. Define the magnitude of the jump of discontinuous components of **e** and **j** through the tensors of conductivity.

3. How many external fields are needed to compute all coefficients of two- and three-dimensional conductivity tensors by calculating the energy? Suggest an algebraic procedure to calculate the eigenvalues and eigenvectors of an effective tensor.

4. Derive the effective properties using an external current instead of the external field. Prove that the resulting effective tensor remains the same.

5. Derive the effective properties for the three-dimensional geometry of "coated spheres."

3
Bounds and G-Closures

In structural optimization, the effective properties of layouts are controls: An optimal structure adapts itself to the local fields. The layout is no longer periodic but almost periodic function. Here we introduce the corresponding technique which is the G-convergence of a sequence of linear operators.

In control problems, it is essential to know the range of effective properties. Here we establish some bounds for the effective tensors. We also introduce the notion of the G-closure: the set of effective tensors of a composite with arbitrary microstructures.

3.1 Effective Tensors: Variational Approach

Here we compute the effective tensors from the variational principles and we establish inequalities for these tensors.

3.1.1 Calculation of Effective Tensors

The Energy of a Homogenized Body

Consider the sequence σ_ε of periodic layouts. The solution w_ε of the conductivity equations is a minimizer of a corresponding variational functional (2.1.23):

$$\int_\Omega (W(\sigma_\varepsilon, \nabla w_\varepsilon) + f w_\varepsilon), \qquad (3.1.1)$$

where

$$W(\boldsymbol{\sigma}_\varepsilon, \nabla w_\varepsilon) = \frac{1}{2} \nabla w_\varepsilon \cdot \boldsymbol{\sigma}_\varepsilon \nabla w_\varepsilon.$$

Also, the solution w_0 (2.2.5) minimizes the energy of the homogenized body:

$$\int_\Omega (W(\boldsymbol{\sigma}_*, \nabla w_0) + f w_0). \qquad (3.1.2)$$

The Euler–Lagrange equation for the last functional, $\nabla \cdot \boldsymbol{\sigma}_* \nabla w_0 = f$, coincides with the homogenized equation.

The minimizer w_ε of the variational problem (3.1.1) tends to the minimizer w_0 of the homogenized medium when $\varepsilon \to 0$. Hence, the sequence of Lagrangians $\{W(\boldsymbol{\sigma}_\varepsilon, \nabla w_\varepsilon) - f w_\varepsilon\}$ tends to the Lagrangian $W(\boldsymbol{\sigma}_*, \nabla w_0) - f w_0$. In other words, the average of the energy over a small region in an inhomogeneous body is arbitrarily close to the energy of an equivalent homogeneous material[1].

The sequence of energies weakly converges (in $L_1(\Omega)$) to the energy of the homogenized material

$$\langle W(\boldsymbol{\sigma}_\varepsilon, \nabla w_\varepsilon) \rangle \rightharpoonup W(\boldsymbol{\sigma}_*, \nabla w_0). \qquad (3.1.3)$$

The last relationship can be rewritten as either

$$\langle \mathbf{e}_\varepsilon \cdot \boldsymbol{\sigma}_\varepsilon \mathbf{e}_\varepsilon \rangle \rightharpoonup \langle \mathbf{e}_0 \rangle \cdot \boldsymbol{\sigma}_* \langle \mathbf{e}_0 \rangle$$

or

$$\langle \mathbf{j}_\varepsilon \cdot \boldsymbol{\sigma}_\varepsilon^{-1} \cdot \mathbf{j}_\varepsilon \rangle \rightharpoonup \langle \mathbf{j}_0 \rangle \cdot \boldsymbol{\sigma}_*^{-1} \cdot \langle \mathbf{j}_0 \rangle.$$

Essentially, these formulas introduce the effective tensor $\boldsymbol{\sigma}_*$. One can check that this definition is equivalent to the earlier definition of the effective tensor as the proportionality coefficients between the averaged current and field (Bensoussan et al., 1978; Jikov et al., 1994).

Remark 3.1.1 *The symmetric form $\mathbf{e}_\varepsilon \cdot \mathbf{j}_\varepsilon$ of the energy deals explicitly only with the currents and fields but not with the properties. The limiting equality (3.1.3) takes the form:*

$$\langle \mathbf{e}_\varepsilon \cdot \mathbf{j}_\varepsilon \rangle \rightharpoonup \langle \mathbf{e}_0 \rangle \cdot \langle \mathbf{j}_0 \rangle.$$

This representation looks surprising because the operation of integration (averaging) commutes with the scalar product operation. This relation follows from the variational principle; it will be analyzed and explained later using the theory of compensated compactness(see Chapter 7).

---

[1] Generally speaking, these energies can differ by a *null-Lagrangian*, that is, by a term for which the Euler–Lagrange equation is identically zero (see the discussion in Chapters 5, 7, and 12).

Calculation of the Effective Tensor Using Variational Approach

We use the variational principle to compute the effective tensor because a cell of periodicity Ω in an inhomogeneous medium stores the same amount of energy as the effective material:

$$\langle \mathbf{e} \cdot \boldsymbol{\sigma}\mathbf{e} \rangle = \langle \mathbf{e} \rangle \cdot \boldsymbol{\sigma}_* \langle \mathbf{e} \rangle.$$

This equation can be used to determine the effective properties tensor itself. For example, applying a field $\mathbf{e} = \mathbf{i}_1$ of unit magnitude and calculating the energy in the unit cell, we find that this energy is equal to the upper-left element σ_{11}^* of the tensor $\boldsymbol{\sigma}_*$.

This element is the cost of the variational problem (3.1.2):

$$(\sigma_*)_{11} = \min_{\mathbf{e} \in \mathcal{E}} \langle \mathbf{e} \cdot \boldsymbol{\sigma}\mathbf{e} \rangle, \tag{3.1.4}$$

where

$$\mathcal{E} = \{ \mathbf{e} : \nabla \times \mathbf{e} = 0, \quad \langle \mathbf{e} \rangle = \mathbf{i}_1, \quad \mathbf{e} \text{ is 1-periodic} \}. \tag{3.1.5}$$

Repeating this procedure several times with differently oriented external fields $\mathbf{e}$, one can calculate all elements of $\boldsymbol{\sigma}_*$.

3.1.2 Wiener Bounds

The variational method allows us to derive the bounds for coefficients of the effective tensor. Indeed, any admissible trial function $\mathbf{e}_{\text{trial}}(\mathbf{x})$ that satisfies (3.1.5) provides an upper bound for a diagonal coefficient of $\boldsymbol{\sigma}_*$ due to (3.1.4).

The simplest bound is given by a *constant* trial function

$$\mathbf{e}_{\text{trial}}(\mathbf{x}) = \text{constant}(\mathbf{x}) = \mathbf{i}_1 \quad \forall \mathbf{x} \tag{3.1.6}$$

that obviously belongs to the set $\mathcal{E}$ (see (3.1.5)). If we substitute $\mathbf{e}_{\text{trial}}$ into (3.1.4) and recall that $\boldsymbol{\sigma}(\mathbf{x}) = \sigma(\mathbf{x})I$, we obtain

$$(\sigma_*)_{11} \leq \langle \mathbf{i}_1 \cdot \boldsymbol{\sigma}\mathbf{i}_1 \rangle = \langle \sigma_{11} \rangle.$$

Varying the orientation of the vector of $\mathbf{i}$, we obtain the matrix inequality:

$$\boldsymbol{\sigma}_* \leq \langle \boldsymbol{\sigma} \rangle. \tag{3.1.7}$$

Particularly, the maximal eigenvalue of $\boldsymbol{\sigma}_*$ is bounded from above by the maximal eigenvalue of $\langle \boldsymbol{\sigma} \rangle$.

For a composite assembled from several materials with volume fractions m_i and conductivity tensors $\boldsymbol{\sigma}_i$ we have

$$\langle \boldsymbol{\sigma} \rangle = \sum_{i=1}^{N} m_i \boldsymbol{\sigma}_i = \boldsymbol{\sigma}_a, \tag{3.1.8}$$

where subindex $_a$ denotes the arithmetic mean. The bound (3.1.7) is called the *Reuss bound* (Reuss, 1929) or the *arithmetic mean bound*.

The dual variational principle (Thompson's principle) also determines a bound for the effective tensor $\boldsymbol{\sigma}_*$. The diagonal coefficient β_*^{11} of the inverse tensor $\boldsymbol{\beta} = \boldsymbol{\sigma}^{-1}$ is

$$\beta_*^{11} = \min_{\mathbf{j} \in \mathcal{J}} \langle \mathbf{j} \cdot \boldsymbol{\sigma}^{-1} \mathbf{j} \rangle,$$

where

$$\mathcal{J} = \{\mathbf{j} : \nabla \cdot \mathbf{j} = 0, \quad \langle \mathbf{j} \rangle = \mathbf{i}_1, \quad \mathbf{j} \text{ is 1-periodic}\}.$$

Thompson's principle leads to upper estimates of the coefficients of the inverse tensor $\boldsymbol{\sigma}_*^{-1}$ (which are the *lower* estimates of the tensor $\boldsymbol{\sigma}_*$). Again, using the constant trial function, one obtains the inequality

$$\beta_*^{11} \leq \langle \mathbf{i}_1 \cdot \boldsymbol{\sigma}^{-1} \mathbf{i}_1 \rangle,$$

which leads to

$$\boldsymbol{\sigma}_*^{-1} \leq \langle \boldsymbol{\sigma}^{-1} \rangle = \sum_{i=1}^{N} m_i \boldsymbol{\sigma}_i^{-1} = \boldsymbol{\sigma}_h^{-1},$$

where

$$\boldsymbol{\sigma}_h = \left(\sum_{i=1}^{N} m_i \boldsymbol{\sigma}_i^{-1} \right)^{-1} \tag{3.1.9}$$

denotes the harmonic mean. This bound is called the *Voigt bound* (Voigt, 1928) or the *harmonic mean bound*.

Together, inequalities (3.1.7) and (3.1.9) provide two-sided bounds of the range of variation of the effective properties tensor:

$$\boldsymbol{\sigma}_h \leq \boldsymbol{\sigma}_* \leq \boldsymbol{\sigma}_a. \tag{3.1.10}$$

The range $[\boldsymbol{\sigma}_h, \boldsymbol{\sigma}_a]$ is called the *Wiener box*. It depends only on the properties of the initial materials and their fractions in the composite. The inequalities (3.1.10) are valid for any composite regardless of its geometry; we call them geometrically independent bounds. These inequalities are also called *Wiener inequalities* (Wiener, 1912).

Remark 3.1.2 *Similar bounds can be established for other equilibria that satisfy a minimum variational principle. Indeed, the constant trial function similar to (3.1.6) trivially satisfies any linear differential restrictions.*

Note that the Wiener bounds are invariant to interchanging the properties tensors with their inverses:

$$\left(\boldsymbol{\sigma}^{-1} \right)_h \leq \boldsymbol{\sigma}_*^{-1} \leq \left(\boldsymbol{\sigma}^{-1} \right)_a.$$

The equivalence follows from obvious identities

$$\left(\boldsymbol{\sigma}^{-1} \right)_h = \left(\boldsymbol{\sigma}_a \right)^{-1}, \quad \left(\boldsymbol{\sigma}^{-1} \right)_a = \left(\boldsymbol{\sigma}_h \right)^{-1}.$$

They demonstrate that the upper bound for the "direct" tensor $\boldsymbol{\sigma}$ becomes the lower estimate for the inverse tensor $\boldsymbol{\sigma}^{-1}$ and vice versa.

Bounds on Composites' Properties

The derived Wiener bounds are the simplest examples of the bounds on effective properties. More complicated procedures take into account the differential properties of the acting fields like the curlfree nature of the fields **e**. In this book, we will develop several methods of this kind. However, a number of the approaches is not discussed because our main focus is structural optimization. Instead, we refer to the collections and monographs (Hashin, 1970b; Christensen, 1979; Berdichevsky, 1983; Nemat-Nasser and Hori, 1993; Berdichevsky et al., 1999; Markov and Preziosi, 1999; Markov and Inan, 1999) where the reader can find these approaches.

A number of papers deals with bounds on the overall properties of composites from nonlinear materials. We mention (Hashin, 1983; Talbot and Willis, 1985; Ponte Castañeda and Willis, 1988; Bergman, 1991; Hashin, 1992; Talbot and Willis, 1992; Bourgeat et al., 1995; Khruslov, 1995; Oleĭnik, Yosifian, and Temam, 1995; Talbot, Willis, and Nesi, 1995; Talbot and Willis, 1995; Telega, 1995; Zhikov, 1995; Ponte Castañeda, 1996; Ponte Castañeda, 1997; Talbot and Willis, 1997; Milton and Serkov, 1999; Torquato, 1999) where a number of bounding methods is developed.

3.2   *G*-Closure Problem

3.2.1   G-convergence

Definition

Generalization of the homogenization procedure for linear operators leads to the introduction of the *G*-convergence. The theory of *G*-convergence studies the behavior of sequences of linear operators L^s and of corresponding solutions w^s of the boundary value problems:

$$L^s w^s = f, \quad \text{in } \Omega, \quad w^s|_{\partial\Omega} = \rho. \tag{3.2.1}$$

The family of the conductivity operators in inhomogeneous media

$$L^s = \nabla \cdot \boldsymbol{\sigma}(\chi^s)\, \nabla,$$

where χ^s is periodic in the cube Ω^s with side $\frac{1}{2^s}$, gives an example of such an operator sequence. The almost periodic layout gives another example.

Consider a sequence $\{L^s\}$ of the operators (3.2.1) and the sequence of their solutions $\{w^s = (L^s)^{-1} f\}$. Suppose that the sequence of the solutions converges weakly (in H^1) to a function w_0:

$$w^s \rightharpoonup w_0 \quad \text{weakly in } H^1(\Omega).$$

Definition 3.2.1 The weak convergence of solutions $w^s = (L^s)^{-1}f$ implies a certain convergence of the operator's sequence, which is called *G-convergence*:

$$L^s \xrightarrow{G} L_* \quad \text{if } (L^s)^{-1}f \rightharpoonup L_*^{-1}f \quad \forall f \in H^{-1}(\Omega).$$

The limiting operator L_* exists for a family of linear elliptic coercive operators L^s if their solutions weakly converge, and this limit is an elliptic operator of the same order as the operators in the sequence, (Marino and Spagnolo, 1969; Bensoussan et al., 1978; Jikov et al., 1994).

More exactly, the sequence $L^s = \nabla \cdot \boldsymbol{\sigma}^s \nabla$ of the conductivity operators G-converges to an operator $L_* = \nabla \cdot \boldsymbol{\sigma}_* \nabla$,

$$L^s = \nabla \cdot \boldsymbol{\sigma}^s \nabla \xrightarrow{G} \nabla \cdot \boldsymbol{\sigma}_* \nabla = L_*, \tag{3.2.2}$$

if the eigenvalues of tensors $\boldsymbol{\sigma}^s$ are constrained,

$$\|\boldsymbol{\sigma}^s\| \le c_1, \quad \|\boldsymbol{\sigma}^{s-1}\| \le c_2, \quad c_1 > 0, c_2 > 0.$$

These conditions mean that the mixed materials are not ideal conductors of insulators. They guarantee that the G-limit of a sequence of the conductivity operators is also a conductivity operator.

The G-convergence of operators is a more general type of convergence than homogenization, but it includes homogenization. Particularly, we can view the limiting operator L_* as the conductivity operator corresponding to an inhomogeneous medium with infinitely fine-scale oscillating properties. The weak limit w_0 of solutions w^s is the averaged potential and the G-limiting operator is the homogenized conductivity operator that depends on the effective conductivity $\boldsymbol{\sigma}_*$.

Instead of a convergence of the conductivity operators we may consider a convergence of the layouts $\{\boldsymbol{\sigma}^s\}$ that define these operators. The notion of G-convergence can be applied to the sequence $\{\boldsymbol{\sigma}^s\}$.

Definition 3.2.2 We say that the sequence of the layouts $\{\boldsymbol{\sigma}^s\}$ *G-converges* to the effective layout $\boldsymbol{\sigma}_*$ if the corresponding sequence $L(\boldsymbol{\sigma}^k)$ G-converges to the conductivity operator $L(\boldsymbol{\sigma}_*)$, $L(\boldsymbol{\sigma}^s) \xrightarrow{G} L(\boldsymbol{\sigma}_*)$; see (3.2.2).

Also, we call the layout $\boldsymbol{\sigma}_*$ the G-limit of the sequence $\{\boldsymbol{\sigma}^s\}$:

$$\boldsymbol{\sigma}^s \xrightarrow{G} \boldsymbol{\sigma}_*.$$

The homogenization procedure corresponds to the case where a G-limiting tensor is independent of $\mathbf{x}$. The G-limiting tensor $\boldsymbol{\sigma}_*$ describes the conductivity of the homogenized media.

Various generalizations for the concept of G-convergence is discussed in (Tartar, 1990; Bensoussan, Boccardo, and Murat, 1992; Dal Maso, 1993; Pedregal, 1997; Chiheb and Panasenko, 1998); see also references therein, and in (Raĭtum, 1999).

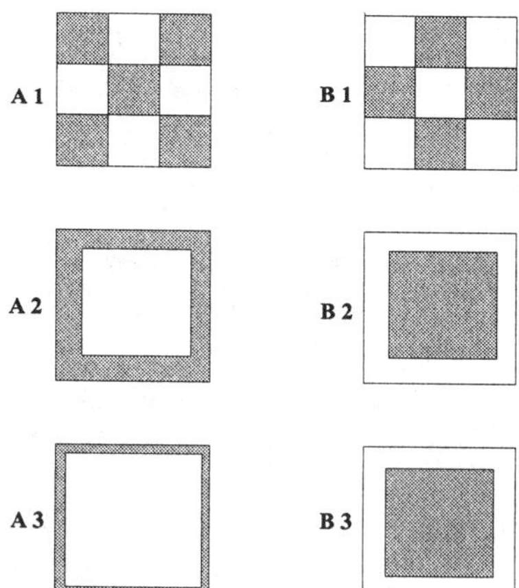

FIGURE 3.1. Various limits in the description of the materials layouts.

G-Convergence and Other Types of Convergence

The following examples (Figure 3.1) illustrate relationships between *G*-convergence and other types of convergence of sequences of materials' layouts.

Example 3.2.1 First, we comment on the relation between the *G*-convergence and strong convergence.

Suppose that an optimal layout $\mathcal{R}_\varepsilon$ of conducting materials is given by a checkerboard structure with squares of size ε made of "white" and "black" materials with conductivities σ_1 and σ_2 ($\sigma_1 < \sigma_2$), respectively (Figure 3.1, **A**1). The structure fills in a domain much larger than a square of the checkerboard and is submerged into a uniform external electrical field. The structure can be replaced with a homogeneous material with isotropic effective conductivity σ_*.

Interchange the materials in the fields and call the new structure $\mathcal{R}'_\varepsilon$ (Figure 3.1, **B**1). Consider the conductivity of the structure $\mathcal{R}'_\varepsilon$ in the same domain and external field. The difference $w' - w$ of the solutions to the corresponding conductivity problems will be as small as the scale of the board is. In the limit, these solutions coincide:

$$w - w' \to 0 \quad \text{as} \quad \varepsilon \to 0.$$

The *G*-convergence does not distinguish between these two layouts that lead to equal solutions to the conductivity problem,

$$\sigma_*(\mathcal{R}_\varepsilon) = \sigma_*(\mathcal{R}'_\varepsilon).$$

However, the pointwise tensor properties of these two layouts are extremely different. The norm of the difference is maximal,

$$|\sigma(\mathcal{R}_\varepsilon) - \sigma(\mathcal{R}'_\varepsilon)| = \sigma_2 - \sigma_1 \quad \forall \mathbf{x} \in \Omega,$$

because the material is switched in each point of the domain.

Therefore, G-convergence does not imply the strong convergence. However, the strong convergence does imply the G-convergence (the consideration is left to the reader).

Example 3.2.2 Consider the relation between weak convergence (averaging) of the materials' layouts and G-convergence. We demonstrate first that the weak limit does not define the G-limit.

Consider a conducting plane of a good conductor σ_2 with periodic square inclusions of a bad conductor (insulator) σ_1 (structure $\mathcal{R}_A$) and suppose that the volume fraction of inclusions is equal to one-half (Figure 3.1, **A2**). Again, consider a sequence of structures in which the size of the periodicity element tends to zero.

The average value of conductivity $\langle \sigma \rangle$ (that is, the weak limit of the conductivity layout) of the structure is $\langle \sigma(\mathcal{R}_A) \rangle = \frac{\sigma_1 + \sigma_2}{2}$. The structure has an isotropic effective conductivity $\boldsymbol{\sigma}_*(\mathcal{R}_A) = \sigma_*(\mathcal{R}_A)I$ due to its symmetry. Physically, it is clear that the effective conductivity of the plane σ_* will remain close to σ_2 ($\sigma_*(\mathcal{R}_A) \approx \sigma_2$), because the conductance is mainly provided by the material σ_2 in the connected phase.

Interchange materials in the composite and call the resulting structure $\mathcal{R}_B$ (Figure 3.1, **B2**). The average conductivity of the structures $\mathcal{R}_A$ and $\mathcal{R}_B$ stays the same, because the same amounts of the materials is used, but the effective conductivity of the structure $\mathcal{R}_B$ is lower; $\sigma_*(\mathcal{R}_B) < \sigma_*(\mathcal{R}_A)$ because its conductance is now mainly determined by the first material ($\sigma_*(\mathcal{R}_B) \approx \sigma_1$) that forms the connected phase.

These two structures have the same mean value of conductivity but different G-limits:

$$\langle \sigma(\mathcal{R}_A) \rangle = \langle \sigma(\mathcal{R}_B) \rangle, \quad \text{but} \quad \sigma_*(\mathcal{R}_B) < \sigma_*(\mathcal{R}_A).$$

Example 3.2.3 On the other hand, the G-limit does not determine the weak limit either. Let us demonstrate the structures (Figure 3.1, **A3, B3**) that have the same G-limit of conductivity but different mean conductivities.

Consider again the configuration $\mathcal{R}_A$ (Figure 3.1, **A2**) with square inclusions occupied by the bad conductor σ_1. Let us increase the fraction m of the inclusions in the element of periodicity from $\frac{1}{2}$ to 1 and let us call the structures obtained $\mathcal{R}_A(m)$. The structure $\mathcal{R}_B$ (Figure 3.1) corresponds to the volume fraction $\frac{1}{2}$ and is denoted $\mathcal{R}_B\left(\frac{1}{2}\right)$.

We already mentioned that

$$\sigma_*\left(\mathcal{R}_A\left(\frac{1}{2}\right)\right) > \sigma_*\left(\mathcal{R}_B\left(\frac{1}{2}\right)\right).$$

Following the increase of m, $\mathcal{R}_A(m)$ continuously decreases down to the value $\sigma_*(\mathcal{R}_A(1)) = \sigma_1$, which is obviously less than $\sigma_*(\mathcal{R}_B)\left(\frac{1}{2}\right)$. Therefore

$\mathcal{R}_A(m)$ meets the effective conductivity $\sigma_* \left(\mathcal{R}_B \left(\frac{1}{2} \right) \right)$ of the configuration $\mathcal{R}_B$ (Figure 3.1, **B2**) somewhere during this process (Figure 3.1, **A3**):

$$\exists \; m_0 \in \left[\frac{1}{2}, \, 1 \right] : \quad \sigma_*(\mathcal{R}_A(m_0)) = \sigma_* \left(\mathcal{R}_B \left(\frac{1}{2} \right) \right).$$

The two composites (Figure 3.1, **A3, B3**) have the same effective conductivity but different mean values of the conductivities, whereas different relative amounts of materials are needed to obtain the same effective conductivity in the configurations $\mathcal{R}_A$ and $\mathcal{R}_B$, $m_0 \neq \frac{1}{2}$.

Thus, the weak limit cannot be determined by the *G*-limit, either.

Moreover, the *G*-limit of an asymmetric structure such as a laminate depends on the direction of the applied field, but the weak limit does not. Thus, the *G*-limit cannot be determined by the weak limit. However, the range of *G*-limits may depend on it.

3.2.2 *G*-Closure: Definition and Properties

Here we introduce the central idea of the *G*-closure of a set of material properties. The *G*-closure is the set of effective properties of all possible composites assembled from given materials. The problem of its description was addressed at the turn of the twentieth century, when the bounds of all possible effective tensors were established in (Wiener, 1912). Hashin and Shtrikman came out with the exact description of isotropic points of *G*-closure in (Hashin and Shtrikman, 1962a). Their work has demonstrated that the bounds for the *G*-closure corresponds to simple explicit formulas. Another simple example was built in (Tartar, 1975; Raĭtum, 1978): we discuss it in the next Section.

The concept of the *G*-closure and the term itself was introduced in (Lurie and Cherkaev, 1981a; Lurie and Cherkaev, 1981c) as the problem of completeness of the *G*-limits. This consideration was motivated by the problem of existence of an optimal layout (Lurie and Cherkaev, 1981a; Armand, Lurie, and Cherkaev, 1984). Here we use the results of the review article (Lurie and Cherkaev, 1986a), where the properties of *G*-closures are systematically studied.

Definitions

Consider a family of materials with known properties D_i, where $i = 1, \ldots,$ N is a parameter of the family[2], and let us call this set $\mathcal{U} = \{D_i\}$.

Consider a composite assembled from these materials. Suppose that the materials are presented in the composite with volume fractions m_i. This

---

[2]The notation D for the materials' properties emphasizes that the linear material may correspond to an equilibrium different from conductivity. For example, elastic materials may be considered with proper exchange in the notation.

composite material is equivalent in the sense of G-convergence to a uniform medium with tensor of effective properties D_*. We recall that the tensor $\mathcal{D}_*$ is independent of external fields. It is determined only by properties of the mixed materials and by the geometrical structure of the composite.

G_m-Closure. We call the G_m-closure of the set $\mathcal{U}$ the set of all possible values of the effective tensors $\mathcal{D}_*$ that correspond to arbitrary microstructures with the fixed volume fractions of materials. We denote the G_m-closure of $\mathcal{U}$ by $G_m\mathcal{U}$. It depends only on the set $\mathcal{U}$ of the properties of those materials and on their volume fractions m_i in a composite:

$$G_m\mathcal{U} = G_m(D_i, m_i).$$

Any tensor $D_* \in G_m\mathcal{U}$ is characterized by angles of orientation of the coordinate system and by rotationally invariant parameters such as the eigenvalues. The G_m-closure set depends only on these invariants, and it is represented as a domain in a corresponding finite-dimensional space. Each microstructure corresponds to a point in this domain.

G-Closure. We define the G-closure of the set of properties of the materials, that is, the set of possible values of the tensor $\mathcal{D}_*$ corresponding to an arbitrary microstructure and arbitrary volume fractions of the materials. The G-closure depends only on the properties of the materials in the set $\mathcal{U}$:

$$G\mathcal{U} = \bigcup_{m_i \in \mathbf{m}} G_m\mathcal{U}, \quad G\mathcal{U} = G\mathcal{U}(D_i).$$

The G-closures are of special interest for the study of polycrystals, where they naturally represent a variety of all composites made from differently oriented fragments of an anisotropic material.

Where Is the Description of G_m-Closure Used?

The following problems are examples where G_m-closures are needed:

- G_m-closures provide a priori bounds for calculation of the effective properties of any prescribed structure. It is useful to know G_m-closures dealing with structures that are either unknown or random.

- It is necessary to know the G_m-closure if a structure of a composite is to be chosen to improve its properties.

- In structural optimization, G_m-closures describe the set of admissible controls, because it is not known a priori what composite is the most effective at a specific point of a construction.

Remark 3.2.1 *For some optimization problems it is enough to find only some components of the G_m-closures. For example, we could be interested in structures of composites that store the minimal energy in an arbitrary external field.*

We also notice that a description of the closures is often presented in an explicit form; they are described by rather simple inequalities that connect invariants of any possible effective tensor. On the other hand, the problem of calculating the effective properties of a given structure typically can only be solved numerically.

G-Closeness of Sets of Materials

Most applications deal with sets of available materials $\mathcal{U}$ that are not *G*-closed, i.e., they do not coincide with their *G*-closure. We cite a few examples:

1. Discrete set that consists of several materials (the composites have intermediate effective properties).

2. Arbitrary set of isotropic media (laminates of isotropic materials are generally not isotropic).[3]

3. Set of anisotropic crystals that differ only in the orientation of their principal axes (a polycrystal composite could be isotropic).

Finally, let us give an example of the *G*-closed set of conducting materials. It is the set of anisotropic materials $\boldsymbol{\sigma}$ with the eigenvalues λ_i, $i = 1, \ldots d$ that are restricted by two constants a and b:

$$0 < a \leq \lambda_i \leq b < \infty.$$

Proof of the *G*-closeness is left to the reader.

Notice that this example is not very natural. It is much easier to find not-*G*-closed sets of materials than to find a *G*-closed set.

Properties of G-Closures and G_m-Closures

Finiteness, Connectedness

Consider the *G*-closure of a set of conductivity tensors $\boldsymbol{\sigma}_*$. Each tensor is characterized by its eigenvalues λ_i, $i = 1, \ldots, d$, and by angles of orientation of the tensor in space. We are interested in a description of the set of eigenvalues only, because the orientation of an effective tensor can be arbitrarily chosen by an orientation of the periodic structure as a whole. It is easy to find that the G_m-closure is a closed, simply connected, and bounded set in the space of invariants of tensor properties.[4]

Indeed, it is bounded by the Wiener inequalities

$$\sigma_h I \leq \boldsymbol{\sigma}_* \leq \sigma_a I,$$

---

[3]An exclusive counterexample of isotropic *G*-closure is discussed in Chapter 15.

[4]The properties discussed are valid for the *G*-closures of the set of linear materials with arbitrary, not only conducting, properties. Additional consideration is needed to describe the proper invariants of the materials' characteristics.

which imply that every eigenvalue belongs to the interval

$$\lambda_i \in [\sigma_h, \sigma_a] \, .$$

Therefore the G-closure is bounded.

A G-closure is connected. Indeed, any two points σ_A, σ_B of a G-closure can be linked by a family of continuous curves that also belong to the G-closure. These curves correspond, for example, to the effective properties of laminates assembled from the materials σ_A and σ_B or to another family of microstructures with variable volume fractions. Obviously, the properties of a structure continuously depend on the volume fractions. Different curves correspond to different orientations of the normal to the layers in the laminates. Motion along the curve corresponds to varying the fractions of materials σ_A and σ_B in the composite, and the ends of the curve correspond to the vanishing of one of the materials in the composite.

A G_m-closure set is connected, too. If σ_A and σ_B represent composites with equal concentration of some initial materials, then a composite of σ_A and σ_B obviously has the same concentrations of these materials, which means that any such composite belongs to G_m-closure.

Both the G-closure and the G_m-closure contain a family of curves that link any two points in it and that correspond to different microstructures with different properties. Generally (but not always), the G-closures are sets with nonempty interior in the space of eigenvalues of σ_*.

Other Properties

We notice some properties of the G-closure of a set $\mathcal{U}$ that are similar to properties of convex envelopes:

1. The envelope rule: Each set $\mathcal{U}$ belongs to its G-closure $G\mathcal{U}$:

$$\mathcal{U} \in G\mathcal{U}.$$

2. The closure rule: The G-closure of a G-closed set coincides with the set:

$$G(G\mathcal{U}) = G\mathcal{U}.$$

3. The junction rule: The union of the G-closures of two sets is smaller than or equal to the G-closure of the union of these sets:

$$G(\mathcal{U}_1 \cup \mathcal{U}_2) \supset G(\mathcal{U}_1) \cup G(\mathcal{U}_2).$$

4. The swallow rule: If a set M belongs to the G-closure of the set $\mathcal{U}$ (but not necessarily to $\mathcal{U}$ itself), then the G-closure of the set $\mathcal{U} \cup M$ is equal to the G-closure of $\mathcal{U}$:

$$M \in G\mathcal{U} \quad \Rightarrow \quad G(\mathcal{U} \cup M) = G\mathcal{U}.$$

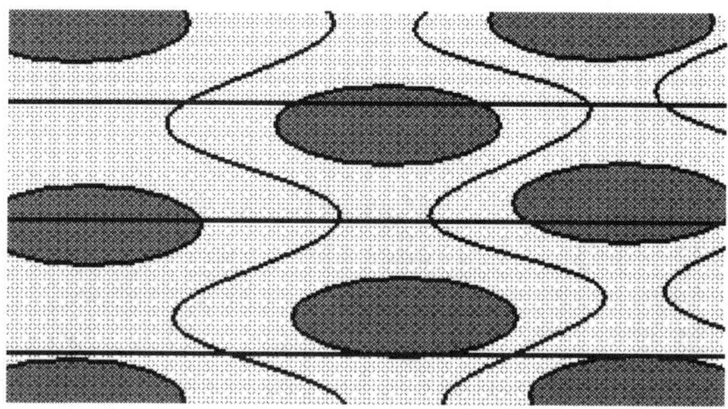

FIGURE 3.2. Illustration of the conservation property of *G*-closure. The phases have a common conductivity λ_0 in the horizontal direction. The applied homogeneous horizontal field causes a constant field everywhere. The applied homogeneous vertical field causes a variable field inside the structure.

These properties are physically obvious; for example, the last one means that if a material from the set M is in the *G*-closure of $\mathcal{U}$, then it could be replaced by a composite of materials from $\mathcal{U}$, and therefore adding this material to the set $\mathcal{U}$ does not change the *G*-closure. The formal proofs of these properties are left to the reader.

The Conservation Property of the G-Closure

Consider the case where the mixed anisotropic materials are represented by the tensors σ_i that all have a common eigenvalue and common eigenvector. Let us denote the common eigenvalue by λ_0 and the common eigenvector by $\mathbf{a}$. The conductivity tensors σ_i of mixing materials are of the form

$$\sigma^i = \lambda_0\,(\mathbf{a} \otimes \mathbf{a}) + \sum_{j=2}^{d} \lambda_j^i (\mathbf{a}_j^i \otimes \mathbf{a}_j^i), \qquad (3.2.3)$$

where j is the number of an eigenvalue, i is the number of a material, and $\otimes$ denotes the *dyadic product* as follows: $C = \{c_{ij} = \mathbf{a} \otimes \mathbf{b}$ if $c_{ij} = a_i b_j\}$.

Let us demonstrate that any matrix of material properties σ_* from the *G*-closure has the same eigenvalue and eigenvector:

$$\sigma_* = \lambda_0\,(\mathbf{a} \otimes \mathbf{a}) + \sum_{j=2}^{d} \lambda_{*j}(\mathbf{a}_{*j} \otimes \mathbf{a}_{*j}) \; \forall \sigma_* \in G\text{-closure.}$$

Indeed, consider a composite with arbitrary shapes of the fragments (see Figure 3.2) and calculate the fields in the composite in the response to the external field $\mathbf{e}_0 = \gamma \mathbf{a}$ applied in the direction $\mathbf{a}$. The pair of the uniform field

$$\mathbf{e}_0(\mathbf{x}) = \gamma \mathbf{a} = \text{constant}(\mathbf{x})$$

and the uniform current

$$\mathbf{j}(\mathbf{x}) = \lambda_0 \mathbf{e}_0(\mathbf{x}) = \lambda_0 \gamma \mathbf{a} = \text{constant}(\mathbf{x})$$

represents a solution to the problem: These constant fields trivially satisfy differential constraints, they satisfy the constitutive equations, and the boundary conditions do not imply any discontinuities in the fields because the only property λ_0 involved in the conductance has the same value in all fragments of the structure. Therefore, the continuity of the normal component of the current $[\mathbf{j}] = 0$ implies the continuity of this component of the field: $[\mathbf{j}] = \lambda_0[\mathbf{e}] = 0$. The current $\mathbf{j}$ that corresponds to the applied field $\mathbf{e}$ is constant everywhere and is aligned with $\mathbf{e}$. Informally speaking, the fragments of the microstructure become "clear" or "invisible" in that field. However, the microstructure manifests itself if any other field $\mathbf{e}_1(\infty)$ is applied. This time, the current $\mathbf{j}_1(\mathbf{x}) = \boldsymbol{\sigma}(\mathbf{x})\mathbf{e}_1(\mathbf{x})$ is inhomogeneous and so is the field $\mathbf{e}_1(\mathbf{x})$ (see Figure 3.2).

Remark 3.2.2 *The conservation property can also be established for elastic materials. We discuss an example in Chapter 15. Moreover, it is valid even for nonlinear composites if their property λ_0 in a direction depends on the field: $\lambda_0 = \lambda_0(\mathbf{e})$ but is constant in all fragments. The reason is the same: The applied constant field $\mathbf{e}$ corresponds to the aligned current $\mathbf{j}$ that is constant everywhere.*

The investigation of the conservation property of G-closures can be formulated as the search for properties of composites that are "stable under homogenization" (Grabovsky and Milton, 1998). Namely, one can ask what sets of material properties $\mathcal{U}$ lead to the set $G\mathcal{U}$ with empty interior. Such G-closures are characterized by equalities called *exact relations* rather than by inequalities. The G-closure of the materials with a common eigenvalue and eigenvector is an example of a set with empty interior, whereas one of the eigenvalues of the G-limit is fixed.

Several examples of G-closures with empty interior are discussed in Chapters 10 and 14. Generally, the conserved property may correspond not to a chosen direction of an external field but to a combination of the applied fields. For example, we demonstrate in Chapter 10 that the determinant of a two-dimensional polycrystal is constant independent of the structure. To obtain this conservation property we consider the mutual dependence of the currents corresponding to two applied fields.

We also refer to (Milgrom and Shtrikman, 1989; Bruno, 1991; Cherkaev and Gibiansky, 1992; Benveniste, 1994; Benveniste, 1995), where various exact relations on G-closures have been found. Recent papers (Grabovsky and Milton, 1998; Grabovsky and Sage, 1998; Grabovsky, 1998) treat this problem generally and suggest algebraic algorithms for a systematic search of exact relations.

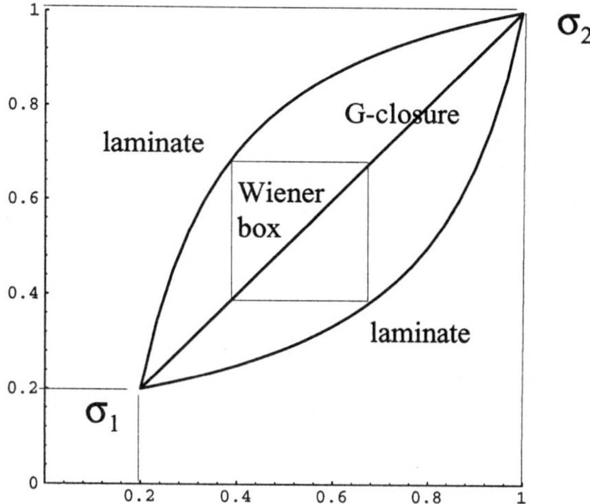

FIGURE 3.3. *G*-closure set of two isotropic conductors in two dimensions.

3.2.3 Example: The G-Closure of Isotropic Materials

We construct the *G*-closure of a set of isotropic conductors in two dimensions by using only the simplest properties of *G*-closures. This set was constructed in (Tartar, 1975; Raĭtum, 1978).

Consider a composite of two isotropic materials with conductivities σ_1 and σ_2 ($0 < \sigma_1 \leq \sigma_2 < \infty$) mixed in an arbitrary proportion. The conductivity of the composite is described by the effective properties tensor σ_*. The material properties of the effective conductivity are presented by the pair λ_1, λ_2 of its eigenvalues, $\lambda_1 \leq \lambda_2$. Let us describe the domain in the λ_1, λ_2–plane that corresponds to the *G*-closure.

The greater eigenvalue λ_2 of an effective tensor of a composite is less than the arithmetic mean of the materials' conductivities (see (3.1.8)), and the smaller eigenvalue of an effective tensor is greater than the harmonic mean of them:

$$\lambda_2 \leq m_1\sigma_1 + m_2\sigma_2, \quad \lambda_1 \geq \frac{1}{\frac{m_1}{\sigma_1} + \frac{m_2}{\sigma_2}}.$$

If we exclude the volume fractions $m_1 \geq 0$, $m_2 = 1 - m_1 \geq 0$ from the last two inequalities, we obtain the bound

$$\sigma_1 \leq \lambda_1 \leq \frac{\sigma_1\sigma_2}{\sigma_2 + \sigma_1 - \lambda_2} \leq \lambda_2 \leq \sigma_2. \tag{3.2.4}$$

The last inequalities provide a complete characterization of the *G*-closure (see Figure 3.3). Indeed, we can demonstrate the specific composite corresponding to each point of its boundary: It is a laminate with a properly chosen volume fraction. The set of laminates corresponds to the equality (see Figure 3.3)

$$\lambda_1 = \frac{\sigma_1\sigma_2}{\sigma_2 + \sigma_1 - \lambda_2}$$

because the eigenvalue corresponding to the normal component is averaged as a harmonic mean, and the eigenvalue corresponding to the tangent component as an arithmetic mean.

The geometric interpretation of this result is as follows. The G_m-closure set lies inside the Wiener box, which parametrically depends on m. Hence the G-closure lies inside the union of all rectangles corresponding to all volume fractions $m \in [0, 1]$ (see Figure 3.3). The boundary of this set is drawn by the motion of two symmetric nondiagonal vertices of those rectangles with coordinates σ_a, σ_h and σ_h, σ_a, respectively. Only the coordinates of these nondiagonal corners of G_m-closure are of importance. Fortunately, these points correspond to the effective properties of the known (laminate) structure. Therefore, laminates form at least a part of the boundary of the G-closure.

In the two-dimensional case, the G-closure is a domain in the plane of eigenvalues of $\boldsymbol{\sigma}_*$. The laminates describe the entire boundary of the G-closure because the set of their properties corresponds to a closed curve in that plane.

Simple-Connectedness

To conclude, we demonstrate that a G-closure is simply connected (it does not contain "holes" inside). The simplest way to demonstrate this is to build a class of microstructures that cover all points inside the domain (3.2.4). We use a two-step procedure to imitate a conductivity tensor with eigenvalues $\sigma', \sigma'' \in G$-closure. First, we build isotropic composites $\boldsymbol{\sigma}_{\text{is}}$ with all intermediate properties $\sigma' \in [\sigma_1, \sigma_2]$ (they correspond, for example, to a class of symmetric microstructures like checkerboards with the volume fraction of one of the materials varying from zero to one). Second, we build a laminate $\boldsymbol{\sigma}_{\text{lam}}$; we choose the volume fraction of materials in that laminate so that one of its eigenvalues becomes equal to σ' (the other eigenvalue σ_{l2} is equal to $\sigma_{l2} = \frac{\sigma_1 \sigma_2}{\sigma_1 + \sigma_2 - \sigma'}$). Now mix the materials $\boldsymbol{\sigma}_{\text{is}}$ and $\boldsymbol{\sigma}_{\text{lam}}$. Note that one of the eigenvalues of $\boldsymbol{\sigma}_{\text{lam}}$ is equal to the eigenvalues of $\boldsymbol{\sigma}_{is}$. By the conservation property, one eigenvalue of the composite is equal to σ' (see (3.2.3)), and the other varies in the interval $[\sigma', \sigma_{l2}]$ when the volume fraction of the isotropic phase changes from one to zero. Particularly, we can choose this fraction to make this eigenvalue equal to the given parameter σ''. Therefore, the set of composites of this kind imitates all points of the G-closure (see Figure 3.3).

Three-Dimensional Case

In the three-dimensional case, the laminates correspond to curves on the boundary surface of the G-closure. We leave the complete description of three-dimensional G-closure for Chapter 10, because it requires a special technique.

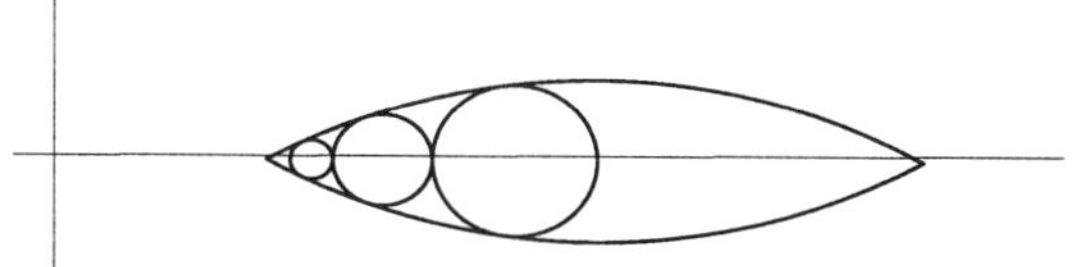

FIGURE 3.4. The domain of attainability of the current j_{x_1}, j_{x_2}. The circles correspond to the fixed volume fractions of materials, the lens corresponds to the *G*-closure.

3.2.4 Weak G-Closure (Range of Attainability)

Two-Dimensional Case

Another way to characterize the *G*-closure is to observe the range of the currents $\mathbf{j} = [j_{x_1},\ j_{x_2}]$ that corresponds to the unit field $\mathbf{e} = [1,\ 0]$ and all possible composites. We use the relation $\mathbf{j} = \boldsymbol{\sigma}_* \mathbf{e}$. The current is equal to $\mathbf{j} = [\boldsymbol{\sigma}_{11},\ \boldsymbol{\sigma}_{12}]$. We express the elements of the tensor through its eigenvalues λ_1 and λ_2, and the orientation of an eigenvector Φ:

$$j_{x_1} = \frac{\lambda_1 + \lambda_2}{2} + \frac{\lambda_1 - \lambda_2}{2}\cos 2\Phi, \quad j_{x_2} = \frac{\lambda_1 - \lambda_2}{2}\sin 2\Phi.$$

The set of all possible composites with the fixed volume fractions of materials corresponds to the vector $\mathbf{j}$ that belongs to the disk:

$$F(\lambda_1), \lambda_2, m) = \left(\mathbf{j}_1 - \frac{\lambda_1 + \lambda_2}{2}\right)^2 + \mathbf{j}_2^2 - \left(\frac{\lambda_1 - \lambda_2}{2}\right)^2 \leq 0;$$

where the eigenvalues λ_1 and λ_2 take the extreme values $\lambda_1 = (m\sigma_1^{-1} + (1 - m)\sigma_2^{-1})^{-1}$ and $\lambda_2 = m\sigma_1 + (1 - m)\sigma_2$ equal, respectively, to the arithmetic and harmonic means of the mixed materials. The differently oriented laminates correspond to the circumference, and the other structures correspond to the inner points of the disk (see Figure 3.4).

When the volume fraction m varies, the family of circles forms a domain of attainability. This domain is just the envelope of the family of circles. We find it by solving the equation $\frac{\partial}{\partial m} F(\lambda_1, \lambda_2, m) = 0$ and excluding $m \in [0, 1]$.

The equation of the domain of attainability is:

$$|j_{x_2}| \leq \sqrt{j_{x_1}(\sigma_1 + \sigma_2 - j_{x_1})} - \sqrt{\sigma_1 \sigma_2}.$$

This domain of attainability is shown in Figure 3.4. It is shaped like a lens; the vertices correspond to the pure materials σ_1 and σ_2.

The Three-Dimensional Case

The three-dimensional case is considered similarly. In dealing with the range of currents, we notice that the maximal range of $\mathbf{j}$ corresponds to the situation where the plane of the maximal and minimal eigenvalues

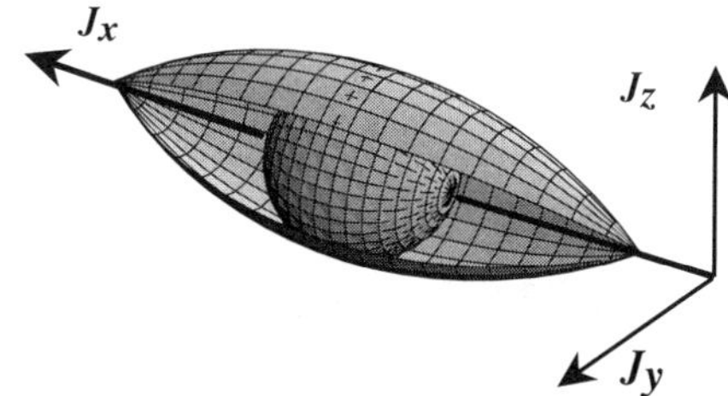

FIGURE 3.5. The domain of attainability of the current in the three-dimensional case. The inner sphere corresponds to the fixed volume fractions; the exterior surface corresponds to the weak G-closure.

of $\boldsymbol{\sigma}_*$ includes both vectors $\mathbf{e}$ and $\mathbf{j}$. In that plane the problem is two-dimensional. Therefore, all the previous results are valid. The value of the intermediate eigenvalue is irrelevant. The domain of attainability in the three-dimensional case is a surface of revolution (Figure 3.5)

$$\sqrt{j_{x_2}^2 + j_{x_3}^2} = \sqrt{j_{x_1}(\sigma_1 + \sigma_2 - j_{x_1})} - \sqrt{\sigma_1 \sigma_2},$$

where x_1 is the direction of the given field.

Notice that the boundary points of the domain correspond to laminate structures. Hence, the class of laminates is sufficient for the solution to a class of structural optimization problems that correspond to optimization of the behavior of electrical fields. A similar concept was used in (Raĭtum, 1989) to prove the existence of the optimal solution in the class of controls that consists of initial materials and their laminates. The set of layouts corresponding to attainability of currents is called the *weak G-closure*.

The advanced generalization of the concept of the weak G-closure to the nonlinear materials and additional references can be found in (Milton and Serkov, 1999).

3.3   Conclusion and Problems

We established simple bounds on the effective properties tensor and introduced the G-closures: sets of all possible effective tensors that correspond to arbitrary microstructures of a composite assembled of material with fixed properties. Their topological properties were studied, and an example was presented.

Now we are prepared to discuss structural optimization problems for conducting media.

Problems

1. How many external fields are needed to compute all coefficients of two- and three-dimensional conductivity tensors by calculating the energy? Suggest an algebraic procedure to calculate the eigenvalues and eigenvectors of an effective tensor.

2. Show that the G-closure is bounded if the mixed materials have finite conductivities.

3. Prove the topological properties of G-closures.

4. Describe the G-closure for the set of two anisotropic materials with conductivities

$$\sigma_1 = \begin{pmatrix} \lambda & 0 \\ 0 & \lambda_1 \end{pmatrix} \quad \text{and} \quad \sigma_2 = \begin{pmatrix} \lambda & 0 \\ 0 & \lambda_2 \end{pmatrix}.$$

Part II

Optimization of Conducting Composites

4
Domains of Extremal Conductivity

In this chapter we consider the simplest optimal design problem for the best structure of a two-component conducting body of minimal or maximal total conductivity. This problem was used as a testing ground for various methods of structural optimization. To solve it, we introduce several different approaches, which are driven by different arguments but lead to similar results. Each approach has an analogue for the one-dimensional variational problem discussed in Chapter 1, and each approach will be developed for more general multidimensional problems.

The relaxation of an optimal design by means of composites was suggested in (Lurie and Cherkaev, 1978), where the Weierstrass conditions were used for the relaxation and the numerical results were obtained in (Lavrov, Lurie, and Cherkaev, 1980). The problem of an elastic bar of the extremal torsion stiffness was considered. This problem is formally equivalent to the problem of the domain of extremal conductivity.

The relaxation (G-closure) approach was applied to specific design problems in (Lurie, Fedorov, and Cherkaev, 1980b; Lurie, Fedorov, and Cherkaev, 1980a) and (Murat and Tartar, 1985a; Lurie and Cherkaev, 1986a) following earlier research (see, for example, (Lurie, 1975; Murat, 1977)); the convexification of the corresponding nonconvex functional was suggested in (Goodman, Kohn, and Reyna, 1986; Kohn and Strang, 1986a); numerical schemes were developed in (Lavrov et al., 1980; Gibiansky and Cherkaev, 1984; Goodman et al., 1986; Kohn and Strang, 1986a; Dvořák, 1994; Haslinger and Dvořák, 1995; Burns and Cherkaev, 1997) and other papers. The multicomponent optimal design was considered in (Burns and Cherkaev, 1997).

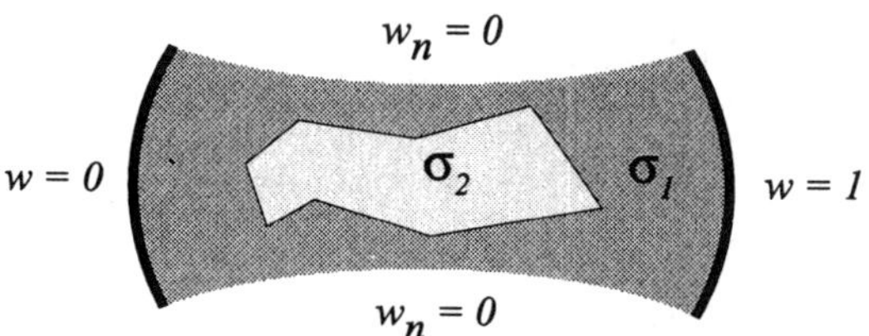

FIGURE 4.1. The problem of optimal conductivity of a domain.

4.1   Statement of the Problem

The Total Conductivity of a Domain

Suppose that two materials with different conductivities σ_1 and σ_2, $\sigma_1 < \sigma_2$, are available. We want to displace these materials in a body that occupies a given domain Ω. The conductivity $\sigma(\mathbf{x})$ at a point $\mathbf{x}$ is equal to

$$\sigma(\mathbf{x}) = \chi(\mathbf{x})\sigma_1 + (1 - \chi(\mathbf{x}))\sigma_2,$$

where $\chi(\mathbf{x})$ is the characteristic function of the domain occupied by the material σ_1.

The boundary conditions are fixed. Suppose for definiteness that the boundary $\partial\Omega$ consists of three components $\partial\Omega_i$ (see Figure 4.1) and that the boundary conditions are

$$w \in \mathcal{W}, \quad \mathcal{W} = \{w : \ w|_{\partial\Omega_1} = 0, \ w|_{\partial\Omega_2} = 1\}; \quad \mathbf{j} \cdot \mathbf{n}|_{\partial\Omega_3} = 0. \qquad (4.1.1)$$

Two components $\partial\Omega_1$ and $\partial\Omega_2$ of the boundary are kept at potential values $w = 0$ and $w = 1$, respectively, and $\partial\Omega_3$ is insulated.

Recall that the energy stored in Ω is equal to the work of the exterior forces (potentials) applied on the boundary of the body. In particular, if the difference of potentials between two parts of the boundary is given by (4.1.1), then the stored energy is equal to the integral of the normal component of the current $\mathbf{j} \cdot \mathbf{n}$ that passes through the boundary of the domain:

$$I_\chi(\chi) = \int_{\partial\Omega_1} (\mathbf{j} \cdot \mathbf{n}) = \min_{w \in \mathcal{W}} \frac{1}{2} \int_\Omega \sigma(\chi)(\nabla w)^2. \qquad (4.1.2)$$

We call I_χ the *total conductivity* of the domain Ω. It is naturally defined as the ratio between the total current and the difference in potentials on the boundary components $\partial\Omega_1$ and $\partial\Omega_2$. Functional I_χ depends on the layout of the materials in Ω described by the characteristic function χ.

We will keep the definition (4.1.2) of the total conductivity in general setting of the Dirichlet boundary conditions on the components $\partial\Omega_1$ and $\partial\Omega_2$ of the boundary.

Optimal Design Problem

Consider the following optimal design problem: Find a layout $\chi(\mathbf{x})$ that minimizes the total conductance I_χ of the domain Ω:

$$\inf_\chi I_\chi(\chi), \qquad (4.1.3)$$

where the cost I is the minimal conductivity of Ω. In the absence of additional constraints, the solution to this problem is trivial; the material with minimal conductivity σ_1 is placed everywhere.

To make the problem nontrivial, we assume that the total mass M_0 of the first material is fixed:

$$\int_\Omega \chi = M_0. \qquad (4.1.4)$$

Constraint (4.1.4) is considered in the standard way by adding (4.1.4) with the Lagrange multiplier γ to the doubled energy (4.1.3) (we double the energy to avoid repeatedly writing the factor $\frac{1}{2}$ in front of the quadratic form (4.1.2)). The problem becomes

$$\min_\chi \max_\gamma \left[\int_\Omega \left(\sigma(\chi)(\nabla w)^2 \right) + \gamma\chi \right) - \gamma M_0 \right]. \qquad (4.1.5)$$

Next, we fix the value of the constant γ and solve the problem for the augmented functional J_a that differs from (4.1.5) by a constant term $-\gamma M_0$:

$$I = \min_\chi \min_{w \in \mathcal{W}} J_a, \qquad J_a = \int_\Omega [\sigma(\chi)(\nabla w)^2 + \gamma\chi]. \qquad (4.1.6)$$

The augmented Lagrangian J_a depends on the Lagrange multiplier γ as on a parameter. Different values of γ correspond to different fixed amounts $M_0 \in [0, \, |\Omega|]$ of the first material. After the solution $w = w(\gamma)$, $\chi = \chi(\gamma)$ of (4.1.6) is obtained, we use the constraint (4.1.4) to determine γ.

Multiplier γ can be interpreted as the difference between the costs of the two materials. Problem (4.1.6) asks for the minimization of the sum of the total conductivity on the domain and its cost. Notice that $\gamma > 0$ or the solution is trivial: $\sigma_{\text{opt}} = \sigma_1$ everywhere. In other words, we assume that the more expensive material is also less conducting.

4.2   Relaxation Based on the G-Closure

4.2.1   Relaxation

First we describe the relaxation technique based on the completeness of the set of controls.

In solving the optimization problem (4.1.6) one should take into account possible fine-scale oscillations of the control σ. These oscillations physically mean that the optimally designed body may tend to become a *composite*. Let us admit that the optimal layout of materials may be characterized by composite zones. The composites enlarge the set of admissible controls because they represent limits of rapidly oscillating sequences of the original controls (layouts).

We use the homogenization approach to effectively describe fine-scale oscillations of layouts. In other words, we describe a composite by its effective tensor. This approach replaces the set $U = \{\sigma_1, \sigma_2\}$ of admissible materials with the G_m-closure of this set. This way, we take into account all possible fine-scale oscillations of $\chi(\mathbf{x})$.

In dealing with composites we must determine the best microstructures. The best structure of a composite is obtained from the solution to a so-called *local problem*. This is a variational problem of structural optimization in an infinitesimally small neighborhood of a point of the designed body.

The energy of a highly inhomogeneous medium σ_ε in a small regular domain ω is replaced by the equal energy of an equivalent homogenized medium as follows:

$$\langle \sigma_\varepsilon (\nabla w_\varepsilon)^2 \rangle = \mathbf{e} \cdot \boldsymbol{\sigma}_* \mathbf{e} + o(\|\omega\|),$$

where $\mathbf{e} = \langle \nabla w_\varepsilon \rangle$ and $\langle \; \rangle$ is the averaging operator (2.2.2).

The total amount of the first material is constrained by (4.1.4). The constraint (4.1.4) can be replaced by an equivalent integral constraint on the volume fraction m of the first material in the composite, $m = m(\mathbf{x}) \in \mathcal{M}$, where

$$\mathcal{M} = \left\{ m(\mathbf{x}) \in [0, 1], \quad \int_\Omega m(\mathbf{x}) = M \right\}. \tag{4.2.1}$$

The averaged functional in (4.1.6) becomes

$$J_a = \int_\Omega \left(\langle \sigma(\chi)(\nabla w)^2 \rangle + \gamma \langle \chi \rangle \right) = \int_\Omega (\mathbf{e} \cdot \boldsymbol{\sigma}_*(m)\mathbf{e} + \gamma m) + o(\|\omega\|).$$

The effective tensor $\boldsymbol{\sigma}_*(m)$ of an optimal composite may vary from point to point together with the field $\mathbf{e}$, but its value belongs to the G_m-closure: $\boldsymbol{\sigma}_*(m) \in G_m U$.

The optimization problem (4.1.6), rounded to $\|\omega\|$, becomes:

$$I = \min_{m \in \mathcal{M}} \; \min_{\mathbf{e} \in \mathcal{E}} \; \min_{\boldsymbol{\sigma}_* \in G_m U} \int_\Omega (\mathbf{e} \cdot \boldsymbol{\sigma}_* \mathbf{e} + \gamma m) \tag{4.2.2}$$

where

$$\mathcal{E} = \{\mathbf{e} : \; \mathbf{e} = \nabla w, \quad w \in \mathcal{W}\}.$$

It is called the *relaxed problem*. Note that the relaxed problem does not have rapidly oscillating minimizing sequences of layouts because the G-limits of these sequences are already included in the set of admissible controls.

The inner operation $\min_{\boldsymbol{\sigma}_* \in G_m U}$ asks for the best structure of a composite with fixed fraction m submerged into a fixed field $\mathbf{e}$. The next operation $\min_{\mathbf{e}_0 \in E}$ defines the field in the domain Ω if the structure is chosen optimally but the layout of the volume fraction $m(\mathbf{x})$ is somehow assigned. The last operation $\min_{m \in \mathcal{M}}$ determines the layout $m(\mathbf{x})$ subject to the integral constraint (4.2.1). The order of the minimal operations can be chosen arbitrarily.

4.2.2 Sufficient Conditions

The Local Problem: Lower Bound. We start with the inner minimization problem in the infinitesimal neighborhood $\omega(\mathbf{x})$ of a point $\mathbf{x}$:

$$\min_{\boldsymbol{\sigma}_* \in G_m U} \mathbf{e} \cdot \boldsymbol{\sigma}_* \mathbf{e}. \tag{4.2.3}$$

We have not described the G_m-closure set for the conductivity problem.[1] Fortunately, problem (4.2.3) can be solved without the complete description of that set (Lurie and Cherkaev, 1978; Kohn and Strang, 1986a).

First, notice that the orientation of the effective tensor $\boldsymbol{\sigma}_*$ is arbitrary because an optimal structure can be arbitrarily rotated. The optimal orientation of $\boldsymbol{\sigma}_*$ is realized when the eigenvector that corresponds minimal eigenvalue $\lambda_{\min}$ is codirected with $\mathbf{e}$. The quadratic form $\mathbf{e} \cdot \boldsymbol{\sigma}_* \mathbf{e}$ becomes

$$\min_{\text{orientation}} \mathbf{e} \cdot \sigma_* \mathbf{e} = \lambda_{\min} \mathbf{e}^2.$$

Next, the optimal structure must possess the minimal value of $\lambda_{\min}$ among all microstructures. Recall (see (2.2.13)) that all eigenvalues of $\boldsymbol{\sigma}_*$ vary in the interval $[\sigma_h, \sigma_a]$. Particularly, the minimal eigenvalue does not exceed the harmonic mean σ_h of mixed conductivities:

$$\lambda_{\min} \geq \sigma_h, \quad \sigma_h = \frac{\sigma_1 \sigma_2}{m \sigma_2 + (1 - m) \sigma_1}.$$

Therefore, the minimum in (4.2.3) in any infinitely small region ω is bounded from below:

$$\min_{\boldsymbol{\sigma}_* \in G_m U} \mathbf{e} \cdot \boldsymbol{\sigma}_* \mathbf{e} \geq \sigma_h \mathbf{e}^2. \tag{4.2.4}$$

The last inequality demonstrates the *sufficient optimality conditions* for the stored energy.

Attainability of the Bound. The bound (4.2.4) is attainable: It corresponds to a laminate structure where laminates are oriented along the field. Indeed, the harmonic mean of the conductivities is exactly the effective conductivity of laminates in that direction. This says that optimal structures can be imitated by properly oriented laminates.

---

[1]Actually, this set is described in Chapter 11.

The Relaxed Problem in Large. We obtain the formulation of the relaxed problem by substitution of the relaxed Lagrangian (4.2.4) into the minimization problem (4.2.2):

$$I = \min_{m \in \mathcal{M}} \min_{e = \nabla w} \int_{\Omega} \left(\sigma_h(m) e^2 + \gamma m \right), \quad \sigma_h(m) = \frac{\sigma_1 \sigma_2}{m \sigma_2 + (1 - m)\sigma_1}.$$
(4.2.5)

The isotropy of the relaxed problem is expected because the optimal structure is chosen among all structures of arbitrary orientation, and the bound (4.2.4) therefore is independent of direction.

Remark 4.2.1 *The minimal energy stored in anisotropic laminate structures is equal to the energy of an isotropic material with conductivity σ_h. This equivalence was used in a numerical scheme (Lavrov et al., 1980) to simplify the calculations. Namely, we replaced the optimally oriented anisotropic composite by the isotropic material with conductivity $\sigma_h(m)$ and numerically found the best layout of m. After the numerical solution was found, we easily determined the laminate composite with the same energy and used the same amount of materials as the isotropic medium σ_h.*

The relaxation is successful due to

1. the available geometrically independent bound (the harmonic mean bound), and

2. the known optimal structure (laminates) that realizes the bound.

Solution to the Relaxed Problem

Lagrangian. The relaxed problem can be solved by a standard technique of the calculus of variations. First, we establish necessary conditions of optimality. We change the sequence of minimal operations and minimize the integrand $\Phi(e)$ of (4.2.5) over m with the "frozen" field e:

$$\Phi(e) = \min_{m \in [0,1]} \left(\frac{\sigma_1 \sigma_2}{m \sigma_2 + (1 - m)\sigma_1} e^2 + \gamma m \right).$$

The optimal value m_0 of m is expressed through the field e and is equal to

$$m_0 = \begin{cases} 0 & \text{if } |e| \leq \frac{C}{\sigma_2}, \\ -\frac{\sigma_1}{\sigma_2 - \sigma_1} + \frac{C}{\gamma}|e| & \text{if } |e| \in \left[\frac{C}{\sigma_2}, \frac{C}{\sigma_1}\right], \\ 1 & \text{if } |e| \geq \frac{C}{\sigma_1}, \end{cases} \tag{4.2.6}$$

where

$$C = \sqrt{\frac{\gamma \sigma_1 \sigma_2}{(\sigma_2 - \sigma_1)}}.$$

This condition says that the volume fraction of σ_1 decreases when the density of the field increases until it reaches the boundaries of its range. The Lagrangian $\Phi(\mathbf{e})$ is

$$\Phi(\mathbf{e}) = \begin{cases} \sigma_1 \mathbf{e}^2 + \gamma & \text{if} \quad |\mathbf{e}| \leq \frac{C}{\sigma_2}, \\ -\gamma \frac{\sigma_1}{\sigma_2 - \sigma_1} + 2C|\mathbf{e}| & \text{if} \quad |\mathbf{e}| \in \left[\frac{C}{\sigma_2}, \frac{C}{\sigma_1}\right], \\ \sigma_2 \mathbf{e}^2 & \text{if} \quad |\mathbf{e}| \geq \frac{C}{\sigma_1}. \end{cases} \qquad (4.2.7)$$

Remark 4.2.2 *The energy $\Phi(\mathbf{e})$ in the composite zone is an affine function of $\mathbf{e}$. This property deserves a physical explanation, because the energy of a linear composite is a quadratic function of $\mathbf{e}$. To explain the linearity of the energy of an optimal composite we observe that the increase of the magnitude of the field $\mathbf{e}$ leads to a change in the structure of the optimal composite (here, to a decrease in the volume fraction m). The variation of the conductivity of the optimal composite partly compensates the increase of the energy with the magnitude of the field.*

We also find from (4.2.7) that the magnitude $|\mathbf{j}|$ of the current

$$\mathbf{j} = \frac{\partial \Phi}{\partial \mathbf{e}} = \sigma_h \mathbf{e}$$

is constant in the composite zone: $|\mathbf{j}| = 2C$. This condition expresses a qualitative property of an optimal design: it evenly distributes acting fields throughout the domain.

Optimal Solution. The optimal solution w ($\nabla w = \mathbf{e}$) is the solution of the Euler equation to the variational problem

$$I = \min_w \int_\Omega \Phi(\nabla w),$$

where w also satisfies the boundary conditions. If m reaches its bounds ($m = 0$ or $m = 1$), then the composite becomes pure materials and w satisfies the Laplace equation

$$\Delta w = 0 \begin{cases} \text{if} \quad |\nabla w| \leq \frac{C}{\sigma_2} \gamma, \\ \text{if} \quad |\nabla w| \geq \frac{C}{\sigma_1} \gamma. \end{cases}$$

In the composite zone, $m \in (0, 1)$, the Lagrangian $\Phi(\mathbf{e})$ is an affine function of $|\mathbf{e}|$, (4.2.7). The second-order Euler–Lagrange equation degenerates into the system of two nonlinear first-order equations.

1. The current has the constant magnitude and is divergencefree

$$|\mathbf{j}| = 2C, \quad \nabla \cdot \mathbf{j} = 0.$$

The first equation implies that $\mathbf{j}$ can be represented through a scalar function ϕ as

$$\mathbf{j}(\phi) = 2C(\cos \phi, \ \sin \phi)$$

and the second equation states that ϕ satisfies the first-order partial differential equation

$$-\sin\phi\frac{\partial\phi}{\partial x_1} + \cos\phi\frac{\partial\phi}{\partial x_2} = 0,$$

which follows from $\nabla \cdot \mathbf{j} = 0$.

2. The constitutive relation $\mathbf{e} = \frac{1}{\sigma_h}\mathbf{j}$ expresses the curlfree field $\mathbf{e} = \nabla w$ ($\nabla \times \mathbf{e} = 0$) through the control σ_h.

$$\nabla \times \left(\frac{\mathbf{j}(\phi)}{\sigma_h}\right) = 0.$$

This relation serves to find the control.

Observe that the second-order elliptical equation of conductivity splits into two nonlinear first-order equations for ϕ and m. This says that the equation for the optimal conductor reaches the boundary of the ellipticity.

In solving these equations we obtain a solution $w(\gamma), \sigma_h(\gamma)$ that depends only on γ. Finally, we choose γ to satisfy the integral constraint (4.1.4) on the available amounts of materials.

G_m-Closures and the Optimal Composites. The optimal composite has been determined without direct reference to the complete description of the G_m-closure. Instead, structures have been found that minimize the stored energy in a given mean field. In other words, a special part of the boundary of the G_m-closure set has been found, which corresponds to the minimization of the form

$$\min_{\boldsymbol{\sigma}_* \in G_m U} \{\mathbf{e} \cdot \boldsymbol{\sigma}_* \mathbf{e} \quad \forall \mathbf{e} : \|\mathbf{e}\| = 1\}.$$

This problem requires the minimization of only one component of the tensor $\boldsymbol{\sigma}_*$, namely, the minimal eigenvalue of the conductivity tensor. The values of the other components of $\boldsymbol{\sigma}_*$ are irrelevant to the problem. This makes the problem of structural optimization simpler than the problem of the G_m-closure.

Remark 4.2.3 *We could apply the technique to more complicated cases, for example, to the elasticity problem. The Wiener bounds are valid for elasticity as well, so we can estimate from below the energy of an elementary cell by the energy of an effective medium that has the effective stiffness equal to the harmonic mean of the stiffness of components. By doing this, we could obtain a lower bound of energy stored in a structure. Also, one can compute the elastic energy stored in optimally oriented laminates. But this time the harmonic mean bound does not coincide with the energy of any laminate. One can expect that either the bound is not exact or the laminates are not a suitable class of microstructures. Actually, both reasons are correct. In the next chapters we will improve the bounds, and we investigate more complicated structures.*

4.2.3 A Dual Problem

The problem of a domain with maximal total conductivity is solved in the same way (Lurie and Cherkaev, 1978; Lavrov et al., 1980). Suppose that the total current through the domain is given and we want to minimize the difference in potentials on two components of its boundary. The dual (Thomson) variational principle is used, and the problem becomes

$$\min_{\chi} \ \min_{\mathbf{j} \in \mathcal{J}} \left(\int_{\Omega} (\mathbf{j} \cdot \sigma^{-1}(\chi)\mathbf{j} + \gamma\chi) \right),$$

where γ is the Lagrange multiplier corresponding to constraint (4.1.4) and

$$\mathcal{J} = \left\{ \mathbf{j} : \ \nabla \cdot \mathbf{j} = 0, \ \int_{\partial\Omega_1} j_n = J_0, \ \mathbf{j} \cdot \mathbf{n}|_{\partial\Omega_3} = 0 \right\}.$$

The problem asks for the most conducting domain: The fixed current corresponds to minimum in the difference of potentials.

Relaxation. The relaxation of this problem is performed the same way as in the previous case. Assuming the possibility of formation of composite zones in optimal design, we replace the problem with its relaxed form:

$$I = \min_{m \in \mathcal{M}} \min_{\mathbf{j} \in \mathcal{J}} \left(\min_{\sigma_* \in G_m U} \int_{\Omega} (\mathbf{j} \cdot \sigma_*^{-1}\mathbf{j} + \gamma m) \right).$$

To obtain the relaxed solution, we solve the interior minimum problem of optimal structure,

$$\min_{\sigma_* \in G_m U} \mathbf{j} \cdot \sigma_*^{-1}\mathbf{j}, \tag{4.2.8}$$

and we express the result as a function $F_*(m, \mathbf{j})$ of the volume fraction m and the mean field $\mathbf{j}$.

Lower Bound. To obtain the lower bound of (4.2.8) we first optimize the orientation of σ_*. This time, we direct the eigenvector corresponding to the maximal eigenvalue along $\mathbf{j}$:

$$\mathbf{j} \cdot \sigma_*^{-1}\mathbf{j} \geq \frac{\mathbf{j}^2}{\lambda_{\max}}.$$

Again, we do not need to know the exact bounds on the G_m-closure, only the exact upper bound of the maximal eigenvalue $\lambda_{\max}$ of σ_*: $\lambda_{\max} \leq \sigma_a = m\sigma_1 + (1-m)\sigma_2$. The cost of problem (4.2.8) is bounded by the inequality

$$\min_{\sigma_* \in G_m U} \mathbf{j} \cdot \sigma_*^{-1}\mathbf{j} \geq \sigma_a^{-1}\mathbf{j}^2. \tag{4.2.9}$$

This bound is attainable, too: It corresponds to the laminate structure oriented across the current. The arithmetic mean of the conductivities is the effective conductivity of laminates in that direction. We see again that the relaxation requires only laminates.

Remark 4.2.4 *An anisotropic laminate structure is simultaneously the structure of the greatest and the least conductivity, depending on its orientation.*

The relaxed problem takes the form

$$I = \min_{m \in \mathcal{M}} \min_{\mathbf{j}_0} \int_{\Omega} G(m, \mathbf{j}), \quad G = \frac{1}{m\sigma_1 + (1-m)\sigma_2}\mathbf{j}^2 + \gamma m.$$

Again we interchange the sequence of minimal operations and minimize over m the value of the integrand with the "frozen" current $\mathbf{j}$. We compute $\Phi(\mathbf{j}) = \min_m G(m, \mathbf{j})$ and observe that $\Phi(\mathbf{j})$ linearly depends on $|\mathbf{j}|$ in the composite zone and that the absolute value of the vector $\mathbf{e} = \sigma_a^{-1}\mathbf{j}$ is constant in this zone. The Euler–Lagrange equations are similar to the one in the previous problem; we leave them to the reader.

4.2.4 Convex Envelope and Compatibility Conditions

Reduction to a Nonconvex Problem

Here we analyze the results of relaxation from a different perspective, following the approach of (Goodman et al., 1986; Kohn and Strang, 1986a). Instead of referring to composites, we treat the optimal design problem as a variational problem, and we concentrate our attention on the properties of the minimizers.

The minimization problem (4.1.6) requires the minimization over the field w and the control χ. Recall that the optimal design problem (4.1.6) has the form

$$J = \inf_{\chi} \min_{w \in \mathcal{W}} \left\{ \int_{\Omega} \left(\chi(\sigma_1(\nabla w)^2 + \gamma_1) + (1-\chi)(\sigma_2(\nabla w)^2 + \gamma_2) \right) \right\},$$

where γ_1 and γ_2 are the costs of the materials used.[2]

Let us exclude the control χ so that the problem is transformed into a variational problem for the potential w. We interchange the sequence of minimal operations $\inf_\chi$ and $\min_w$. The inner minimization problem can be easily solved. Indeed, $\chi(\mathbf{x})$ takes only values: zero or one. One must simply choose the value of χ which provides a smaller value of the integrand. The problem becomes:

$$J = \inf_{w \in \mathcal{W}} \int_{\Omega} F(\nabla w), \tag{4.2.10}$$

---

[2]Note that the previously introduced Lagrange multiplier γ corresponds to the following assignment of the cost of the materials: $\gamma_1 = \gamma$, $\gamma_2 = 0$. Here we use two Lagrange multipliers to make the expressions symmetric. Of course, the results depend only on the difference $\gamma_1 - \gamma_2$.

where

$$F(\nabla w) = \min\left\{\sigma_1(\nabla w)^2 + \gamma_1, \sigma_2(\nabla w)^2 + \gamma_2\right\}. \qquad (4.2.11)$$

Here F depends on w and the parameters γ_1 and γ_2. The optimal design problem becomes a variational problem for the new "energy" $F(\nabla w)$.

Note that F is a nonconvex function of ∇w. The lack of convexity may cause fine-scale oscillations in the solution to (4.2.10) (see Chapter 1); they physically correspond to the structure from infinitely fast interchanging materials in an optimal project.

We will develop a systematic procedure that automatically selects the right microstructures and homogenizes them.

Convex Envelope

Let us examine the expression for the extended (averaged) Lagrangian given by (4.2.7). One can see that the extended Lagrangian coincides with the convex envelope of the original nonconvex Lagrangian (4.2.11). To demonstrate this, we use the technique of Example 1.3.6; one can show, that the convex envelope of $F(\nabla w)$ has the form (4.2.7).

Because $CF \leq F$, the new problem has a cost that is less than or equal to the cost of the original problem. Similarly to one-dimensional problems (Chapter 1), the solution of a variational problem with a convex Lagrangian is stable against fine-scale perturbations.

The Compatibility Conditions. Relaxation is analogous to the relaxation of a one-dimensional variational problem: we simply convexify the Lagrangian. However, there is a difference: In the one-dimensional problem, the argument of the convexification (the derivative u') is free from any differential constraints. On the other hand, the rapidly varying variable $\mathbf{e} = \nabla w$ (an analogue of u') is subject to differential constraints: It is curlfree; $\nabla \times \mathbf{e} = 0$. This constraint is not taken into account in the relaxation by the convex envelope; the argument $\mathbf{e}$ is treated as a free vector.

The convex envelope of the two-well Lagrangian (4.2.11) corresponds to a minimizing sequence of the rapidly oscillating vector $\mathbf{e}(\mathbf{x})$. This vector takes two values $\mathbf{e}_1$ and $\mathbf{e}_2$ that belong to different wells of Lagrangian. The values $\mathbf{e}_1$ and $\mathbf{e}_2$ are the supporting points of the convex envelope, i.e., the points of the common tangent to the wells in the Lagrangian. Physically, $\mathbf{e}_1$ and $\mathbf{e}_2$ represent fields in the first and second materials, respectively.

The supporting vectors $\mathbf{e}_1$ and $\mathbf{e}_2$ coexist within an optimal microstructure, and hence they must share a common boundary. On the other hand, the constraint $\nabla \times \mathbf{e} = 0$ requires the continuity of the tangent component(s) $\mathbf{t}$ of $\mathbf{e}$ on the dividing line between the zones with the materials:

$$(\mathbf{e}_1 - \mathbf{e}_2) \times \mathbf{t} = 0. \qquad (4.2.12)$$

This requirement poses an additional constraint that might restrict the choice of e_1 and e_2, which are determined independently of the continuity consideration.

However, equation (4.2.12) can be satisfied for arbitrary e_1 and e_2 by the proper choice of the tangent t. Indeed, $d-1$ linearly independent equations (4.2.12) can always be solved for the d-dimensional vector t, which plays the role of additional control. The differential constraint $\nabla \times e = 0$ restricts the geometry of the optimal structure (the tangent t), but it is unable to change the cost of the problem.

Remark 4.2.5 *The more general optimization problems discussed in the next chapters cannot be related by passing to the convex envelope of the Lagrangian. For them, the convex envelope does not correspond to any minimizing sequence because the requirements of the type of (4.2.12) become contradictory. For these problems a more delicate technique of relaxation is developed; see Chapters 6–9.*

Remark 4.2.6 *These results can be generalized to nonquadratic energies, i.e., to the nonlinear materials. Examples of optimization of composites from nonlinear materials can be found in (Cherkaev and Gibiansky, 1988).*

4.3   Weierstrass Test

4.3.1   Variation in a Strip

Let us introduce the technique of necessary conditions of optimality and minimal extensions, and apply it to the discussing problem of conducting bodies with extremal total conductivity. In doing this analysis we obtain more detailed information on the structure of optimal fields in the solution. We also start to develop a general technique of necessary conditions that is applicable to a broad class of optimization problems. We follow here the approach suggested in (Lurie, 1963; Lurie, 1967) and summarized in (Lurie, 1975). The approach is the multivariable analogue of Pontryagin's maximum principle (Pontryagin, Boltyanskii, Gamkrelidze, and Mishchenko, 1964; Rozonoèr, 1959) of the theory of optimal control. The considered problem was analyzed using this technique in (Lavrov et al., 1980; Lurie, Cherkaev, and Fedorov, 1982).

The Scheme. The variational problems of structural optimization have a characteristic function χ among the minimizers. This function takes only values zero and one; therefore any variation of χ has unit magnitude. However, the size of the domain of variation can be made arbitrary small. These variations with a finite magnitude and an infinitesimal support are called the *Weierstrass variations.*

The idea of a Weierstrass-type test in optimal design problems is quite simple: If a layout χ is optimal, then it must be stable against any interchanging of materials. The Weierstrass-type condition requires the nonnegativity of the increment of the functional due to inserting of an infinitesimal inclusion of a material into a domain occupied by another material. One should choose the shape of inclusions to test the optimality, because the result of variation depends on it. We are interested in choosing a domain that provides the strongest variations.

Remark 4.3.1 *The dependence of the increment on the shape of the domain of variation reflects the dependence of the effective tensor of a composite on its microstructure. Indeed, the variation by interchange of materials can start formation of "seeds of composite" with infinitely small volume fractions of inclusion material.*

Denote the cost of the optimization problem as I:

$$I(\sigma_0, \mathbf{e}) = \min_{\sigma} I(\sigma, \mathbf{e}).$$

We suppose that the material with conductivity σ_0 is optimal at a point $\mathbf{x}_0 \in \Omega$ where the field is equal to $\mathbf{e}$.

Let us introduce the variation. We replace the material σ in a neighborhood ω of $\mathbf{x}_0$ by another material with conductivity σ'. The new layout of materials is called $\sigma_0 + \delta\sigma$:

$$\sigma_0 + \delta\sigma = \begin{cases} \sigma' & \text{if } \mathbf{x} \in \omega, \\ \sigma_0 & \text{if } \mathbf{x} \notin \omega. \end{cases}$$

Due to this variation, the field $\mathbf{e}$ varies and becomes $\mathbf{e}' = \mathbf{e} + \delta\mathbf{e}$. If σ_0 is an optimal layout, then the increment of the functional due to the variation is nonnegative:

$$\delta I(\sigma_0 + \delta\sigma, \mathbf{e} + \delta\mathbf{e}) - I(\sigma_0, \mathbf{e}) \geq 0.$$

The variation $\delta\mathbf{e}$ of the field depends on the properties of the host and the injected materials, on the field, and on the shape ω of the domain of variation: $\delta\mathbf{e} = \delta\mathbf{e}(\sigma_0, \sigma', \mathbf{e}, \omega)$. To find the variation $\delta\mathbf{e}$ one must find the field in a perturbed system. The perturbation is caused by an inclusion ω that is filled with σ' and inserted into an infinite homogeneous domain filled with σ_0.

Variation in a Strip

Suppose that the chosen domain of variation ω is a strip (Figure 4.2). At this point we cannot justify this choice; this is done a posteriori by examination of the result of the variation. Now we can only appeal to the simplicity of the resulting problem.

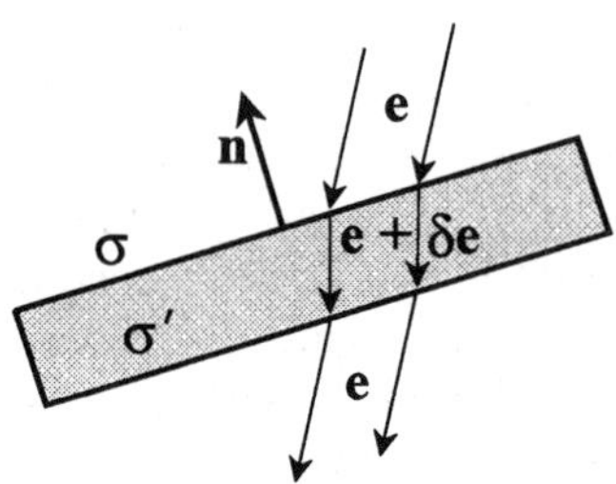

FIGURE 4.2. Variation in a strip.

Variation of the Field. Let us calculate $\delta\mathbf{e}$. To perform the rigorous computation, one can consider (Lurie, 1975) the field in an infinite domain with an elliptical inclusion submerged into a homogeneous field and then pass to the limit when the eccentricity of the ellipse is maximal. Here, we show a less rigorous derivation (Lurie et al., 1980b) that leads to the same results.

The field $\mathbf{e}' = \mathbf{e} + \delta\mathbf{e}$ inside the strip is calculated through the field $\mathbf{e}$ outside the strip. The continuity conditions (4.2.12) show that

$$\mathbf{e}' \cdot \mathbf{t} = \mathbf{e} \cdot \mathbf{t} \quad \mathbf{e}' \cdot \mathbf{n} = \frac{\sigma_0}{\sigma'}\mathbf{e} \cdot \mathbf{n}, \tag{4.3.1}$$

where $\mathbf{t}$ and $\mathbf{n}$ are the normal and the tangent to the strip.

Assuming that the strip is infinitely small, we conclude that the field $\mathbf{e}$ outside the strip tends to the field that has been in the region in the absence of the strip and that the field $\mathbf{e}'$ inside the strip stays constant within the strip (Lurie, 1975). Conditions (4.3.1) say that the field $\mathbf{e}'$ in the varied system is

$$\mathbf{e}' = \begin{cases} \mathbf{e} + \frac{\sigma_0}{\sigma'}\mathbf{n} \cdot \mathbf{e} + O\|\omega\| & \text{if } \mathbf{x} \in \omega, \\ \mathbf{e} + O\|\omega\| & \text{if } \mathbf{x} \notin \omega. \end{cases} \tag{4.3.2}$$

The Increment. The increment $\delta I = \delta I_1 + \delta I_2$ of the cost of problem (4.1.6) consists of two terms. The first term δI_1 describes the change of the cost of used materials due to increasing the amount of material σ' with cost γ' and decreasing the amount of material σ with cost γ. It is equal to

$$\delta I_1 = |\omega|(\gamma' - \gamma).$$

The second term δI_2 is the increment of the energy:

$$\begin{aligned}
\delta I_2 &= \int_\Omega \left((\sigma_0 + \delta\sigma)(\mathbf{e} + \delta\mathbf{e})^2 - \sigma_0\mathbf{e}^2\right) \\
&= \int_\omega \delta\sigma(\mathbf{e} + \delta\mathbf{e})^2 + \int_\Omega \sigma_0\left((\mathbf{e} + \delta\mathbf{e})^2 - \mathbf{e}^2\right),
\end{aligned}$$

where $\mathbf{e}$ and $\mathbf{e} + \delta\mathbf{e}$ are the field in the tested and perturbed configurations, respectively.

Using equalities (4.3.2), we exclude $\delta \mathbf{e}$. The increment becomes

$$\delta I_2 = |\omega| \left[(\sigma' - \sigma_0)\mathbf{e}^2 - \frac{(\sigma' - \sigma_0)^2}{\sigma'}(\mathbf{e} \cdot \mathbf{n})^2 \right] + O(|\omega|). \qquad (4.3.3)$$

Optimal Orientation of the Strip. Formula (4.3.3) shows that the increment $\delta I = \delta I_2(\mathbf{n})$ depends on the orientation of the normal $\mathbf{n}$ of the strip. We choose the orientation to maximize the effect of the variation, that is, to minimize δI_2. We compute

$$\Delta I_2 = \min_{\mathbf{n}} \delta I_2(\mathbf{n}).$$

The increment $\Delta I_2 + \delta I_1$ must be nonnegative for all orientations of the strip.

From (4.3.3) we observe that the minimum of $\delta I_2(\mathbf{n})$ is achieved when the normal $\mathbf{n}$ is aligned with the field $\mathbf{e}$: $\mathbf{n} = \frac{\mathbf{e}}{|\mathbf{e}|}$. In this case, we have $(\mathbf{e} \cdot \mathbf{n})^2 = \mathbf{e}^2$ and

$$\Delta I_2(\sigma_0, \sigma', \mathbf{e}) = \min_{\mathbf{n}} \delta I_2(\mathbf{n}) = |\omega| \frac{(\sigma' - \sigma_0)\sigma_0}{\sigma'} |\mathbf{e}|^2. \qquad (4.3.4)$$

Finally, the increment of the cost is

$$\delta I = \delta I_1 + \Delta I_2 = |\omega| \left((\gamma' - \gamma) + \frac{(\sigma' - \sigma_0)\sigma_0}{\sigma'} |\mathbf{e}|^2 \right).$$

Weierstrass Condition. If the layout is optimal, the increment δI is positive everywhere in the domain occupied by the material σ; the injected material σ' can be any of the admissible materials. This argument leads to the following Weierstrass condition:

> The material σ is optimal if the field $\mathbf{e}$ in it satisfies the inequality
>
> $$(\gamma' - \gamma) + \frac{(\sigma' - \sigma)\sigma}{\sigma'} |\mathbf{e}|^2 \geq 0$$
>
> for any admissible material σ'.

Weierstrass Test for Two-Component Composite

Let us analyze the problem of a two-component optimal composite. The Weierstrass condition shows that optimal fields $\mathbf{e}_1$ and $\mathbf{e}_2$ in the first and second materials satisfy the inequalities

$$|\mathbf{e}_1| \geq \frac{C}{\sigma_1}, \quad |\mathbf{e}_2| \leq \frac{C}{\sigma_2}, \qquad (4.3.5)$$

where $C = \sqrt{\frac{\sigma_1 \sigma_2 (\gamma_1 - \gamma_2)}{\sigma_2 - \sigma_1}}$. This formula is in accord with (4.2.7), which shows the optimality of the pure phases in the extension procedure.

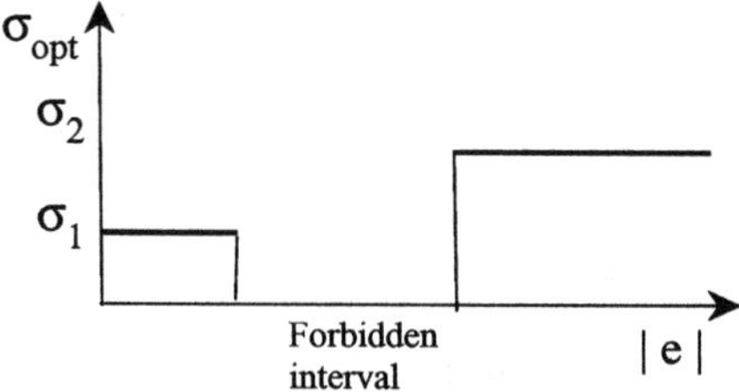

FIGURE 4.3. Forbidden interval.

Forbidden Region

The Weierstrass conditions (4.3.5) show that the magnitudes of fields in the materials in the optimal structures are separated. We observe that σ_1 is optimal in the zone of low field densities, and σ_2 is optimal in the zone of high field densities. Between these two zones lies the *forbidden interval* $\mathcal{E}_{\mathrm{f}}$ of field densities,

$$\mathcal{E}_{\mathrm{f}} = \left[\frac{C}{\sigma_1}, \frac{C}{\sigma_2} \right], \tag{4.3.6}$$

where none of the fields is optimal; see Figure 4.3. These fields must be avoided everywhere in an optimal design:

$$|\mathbf{e}(\mathbf{x})| \notin \mathcal{E}_{\mathrm{f}} \quad \forall\, \mathbf{x} \in \Omega. \tag{4.3.7}$$

Variation in an Optimally Oriented Strip is the Strongest. Let us find whether it is possible to obtain a structure in which the fields do not belong to the forbidden interval $\mathcal{E}_{\mathrm{f}}$.

The material σ_1 is optimal in the locations where the magnitude of the applied field is below the lower boundary of the forbidden interval $\mathcal{E}_{\mathrm{f}}$. The material σ_2 is optimal where the magnitude of the field is above the upper boundary of $\mathcal{E}_{\mathrm{f}}$.

Suppose now that the mean applied field belongs to $\mathcal{E}_{\mathrm{f}}$. Whereas the field cannot belong to $\mathcal{E}_{\mathrm{f}}$ in any point, this mean field is an average of the fields that lie either above or below this interval. This case corresponds to a fine-scale mixture of σ_1 and σ_2. This is how the optimality requirement (4.3.7) leads to oscillating solutions.

Consider the boundary Γ between regions filled with materials σ_1 and σ_2 and denote its normal and tangent by $\mathbf{n}$ and $\mathbf{t}$, respectively. The jump conditions (see (2.1.14), (2.1.15)) are

$$\sigma_2(\mathbf{e}_2 \cdot \mathbf{n}) = \sigma_1(\mathbf{e}_1 \cdot \mathbf{n}), \quad (\mathbf{e}_2 \cdot \mathbf{t}) = (\mathbf{e}_1 \cdot \mathbf{t}),$$

where $\mathbf{e}_1$ and $\mathbf{e}_2$ are the fields at the neighbor points on the opposite sides of Γ. These conditions give

$$\sigma_2|\mathbf{e}_2| = \sqrt{\sigma_1^2|\mathbf{e}_1|^2 + (\sigma_2^2 - \sigma_1^2)(\mathbf{e}_1 \cdot \mathbf{t})^2}. \tag{4.3.8}$$

On the other hand, Weierstrass test (4.3.5) requires that

$$\sigma_2|\mathbf{e}_2| \leq C \leq \sigma_1|\mathbf{e}_1|.$$

Comparing this inequality with (4.3.8) and recalling that $\sigma_1 < \sigma_2$ we see that the Weierstrass test can be satisfied on both sides of the boundary only if its tangent is orthogonal to the direction of the field everywhere:

$$\mathbf{e}_1 \cdot \mathbf{t} = 0. \tag{4.3.9}$$

This condition proves that the chosen domain of variation (the strip) leads to the strongest possible local variation. Indeed, the forbidden interval $\mathcal{E}_f$ obtained from the variation in a strip cannot be made wider. If this were possible, then the zones of optimality of the first and second materials could not have a common boundary, because the continuity conditions would not allow the field $\mathbf{e}$ to jump over the forbidden interval. Hence the materials σ_1 and σ_2 could not coexist in a structure.

For these reasons, one cannot expect to find a stronger local necessary condition. On the other hand, one could try other types of variations, say, the variation in a circle instead of in an oriented strip. These variations lead to smaller forbidden intervals and the condition for the normal to the boundary admits multiple solutions. This says that the variation in a circle is not the strongest one.

Weierstrass Test and Composites. The analysis of the necessary conditions leads to the conclusion of the necessity of composite zones in an optimal project. Observe that the boundary curve Γ is obtained from the Weierstrass conditions (4.3.5). At the same time, its normal must satisfy the requirements (4.3.9). These requirements overdetermine the boundary curve Γ. Indeed, the curve itself and its normal satisfy two different equations, which are not generally consistent. This observation has made in (Klosowicz and Lurie, 1971). This phenomenon suggests that the boundary between zones of pure materials is a generalized curve with infinitely fast wriggles. A generalized curve decouples the direction of the normal and the position of the curve itself. This curve densely covers a region of nonzero measure. Inside the region, the optimal layout corresponds to composites. The effective properties of composites correspond to a homogenized conductivity in the region of the chattering boundary Γ.

This consideration again points to the necessity of enlarging the class of available materials by including composites.

Optimal Fields and Optimal Currents

Let us examine fields and currents in an optimal structure before and after relaxation. The current $\mathbf{j}$ in a material is found to be

$$\mathbf{j}_{\mathrm{opt}} = \frac{\partial}{\partial \mathbf{e}} F_{\mathrm{opt}}(\mathbf{e}).$$

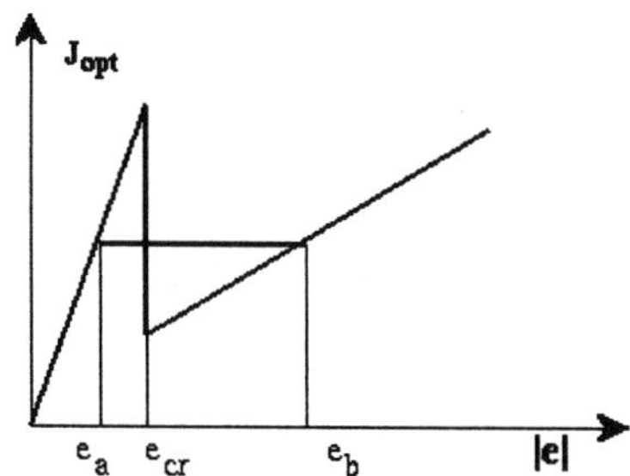

FIGURE 4.4. The constitutive relations in the optimal medium.

The constitutive relations in the optimal medium (without the relaxation)
is

$$
\mathbf{j}_{\mathrm{opt}} = \begin{cases} \sigma_2 \mathbf{e} & \text{if } |\mathbf{e}| \in [\, 0,\, e_{\mathrm{cr}}\,], \\ \sigma_1 \mathbf{e} & \text{if } |\mathbf{e}| \in [\, e_{\mathrm{cr}},\, \infty\,], \end{cases} \qquad e_{\mathrm{cr}} = \sqrt{\dfrac{\gamma}{\sigma_2 - \sigma_1}};
$$

see Figure 4.4. These relations are the Euler equations of the initial variational problem (4.2.10), (4.2.11).

Observe that the inequalities

$$
\mathbf{j}_1 \geq \mathbf{j}_2, \quad \mathbf{e}_1 \leq \mathbf{e}_2
$$

hold in the forbidden interval.

Outside of the forbidden interval, the current monotonically depends on
the field. However, the monotonicity is lost in the forbidden interval, where
a smaller field multiplied by greater conductivity leads to a greater current,
and vice versa: A greater field multiplied by smaller conductivity leads to
a smaller current. This phenomenon is the physical cause of the instability.

On the other hand, the current in the relaxed problem is a monotone
function of the field:

$$
\mathbf{j}_{\mathrm{opt}} = \begin{cases} \sigma_2 \mathbf{e} & \text{if } |\mathbf{e}| \in [0,\, \frac{C}{\sigma_1}], \\[2mm] 2\gamma C \dfrac{\mathbf{e}}{|\mathbf{e}|} & \text{if } |\mathbf{e}| \in \left[\dfrac{C}{\sigma_1},\, \dfrac{C}{\sigma_2}\right], \\[2mm] \sigma_1 \mathbf{e} & \text{if } |\mathbf{e}| \in \left[\dfrac{C}{\sigma_2},\, \infty\right] \end{cases} \tag{4.3.10}
$$

(see Figure 4.4). The absolute value of the current $\mathbf{j}$ is constant in the
forbidden interval.

Optimality of Local Fields

Let us demonstrate how the necessary conditions are satisfied in the optimal
laminates oriented across the field at each point of the domain. Due to the
optimal orientation of the layers, the mean field is aligned with the local
fields in the optimal laminates everywhere. The fields $\mathbf{e}_1$ and $\mathbf{e}_2$ in the
layers of the first and second materials are expressed through the mean

field $\mathbf{e}$ as follows:

$$\mathbf{e}_1(m) = \frac{\sigma_2}{\sigma_2 m + \sigma_1(1-m)}\mathbf{e}, \quad \mathbf{e}_2(m) = \frac{\sigma_1}{\sigma_2 m + \sigma_1(1-m)}\mathbf{e}.$$

The optimal volume fractions m^{opt} calculated by (4.2.6) correspond to the fields $\mathbf{e}_1$, $\mathbf{e}_2$:

$$\left|\mathbf{e}_1\left(m^{\mathrm{opt}}\right)\right| = \frac{C}{\sigma_1}, \quad \left|\mathbf{e}_2\left(m^{\mathrm{opt}}\right)\right| = \frac{C}{\sigma_2}. \tag{4.3.11}$$

Each field has a constant magnitude everywhere in the composite zone. The mean field $\mathbf{e}$ varies only due to the variation of the volume fractions and variation of the orientation of the normal to the layers. The pointwise fields $\mathbf{e}_1$ and $\mathbf{e}_2$ everywhere belong to the boundaries of the forbidden interval(4.3.11). Therefore, the Weierstrass conditions are satisfied as equalities everywhere in the composite zone.

The average field $\mathbf{e}$ varies only due to variation of the volume fraction m^{opt} of materials in the structure. The constancy of the magnitudes of the fields in each material is a manifestation of the optimality principle: The materials are equally loaded everywhere in the composite zone.

4.3.2 The Minimal Extension

A relaxation of the optimal design problem can be obtained from pure variational arguments without referring to the G-closure problem. Here we obtain the relaxation, considering only the stability against the Weierstrass variations. As in Chapter 1, we perform the minimal extension of the Lagrangian, changing it in such a way that the new extended problem has the same cost as the original one and satisfies the Weierstrass conditions everywhere.

The extended Lagrangian $\mathcal{S}F(\mathbf{e})$ is defined as the maximum of the functions $f(\mathbf{e})$ that are not greater than $F(\mathbf{e})$ and that satisfy the Weierstrass test everywhere:

$$\mathcal{S}F(\mathbf{e}) = \max\left\{f(\mathbf{e}): \quad f(\mathbf{e}) \le F(\mathbf{e}), \ \mathcal{W}(f(\mathbf{e})) \ge 0 \ \forall \mathbf{e}\right\}$$

where $\mathcal{W}(f(\mathbf{e}))$ is the increment of the functional due to the Weierstrass variation. The Lagrangian $\mathcal{S}F_u(\mathbf{e})$ leads to the variational problem, stable against Weierstrass variations.

Outside of the forbidden region (4.3.5), the Weierstrass test is satisfied, $\mathcal{W}(F(\mathbf{e})) \ge 0$ and the extended Lagrangian coincides with the original one:

$$\mathcal{S}F(\mathbf{e}) = F(\mathbf{e}) \quad \text{if } \mathbf{e} \notin \mathcal{E}_{\mathrm{f}}.$$

In the forbidden region we define an extended Lagrangian $\mathcal{S}F(\mathbf{e}) < F(\mathbf{e})$ so that the Weierstrass test is satisfied as an equality.

Without loss of generality we can represent the function $\mathcal{S}F(\mathbf{e})$ in the form

$$\mathcal{S}F(\mathbf{e}) = \sigma_*(\mathbf{e})\mathbf{e}^2 + \gamma_*, \qquad (4.3.12)$$

which allows us to treat the problem as an extension of the materials' properties: We add the "materials" with conductivity σ_* (which, however, may depend on $\mathbf{e}$) to the given set $U = \{\sigma_1, \sigma_2\}$ of materials. The constant γ_* represents the cost of the added material. We assume the representation

$$\gamma_* = m\gamma_1 + (1 - m)\gamma_2 \qquad (4.3.13)$$

for this parameter, which introduces the variable m. This representation says that the cost of the composite σ_* is the sum of costs of the mixing materials.

The maximum nature of $\mathcal{S}F(\mathbf{e})$ assumes neutrality to the Weierstrass test:

$$\mathcal{W}(\mathcal{S}F(\mathbf{e})) = 0 \quad \forall \, \mathbf{e} \in \mathcal{E}_{\mathrm{f}}.$$

This equality determines the "properties" σ_* throughout the field $\mathbf{e}$ and determines the extension itself.

Examine the Weierstrass test that is performed by interchanging materials σ_1 and $\sigma_*(\mathbf{e})$. We suppose that the field $\mathbf{e}_1$ in the material σ_1 varies in the interval of optimality of this material (4.3.5), and the field $\mathbf{e}$ belongs to the forbidden interval. The necessary condition (4.3.5) applied to pair $(\sigma_*(\mathbf{e}), \, \mathbf{e})$ and $(\sigma_1, \, \mathbf{e}_1)$ leads to the inequalities

$$|\mathbf{e}| \le \frac{C_*}{\sigma_*}, \quad |\mathbf{e}_1| \ge \frac{C_*}{\sigma_1},$$

where the constant C_* is computed as in (4.3.5):

$$C_* = \sqrt{\frac{(\gamma_1 - \gamma_*)\sigma_*\sigma_1}{\sigma_* - \sigma_1}}.$$

To derive the equation for the minimal extension or for the coefficient σ_*, we require the Weierstrass conditions (4.3.5) be satisfied as limiting equalities,

$$\sigma_*|\mathbf{e}| = C_*, \quad \sigma_1\mathbf{e}_1^2 = C_*,$$

or

$$\sigma_*|\mathbf{e}| = \sigma_1|\mathbf{e}_1|, \qquad (4.3.14)$$

which says that the magnitude of the current in the forbidden interval remains constant.

The field $\mathbf{e}$ in the forbidden interval is the composition of the codirected the fields $\mathbf{e}_1$ and $\mathbf{e}_2$:

$$|\mathbf{e}| = m|\mathbf{e}_1| + (1 - m)|\mathbf{e}_2|. \qquad (4.3.15)$$

Then, from (4.3.14), (4.3.15),

$$\sigma_* = \frac{|\mathbf{e}_1|}{m|\mathbf{e}_1| + (1-m)|\mathbf{e}_2|}\sigma_1.$$

Recall that $|\mathbf{e}_1|$ and $|\mathbf{e}_2|$ vary in their intervals $\mathcal{E}_1$ and $\mathcal{E}_2$, respectively (see (4.3.5)), that were determined by the Weierstrass variation.

The condition in the forbidden interval is satisfied if

$$\sigma_* \leq \min_{\mathbf{e}_1 \in \mathcal{E}_1} \min_{\mathbf{e}_2 \in \mathcal{E}_2} \frac{1}{m + (1-m)\frac{|\mathbf{e}_2|}{|\mathbf{e}_1|}}\sigma_1$$

or, due to (4.3.5),

$$\sigma_* \leq \frac{\sigma_1 \sigma_2}{m\sigma_2 - (1-m)\sigma_1}.$$

Finally we choose the maximal possible σ_* that is consistent with the Weierstrass conditions:

$$\sigma_* = \frac{\sigma_1 \sigma_2}{m\sigma_2 - (1-m)\sigma_1} = \left(\frac{m}{\sigma_1} + \frac{1-m}{\sigma_2}\right)^{-1}. \tag{4.3.16}$$

We find again that the extremal conductivity in the extension is a harmonic mean of the conductivities in the composite; the parameter m is the volume fraction of σ_1.

Formulas (4.3.12), (4.3.13), and (4.3.16) determine the minimal extension of the variational problem based on stability against special fine-scale perturbations (against the variation in a strip).

Remark 4.3.2 *A supplementary test of interchanging the pair σ_2 and σ_* (instead of σ_1 and σ_*) leads to the same extension. This is clear from the symmetry of the extension (4.3.16). Also, the extension remains the same if we interchange any two of the introduced "materials" $\sigma_*(\mathbf{e})$ that correspond to two different values of $\mathbf{e}$.*

4.3.3 Summary

We have examined several approaches to the extension of the optimization problem based on necessary and sufficient conditions of optimality:

- The enlarging of the set of layouts to its G-closure yields to the well-posed optimization problem. The boundary of the G-closure corresponds to optimal mixtures.

- A sufficient condition is given by the convex envelope $CF(\mathbf{e})$ of the original nonconvex energy. This condition bounds the optimal energy from below,

$$CF(\mathbf{e}) \leq F_{\text{opt}}(\mathbf{e}),$$

because it leads to a stable variational problem with a cost that is less than or equal to the cost of the original problem. For composite problems, the convex envelope corresponds to the harmonic mean of the tensors of properties of mixed materials. This bound is valid for all composites, although in the general case it is not known whether this bound is realizable by any structures.

- The first type of necessary condition is realized as the minimum among a special class of microstructures: the laminates. This class is chosen essentially by guessing. The effective properties of laminates are explicitly computed, and the best orientation corresponds to minimal energy in the class of these structures. This extension $\mathcal{L}F(\mathbf{e})$ gives the minimal value of the functional in the class of laminates. It provides an upper bound of the optimal energy:

$$\mathcal{L}F(\mathbf{e}) \geq F_{\mathrm{opt}}(\mathbf{e}),$$

because it is generally not known whether a laminate structure is the best possible.

- The second approach is based on the Weierstrass text. The extension requires stability against Weierstrass variations. Extension due to the Weierstrass test also leads to an upper bound $\mathcal{S}F(\mathbf{e})$ for the optimal energy:

$$\mathcal{S}F(\mathbf{e}) \geq F_{\mathrm{opt}}(\mathbf{e}).$$

Note that the Weierstrass variation in an optimally oriented strip points to the right microstructure, which consists of alternation of such strips.

This extension does not guarantee that the resulting energy $\mathcal{S}F(\mathbf{e})$ is stable against all variations, because special local variations are used in the test.

- For the considered problem, the upper and lower extensions coincide, and they determine the final extension:

$$\mathcal{L}F(\mathbf{e}) = \mathcal{S}F(\mathbf{e}) = \mathcal{C}F(\mathbf{e}) = F_{\mathrm{opt}}(\mathbf{e}).$$

The forbidden interval (4.3.6) in the Weierstrass variation is exactly the interval of nonconvexity of the Lagrangian. The coincidence of the extensions proves that (i) the lower bound is exact, (ii) the Weierstrass variation is the strongest, and (iii) the energy minimum achieved on laminates is actually the global minimum among all structures.

4.4 Dual Problem with Nonsmooth Lagrangian

Reduction to a Nonsmooth Variational Problem

The same problem of minimization of the conductivity of a domain can be formulated as a minimax variational problem. This time we use the dual variational principle that determines the resistance R of the domain as follows:

$$R(\chi) = \min_{\mathbf{j} \in \mathcal{J}} \int_{\Omega} \frac{\mathbf{j}^2}{\sigma}, \tag{4.4.1}$$

where

$$\mathcal{J} = \{\mathbf{j}: \ \nabla \cdot \mathbf{j} = 0, \ \text{in } \Omega, \ \mathbf{n} \cdot \mathbf{j} = \rho_2 \ \text{on } \partial\Omega_2\}.$$

For simplicity in notation, we suppose here that $\rho_1 = 0$.

We may replace the problem of minimization of the conductivity with the problem of maximization of the resistance R. Namely, we consider the following optimization problem:

$$\max_{\chi} R(\chi) \quad \text{if} \quad \int_{\Omega} \chi = M.$$

Adding the integral constraint with the Lagrange multiplier to the doubled functional, we obtain the problem

$$I = \max_{\chi} J_r(\chi), \tag{4.4.2}$$

where

$$J_r(\chi) = \inf_{\mathbf{j} \in \mathcal{J}} \int_{\Omega} \left[\chi \left(\frac{1}{\sigma_1} \mathbf{j}^2 - \gamma \right) + (1 - \chi) \frac{1}{\sigma_2} \mathbf{j}^2 \right]. \tag{4.4.3}$$

Instead of looking for a layout χ that minimizes the energy of the domain, we are looking for a layout that maximizes the dual form of the energy. It is physically clear that the solutions to both problems must coincide. This problem has been analyzed in (Cherkaev and Gibiansky, 1988).

If we interchange the sequence of operations $\sup_{\chi} \inf_{\mathbf{j}}$, then the problem can be solved by a necessary condition of optimality. However, this time the interchange leads to an inequality,

$$\sup_{\chi} \inf_{\mathbf{j}} f(\chi, \mathbf{j}) \leq \inf_{\mathbf{j}} \sup_{\chi} f(\chi, \mathbf{j}), \tag{4.4.4}$$

according to the maximin theorem (see, for example, (Dem'yanov and Malozëmov, 1990; Rockafellar, 1997)). Therefore, the cost of the variational problem I (equation (4.4.3)) is bounded from above by the cost I^d of the problem (4.4.4):

$$I \leq I^d, \quad I^d = \inf_{\mathbf{j} \in \mathcal{J}} \int_{\Omega} F^d(\mathbf{j}), \tag{4.4.5}$$

FIGURE 4.5. Optimal complementary energy, dependence on $|\mathbf{j}|$. Notice the convexity and nonsmoothness of the energy.

where

$$F^d(\mathbf{j}) = \max\left\{\left(\frac{1}{\sigma_1}\mathbf{j}^2 - \gamma\right), \frac{1}{\sigma_2}\mathbf{j}^2\right\}. \qquad (4.4.6)$$

To obtain (4.4.6), we take the maximum over χ of the integrand in (4.4.3).

The variational problem I^d has a convex Lagrangian F^d (Figure 4.5). However, F^d is not smooth; the derivative $\frac{\partial}{\partial|\mathbf{j}|}F^d$ is discontinuous at the point $\mathbf{j}_c = \sqrt{\frac{\sigma_1\sigma_2}{\gamma(\sigma_2-\sigma_1)}}$, where

$$\frac{1}{\sigma_1}\mathbf{j}_c^2 - \gamma = \frac{1}{\sigma_2}\mathbf{j}_c^2.$$

Because of the convexity of the Lagrangian one concludes that the solution to (4.4.5) is stable against fine-scale perturbations. On the other hand, we should also expect that solutions to the problems of minimization of the conductance and maximization of the resistance coincide. Hence, both optimal solutions correspond to an infinitely rapidly oscillating layout χ. The corresponding laminate structures that are optimal for the minimal problem must also be optimal for the minimax problem.

Let us analyze this contradiction, that is to systematically detect instabilities (composite zones) in a solution to a multidimensional variational problem with a convex but nonsmooth Lagrangian, and to describe them using the relaxation technique.

Euler–Lagrange Equation and the Weierstrass Test

The Euler–Lagrange equation of problem (4.4.5) matches that of the minimization problem:

$$\begin{aligned}
\nabla w = \tfrac{1}{\sigma_2}\mathbf{j}, \quad & \nabla\cdot\mathbf{j} = 0 \quad && \text{if } |\mathbf{j}| \in [0, |\mathbf{j}_c|], \\
\nabla w = \tfrac{1}{\sigma_1}\mathbf{j}, \quad & \nabla\cdot\mathbf{j} = 0 \quad && \text{if } |\mathbf{j}| \in [|\mathbf{j}_c|, \infty].
\end{aligned}$$

We could derive the Weierstrass test following the procedure just described. Namely, we calculate the increment of the energy caused by injection of a small strip of material σ' into the environment of material σ. This time we suppose that a homogeneous current $\mathbf{j}$ is fixed at infinity.

Relations between Divergencefree and Curlfree Vectors in Two Dimensions.
We obtain the result immediately in the plane problem if we notice that
any current (a divergencefree vector) is transformed into a field (a curlfree
vector) by the rotation through a right angle and vice versa (Courant and
Hilbert, 1962). In other words, any current can be represented as a field
rotated by 90° and vice versa.

Indeed, the components of a divergencefree current vector $\mathbf{j}$ satisfy the
constraints

$$\frac{\partial j_1}{\partial x_1} + \frac{\partial j_2}{\partial x_2} = 0. \tag{4.4.7}$$

The rotation matrix is

$$R = \begin{pmatrix} 0 & 1 \\ -1 & 0 \end{pmatrix}. \tag{4.4.8}$$

Hence the components of the rotated vector $\mathbf{x}' = R\mathbf{x}$ are $x_1' = x_2$, $x_2' = -x_1$.

The constraint (4.4.7) becomes

$$0 = \frac{\partial j_1}{\partial x_1} + \frac{\partial j_2}{\partial x_2} = \frac{\partial j_1}{\partial x_2'} - \frac{\partial j_2}{\partial x_1'} = \nabla_{x'} \times \mathbf{j}.$$

It says that the divergencefree vector $\mathbf{j}$ becomes a curlfree vector in the
rotated coordinates. In the same way, one can check that the divergence of
a rotated gradient is zero.

The Weierstrass Test. This observation allows us to immediately extend
the results obtained for the fields $\mathbf{e}$ to the currents $\mathbf{j}$. We simply replace in
(4.3.4)

$$\mathbf{e} \text{ by } \mathbf{j}, \quad \sigma \text{ by } \frac{1}{\sigma}, \quad \text{and} \quad \mathbf{n} \text{ by } \mathbf{t}.$$

The increment of the energy $\delta I_2^d(\mathbf{t})$ of I_2^d (see (4.4.5)) becomes

$$\delta I_2^d(\mathbf{t}) = |\Omega_{\text{strip}}| \left[\frac{\sigma - \sigma'}{\sigma'\sigma} \mathbf{j}^2 - \left(\frac{\sigma - \sigma'}{\sigma} \right)^2 (\mathbf{j} \cdot \mathbf{t})^2 \right].$$

For the considered problem (4.4.5), an optimal layout corresponds to the
maximum of the functional; therefore the increment $\delta I_2^d(\mathbf{t})$ must be nonpositive for all orientations of the trial strip. The most sensitive orientation
that yields the maximum of the increment corresponds to an orientation of
the strip such that $\mathbf{j} \cdot \mathbf{t} = 0$. Hence, $\mathbf{j}$ is parallel to the normal $\mathbf{n}$ and therefore $\mathbf{e}$ is parallel to $\mathbf{n}$, in accordance with (4.3.9). The maximal increment
$\Delta^{\text{dual}}(\sigma, \sigma', \mathbf{j})$ is

$$\Delta^{\text{dual}}(\sigma, \sigma', \mathbf{j}) = \max_{\mathbf{t}} \delta^{\text{dual}} I_2(\mathbf{t}) = |\Omega_{\text{strip}}| \frac{(\sigma' - \sigma)}{\sigma'\sigma} (\mathbf{j})^2. \tag{4.4.9}$$

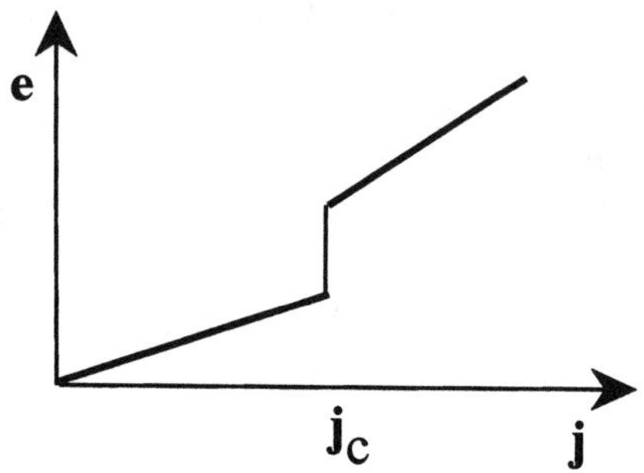

FIGURE 4.6. The constitutive equation for the optimal medium.

Interchanging two strips of the materials σ_1 and σ_2 leads to the optimality condition

$$\Delta^{\text{dual}}(\sigma_1, \sigma_2, \mathbf{j}_1) + \Delta^{\text{dual}}(\sigma_2, \sigma_1, \mathbf{j}_2) \geq 0, \qquad (4.4.10)$$

where $\mathbf{j}_1$ and $\mathbf{j}_2$ are currents at two arbitrary points of the first and second materials, respectively. Assuming as before $\sigma_1 < \sigma_2$, we obtain from (4.4.9), (4.4.10) that

$$(\mathbf{j}_1)^2 - (\mathbf{j}_2)^2 \leq 0 \quad \text{if} \quad \sigma_1 \text{ is optimal.}$$

This time the Weierstrass condition does not lead to a forbidden interval. The condition simply states that smaller conductivity corresponds to a current with smaller magnitude. The switching point β^* corresponds to the critical point j_0 in the stationary solution (4.4.1).

The problem has a convex integrand, and therefore it has a differentiable solution $\mathbf{j}$. One can see that the magnitude $|\mathbf{e}|$ of the field $\mathbf{e} = \frac{\partial}{\partial \mathbf{j}} F^d$ is a monotonic function of the current $|\mathbf{j}|$ everywhere (see Figure 4.6). We observe also that $|\mathbf{e}|$ experiences a jump in the interval $\left[\frac{j_c}{\sigma_2}, \frac{j_c}{\sigma_1}\right]$ that corresponds to the discontinuity of the derivative of the Lagrangian F^d. Actually, the dependence coincides with (4.3.10) when the last is solved for $\mathbf{e}$, as expected.

Instabilities

Recall that the nonconvexity of a Lagrangian points to instability of a solution to the minimization problem. In the minimax problem under discussion, the same instability of the solution is hidden in the interchanging of the extremal operations max min and min max. After this interchanging, we obtain a well-defined problem. But the interchanging leads to inequality (4.4.4). The exactness of this inequality needs to be examined. Namely, we need to find a minimizing sequence in the original problem that corresponds to a solution to the problem with inverse sequence of extremal operations. These optimal sequences may correspond to the so-called mixed strategies, that is, to alternation among several equally optimal controls (von

Neumann and Morgenstern, 1980). The alternation (mix) of the strategies again corresponds to fine-scale inhomogeneous layouts.

Technically, the instabilities can be detected from the analysis of optimality of the discontinuous field $\mathbf{e} = \frac{\partial}{\partial \mathbf{j}} F^d$ at the point $\mathbf{j} = \mathbf{j}_c$.

Continuity Conditions and Composite Zones. Let us demonstrate that the absolute value of the current typically takes its critical value $\mathbf{j}_c$ in a domain of nonzero measure, not at a curve. The previous consideration ensures that this must be true. We have already shown that the solution to the first problem (Sections 4.1 and 4.2) with a nonconvex Lagrangian has a composite zone and that the absolute value of the current is constant within this zone. The solution to the current problem with a convex but not smooth Lagrangian must have this feature too, because the solutions to both problems are identical.

Let us find a formal test to detect the composite zone. Consider the zones in Ω occupied by the first and second materials and suppose that they are divided by a curve Γ along which the absolute value of the current is constant:

$$|\mathbf{j}(\mathbf{x})|_{\mathbf{x} \in \Gamma} = \text{constant}(\mathbf{x}). \qquad (4.4.11)$$

Compare the jump conditions with the Weierstrass test, as we have done in (4.3.8) in solving the previous problem. As before, we denote the normal and tangent to that curve Γ as $\mathbf{n}$ and $\mathbf{t}$, respectively. The jump condition is:

$$\begin{aligned}
(\mathbf{j}_1)^2 &= (\mathbf{j}_1 \cdot \mathbf{t})^2 + (\mathbf{j}_1 \cdot \mathbf{n})^2 \\
&= (\mathbf{j}_2 \cdot \mathbf{n})^2 + \frac{\sigma_1}{\sigma_2}(\mathbf{j}_2 \cdot \mathbf{t})^2 = (\mathbf{j}_2)^2 - \frac{\sigma_2 - \sigma_1}{\sigma_2}(\mathbf{j}_2 \cdot \mathbf{t})^2.
\end{aligned}$$

The continuity of currents required by the Weierstrass test poses an additional condition on the curve Γ:

$$\mathbf{j}_2 \cdot \mathbf{t} = 0. \qquad (4.4.12)$$

Conditions (4.4.11) and (4.4.12) are independent of each other; therefore they overdetermine the boundary Γ: The first condition determines the curve, and the second independently fixes its normal. If there exists a curve that divides two subdomains, then the continuity conditions and optimality requirements would overdetermine the equation for finding such a curve. To solve this inconsistency, we suggest that the point (4.4.11) in the space of parameters corresponds to a domain in the $\mathbf{x}$ plane.

Remark 4.4.1 *The equality of the cost of the maximin and the minimax problems (4.4.2), (4.4.5) should be proved in each problem. It mirrors the attainability of the convex envelope in the dual minimization problem.*

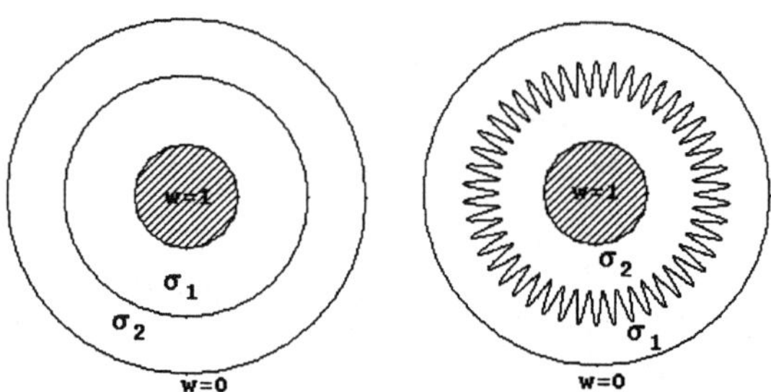

FIGURE 4.7. The optimal annular conductors of the minimal (problem A) and maximal (problem B) conductivity.

4.5   Example: The Annulus of Extremal Conductivity

Here we consider the following problem: Find the structure of a composite annulus $\Omega : (r_0 \leq r \leq 1)$ assembled from given amounts of two materials with conductivities σ_1, σ_2 that minimizes (problem A) or maximizes (problem B) the total current through it. We assume that the potential of the inner boundary $\partial\Omega_0$ is equal to one, the potential of the exterior boundary $\partial\Omega_1$ is zero, and the amount of material σ_1 is fixed:

$$M = \pi \int_{r_0}^{1} m(r) r \, dr. \tag{4.5.1}$$

The similar example was described in (Murat and Tartar, 1997).

The problem has axial symmetry. This symmetry calls for an axisymmetric layout of composites, $\boldsymbol{\sigma}_* = \boldsymbol{\sigma}_*(r)$, and for axisymmetric currents $\mathbf{j} = \mathbf{j}(r)$ and fields $\mathbf{e} = \mathbf{e}(r)$.

The symmetry assumes that the current depends on the radius only and is directed radially, $\mathbf{j} = j_r(r)\mathbf{i}_r$, where $\mathbf{i}_r$ is the unit vector directed along the radius. Its radial component $j_r(r)$ is calculated independently of material properties:

$$\nabla \cdot \mathbf{j} = 0 \quad \Rightarrow \quad j_r(r) = \frac{J}{r},$$

where constant J is equal to the total current. The radial current monotonically decreases to the periphery of the ring independently of the materials' layout.

Minimal Conductivity

Consider problem A. Contrary to our previous arguments, the solution to this problem does not contain composite zones! Indeed, the layout

$$\sigma(r) = \begin{cases} \sigma_1 & \text{if} \quad r \in [r_0, r_*], \\ \sigma_2 & \text{if} \quad r \in [r_*, r_1], \end{cases}$$

where r_* is defined such that (4.5.1) is fulfilled (see Figure 4.6), satisfies all necessary conditions. The contradiction in the necessary conditions does not occur due to axial symmetry of the problem. Recall that the contradiction in the general case occurs because the normal to a boundary between zones of different materials is independent of the boundary itself. Here, however, the circumferential boundary line automatically satisfies both equations. The current is directed along the radius (that is, along the normal of the dividing curve) due to the symmetry of the problem, which makes the circumferential component equal to zero. The optimal control

$$m(r) = \begin{cases} 1 & \text{if} \quad r \in [r_0, r_*] \\ 0 & \text{if} \quad r \in [r_0, r_*] \end{cases}$$

satisfies necessary conditions. In the optimal project, the better conducting material σ_2 is placed outside, where the current is smaller; see Figure 4.7.

Remark 4.5.1 *This example should not discourage us to look for a composite extension of materials' layout. It shows that the extension leads to the classical solution if this solution exists due to some special circumstances. The extension cannot hurt.*

Maximal Conductivity, Minimal Resistivity

Consider problem B. This problem asks for the structure of the annulus of the minimal total resistance that is determined as the ratio of the potential difference and the current. Let us assume that the total current J is fixed, and minimize the potential of the exterior circle that is proportional to the energy of the annulus. In order to minimize the resistance we need to minimize the energy stored in the body when the current is fixed.

According to our analysis (4.2.9), we see that the optimal structures are laminates oriented along the radius (see Figure 4.7). The volume fractions m and $1-m$ of materials in optimal laminates varies to fulfill the optimality requirement $|\nabla w| = c$, where c is a constant.

Ohm's law becomes

$$\frac{J}{r} = [\sigma_2 + m(\sigma_1 - \sigma_2)]c.$$

The last condition allows us to determine the size of the composite zone and its location. Indeed, suppose that the composite zone occupies a ring $r_1 < r < r_2$. At the points r_1, r_2 the following conditions hold:

$$\frac{J}{r_1} = \sigma_1 c, \quad \frac{J}{r_2} = \sigma_2 c \quad \Rightarrow \quad \frac{r_1}{r_2} = \frac{\sigma_2}{\sigma_1}.$$

The optimal layout of $m(r)$ (see Figure 4.7) is

$$m(r) = \begin{cases} 1 & \text{if} \quad r \in [r_0, \, r_1], \\ \dfrac{J - \sigma_2 c\, r}{c\, r(\sigma_1 - \sigma_2)]} & \text{if} \quad r \in [r_1, \, r_2], \\ 0 & \text{if} \quad r \in [r_2, \, 1]. \end{cases}$$

The integral restriction (4.5.1) allows us to determine the parameter c.

The optimal design consists of three zones. The intermediate zone is occupied by radial laminates with variable volume fractions of materials.

4.6   Optimal Multiphase Composites

4.6.1   An Elastic Bar of Extremal Torsion Stiffness

In this section, we analyze the optimal layout of several materials and introduce another physical setting for the problem of the domain of extremal conductivity. This problem is mathematically equivalent to the problem of the torsioned elastic bar of extremal torsion stiffness; see, for example, (Sokolnikoff, 1983).

Consider a problem of an elastic bar with the extremal torsion stiffness, made from two materials with the different shear moduli μ_1 and μ_2. We assume that the bar is cylindrical, infinitely long, and that its cross section is an arbitrary simply connected domain Ω. The stiffness of the bar is determined as the ratio between the angle of torsion of the cross section per unit length and the applied torsion moment $\mathcal{J}$. Equivalently, it can be determined as the specific energy stored in the portion of the bar of unit length if the torsion moment $\mathcal{J}$ is applied.

The torsion of an elastic bar is described by the equations that coincide with the conductivity equation, where the role of conductivity plays the inverse of the shear modulus $\sigma \to \frac{1}{\mu}$. The potential w is the so-called Prandtl function; the partial derivatives of the potential represent the shear stresses at a point of the cross section of the cylinder.

The following variational problem describes the stiffness of an elastic bar

$$I(\mu) = \min_{w:w|_{\partial\Omega}=0} \int_\Omega \left(\frac{1}{2\mu} (\nabla w)^2 - w \right),$$

where μ is the shear modulus of the material and I is the variation of the angle of the rotation of the bar's cross section per unit length of the bar. It is assumed that the applied torsion moment is equal to one: $\mathcal{J} = 1$.

The optimization problem asks for the layout of the materials in the cross section of the bar that maximizes or minimizes its stiffness. This problem for a bar under torsion differs from those already discussed mainly in notation. It was studied in (Klosowicz and Lurie, 1971; Lurie and Cherkaev, 1978; Lavrov et al., 1980; Goodman et al., 1986) for two-material composite.

Depending on the values of $|\nabla w|$, the optimal layout consists of the zone of the stiff material, the laminate zone, and the zone of weak material. Function $|\nabla w(x_1, x_2)|$ and the volume fractions of the laminates is shown in the Figure 4.8 for the square cross section. One can see that the weak material is placed in the middle and near the corners where $|\nabla w|$ is small;

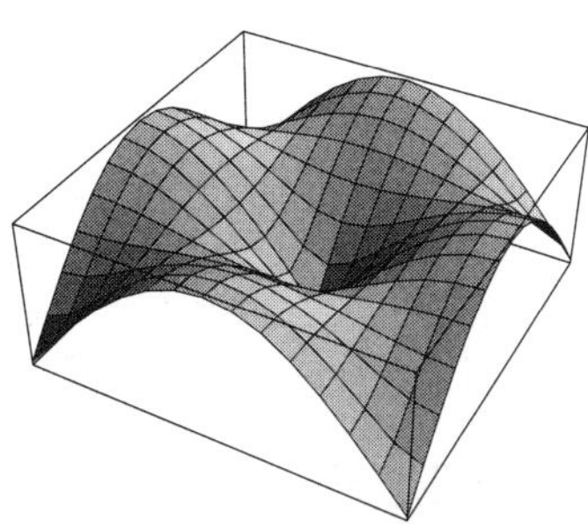 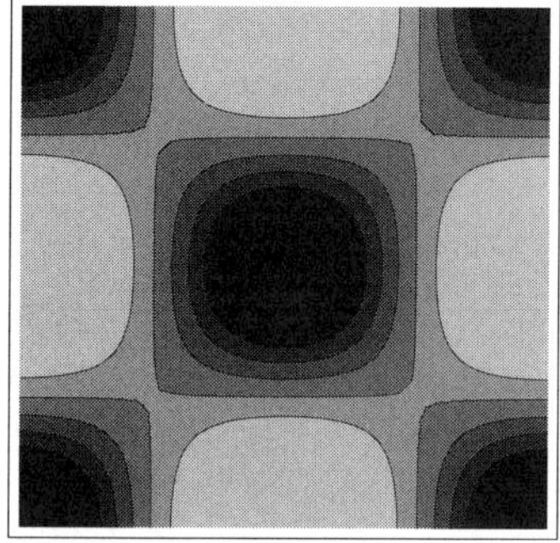

FIGURE 4.8. The cross section of a bar of maximal torsion stiffness: (Left) The graph of $|\nabla w(x_1, x_2)|$. (Right): The volume fraction of the stiff material in the optimal laminate. Black and white zones show the solid weak and stiff material, respectively.

the stiff material is located near the middle of the sides where $|\nabla w|$ is large. The rest is filled with laminates with variable volume fraction. The laminates are directed along the level lines of w.

4.6.2   Multimaterial Design

The optimization problems we have dealt with so far have been formulated for two-material composites; we now extend the results to the multimaterial case. The extension procedure is the same: We build the convex envelope of the nonconvex multiwell Lagrangian. We follow (Burns and Cherkaev, 1997).

Consider again the problem of a layout of material in a conducting body that minimizes its total energy. For definiteness let us consider the problem of optimal layout of three isotropic conducting materials with conductivities $\sigma_1 < \sigma_2 < \sigma_3$,

$$\sigma(\mathbf{x}) = \sum_{i=1}^{3} \chi_i(\mathbf{x})\sigma_i,$$

whose total volumes are fixed:

$$\int_\Omega \chi_i = M_i, \quad \sum_{i=1}^{3} M_i = \|\Omega\|. \tag{4.6.1}$$

We formulate the optimization problem (compare with (4.2.10), (4.2.11)) as

$$J = \inf_w \int_\Omega F(|\nabla w|, \gamma_i), \tag{4.6.2}$$

$$F(v, \gamma_i) = \min_i \{\sigma_i v^2 + \gamma_i\},$$

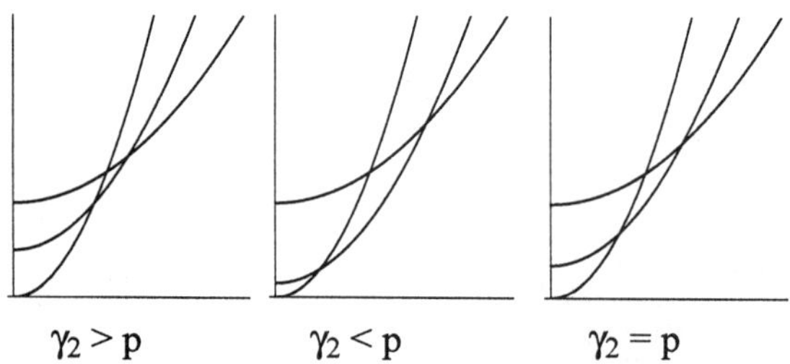

FIGURE 4.9. Multiwell Lagrangians.

where we use the notation $v = |\nabla w|$. The Lagrange multipliers γ_i correspond to the constraints (4.6.1). The multiplier γ_i can also be viewed as the cost of the corresponding material. We can always assume that $\gamma_3 = 0$. To avoid trivial degenerations we will expect that the better material is more expensive:

$$\gamma_3 < \gamma_2 < \gamma_1.$$

Otherwise, one or two of the materials have zero fraction in the optimal construction.

Again we are dealing with the nonconvex variational problem (4.6.2) for a multiwell Lagrangians. We relax it by passing to the convex envelope $CF(v)$ of $F(v)$ ($v = |\nabla w|$). The relaxed problem has the same cost J, and possesses a classical solution.

The picture of the envelope is more sophisticated than before. Depending on the values of γ_i, one can meet different cases of relative positions of the parabolas $\sigma_i v^2 + \gamma_i$ (see Figure 4.9) that represent the convex wells for the multiwell problem:

1. The convex envelope is supported by the first and third parabolas only. This is the easiest case; the second material is not presented in the composite. The corresponding restriction on the second material is trivial: $M_2 = 0$.

2. The convex envelope is supported by the first and second and by the second and third parabolas. The equations of the envelope are

$$CF(v) = \begin{cases} \sigma_1 v^2 + \gamma_1 & \text{if} & 0 \le v \le \alpha_{12}, \\ a_{12}v + b_{12} & \text{if} & \alpha_{12} \le v \le \beta_{12}, \\ \sigma_2 v^2 + \gamma_2 & \text{if} & \beta_{12} \le v \le \alpha_{23}, \\ a_{23}v + b_{23} & \text{if} & \alpha_{23} \le v \le \beta_{23}, \\ \sigma_3 v^2 + \gamma_3 & \text{if} & \beta_{23} \le v \le \infty, \end{cases}$$

where

$$\alpha_{ij} = \sqrt{\frac{\sigma_j(\gamma_i - \gamma_j)}{\sigma_i(\sigma_i - \sigma_j)}}, \quad \beta_{ij} = \sqrt{\frac{\sigma_i(\gamma_i - \gamma_j)}{\sigma_j(\sigma_i - \sigma_j)}},$$

and

$$a_{ij} = 2\sqrt{\frac{\sigma_i\sigma_i(\gamma_1 - \gamma_j)}{\sigma_i - \sigma_j}}, \quad b_{ij} = a_{ij}^2 + \gamma_i.$$

Here the optimal project contains zones of all pure materials and laminates made from the first and second materials and from the second and third materials.

The volume fraction M_2 of the second material must be large enough (greater than a positive constant $p > 0$),

$$M_2 \geq p, \tag{4.6.3}$$

to provide the possibility of this regime because all composite zones use the second material.

3. The convex envelope is a straight line supported by all three parabolas that have the common tangent:

$$a_{12} = a_{23} = a_{13}, \quad b_{12} = b_{23} = b_{13}.$$

The equations are

$$CF(v) = \begin{cases} \sigma_1 v^2 + \gamma_1 & \text{if} \quad 0 \leq v \leq \alpha_{13}, \\ a_{13}v + b_{13} & \text{if} \quad \alpha_{13} \leq v \leq \beta_{13}, \\ \sigma_3 v^2 + \gamma_3 & \text{if} \quad \beta_{13} \leq v \leq \infty. \end{cases}$$

This regime of control corresponds to a special set of costs γ_i. The optimal project contains zones of the first and third pure materials and of laminates that are made of all three materials. Laminates are oriented across the field and have the conductivity

$$\sigma_h = \left[\frac{m_1}{\sigma_1} + \frac{m_2}{\sigma_2} + \frac{m_1}{\sigma_3}\right]^{-1}$$

along the field.

Nonuniqueness. A characteristic feature of this minimizing layout is that the values of σ_h and of the cost do not determine completely the volume fractions of presented materials. Indeed, one can replace a portion of the intermediate material σ_2 with the laminates made of the first and third materials that show equal conductivity across the layers; these replacements do not change either the local resistance or the cost of materials used.

The volume fraction μ_1 of σ_1 in the laminate that replaces the second material is found from

$$\frac{1}{\sigma_2} = \frac{\mu}{\sigma_1} + \frac{1-\mu}{\sigma_3},$$

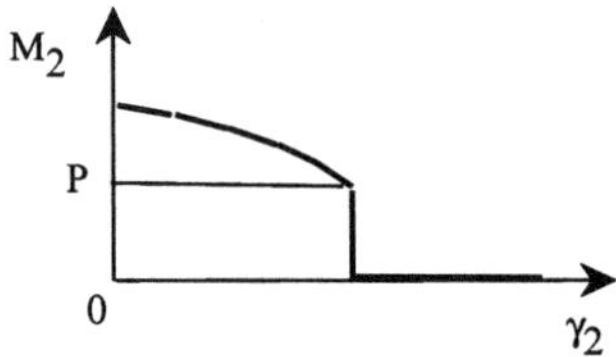

FIGURE 4.10. Dependence of the volume M_2 of σ_2 on the cost γ_2.

and it equals

$$\mu = \frac{(\sigma_3 - \sigma_2)\sigma_1}{(\sigma_3 - \sigma_1)\sigma_2}.$$

The total cost of the materials used remains the same ($\mu\gamma_1 + (1 - \mu)\gamma_3 = \gamma_2$). Therefore, the cost of the optimal problem remains invariant under this replacement.

This regime corresponds to a range of volume fraction M_2 of the second material. Let us compute the range of M_2 that can be used in a composite with a given effective conductivity σ_h. Clearly, it varies from zero to a maximal value $\nu = \nu(\sigma_h)$ that equals the amount of this material in a two-component composite with the same conductivity when one of the outer materials is completely replaced. This value is

$$\nu = \min\left\{ \frac{(\sigma_3 - \sigma_h)\sigma_2}{(\sigma_3 - \sigma_2)\sigma_h}, \quad \frac{(\sigma_1 - \sigma_h)\sigma_2}{(\sigma_1 - \sigma_2)\sigma_h} \right\}.$$

Given a layout $\sigma_h(\mathbf{x})$, one can place any amount $m_2(\mathbf{x}) \in [0, \nu]$ of σ_2 allowed by the restriction (4.6.1) in it. The maximal amount σ_2 used in the design is less than the constant p (see (4.6.3)):

$$0 \leq M_2 \leq \int_\Omega \nu \leq p.$$

One could draw an arbitrary contour in the region of optimal design filled with composite and then change the volume fraction of materials by adding some quantity of the second material in the marked region in exchange for the equivalent composite of the first and third materials.

The suggested analysis implies an interesting dependence of the volume M_2 (see Figure 4.10) on the cost γ_2 of this material. This dependence is characterized by the horizontal component $M_2 = 0$ (case 1), by the jump in the critical point (case 3), and by the monotonic part (case 2). The critical point is characterized by a "phase transition": An infinitely small decrease in the cost γ_2 is followed by a finite variation of the optimal fraction M_2 from zero to a positive value P

$$P = \int_\Omega \nu(\sigma_h)$$

that depends on the global solution via layout of $\sigma_h = \sigma_h(\mathbf{x})$ throughout the designed body. Of course, the cost of the problem remains a continuous function of γ_2. because at the point $\gamma_2 = p$ it is independent on M_2. This analysis was implemented in (Burns and Cherkaev, 1997) in a numerical algorithm that selects the case and chooses the optimal volume fraction.

4.7 Problems

Consider the problem of minimal conductivity of a domain filled with two anisotropic materials with the property tensors

$$\boldsymbol{\sigma}_1 = \begin{pmatrix} \lambda_{11} & 0 \\ 0 & \lambda_{12} \end{pmatrix}, \quad \lambda_{11} \leq \lambda_{12},$$

and

$$\boldsymbol{\sigma}_2 = \begin{pmatrix} \lambda_{21} & 0 \\ 0 & \lambda_{22} \end{pmatrix}, \quad \lambda_{21} \leq \lambda_{22}.$$

Assume one can rotate the tensors $\boldsymbol{\sigma}_1$ and $\boldsymbol{\sigma}_2$ in a microstructure.

1. Derive the Weierstrass condition and the forbidden interval. Discuss the dependence of optimal project on λ_{ij}.

2. Is it possible to change some of the parameters λ_{ij} without changing the cost of the optimal project?

3. Show that the optimal energy is isotropic. Draw the graph of the optimal energy and its convex envelope.

4. Consider the dual problem of minimal resistivity. Derive the Weierstrass condition and describe the forbidden interval. Discuss the dependence of the optimal project on λ_{ij}. Is it possible to change some of the parameters λ_{ij} without changing the cost of the optimal project?

5. What is the relation between these two problems? Analyze various orderings of λ_{12} and λ_{22}.

6. Assume that the eigenvectors of the tensors $\boldsymbol{\sigma}_1$ and $\boldsymbol{\sigma}_2$ are fixed and codirected. Assume also that $\lambda_{12} = \lambda_{22}$.

   Derive the Weierstrass conditions for minimum and maximum conductivity. Define the forbidden region. Draw a picture of this region in the plane of parameters $\frac{\partial w}{\partial x_1}$ and $\frac{\partial w}{\partial x_2}$.

7. Derive the Weierstrass condition for the three-dimensional conductivity problem. Use variation in a disk-shaped domain. Compare the results with the two-dimensional case.

8. Consider the problem of minimal conductivity for four isotropic materials. What are the relations between cost and conductivities of material if the optimal project:

- consists of materials 1 and 4?
- consists of materials 1, 3, and 4?
- consists of materials 2, and 4?

Suggest an algorithm for the optimization problem.

5
Optimal Conducting Structures

In this chapter we consider minimization of functionals that depend on the solution of the stationary conductivity problem. Energy minimization is one such problem. Other examples include minimization of the mean temperature within some area in the body, minimization of a norm of the difference between the desired and actual temperature, or maximization of the total current through a boundary component. Again, microstructures appear in these optimal designs. We describe an approach based on homogenization and demonstrate that the optimal composites are surprisingly simple: Laminates are the optimal structures for a large class of cost functionals.

5.1 Relaxation and G-Convergence

5.1.1 Weak Continuity and Weak Lower Semicontinuity

First, we describe the type of functionals that can be minimized by homogenization methods. In minimizing the functional, we likely end up with a highly inhomogeneous sequence of materials layout. The homogenization replaces these highly inhomogeneous materials with effective materials. The question is: How does this replacement effect a cost functional?

Formulation

Consider again a domain Ω with a smooth boundary $\partial\Omega$ filled with a two-phase inhomogeneous material of conductivity $\sigma(\chi)$, where $\chi = \chi(\mathbf{x})$ is the characteristic function of the subdomain Ω_1 occupied with the material σ_1. The rest of the domain is filled with material σ_2.

Consider the conductivity equations (2.1.2), (2.1.3), (2.1.4), and (2.1.7) in Ω. We rewrite them for convenience:

$$\left. \begin{array}{r} \nabla \cdot \mathbf{j} = q \\ \sigma(\chi)\nabla w = \mathbf{j} \end{array} \right\} \text{ in } \Omega, \qquad \begin{array}{l} w = \rho_1 \text{ on } \partial\Omega_1, \\ j_n = \rho_2 \text{ on } \partial\Omega_2, \end{array} \tag{5.1.1}$$

where w is a potential, $\mathbf{j}$ is the current, q is the density of sources, ρ_1 and ρ_2 are the boundary data imposed on the supplementary components $\partial\Omega_1$, $\partial\Omega_2$ of the boundary $\partial\Omega$, and $j_n = \mathbf{j} \cdot \mathbf{n}$ is the normal component of $\mathbf{j}$. We assume that $\rho_1(s)$, $\rho_2(s)$, and $q(\mathbf{x})$ are differentiable, where s is the coordinate on the surface $\partial\Omega$.

We suppose that problem (5.1.1) has a unique solution for the given q, ρ_1, ρ_2 and for arbitrary layout χ, and this solution is bounded,

$$\|w\|_{H^1(\Omega)} = \int_\Omega (\nabla w^2 + w^2) < C.$$

Consider the minimization problem

$$\min_{w \text{ as in } (5.1.1)} I(w); \quad I(w) = \int_\Omega F(w, \nabla w), \tag{5.1.2}$$

where w is the solution to (5.1.1). Equation (5.1.1) is treated as a constraint on $w(\chi)$, and χ is the control. Therefore, the functional I is determined by the control, too: $I = I(w(\chi))$. We denote by $\mathcal{U}$ the set of conductivities of the available materials. Thus we have formulated a restricted variational problem (the so-called Mayer-Bolza problem; see, for example, (Lurie, 1967)).

The scheme of the dependence of the functional on the control is as follows:

$$\chi \implies w(\chi) \implies I(w).$$

Stability against Homogenization

In dealing with structural optimization problems we expect that a minimizer $\chi(\mathbf{x})$ is characterized by fine-scale oscillations. Homogenization can be used to describe such oscillatory solutions. Let us find a class of functionals that can be minimized by this approach. The corresponding mathematical technique is the theory of sequentially weak lower semicontinuity of functionals. We give an informal introduction to the use of this theory in homogenization, and we refer the reader to a rigorous exposition in (Dacorogna, 1982; Jikov et al., 1994; Dal Maso, 1993; Pedregal, 1997).

Compare the potential w_ε associated with the problem for the conductivity operator $\nabla\sigma_\varepsilon\nabla$ with fast oscillating coefficients σ_ε and the potential w_0 associated with the homogenized conductivity operator $\nabla\sigma_*\nabla$ with smooth coefficients σ_*. Recall that w_ε tends to w_0,

$$\lim_{\varepsilon \to 0} \int_\Omega (w_\varepsilon - w_0)^2 = 0, \quad \text{or} \quad w_\varepsilon \to w_0 \text{ strongly in } L_2. \qquad (5.1.3)$$

However, ∇w_0 and ∇w_ε are not close pointwise:

$$\lim_{\varepsilon \to 0} \int_\Omega (\nabla w_\varepsilon - \nabla w_0)^2 > 0$$

because ∇w_ε is a discontinuous function that has finite jumps on the boundary of the regions of different materials and ∇w_0 is a continuous function. The limit ∇w_0 represents the mean value of ∇w_ε over an arbitrary regular small region Ω_ε when the frequency of oscillations $\frac{1}{\varepsilon}$ goes to infinity,

$$\nabla w_0 = \lim_{\varepsilon \to 0} \langle \nabla w_\varepsilon \rangle; \qquad (5.1.4)$$

∇w_ε weakly converges to ∇w_0 in L_2.

Weakly Continuous Functionals

Some functionals $I(w)$ are stable under homogenization, while others can significantly change their value: $I(w_0) - I(w_\varepsilon) = 0(1)$. In the latter case, the solution to the homogenized problem may have nothing in common with the solution to the original problem. Therefore, it is important to distinguish these types of functionals.

The functional $I(w)$ is called *weakly continuous*, if

$$I(w_0) = \lim_{w_s \rightharpoonup w_0} I(w_s)$$

where $\rightharpoonup$ means the weak convergence in $H^1(\Omega)$. Weakly continuous functionals are stable under homogenization.

For example, the functional

$$I_1(w) = \int_\Omega (F(w) + \mathbf{A} \cdot \nabla w),$$

is weakly continuous if F is a continuous function and $\mathbf{A}$ is a constant or smoothly varying vector. Indeed, (5.1.3), (5.1.4) imply that

$$I_1(w_\varepsilon) - I_1(w_0) = \int_\Omega (F(w_\varepsilon) - F(w_0)) + \int_\Omega (\mathbf{A} \cdot \nabla(w_\varepsilon - w_0)) \to 0$$

when $\varepsilon \to 0$. The first integral goes to zero because $\|w_0 - w_\varepsilon\|_{L_2}$ is arbitrary near to zero and the function F is continuous, and the second integral goes to zero because ∇w_ε goes to ∇w_0 weakly in L_2. Therefore, $I_1(w)$ is stable under homogenization.

Generally, the functional $I(w) = \int_\Omega F(\nabla w)$, where F is a nonlinear function, does not keep its value after homogenization. The limit

$$I^0 = \lim_{w_s \to w_0} I(w_\varepsilon),$$

where w_s is a fine-scale oscillatory function, can generally be either greater or smaller than $I(w_0)$ depending on the minimizing sequence $\{w_s\}$. These functionals are called *weakly discontinuous.*

Weakly Lower Semicontinuous Functionals

The weak continuity is sufficient but not necessary to pass to the weak limit of the minimizer in the variational problem. The property needed is called the *weak lower semicontinuity* (Dacorogna, 1982):

$$I^0 = \lim_{\varepsilon \to 0} I(w_\varepsilon) \geq I(w_0), \qquad w_\varepsilon \to w_0.$$

For these functionals, the limit only decreases when the minimizer coincides with the weak limit of the minimizing sequence.

Each functional of the type

$$I_2(w) = \int_\Omega F(w, \nabla w), \tag{5.1.5}$$

where $F(w, \nabla w)$ is a continuous function of w and a convex function of ∇w, is weakly lower semicontinuous.

Example 5.1.1 The functional

$$I_3(w) = \int_\Omega (\nabla w)^2$$

is weakly lower semicontinuous, because

$$I_3^0 = \lim_{w_\varepsilon \to w_0} I(w_\varepsilon) = I_3(w_0) + \int_\Omega (\nabla w_\varepsilon - \nabla w_0)^2 \geq I_3(w_0),$$

when $w_\varepsilon \to w_0$ (to compute I_3^0 we use the limit $\nabla(\langle w_\varepsilon \rangle - w_0) \to 0$).

Dependency on χ. A broader class of optimization problems deals with functionals

$$I_4(w, \chi) = \int_\Omega F(w, \nabla w, \chi)$$

that explicitly depend on the characteristic function χ. The functional I_4 depends on the amount (or cost) of the materials used in the design. For these problems, weak lower semicontinuity is formulated as

$$\lim_{\varepsilon \to 0} I_4(w_\varepsilon, \chi_\varepsilon) \geq I(w^0, m), \quad \text{when} \begin{cases} \chi_\varepsilon \to m, \\ w_\varepsilon \to w_0. \end{cases} \tag{5.1.6}$$

Analogously to (5.1.5), the functional I_4 is weakly lower semicontinuous if $F(w, \nabla w, \chi)$ is a continuous function of w and a convex function of ∇w and χ.

Remark 5.1.1 *Dealing with relaxation of the nonconvex Lagrangians, we consider the inverse problem: What property of the Lagrangian is necessary and sufficient for the weakly lower semicontinuity? This question is discussed in the next chapter. Here, we only mention that the convexity of $F(w, \nabla w, \chi)$ with respect to ∇w and χ and the continuity with respect to w is sufficient for the weakly lower semicontinuity.*

In summary, the homogenization technique is directly applicable to the weakly lower semicontinuous functionals that do not increase their values by homogenization.

5.1.2 Relaxation of Constrained Problems by G-Closure

G-Closed Sets of Control

Here we are dealing with a variational problem with differential constraints that expresses the equilibrium (the Mayer-Bolza problem): w is the solution of an equilibrium problem defined by the control.

Consider the minimization of a weakly lower semicontinuous functional $I(w)$, where w is the solution to a boundary value problem (5.1.1). Consider a sequence of solutions $w_\varepsilon = w(\sigma_\varepsilon)$ that minimizes the functional $I(w)$:

$$I(w_\varepsilon) \;\to\; I_0 = \inf I(w).$$

Because the functional is weakly lower semicontinuous, the minimizing sequence $\{w_\varepsilon\}$ weakly converges to w_0,

$$w_\varepsilon \rightharpoonup w_0 : \quad I_0 = I(w_0).$$

Let us find out what happens to the corresponding sequence of materials' layouts σ_ε. Recall the definition of G-convergence (Chapter 3): A sequence $\{\sigma_\varepsilon\}$ G-converges to a tensor-valued function σ_* if the sequence $\{w_\varepsilon\} : w_\varepsilon = w(\sigma_\varepsilon)$ weakly converges to $w_0 = w(\sigma_*)$. Therefore, an optimal solution w_0 (a weak limit of w_ε) corresponds to a G-limit σ_* of $\{\sigma_\varepsilon\}$:

$$I_0 = \inf_{\sigma_\varepsilon} I(w(\sigma_\varepsilon)) = \min_{\sigma_* \in GU} I(w(\sigma_*)).$$

A minimization problem of a weakly lower semicontinuous functional has a solution in the set U of values of σ if all G-limits σ_* belong to this set or if the set U is G-closed.

Remark 5.1.2 *This approach, called the "homogenization approach," was developed in different forms in many papers (Lurie and Cherkaev, 1978;*

Raĭtum, 1979; Lurie et al., 1982; Tartar, 1994; Lurie, Cherkaev, and Fedorov, 1984; Murat and Tartar, 1985b; Kohn and Strang, 1986a; Kohn and Strang, 1986b; Tartar, 1987; Lurie and Cherkaev, 1986a; Bendsøe and Kikuchi, 1988) We do not mention here the control problems dealing with ordinary differential equations, where these ideas were developed several decades earlier.

G-Closure of the Set of Controls

We mentioned in Chapter 3 that the set of conductivities $\mathcal{U}$ usually does not coincide with its G-closure. In that case, the solution to problem (5.1.2) may not exist. This means that any value $I(w^s)$ that corresponds to a layout of materials $\sigma(\chi^s)$ can be decreased by another layout $\sigma(\chi^{s+1})$ and so on.

Example 5.1.2 Recall the simplest example of the absence of an optimal element in an open set; there is no minimal positive number. The infimum of all positive numbers is zero, but zero is not included in the set. To make the problem of the minimal positive number well-posed, one must enlarge the set of positive numbers by including the limiting point (zero) in it. Similarly, we reformulate (relax) the optimal design problem including all G-limits into the set of admissible controls. An extended problem has a solution equal to the limit of the minimizing sequences of solutions to the original ill-posed problems.

A functional $I_4(w, \chi)$ (5.1.6) that explicitly depends on χ requires similar relaxation. This time we need to consider the convergence of a pair (w^s, χ^s) to the pair (w_0, m), where $m = m(\mathbf{x})$ is the variable volume fraction of the first material in an optimal composite. Accordingly, the G_m-closures (instead of G-closures) are used for the relaxation.

Notion of G-Closure is Sufficient (but Not Necessary) for Relaxation

The problem of existence of optimal controls for lower weakly continuous functionals is trivially solved when one passes from the set $\mathcal{U}$ to its G- or G_m-closure. However, this problem is in fact replaced by another one: how to find the G_m-closure itself. The last problem is by no means simpler than the problem of relaxation. On the contrary, some optimization problems are less complex than the problem of the description of the G-closure. The problem in Chapter 4 is an example of relaxation without complete description of the G_m-closure.

More direct approaches to the relaxation of the optimal problems than the G-closure procedure have been considered in a number of papers, beginning with (Lurie, 1970b; Murat, 1972; Tartar, 1975; Murat, 1977; Raĭtum, 1978). It was noted that the given values of the pair of current and gradient fields in the constitutive relations do not determine the tensor of properties completely. One can consider the equivalence of the class of anisotropic tensors (Raĭtum, 1978) that produces the same current in a given gradient field. This idea (the weak G-closure; see Chapter 3) allows one to prove that only laminates can be used for relaxation.

A different approach (Lurie, 1990b; Lurie, 1994) is based on direct bounds of the value of the minimax augmented functional that replaces the original minimal problem with differential constraints. It was suggested to use the saddle-type Lagrangians for the upper and lower bound of the min-max variational problem for an augmented functional. That approach enables one to immediately find the bounds for a minimax problem.

The Detection of Ill-Posed Extremal Problems. To find whether the optimization problem is ill-posed and needs relaxation, one can use the Weierstrass test. For example, one can consider the variation in a strip and detect the existence of a forbidden region where the test fails. This approach was suggested in early papers (Lurie, 1963; Lurie, 1967) where the Weierstrass-type conditions were derived and the forbidden region was detected in a problem similar to the one considered here.

5.2   Solution to an Optimal Design Problem

We are turning toward a construction of minimizers for problem (5.1.1), (5.1.2). Here, we introduce the formal solution scheme to the problem and demonstrate that optimal designs correspond to the laminate structures. We describe a method suggested in (Cherkaev, 1994): The constrained extremal problem is reduced to a minimal unconstrained variational problem similar to the problem of energy minimization.

For definiteness, we consider a weakly continuous functional of the type

$$I(w) = \int_{\Omega} F(w) + \oint_{\partial\Omega} \Phi(w, w_n), \qquad (5.2.1)$$

where F and Φ are differentiable functions. As an example, one can think of the problem of minimizing a temperature in a region inside the domain Ω.

5.2.1   Augmented Functional

Consider the minimization problem

$$I = \min_{\sigma} \int_{\Omega} F(w) + \oint_{\partial\Omega} \Phi(w, w_n) \qquad (5.2.2)$$

where w is the solution to system (5.1.1). Assume that the functional (5.2.1) is weakly continuous. Therefore, the solution to (5.2.1) exists if the set $\mathcal{U}$ of controls is G-closed. Here we find that all optimal structures are laminates.

In dealing with the constrained optimization problem we use Lagrange multipliers to take into account the differential constraints. We construct the augmented functional I_A. Namely, we add to I the two differential equations (5.1.1) multiplied by the scalar Lagrange multiplier $\lambda = \lambda(\mathbf{x})$ and vector Lagrange multiplier $\boldsymbol{\kappa} = \boldsymbol{\kappa}(\mathbf{x})$, respectively,

$$G = \lambda(\nabla \cdot \mathbf{j} - q) + \boldsymbol{\kappa} \cdot (\nabla w - \boldsymbol{\sigma}^{-1}\mathbf{j}).$$

This way we obtain the augmented functional

$$I_A = \min_{\boldsymbol{\sigma}} \min_{w,\mathbf{j}} \max_{\lambda,\boldsymbol{\kappa}} \int_\Omega (F + G) + \oint_{\partial\Omega} \Phi(w, w_n). \qquad (5.2.3)$$

The boundary conditions (5.1.1) are assumed. The value of the augmented functionalI_A is equal to the value of I; see, for example, (Gelfand and Fomin, 1963).

The Adjoint Problem

To find the Lagrange multipliers λ and $\boldsymbol{\kappa}$ we calculate the variation of the augmented functional (5.2.3) caused by the variation δw of w and the variation $\delta\mathbf{j}$ of $\mathbf{j}$. The multipliers satisfy the stationarity condition $\delta I_A = 0$, where

$$\delta I_A = \int_\Omega \left(\frac{\partial F}{\partial w}\delta w + \lambda(\nabla \cdot \delta\mathbf{j}) + \boldsymbol{\kappa} \cdot (\nabla\delta w - \boldsymbol{\sigma}^{-1}\delta\mathbf{j}) \right)$$
$$+ \oint_{\partial\Omega} \left(\frac{\partial\Phi}{\partial w}\delta w + \frac{\partial\Phi}{\partial j_n}\delta j_n \right).$$

We apply the standard variational technique, transforming the terms on the right-hand side using Green's theorem:

$$\int_\Omega (a\nabla \cdot \mathbf{b} + \mathbf{b} \cdot \nabla a) = \oint_{\partial\Omega} ab_n$$

where $a, \mathbf{b}$ are differentiable scalar and vector functions, respectively; $b_n = \mathbf{b} \cdot \mathbf{n}$ is the normal component of $\mathbf{b}$, and Ω is a domain with a smooth boundary $\partial\Omega$.

We apply this theorem to the two terms on the right-hand side of the expression for δI_A:

$$\int_\Omega \lambda\nabla \cdot \delta\mathbf{j} = -\int_\Omega \delta\mathbf{j} \cdot \nabla\lambda + \oint_{\partial\Omega} \lambda\delta j_n,$$
$$\int_\Omega \boldsymbol{\kappa} \cdot \nabla\delta w = -\int_\Omega \delta w\nabla \cdot \boldsymbol{\kappa} + \oint_{\partial\Omega} \kappa_n\delta w.$$

This brings δI_A to the form

$$\delta I_A = \int_\Omega (\mathcal{A}\delta w + \mathcal{B} \cdot \delta\mathbf{j}) + \oint_{\partial\Omega} (\mathcal{C}\delta w + \mathcal{D}\delta j_n),$$

where
$$\mathcal{A} = -\nabla\cdot(\kappa) + \frac{\partial F}{\partial w}, \qquad \mathcal{B} = -\sigma^{-1}\kappa - \nabla\lambda,$$
$$\mathcal{C} = \frac{\partial \Phi}{\partial w} + \kappa_n, \qquad \mathcal{D} = \frac{\partial \Phi}{\partial j_n} + \lambda.$$

The stationarity $\delta I_A = 0$ of I_A with respect to the variations δw and $\delta \mathbf{j}$, together with the boundary conditions (5.1.1), implies

$$\mathcal{A} = 0 \ \ \text{in} \ \Omega, \qquad \mathcal{B} = 0 \ \ \text{in} \ \Omega,$$
$$\mathcal{C} = 0 \ \ \text{on} \ \partial\Omega_1, \qquad \mathcal{D} = 0 \ \ \text{on} \ \partial\Omega_2.$$

The pair λ, κ satisfies the adjoint problem (compare with (5.1.1))

$$\left.\begin{array}{c} \nabla \cdot \kappa = -\frac{\partial F}{\partial w} \\[4pt] \sigma(\chi)\nabla\lambda = -\kappa \end{array}\right\} \ \text{in} \ \Omega, \qquad \begin{array}{l} \lambda = -\frac{\partial \Phi}{\partial j_n} \ \ \text{on} \ \partial\Omega_1, \\[4pt] \kappa_n = -\frac{\partial \Phi}{\partial w} \ \ \text{on} \ \partial\Omega_2. \end{array} \qquad (5.2.4)$$

To obtain these boundary conditions we notice that $\delta w|_{\partial\Omega_1} = 0$ and $\delta j_n|_{\partial\Omega_2} = 0$.

System (5.2.4) allows us to determine λ for given σ and w. The multipliers λ and κ (5.2.4) are similar to the variables w and $-\mathbf{j}$ (5.1.1), respectively, because they correspond to the same inhomogeneous layout σ but to a different right-hand-side term and boundary conditions.

Self-Adjoint Problem. An important special case of problem (5.2.3) is the coincidence of λ and w. This happens if

$$\frac{\partial F}{\partial w} = q, \qquad \frac{\partial \Phi}{\partial j_n} = -\rho_1, \qquad \frac{\partial \Phi}{\partial w} = \rho_2.$$

The solutions to these problems also coincide:

$$\lambda = w, \quad \mathbf{j} = -\kappa. \qquad (5.2.5)$$

In this case we call the problem *self-adjoint*. The self-adjoint problem corresponds to the special functional (5.2.2):

$$F(w) = qw, \qquad \Phi = \begin{cases} -\rho_1 w & \text{on} \ \partial\Omega_1, \\ \rho_2 j_n & \text{on} \ \partial\Omega_2. \end{cases} \qquad (5.2.6)$$

The reader can check that this functional represents the total energy stored in the body.

Also, we deal with a self-adjoint problem when the problem for a Lagrange multiplier λ differs only by sign from the variable w:

$$\lambda = -w, \quad \mathbf{j} = \kappa. \qquad (5.2.7)$$

This happens when the negative of the value of the work is minimized, or, equivalently, when the work is maximized; in this case both integrands F and Φ are negatives of those given by (5.2.6).

The self-adjoint problems minimize or maximize the energy of a structure; they were discussed in Chapter 4.

5.2.2   The Local Problem

Equations (5.1.1) and (5.2.4) determine two of three unknown functions w, λ, and σ. The remaining problem is to find an additional relation among σ, w, and λ.

Remark 5.2.1 *In the exposition, we will associate the conductivity problem with the potential w assuming that $\mathbf{j}$ is computed from (5.1.1). Similarly, we associate the adjoint problem with the potential λ assuming that κ is computed from (5.2.4).*

We use the homogenization approach. Let us average the augmented functional. We bring problem (5.2.2) to a symmetric form by integration by parts of the term that contains the control σ:

$$I_A = \min_{\sigma} \min_{w} \max_{\lambda} \int_{\Omega} (F(w) - q\lambda - \nabla\lambda \cdot \sigma\nabla w) + \text{boundary terms}, \quad (5.2.8)$$

and we apply the averaging operator (2.2.2) $\langle \ \rangle$ to it.

In performing the averaging we replace (5.2.8) by the homogenized problem

$$I_A^{\varepsilon} = \min_{\sigma} \min_{\langle w \rangle} \max_{\langle \lambda \rangle} P,$$

where

$$P = \int_{\Omega} (\langle F(w) \rangle - \langle q\lambda \rangle - \langle \nabla\lambda\sigma \cdot \nabla w \rangle) + \text{boundary terms}. \quad (5.2.9)$$

The cost of the homogenized problem is arbitrarily close to the cost of the original problem:

$$I_A^{\varepsilon} \to I_A \quad \text{when } \varepsilon \to 0.$$

It is easy to compute the first two terms of the integrand in the volume integral on the right-hand side of (5.2.9), supposing that the functions $F(w)$ and q are continuous:

$$\langle F(w) \rangle = F(\langle w \rangle) + o(\varepsilon) \quad \text{and} \quad \langle q \cdot \lambda \rangle = q \cdot \langle \lambda \rangle + o(\varepsilon).$$

Calculation of the remaining term,

$$\langle \nabla\lambda\sigma \cdot \nabla w \rangle = K(\nabla\lambda, \nabla w),$$

where

$$K(\nabla\lambda, \nabla w) = \frac{1}{\|\omega\|} \min_{\sigma(\mathbf{x})} \min_{w} \max_{\lambda} \int_{\omega} \nabla(-\lambda)(\sigma(\mathbf{x}))\nabla w, \quad (5.2.10)$$

requires homogenization. Term K should be expressed as a function of the averaged fields

$$\mathbf{p} = \langle \nabla w \rangle \quad \text{and} \quad \mathbf{q} = \langle \nabla\lambda \rangle$$

as $K = K(\mathbf{p}, \mathbf{q})$.

This problem is called the *local problem*. It asks for an optimal layout σ in the element of periodicity ω (considered a neighborhood of a point of the body Ω in the large scale).

Fixed Volume Fractions

First, we solve an auxiliary problem. Consider the functional K (5.2.10) and assume that the volume fraction m of the first material in the structure is also given. The cost of (5.2.10) can be expressed through the tensor of effective properties σ_*. It takes the form

$$J(\mathbf{p}, \mathbf{q}, m) = \min_{\sigma_* \in G_m U} \langle \mathbf{p} \cdot \sigma_* \mathbf{q} \rangle$$

and asks for the best structure of a composite with the fixed volume fractions of components. The volume fraction m will be determined later to minimize the cost of (5.2.10). Note that

$$K(\mathbf{p}, \mathbf{q}) = \min_{m \in [0,1]} J(\mathbf{p}, \mathbf{q}, m). \tag{5.2.11}$$

We calculate $J(\mathbf{p}, \mathbf{q}, m)$, referring only to the Wiener bounds; the eigenvalues λ_i of any effective tensor σ are bounded as

$$\sigma_h \leq \lambda_1 \leq \lambda_2 \leq \lambda_3 \leq \sigma_a. \tag{5.2.12}$$

Note that the bounds of the G_m-closure restrict eigenvalues of the effective tensor but not its orientation. Although these bounds are uncoupled and therefore not complete, they provide enough information to solve the optimization problem.

Let us analyze the bilinear form J. It depends only on magnitudes $|\mathbf{p}|$ and $|\mathbf{q}|$ and the angle 2θ between them:

$$J = \min_{\sigma \in G_m U} \mathbf{p} \cdot \sigma \mathbf{q} = J(|\mathbf{p}|, \, |\mathbf{q}|, \theta),$$

where

$$\theta = \frac{1}{2} \arccos \frac{|\mathbf{p} \cdot \mathbf{q}|}{|\mathbf{p}| \, |\mathbf{q}|}. \tag{5.2.13}$$

Next, J is proportional to the magnitudes $|\mathbf{p}|$ and $|\mathbf{q}|$ of both external fields $\mathbf{p}$ and $\mathbf{q}$; it can be rewritten in the form

$$J(\mathbf{p}, \mathbf{q}, m) = |\mathbf{p}||\mathbf{q}| \min_{\sigma_* \in G_m U} \mathbf{i}_p \cdot \sigma_* \mathbf{i}_q, \tag{5.2.14}$$

where the unit vectors $\mathbf{i}_p$ and $\mathbf{i}_q$ show the directions of $\mathbf{p}$ and $\mathbf{q}$:

$$\mathbf{i}_p = \frac{1}{|\mathbf{p}|} \mathbf{p}, \quad \mathbf{i}_q = \frac{1}{|\mathbf{q}|} \mathbf{q}. \tag{5.2.15}$$

Thus (5.2.14) is reduced to the minimization of a nondiagonal pq-component of the tensor $\sigma \in G_m U$ in the nonorthogonal coordinate system of axes $\mathbf{i}_p$ and $\mathbf{i}_q$.

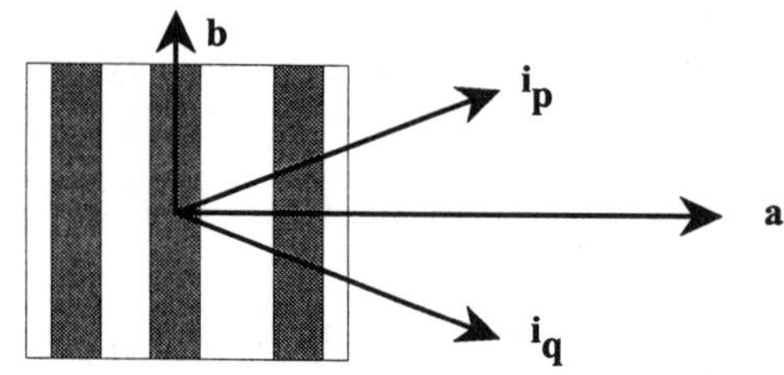

FIGURE 5.1. The local problem.

Diagonalization

To find the optimal orientation of the tensor σ we introduce a pair of vectors

$$\mathbf{a} = \mathbf{i}_p + \mathbf{i}_q, \qquad \mathbf{b} = \mathbf{i}_p - \mathbf{i}_q.$$

Due to normalization (5.2.15), these fields are orthogonal: $\mathbf{a} \cdot \mathbf{b} = 0$.

The magnitudes of $\mathbf{a}$ and $\mathbf{b}$ depend only on θ (see (5.2.13)). They equal $|\mathbf{a}| = 2\cos\theta$, $|\mathbf{b}| = 2\sin\theta$ (see Figure 5.1).

In the introduced notation, the bilinear form J takes the form of the difference of two quadratic functions,

$$J = \min_{\sigma \in G_m U} |\mathbf{p}||\mathbf{q}|[\mathbf{a} \cdot \sigma\mathbf{a} - \mathbf{b} \cdot \sigma\mathbf{b}]. \tag{5.2.16}$$

Optimal Effective Tensor

The minimization of J over $\sigma \in G_m U$ requires the optimal choice of the eigenvectors and eigenvalues of σ. To minimize J, we direct the eigenvectors of σ as follows. The eigenvector of σ that corresponds to the minimal eigenvalue $\lambda_{\min}$ is oriented parallel to $\mathbf{a}$, and the eigenvector of σ that corresponds to the maximal eigenvalue $\lambda_{\max}$ is oriented parallel to $\mathbf{b}$:

$$J \geq |\mathbf{p}||\mathbf{q}|(\lambda_{\min}\mathbf{a}^2 - \lambda_{\max}\mathbf{b}^2).$$

Next, J decreases if $\lambda_{\min}$ coincides with its lower bound, and $\lambda_{\max}$ coincides with its upper bound; the bounds are given by (5.2.12). We have

$$J \geq |\mathbf{p}||\mathbf{q}|(\sigma_h\mathbf{a}^2 - \sigma_a\mathbf{b}^2). \tag{5.2.17}$$

Remark 5.2.2 *The value of the intermediate eigenvalue is irrelevant. This feature recalls the definition of the weak G-closure (Chapter 3). The weak G-closure is adequate for this problem because it deals with an arbitrary pair of the field and the current.*

Formula (5.2.17) displays the basic qualitative property of an optimal composite:

> The optimal effective tensor σ_* possesses maximal difference between weighted maximal and minimal eigenvalues. The optimal structures are extremely anisotropic.

Substituting the values of the original fields $\mathbf{a}$ and $\mathbf{b}$ into (5.2.17), we find that J is bounded from below as follows:

$$J \geq 2[(\sigma_a + \sigma_h)(|\mathbf{p}||\mathbf{q}| - (\sigma_a - \sigma_h)\mathbf{p} \cdot \mathbf{q}). \tag{5.2.18}$$

Inequality (5.2.18) is valid for all composites independent of their structure.

The fields $\mathbf{p}$ and $\mathbf{q}$ lie in the plane of maximal and minimal eigenvalues of $\boldsymbol{\sigma}_*$. The direction of $\lambda_{\min}$ bisects the angle between the vectors $\mathbf{p}$ and $\mathbf{q}$. Qualitatively speaking, $\mathbf{p}$ shows the attainable direction of currents, $\mathbf{q}$ shows the desired direction of them. The direction of maximal conductivity in the optimal structure bisects this angle. This rule provides a compromise between the availability and the desire.

Remark 5.2.3 *Note that the self-adjoint problem (5.2.5), where* $\mathbf{p} = \mathbf{q}$, *corresponds to* $\mathbf{a} = 0$ *and the problem where* $\mathbf{p} \equiv -\mathbf{q}$ *(5.2.7) corresponds to* $\mathbf{b} = 0$.

Optimal Structures. Inequality (5.2.18) is strict; an appropriately oriented laminate provides the minimal value of J. Indeed, the laminates have simultaneously the maximal conductivity σ_a in a direction(s) (along the layers) and the minimal conductivity σ_h in the perpendicular direction (across the layers).

The optimal laminates are oriented so that the normal $\mathbf{n}$ coincides with the direction of the field $\mathbf{a}$, and the tangent $\mathbf{t}$ coincides with the direction of $\mathbf{b}$. The cost $J(\boldsymbol{\sigma}_{\mathrm{lam}})$ of the local problem for laminate structure $\boldsymbol{\sigma}_{\mathrm{lam}}$ coincides with the bound (5.2.18).

5.2.3 Solution in the Large Scale

The solution to the auxiliary local problem allows us to compute K from (5.2.11). We denote $m_1 = m$, $m_2 = 1 - m$; assume that $\sigma_2 > \sigma_1$; and calculate an optimal value of the volume fractions of materials in the laminates. We have

$$\frac{K}{|\mathbf{p}|\,|\mathbf{q}|} = \min_{m \in [0,1]} \left(\lambda_{\min}(m)\mathbf{a}^2 - \lambda_{\max}(m)\mathbf{b}^2 \right)$$

$$= \min_{m \in [0,1]} \left(\frac{\sigma_1 \sigma_2}{m\sigma_2 + (1-m)\sigma_1} \mathbf{a}^2 + (m\sigma_1 + (1-m)\sigma_2)\mathbf{b}^2 \right).$$

The optimal value m^{opt} of m depends only on the ratio between $|\mathbf{a}|$ and $|\mathbf{b}|$,

$$\frac{|\mathbf{a}|}{|\mathbf{b}|} = \cot \theta, \tag{5.2.19}$$

and is equal to

$$
m^{\text{opt}} = \begin{cases}
0 & \text{if } \cot\theta \le \sqrt{\frac{\sigma_1}{\sigma_2}}, \\
\frac{\sqrt{\sigma_1\sigma_2}}{\sigma_2-\sigma_1}\left(\cot\theta - \sqrt{\frac{\sigma_1}{\sigma_2}}\right) & \text{if } \sqrt{\frac{\sigma_1}{\sigma_2}} \le \cot\theta \le \sqrt{\frac{\sigma_2}{\sigma_1}}, \\
1 & \text{if } \cot\theta \ge \sqrt{\frac{\sigma_2}{\sigma_1}}.
\end{cases}
\tag{5.2.20}
$$

Equation (5.2.20) says that the optimal concentration of materials in the laminates depends only on the angle θ.

We find the optimal value of the functional $K = J(m^{\text{opt}})$:

$$
\frac{K}{|\mathbf{p}|\,|\mathbf{q}|} = \begin{cases}
\sigma_2 \cos 2\theta & \text{if } \cot\theta \le \sqrt{\frac{\sigma_1}{\sigma_2}}, \\
(\sigma_1 + \sigma_2)\sin^2\theta & \text{if } \sqrt{\frac{\sigma_1}{\sigma_2}} \le \cot\theta \le \sqrt{\frac{\sigma_2}{\sigma_1}}, \\
\sigma_1 \cos 2\theta & \text{if } \cot\theta \ge \sqrt{\frac{\sigma_2}{\sigma_1}}.
\end{cases}
\tag{5.2.21}
$$

To complete the solution, it remains to pass to the original notation $\nabla\langle w\rangle = \mathbf{p}$ and $\nabla\langle\lambda\rangle = \mathbf{q}$, substitute the value of the local problem into the functional (5.2.3), and find the Euler–Lagrange equations of the problem:

$$
I_A = \min_{\langle w\rangle}\max_{\langle\lambda\rangle}\int_{\mathcal{O}}[F(\langle w\rangle) + \langle\lambda\rangle q + K].
$$

Note that the equations for $\langle w\rangle$ and $\langle\lambda\rangle$ are coupled because the optimal properties depend on both of them.

Numerical Procedure

Practically, we have used a different procedure for the numerical solution; see (Gibiansky, Lurie, and Cherkaev, 1988). The iterative method has been organized as follows:

1. Given a layout of $\boldsymbol{\sigma}$, we compute the solution w of problem (5.1.1) and the solution λ of the adjoint problem (5.2.4).

2. The optimal layouts $m(\mathbf{x})$ and $\theta(\mathbf{x})$ is found from (5.2.19). Then we return to the first step.

5.3 Reducing to a Minimum Variational Problem

Duality

In this section, we reformulate the local minimax problem (5.2.16) as a minimal variational problem. This way we reduce the problem to the type discussed in Chapter 4. The relaxation is obtained by the convexification. The transformation is done by the Legendre transform.

The Legendre transform replaces the saddle Lagrangian (5.2.16) with a convex Lagrangian. Recall (chapter 1) that the Legendre transform or Young–Fenchel transform $f^*(x^*)$ (1.3.25) of a concave function $f(x)$ is given by (1.3.25),

$$f^*(x^*) = \min_x [x\, x^* - f(x)],$$

where $f^*(x^*)$ is the conjugate to $f(x)$.

Consider the Legendre transform of the function

$$f(b) = \frac{\lambda_1}{2} a^2 - \frac{\lambda_2}{2} b^2,$$

where a is a parameter. We compute

$$f_b^*(a, b^*) = \min_b [b\, b^* - f(a, b)] = \frac{\lambda_1}{2}(a)^2 + \frac{1}{2\lambda_2} b^{*2}, \quad b^* = \lambda_2 b. \qquad (5.3.1)$$

Notice that $f_b^*(a, b^*)$ is a convex function of a and b^*

Consider the normalized local minimax problem (see (5.2.16))

$$J = \min_{\boldsymbol{\sigma}_* \in G_m U} \left\{ \frac{1}{2} \min_{\mathbf{a}} \max_{\mathbf{b}} \left\{ \langle \mathbf{a} \rangle \cdot \boldsymbol{\sigma}_* \langle \mathbf{a} \rangle - \langle \mathbf{b} \rangle \cdot \boldsymbol{\sigma}_* \langle \mathbf{b} \rangle \right\} \right\}, \qquad (5.3.2)$$

where we put $|\mathbf{p}| = |\mathbf{q}| = 1$,

To reduce (5.3.2) to a minimum problem, we perform the Legendre transform on the variable $\mathbf{b}$. We replace the problem of maximization of conductivity in the direction $\mathbf{b}$ with the problem of the minimization of the resistance in this direction. This transform does not include optimization; we simply change the variable $\mathbf{b}$ to its conjugate variable $\mathbf{b}^*$.

The dual to $\mathbf{b}$ variable $\mathbf{b}^* = \mathbf{j}$ is (see (5.3.1)):

$$\mathbf{j} = \boldsymbol{\sigma}\mathbf{b}. \qquad (5.3.3)$$

If we substitute (5.3.3) into (5.3.2), the latter becomes the following minimum problem:

$$R_*(\mathbf{a}, \mathbf{j}) = \min_{\boldsymbol{\sigma}} R(\mathbf{a}, \mathbf{j}, \boldsymbol{\sigma}),$$

$$R(\mathbf{a}, \mathbf{j}, \boldsymbol{\sigma}) = \left\langle \left(-\mathbf{b} \cdot \mathbf{j} + \frac{1}{2} \mathbf{a} \cdot \boldsymbol{\sigma}\mathbf{a} + \frac{1}{2}\mathbf{j} \cdot \boldsymbol{\sigma}^{-1}\mathbf{j} \right) \right\rangle. \qquad (5.3.4)$$

Recall that the eigenvectors of $\boldsymbol{\sigma}$ are oriented along $\mathbf{a}$ and $\mathbf{b}$. Therefore, the vectors $\mathbf{j}$ and $\mathbf{b}$ are parallel, and $\mathbf{j}$ and $\mathbf{a}$ are orthogonal.

The reformulated local problem (5.3.4) asks for a medium that stores *the minimal sum of the energy and the complementary energy* caused by two orthogonal external fields.

Minimal Variational Problem

The function R is a solution to a variational problem for unknown fields and layout, similar to the variational problem of Chapter 4.

This local variational problem can be rewritten as:

$$R(\mathbf{a}, \mathbf{b}, m) = \min_{\chi} \min_{\alpha \in \mathcal{A}} \min_{\beta \in \mathcal{B}} J_R(\alpha, \beta, \chi),$$

where χ is subject to the constraint $\langle \chi \rangle = m$, and

$$J_R(\alpha, \beta, \chi) = \langle -\alpha \cdot \beta + \frac{1}{2}\alpha \cdot \sigma(\chi)\alpha + \frac{1}{2}\beta \cdot \sigma(\chi)^{-1}\beta \rangle, \qquad (5.3.5)$$

$\mathcal{A}$ is the set of periodic gradient fields with mean value $\mathbf{a}$,

$$\mathcal{A} = \{\alpha : \ \nabla \times \alpha = 0, \ \langle \alpha \rangle = \mathbf{a}, \ \alpha(\mathbf{x}) \text{ is } \Omega - \text{periodic}\},$$

and $\mathcal{B}$ is the set of divergencefree periodic vectors with the mean value $\mathbf{b}$,

$$\mathcal{B} = \{\beta : \ \nabla \cdot \beta = 0, \ \langle \beta \rangle = \mathbf{j}, \ \beta(\mathbf{x}) \text{ is } \Omega - \text{periodic}\}.$$

Note that variational problem (5.3.5) does not contain differential constraints. As with the problem in Chapter 4, this problem can be analyzed by classical variational methods.

Remark 5.3.1 *The minimal form J_R of the Lagrangian is symmetric in the sense that the field $\mathbf{a}$ and the current $\mathbf{j}$ enter into the problem in the same way. One expects this property because the original conductivity problem could be formulated in two equivalent ways: By using a field potential or a current potential; the result is unrelatedto this choice. Both self-adjoint cases correspond to either $\mathbf{a} = 0$ or $\mathbf{j} = 0$. The problem is reduced to minimization of the energy in one of the dual forms, as one would expect.*

The Transferred Problem as a Nonconvex Variational Problem

Problem (5.3.5) depends on χ and it needs a relaxation. The problem can be relaxed by convexification. As in the problem in Chapter 4, we can exclude the control χ by interchanging the minimal operations:

$$R(\mathbf{a}, \mathbf{b}, m) = \min_{\sigma(\chi)} \min_{\alpha \in \mathcal{A}} \min_{\beta \in \mathcal{B}} J_R = \min_{\alpha \in \mathcal{A}} \min_{\beta \in \mathcal{B}} \min_{\sigma(\chi)} J_R.$$

The inner minimum $\min_{\sigma(\chi)} J_R$ can be easily computed, because χ takes only two values: zero and one. The problem becomes

$$R(\mathbf{a}, \mathbf{b}, m) = \min_{\mathbf{a} \in \mathcal{A}} \min_{\mathbf{j} \in \mathcal{B}} J_{RR},$$

where

$$J_{RR}(\alpha, \beta) = \left\langle -\alpha \cdot \beta + \frac{1}{2} \min \left\{ \sigma_1 |\alpha|^2 + \frac{1}{\sigma_1}|\beta|^2, \ \sigma_2 |\alpha|^2 + \frac{1}{\sigma_2}|\beta|^2 \right\} \right\rangle.$$

$$(5.3.6)$$

Notice that $J_{RR}(\boldsymbol{\alpha}, \boldsymbol{\beta})$ is a nonconvex function of its arguments.

To relax problem (5.3.5) we again perform the convexification of the Lagrangian $J_{RR}(\boldsymbol{\alpha}, \boldsymbol{\beta})$. It is a two-well Lagrangian, and therefore its convex envelope is supported by two points that belong to different wells. Following the calculation in Chapter 4, we find these points and show that the convex envelope is equal to

$$\mathcal{C}J_{RR}(\mathbf{a}, \mathbf{j}) = -\mathbf{a} \cdot \mathbf{j} + \frac{1}{2} \min_{m \in [0,1]} \left\{ \sigma_h(m) \, \mathbf{a}^2 + \frac{1}{\sigma_a(m)} \, \mathbf{j}^2 \right\},$$

where $(\cdot)_h$ and $(\cdot)_a$ are again the harmonic and the arithmetic means. To obtain the term $\frac{1}{\sigma_a(m)}$, we use the identity $\left(\frac{1}{\sigma}\right)_h = \frac{1}{\sigma_a}$.

To compute the first term $\langle \boldsymbol{\alpha} \cdot \boldsymbol{\beta} \rangle = \mathbf{a} \cdot \mathbf{j}$ on the right-hand side, we use the property of divergencefree and curlfree fields called *compensated compactness* (see Chapter 8 and (Tartar, 1979a; Murat, 1981)).

The convex envelope of the Lagrangian (5.3.6) is again attainable, this time due to the orthogonality of $\mathbf{a}$ and $\mathbf{j}$. Indeed, the laminates have conductivity σ_h and σ_a in orthogonal directions. The structure can be oriented so that the axis σ_h is directed along $\mathbf{a}$ and the axis σ_a along $\mathbf{j}$.

The Legendre transform is an involution: After the convexification is performed, we perform the transform with respect to $\mathbf{j}$ to bring the problem back to the form (5.2.16).

Remark 5.3.2 *Another approach to these problems was developed in (Lurie, 1990b; Lurie, 1994). Instead of transforming the functional, the author developed a method for finding an saddle-type envelope of the Lagrangian of a minimax problem. Both approaches lead to similar results, as expected.*

Summary of the Method

Let us outline the basic steps of the suggested method of relaxation. We assume that the problem is described by self-adjoint elliptic equations; the shape of the domain, boundary conditions, and external loadings (right-hand-side terms) are fixed; the minimizing functional is weakly lower semicontinuous. To relax the problem, we follow this procedure:

1. Formulate the local problem as a min-min-max problem for bilinear form of the field and gradient of the Lagrange multiplier; the coefficients of the form are the effective properties of the composite.

2. Normalize the bilinear form and transform it to the diagonal form by introducing new potentials. A transformed problem asks for minimization of the difference of energies caused by two orthogonal fields by the layout of the materials.

3. Use the Legendre transform to reduce the problem to a minimal variational problem, ending up with a problem of the minimization of the sum of an energy and a complementary energy.

4. Use convexification to bound the minimized nonconvex functional from below and find a minimizing sequence (i.e., the optimal microstructure) that bounds it from above.

5. Return to the original notation performing the Legendre transform of the convexified problem.

5.4   Examples

Example 5.4.1 Consider a thin circular cylindrical shell of height h and radius r made of two conducting materials σ_1 and σ_2. Suppose that its upper and lower faces are kept by different potentials. Let us find a layout of materials that maximizes the circumferential component of the current; this component is zero for isotropic materials.

Introduce the rectangular coordinates z, Θ on the surface. The surface Ω is the rectangle

$$0 \le z \le h, \quad 0 \le \Theta \le 2\pi r.$$

The potential $w(z, \Theta)$ is the solution to the problem

$$\nabla \cdot \sigma \nabla w = 0 \quad \text{in } \Omega,$$

with the boundary conditions

$$w(0, \Theta) = 0, \ w(h, \Theta) = U, \ w(z, 0) = w(z, \pi), \ \frac{\partial w(z, 0)}{\partial \Theta} = \frac{\partial w(z, \pi)}{\partial \Theta}.$$

The maximizing functional is the circumferential component of the current; it is written as

$$I = \int_\Omega \mathbf{i}_\Theta \cdot \sigma \nabla w,$$

where $\mathbf{i}_\Theta$ is a unit vector that points in the circumferential direction.

Applying the preceding analysis, we find that the Lagrange multiplier λ satisfies the problem (see (5.2.4))

$$\nabla \cdot \sigma (\nabla \lambda + \mathbf{i}_\Theta) = 0 \quad \text{in } \Omega,$$

with boundary conditions

$$\lambda(0, \Theta) = 0, \qquad \lambda(h, \Theta) = 0,$$
$$\lambda(z, 0) = \lambda(z, \pi), \qquad \frac{\partial \lambda(z, 0)}{\partial \Theta} = \frac{\partial \lambda(z, \pi)}{\partial \Theta}.$$

The solution to the averaged problem is easily found. We observe that the constant tensor $\boldsymbol{\sigma}_*$ and constant fields

$$\langle \nabla \lambda \rangle = -\mathbf{i}_\Theta \quad \text{and} \quad \langle \nabla w \rangle = \frac{U}{h} \mathbf{i}_z, \quad \boldsymbol{\sigma}_* = \text{constant}(\mathbf{x}),$$

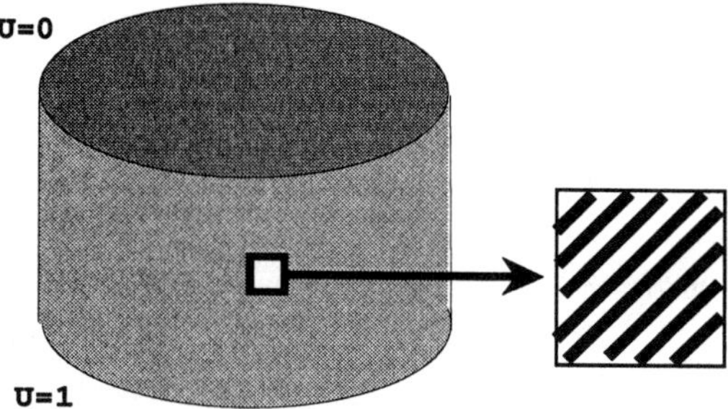

FIGURE 5.2. An optimally conducting cylinder.

satisfy the equations and the boundary conditions due to the symmetry of the domain and the special boundary conditions. The same composite is used at each point of the domain.

The angle θ that bisects the direction of the fields $\langle \nabla \lambda \rangle$ and $\langle \nabla w \rangle$ is equal to $\frac{\pi}{4}$. The optimal volume fraction m_{opt} of the first material (see (5.2.21)) is

$$m_{\mathrm{opt}} = \frac{\sqrt{\sigma_1}}{\sqrt{\sigma_1} + \sqrt{\sigma_2}}.$$

The cost K of the local problem (see (5.2.21)) is equal to the maximal current across the acting field. This cost and the functional I are

$$K = \frac{U}{h}(\sqrt{\sigma_1} - \sqrt{\sigma_2})^2, \quad I = \int_\Omega K = 2\pi r(\sqrt{\sigma_1} - \sqrt{\sigma_2})^2 U.$$

Example 5.4.2 The next problem is more advanced. It deals with an inhomogeneous layout of optimal laminates. The problem has been formulated and solved in (Gibiansky et al., 1988); an exposition of the solution can be found in (Lurie and Cherkaev, 1986a).

Consider a circular cylinder ($0 < r < 1$, $0 < z < h$). Suppose that the constant heat flux $-\mathbf{j} \cdot \mathbf{n}$ is applied to the upper face ($z = h$). The lateral surface ($r = 1$) of the cylinder is heat insulated, and the lower face ($z = 0$) is kept at zero temperature.

The cylinder is filled with two materials with heat conductivities σ_1 and σ_2. The steady state is described by the boundary value problem

$$\begin{aligned}
\mathbf{j} = \sigma(r, z)\nabla T, \quad \nabla \cdot \mathbf{j} &= 0, \quad \text{inside the cylinder,} \\
\mathbf{i}_z \cdot \mathbf{j} &= 1, \quad \text{on the upper face,} \\
T &= 0, \quad \text{on the lower face,} \\
\mathbf{i}_r \cdot \mathbf{j} &= 0, \quad \text{on the lateral surface.}
\end{aligned} \tag{5.4.1}$$

Here T is the temperature; $\sigma(r, z) = \chi(r, z)\sigma_1 + (1 - \chi_1(r, z))\sigma_2$; $\chi(r, z)$ is the characteristic function; $\mathbf{i}_r$ and $\mathbf{i}_z$ are the unit vectors directed along the axes of the cylindrical coordinate system.

It is required to lay out the materials in the domain to minimize the functional over the lower face

$$I = \int_0^1 \rho(r)\mathbf{i}_z \cdot \mathbf{j}|_{z=0}\, r\, dr.$$

Here $\rho(r)$ is a weight function. Notice that I is the boundary integral, and the problem asks for the optimization of boundary currents caused by fixed boundary potentials.

In particular, if the weight function $\rho(r)$ is

$$\rho(r) = \begin{cases} 1 & \text{if } 0 < r < r_0, \\ 0 & \text{if } r_0 < 1, \end{cases} \tag{5.4.2}$$

then the problem is transformed to maximization of the heat flux through a circular "window" of radius r_0 on the lower surface of the cylinder.

Assuming that the set of admissible controls contains the initial materials and the composites assembled from them, we relax the problem. The augmented functional of the problem has the following form:

$$J = \int_0^1 \rho(r)\mathbf{i}_z \cdot \mathbf{j}|_{z=0} r\, dr + 2\pi \int_0^1 \int_0^h \lambda \nabla \cdot \boldsymbol{\sigma}_* \nabla T\, r\, dr\, dz. \tag{5.4.3}$$

Varying (5.4.3) with respect to T and $\mathbf{j}$ and taking into consideration the boundary conditions (5.4.1), we obtain a boundary value problem for the conjugate variable λ:

$$\begin{aligned} \boldsymbol{\kappa} = \boldsymbol{\sigma}_* \nabla \lambda, \quad \nabla \cdot \boldsymbol{\kappa} &= 0 && \text{inside the cylinder,} \\ \boldsymbol{\kappa} \cdot \mathbf{i}_r &= 0 && \text{on the lateral surface,} \\ \boldsymbol{\kappa} \cdot \mathbf{i}_z &= 0 && \text{on the upper face,} \\ \lambda &= \rho(r) && \text{on the lower face.} \end{aligned} \tag{5.4.4}$$

Problem (5.4.4) describes the "temperature" λ generated by the prescribed boundary values of λ on the lower face $z = 0$ if the upper face and lateral surface of the cylinder are heat insulated.

According to our analysis, an optimal layout of materials is characterized by a zone of material of low conductivity if the directions of gradients ∇T and $\nabla \lambda$ are close to each other, by a zone of material of high conductivity if the angle between directions of gradients ∇T and $\nabla \lambda$ is close to π, and by an anisotropic zone if the directions of gradients form an angle close to $\frac{\pi}{2}$. In the last zone the normal to the layers divides the angle between the gradients in half; the optimal medium tries to turn the direction of the vector of heat flow in the appropriate direction.

The drafts of the vector lines ∇T and $\nabla \lambda$ are shown in Figure 5.3. It is assumed that the function $\rho(r)$ is defined by (5.4.2). All vector lines ∇T begin on the lower face and end on the upper face (Figure 5.3); all vector lines $\nabla \lambda$ begin on the part of the lower face where $\rho = 0$ and end on the

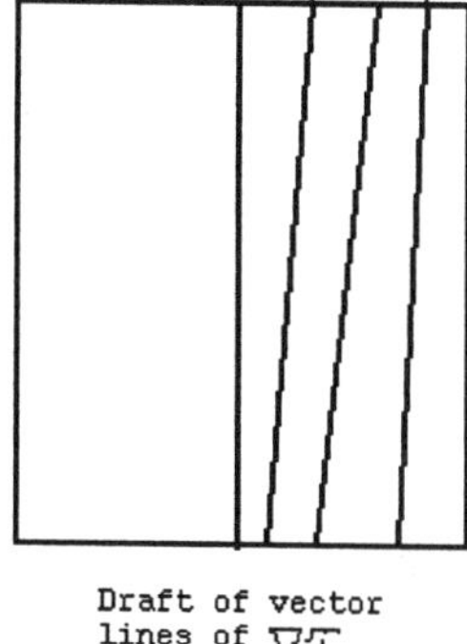

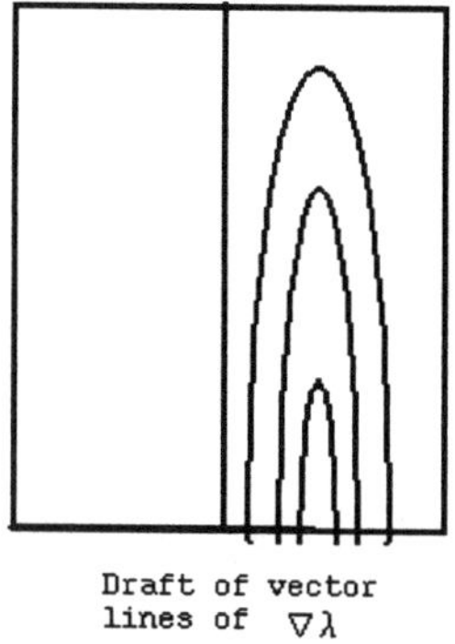

FIGURE 5.3. Draft of the fields ∇w and $\nabla \lambda$.

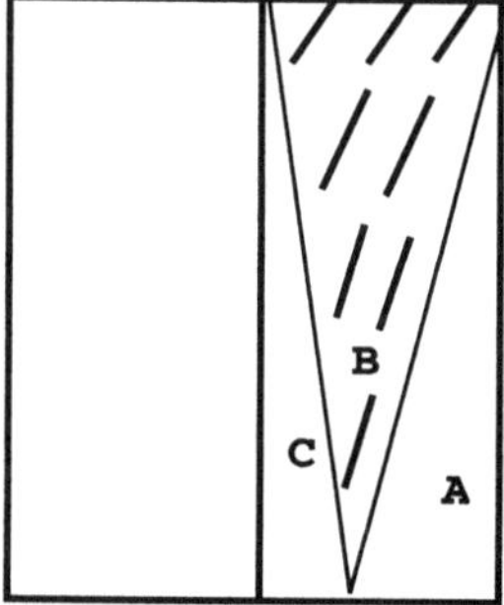

FIGURE 5.4. Draft of the optimal thermolens and the directions of optimally oriented laminates. Zone **A** is filled with an insulator, zone **B** is filled with laminates, and zone **C** is filled with a conductor.

part where $\rho = 1$. The exterior vector line $\nabla \lambda$ passes around the lateral surface and the upper face of the cylinder, returning to the lower face along the axis Oz (Figure 5.3).

It is not difficult to see (Figure 5.4) that the zone **A**, where the angle ψ is close to zero, adjoins the lateral surface of the cylinder. The zone **C** is located near the axes of the cylinder; here this angle is close to $\frac{\pi}{2}$. Zones **A** and **C** join at the endpoint of a "window" through which one should pass the maximal quantity of heat; they are divided by zone **B** where the angle ψ is close to $\frac{\pi}{4}$; this zone adjoins the upper face. In zone **B**, the layered composites are optimal; the directions of the layers are shown by the dotted line.

Physically, the *thermolens* focuses the heat due to the following effects:

1. The heat flow is forced out from zone **A** by the low-conductivity material σ_1.

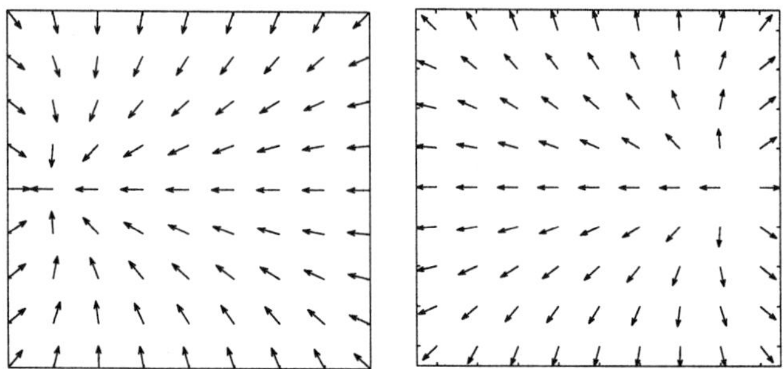

FIGURE 5.5. The fields ∇T and $\nabla \lambda$ in the optimal domain.

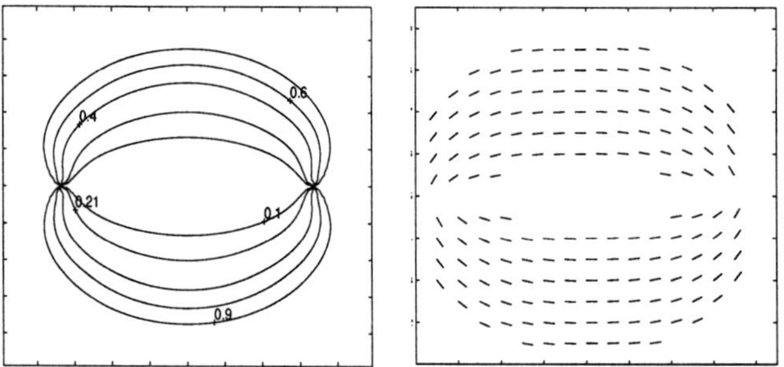

FIGURE 5.6. (Left): the volume fraction of an insulator in the optimal laminates; the periphery is filled by pure insulator and center zone with pure conductor. (Right) Directions of laminates in the composite zone.

2. The heat flow turns in a favorable direction due to refraction in the optimally oriented layers in zone **B**.

3. The heat flow is concentrated in zone **C**, which is occupied by the high-conductivity material σ_2.

The next example of "inhomogeneous heater" demonstrates construction that maximize the temperature in a target point. It was obtained in (Cherkaev and Robbins, 1999) The problem is similar to the previous one.

Example 5.4.3 Consider a domain ($\Omega: \ 0 < x < 1, 0 < y < 1$) filled with two materials with heat conductivities σ_1 and σ_2. Suppose that the boundary $\partial\Omega$ is kept at zero temperature and that the domain has a concentrated source inside.

The equilibrium is described by the boundary value problem

$$\mathbf{j} = \sigma(x,y)\nabla T, \quad \nabla \cdot \mathbf{j} = \delta(x - x_0, y - y_0), \quad \text{in } \Omega$$
$$T = 0, \qquad\qquad\qquad\qquad\qquad\qquad\qquad\quad \text{on } \partial\Omega$$

Here T is the temperature;

$$\boldsymbol{\sigma}(x, y) = \chi(x, y)\sigma_1 + (1 - \chi_1(x, y))\sigma_2;$$

$\delta(x - x_0, y - y_0)$ is the δ-function supported at the point (x_0, y_0) where the source is applied.

It is required to lay out the materials in the domain to maximize the temperature $T(x_t, y_t)$ at a target point $(x_t, y_t) \in \Omega$. The functional is

$$I = \int_\Omega T(x, y)\delta(x - x_t, y - y_t)dx\, dy.$$

Figure 5.5 demonstrates the directions of ∇T and $\nabla\lambda$. Figure 5.6 shows the optimal project.

The optimal layout provides the best conductance between the source and target points and also insulate the boundary. Notice the optimal layout in the proximity of the source and target points: the volume fraction of the insulator varies from zero (in the direction between the point) to one (in the opposite direction).

5.5   Conclusion and Problems

Relaxation and G-Closure

We already mentioned that the solution to an optimal design problem exists if the set of controls is G_m-closures. If this set is known, one could choose the element $\boldsymbol{\sigma}_* \in G_m U$ that provides the minimum of the functional J. However, here we have shown an alternative, straightforward way of solution, so that unnecessary difficulties of a complete description of G_m-closure are avoided.

Recall that the problem of optimal conductivity of a composite was solved using only the simplest bounds of the G_m-closure or by using the weak G-closure (Chapter 3). We deal with a pair of fields $\mathbf{p}$ and $\mathbf{q}$, and we minimize their weighted scalar product $\mathbf{p} \cdot \boldsymbol{\sigma}_* \mathbf{q}$. In this problem, we are looking for a tensor with an extreme element; we are not interested in a description of other elements of it. The extremal tensors belong to a special component of the boundary of $G_m U$; it is enough to describe this component.

Another way to look at this phenomenon is as follows. The constitutive relations in a medium $\langle \mathbf{j} \rangle = \boldsymbol{\sigma}_* \langle \mathbf{e} \rangle$ are completely determined by the projection of the tensor $\boldsymbol{\sigma}_*$ on a plane formed by vectors $\langle \mathbf{j} \rangle$ and $\langle \mathbf{e} \rangle$. Particularly, for any optimization problem only this projection is needed, no matter what the cost functional is. These arguments show again that it is enough to describe only a two-dimensional projection on the plane of eigenvectors of maximal and minimal eigenvalues (the weak G-closure). Moreover, only the "corner" of this two-dimensional set is needed to solve the problem.

Clearly, the results obtained could easily be extended to composites of more than two components, which are considered here for simplicity. The result in the general case is the same: Optimal structures are just laminates that bisect the directions of the fields $\langle \mathbf{p} \rangle$ and $\langle \mathbf{q} \rangle$.

We also could apply this method of relaxation to optimization problems for the processes associated with more general operators than the conductivity operators. The main qualitative result remains the same: The optimal medium has the maximal difference between the energies stored in two different fields. However, we would observe that Wiener bounds are not achievable by a laminate structure. Our next goal, therefore, is to develop a technique for strict bounds and to enlarge the class of available microstructures.

Problems

1. Consider the problem of Section 5.2 for an anisotropic material with variable orientation of the eigenvectors. The G-closure corresponds to all polycrystals. Using the general properties of G-closures, show that the range of eigenvalues of polycrystals lies inside the range of eigenvalues of the original anisotropic material. Analyze the necessary conditions. Do microstructures (polycrystals) appear in an optimal project?

2. Consider the problem of Section 5.2 for three conducting materials. Analyze the necessary conditions. What microstructures are optimal? Does the material with the intermediate conductivity appear in an optimal project?

3. Consider the problem of Section 5.2 with the additional constraint

$$\int_\Omega \chi = M.$$

Derive and analyze the necessary conditions. Can the Lagrange multiplier associated with the constraint change its sign? When does the multiplier equal zero?

4. Investigate the problem of Section 5.2 about the process described by the differential equation

$$\nabla \cdot \sigma(\chi)(\nabla w + \phi(w)) + \mathbf{v} \cdot \nabla w + k(\chi)w = f,$$

where ϕ, k, f are the differentiable functions, and $\mathbf{v}$ is a given vector field. Derive the differential equation for the Lagrange multiplier. What microstructures are optimal? What is the relaxed form of the problem?

5. Consider the example of Section 5.4. Assume that the boundary conditions on the lateral surface are

$$w(h, 1) - \kappa \frac{\partial w(h, r)}{\partial r}\bigg|_{r=1} = 0.$$

Derive the boundary conditions for λ and discuss the dependence of the optimal structure on κ.

6. Qualitatively describe a composite in an optimal conducting rectangular plate $-1 \leq x_1 \leq 1,\quad -1 \leq x_2 \leq 1$, subject to the boundary conditions

$$\frac{\partial w}{\partial x_2}\bigg|_{x_1=-1} = \frac{\partial w}{\partial x_2}\bigg|_{x_1=1} = 0, \quad w|_{x_2=-1} = 0, \quad w|_{x_2=1} = 1$$

with respect to the following optimality requirements:

(a) A domain in the center is kept at lowest temperature

$$\min \int_{\Omega_\varepsilon} T,$$

where Ω_ε is a circle with center at $(0,0)$ and with the radius much smaller than one. Derive the problem for Lagrange multiplier λ and draw a draft of the gradient lines of ∇w and $\nabla \lambda$. Draw a draft of the optimal layout.

(b) The current $\mathbf{j}$ in Ω_ε is directed to minimize the functional

$$\min \int_{\Omega_\varepsilon} \mathbf{j} \cdot \mathbf{t}, \quad \mathbf{t} = \text{constant}.$$

Derive the equation for the Lagrange multiplier λ and draw a draft of ∇w and $\nabla \lambda$. Draw a draft of the optimal layout.

Hint. To draw the draft of an optimal project, sketch the fields ∇w and $\nabla \lambda$ for the isotropic domain and apply necessary conditions to find an approximation to the optimal layout.

Part III

Quasiconvexity and Relaxation

6
Quasiconvexity

6.1 Structural Optimization Problems

We begin the study of optimization problems for equilibria described by a system of linear elliptic partial differential equations. Probably the most practically interesting example is elasticity, but we leave the discussion of this topic for the last part of the book because it deserves special detailed consideration. The exposition of the theory of relaxation of multivariable variational problems can be found in (Ekeland and Temam, 1976; Dacorogna, 1989; Buttazzo, 1989; Pedregal, 1997; Roubíček, 1997). We are dealing with the Lagrangians that depend on the bulk energy; Lagrangians that depend on surface energy were considered in such papers as (Fonseca, 1992; Kohn and Müller, 1994; Klouček and Luskin, 1994; Dolzmann and Müller, 1995; Taylor, 1996; Lipton, 1998).

6.1.1 Statements of Problems of Optimal Design

The problem of a body of extremal conductivity studied in Chapter 4 permits a straightforward generalization. Namely, we may optimize the structure that conducts several substances. For example, we could construct a structure of minimal total heat conductance and with constrained total electrical conductivity or total porosity. The next two examples deal with a conducting medium that is submerged into a number of exterior fields.

Minimizing Sequences

Example 6.1.1 Suppose that a linear medium is composed of two materials, each of which can conduct two substances. Consider the minimization of the energy

$$\int_\Omega W_1(\chi, w_1), \quad W_1(\chi, w_1) = \sigma^1(\chi)(\nabla w_1)^2$$

of one conducting equilibrium when the energy

$$\int_\Omega W_2(\chi, w_2), \quad W_2(\chi, w_2) = \sigma^2(\chi)(\nabla w_2)^2$$

of the second equilibrium and the amounts of materials in the domain are fixed. Here where w_1, w_2 are the potentials of conductivities, $\sigma^1(\chi) = \chi \sigma_1^1 + (1 - \chi)\sigma_2^1$ is the conductivity of the layout with respect to the first process, $\sigma^2(\chi) = \chi \sigma_1^2 + (1 - \chi)\sigma_2^2$ is the conductivity with respect to the second process, and the conductivity of the kth material with respect to the nth process is denoted by σ_n^k. The characteristic function χ defines the structure of the layout.

Adding to the goal functional the constraints with the Lagrange multipliers λ and γ, respectively, we come to the problem of the minimization of a linear combination of two energies over the layout χ:

$$\inf_\chi \min_{w_1, w_2} \left\{ \int_\Omega W_1(\chi, w_1) + \lambda \int_\Omega W_2(\chi, w_2) + \gamma \int_\Omega \chi \right\}. \tag{6.1.1}$$

Two equilibria w_1 and w_2 are coupled through the layout χ. This problem will be investigated in the next chapters, where we will construct a minimizing sequence and a lower geometrically independent bound for its Lagrangian.

We anticipate the following features of its solution:

1. The problem requires a relaxation because its solution is unstable against fine-scale perturbations. Indeed, even its degenerate version $w_2 = 0$ (which has been studied previously) has an unstable solution.

2. Laminate structures cannot provide relaxation for all values of parameters. Indeed, we saw in Chapter 4 that the optimal structure is a laminate oriented across ∇w_1 if $\nabla w_2 = 0$. Similarly, a laminate oriented across ∇w_2 is optimal if $\nabla w_1 = 0$. In a general case, an optimal structure should provide small conductance in both directions simultaneously. We should anticipate that the poor conductor forms a box-type structure that is spread in all directions. Therefore, the optimal structure should be different from laminates.

Similar types of structural optimization problems deal with one physical process of conductivity with N different external loadings.

Example 6.1.2 Construct an inhomogeneous conducting medium that shows the minimal average conductivity when submerged into N external fields.

The energy of the material with an anisotropic conductivity tensor $\boldsymbol{\sigma}$ can be represented in the form[1]

$$W = \mathbf{v}^T \boldsymbol{\sigma} \mathbf{v} = \text{Tr}(\boldsymbol{\sigma}\mathbf{V}), \qquad (6.1.2)$$

where $\mathbf{v} = \nabla w$ and the dyadic product $\mathbf{V} = \mathbf{v} \otimes \mathbf{v}$ is a nonnegative, symmetric, rank-one matrix (recall that $\mathbf{V} = \mathbf{v} \otimes \mathbf{v}$ is the matrix with elements $V_{ij} = v_i v_j$).

The sum of N energies W_k of the potentials w_k caused by different loadings is equal to

$$W = \sum_{k=1}^{N} W_k = \text{Tr}\,\boldsymbol{\sigma}\mathbf{V}_N, \qquad (6.1.3)$$

where $\mathbf{V}_N$ is the matrix

$$\mathbf{V}_N = \sum_{k=1}^{N} \mathbf{v}_i \otimes \mathbf{v}_i$$

such that

$$\mathbf{V}_N \geq 0, \quad \text{rank}(\mathbf{V}_N) = \min\{N, d\}.$$

If $N \geq d$, then $\mathbf{V}_N$ generally is a nondegenerate positive symmetric $d \times d$ matrix. We may diagonalize this matrix and find its eigenvalues μ_k and eigenvectors $\mathbf{h}_k$:

$$\mathbf{V}_N = \sum_{k=1}^{d} \mu_k\,\mathbf{h}_k \otimes \mathbf{h}_k;$$

then the problem (6.1.3) is reduced to the minimization of the sum of energies of the orthogonal fields $\mathbf{h}_k$, $k = 1, \ldots, \text{rank}(\mathbf{V}_N)$.

The nonself-adjoined problems can also be reduced to the similar minimization problem by means of the Lagrange transform, as was shown in Chapter 5; see also Chapters 12 and 17.

Multiwell Lagrangians in Problems of Structural Optimization

Problems of structural optimization deal with such minimizers as currents, potential fields, stresses, and strains that form the vector $\mathbf{v}$, as well as with the vector $\boldsymbol{\chi}(\mathbf{x}) = [\chi_1(\mathbf{x}), \ldots, \chi_N(\mathbf{x})]$ of characteristic functions that describe the material at point $\mathbf{x}$. The problem asks for the optimal $\boldsymbol{\chi}(\mathbf{x})$,

---

[1]In this book, we use the two equivalent formulas $\mathbf{x} \cdot C\mathbf{x}$ and $\mathbf{x}^T C\mathbf{x}$ for quadratic forms. We prefer to use the first form either when $\mathbf{x}$ is a spatial vector as normal or gradient, or when $\mathbf{x}$ is overloaded with superindexes.

that is for optimal microstructure. Let us demonstrate that the problem of minimization of the energy can be represented as the variational problem with a multiwell Lagrangian.

The problems of energy minimization have the form

$$I = \inf_{\chi} \left(\min_{\mathbf{v}} \int_{\Omega} \sum_{i=1}^{N} (\chi_i(W_i(\mathbf{v}) + \gamma_i)) \right), \qquad (6.1.4)$$

where W_i is the energy of the ith material (as in (6.1.2)) and γ_i is its cost.

Remark 6.1.1 *The components of χ are equal to either zero or one; therefore, variations $\delta\chi$ have either unit or zero magnitude at each point. The only small parameter of variation is the measure of the varied function. One cannot use the standard variational technique to investigate problem (6.1.4) because such a technique is based on consideration of variations of infinitely small magnitude.*

One can exclude χ from the variational problem by using necessary conditions of optimality. Namely, we may interchange the sequence of minimizing operations in (6.1.4) and rewrite (6.1.4) as

$$I = \inf_{\mathbf{v}} \int_{\Omega} \min_{\chi} \sum_{i=1}^{N} \chi_i (W_i(\mathbf{v}) + \gamma_i). \qquad (6.1.5)$$

Let us perform the inner minimization. Recall that at each point $\mathbf{x}$, one and only one characteristic function χ_i is equal to one, and the others are equal to zero. Therefore, we have only N terms $W_i(\mathbf{v}) + \gamma_i$ to compare; we choose the minimum of them. The problem becomes

$$I = \min_{\mathbf{v}} \int_{\Omega} F(\mathbf{v}), \quad F(\mathbf{v}) = \min_{1 \leq i \leq N} \{\gamma_i + W_i(\mathbf{v})\}. \qquad (6.1.6)$$

The Lagrangian of (6.1.6) is called a *multiwell Lagrangian* because it is the minimum of several convex functions (wells). The dependence on the constraints M_i–the amounts of the materials–is replaced by the dependence on the Lagrange multipliers γ_i that represent the cost of these materials.

6.1.2 Fields and Differential Constraints

Let us discuss differential constraints on the vector $\mathbf{v}$. Usually, $\mathbf{v}$ consists of either curlfree or divergencefree vectors. They correspond to some potentials and therefore are subject to linear differential constraints that express the integrability. Examples of the constraints were given in Chapter 2 (see Table 2.1). Let us discuss a general form of differential constraints on the minimizers $\mathbf{v}$ to the variational problem. We distinguish three types of variables:

1. Differentiable (smooth) variables $\mathbf{w}$, that represent the potentials, such as temperature. The Lagrangian L explicitly depends on these variables and on gradients of them, $L(\mathbf{w}, \nabla \mathbf{w})$. The L_p norm of all partial derivatives of $\mathbf{w}$ is bounded.

2. Nondifferentiable variables $\mathbf{f}$, such as the density of the sources. These variables belong to L_p spaces, and their partial derivatives are not bounded.

3. Variables $\mathbf{v}$ that represent either curlfree or divergencefree vectors, such as currents or electrical fields. They correspond to some potentials and therefore they are subject to linear differential constraints that express the integrability (see, for example, Table 2.1 and Table 14.1). Namely, some linear combinations of partial derivatives of the components v_i, $i = 1, \ldots n$ of $\mathbf{v}$ are fixed:

$$\sum_{k=1}^{d} \sum_{j=1}^{m} a_{ijk} \frac{\partial v_j}{\partial x_k} = g_i, \ i = 1, 2, \ldots, r. \tag{6.1.7}$$

Here, g_i are given functions (sources), r is the number of differential constraints, and $\mathcal{A} = \{a_{ijk}\}$ is an $r \times m \times d$ third-rank tensor of constants that fixes differential properties of the minimizers.

The tensor form of the constraints is

$$\mathcal{A} : \nabla \mathbf{v} = \mathbf{g}.$$

Here $(:)$ means contraction of two indexes as follows: $\{a_{ijk}\} : \{b_{mn}\} = \sum_{j,k} a_{ijk} \, b_{kj}$ and $\mathbf{g} = (g_1, \ldots, g_r)$.

This type of variable appears only in multidimensional variational problems. They are the focus of our consideration.

Remark 6.1.2 *The first two types of variables also admit the representation (6.1.7) with special $\mathcal{A}$. Namely, each partial derivative of the differentiable variable $\mathbf{w}$ is bounded. No partial derivative of the nondifferentiable variables f are bounded.*

We will often denote the entire argument of the Lagrangian by $\mathbf{v}$. We will not distinguish the types of minimizers unless it is necessary, treating all types of variables as variables of the third type.

The next examples show differential constraints on variables $\mathbf{v}$ in the form (6.1.7).

Example 6.1.3 The divergence of a current field $\mathbf{j}$ is bounded:

$$\sum_{i=1}^{d} \frac{\partial j_i}{\partial x_i} = g,$$

where g is the density of the sources. In this case the constraints (6.1.7) are set as:

$$r = 1, \quad g_1 = q, \quad a_{ijk} = \delta_{1i}\delta_{jk},$$

where δ_{ab} is the Kronecker symbol; $\mathcal{A}$ is the $(1 \times n \times d)$ tensor.

Example 6.1.4 The three-dimensional potential field $\mathbf{e} = \nabla w$ is curlfree:

$$\frac{\partial e_3}{\partial x_2} - \frac{\partial e_2}{\partial x_3} = 0, \quad \frac{\partial e_1}{\partial x_3} - \frac{\partial e_3}{\partial x_1} = 0, \quad \frac{\partial e_2}{\partial x_1} - \frac{\partial e_1}{\partial x_2} = 0.$$

To rewrite the constraints in the form (6.1.7) we set $g_i = 0$ and the $3 \times 3 \times 3$ tensor $\mathcal{A} = \{a_{ijk}\}$ equal to the Levi-Civita tensor. It has the following nonzero elements:

$$a_{132} = a_{213} = a_{321} = 1, \quad a_{123} = a_{231} = a_{312} = -1.$$

Generally, we consider variational problems that depend on a number of divergencefree and/or curlfree fields.

Example 6.1.5 Suppose that a vector $\mathbf{v} = [v_1, \ldots, v_4]$ consists of the elements of a matrix $\nabla \mathbf{y} = (\nabla y_1, \nabla y_2)$, where $y = y(x_1, x_2)$:

$$v_1 = \frac{\partial y_1}{\partial x_1}, \quad v_2 = \frac{\partial y_2}{\partial x_1}, \quad v_3 = \frac{\partial y_1}{\partial x_2}, \quad v_4 = \frac{\partial y_2}{\partial x_2}. \tag{6.1.8}$$

The differential constraints express the equality of the mixed derivatives of y_i or the identity $\nabla \times \nabla \mathbf{y} \equiv 0$. These constraints can be written as:

$$\frac{\partial v_1}{\partial x_2} - \frac{\partial v_3}{\partial x_1} = 0, \quad \frac{\partial v_2}{\partial x_2} - \frac{\partial v_4}{\partial x_1} = 0.$$

If they are written in the form (6.1.7), then the $2 \times 4 \times 2$ tensor $\mathcal{A}$ has the following nonzero coefficients:

$$a_{112} = 1, \quad a_{131} = -1, \quad a_{222} = 1, \quad a_{241} = -1.$$

Remark 6.1.3 *The form (6.1.7) of differential constraints was used for classification of the variational problems starting from (Murat, 1981; Dacorogna, 1982; Tartar, 1986). This form is convenient, but not imperative. At long last, all the fields are linear combinations of elements of gradients of some potentials, because they are linear combinations of some partial derivatives. Therefore, the classical form $L(\mathbf{x}, \mathbf{w}, \nabla \mathbf{w})$ of Lagrangian is equivalent to (6.1.7).*

We could also consider the differential constraints in the form of higher-order differential forms instead of (6.1.7). The analysis of such cases is similar in many respects (see, for example, (Ball, Currie, and Oliver, 1981; Ball and Murat, 1984)); however, the results may be different. An example of these constraints is analyzed in Chapter 15.

6.2 Convexity of Lagrangians and Stability of Solutions

We have learned that a solution to a one-dimensional variational problem is stable against fine-scale perturbations if its Lagrangian is convex. These problems are weakly semicontinuous (Chapter 5). The lack of convexity of the Lagrangian leads to the appearance of rapidly alternating functions in the optimal solution. Here we demonstrate that the stability of a solution of a multidimensional variational problem generally requires less than convexity of the Lagrangian.

6.2.1 Necessary Conditions: Weierstrass Test

Weierstrass-Type Variation. We apply the approaches of Chapter 1 to multidimensional variational problems. For definiteness, here we deal with the special case of differential constraints, $\nabla \times \boldsymbol{\xi} = 0$ that corresponds to the representation $\boldsymbol{\xi} = \nabla \mathbf{w}$. Consider the problem

$$\min_{\mathbf{w}} \int_{\Omega} F(\mathbf{x}, \mathbf{w}, \nabla \mathbf{w}) \tag{6.2.1}$$

where $\mathbf{w} = (w_1, \ldots, w_m)$ is the vector of minimizers, $\mathbf{x}$ is a point in the d-dimensional domain Ω, and F is a twice-differentiable function of its arguments. As in the one-dimensional case, the classical (twice-differentiable) minimizer $\mathbf{w}_0$ of the variational problem corresponds to the Euler–Lagrange equations (2.1.21),

$$\nabla \cdot \frac{\partial F}{\partial \nabla \mathbf{w}} - \frac{\partial F}{\partial \mathbf{w}} = 0.$$

This solution must also satisfy the Weierstrass-type condition. These conditions suggest that a sharp localized perturbation of the extremal does not decrease the functional. Failure to satisfy the Weierstrass test proves the absence of smooth solutions, because these solutions can be improved by adding an oscillatory component.

Let us discuss the Weierstrass-type condition in a multidimensional variational problem.

Definition 6.2.1 By a Weierstrass variation of a multidimensional variational problem we understand a localized perturbation $\delta \mathbf{w}$ of the potential $\mathbf{w}$ that

1. has an arbitrarily small magnitude $|\delta \mathbf{w}| < \varepsilon$;

2. has a finite magnitude of the gradient $|\nabla \mathbf{w}| = O(1)$; and

3. is localized in a small neighborhood $\omega_\varepsilon(\mathbf{x}_0)$ of a point $\mathbf{x}_0$ in the domain Ω: $\delta \mathbf{w}(\mathbf{x}) = 0 \; \forall \; \mathbf{x} \notin \omega_\varepsilon(\mathbf{x}_0)$, where $\omega_\varepsilon(\mathbf{x}_0)$ is a domain in Ω with the following properties: $\mathbf{x}_0 \in \omega_\varepsilon(\mathbf{x}_0)$, $\mathrm{diam}(\omega_\varepsilon) \to 0$ as $\varepsilon \to 0$.

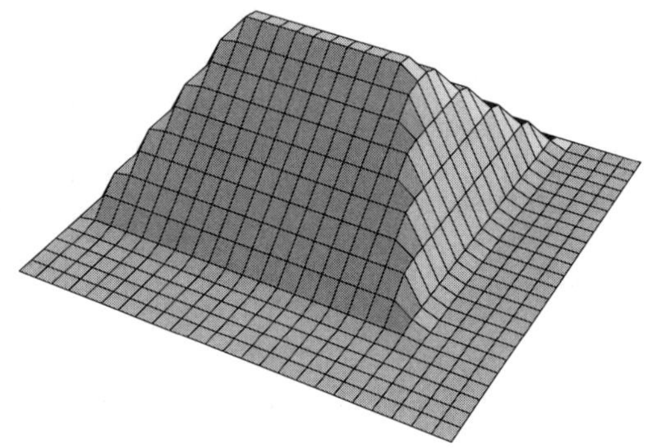

FIGURE 6.1. A strong local perturbation of a potential that keeps it continuous.

There is freedom in choosing the type of Weierstrass variation in a multidimensional problem. First, we choose the shape of ω_ε. It is important that δw is continuous and vanishes on the boundary $\partial\omega_\varepsilon$ of ω_ε.

For example, we may choose ω_ε as a circular domain and consider the trial perturbation δw shaped like a cone or a symmetric paraboloid. For a polygonal domain ω_ε the variation δw can be shaped like a pyramid.

The resulting necessary condition depends on the chosen shape of the variation. We will call the corresponding inequalities the necessary conditions of Weierstrass type or the *Weierstrass conditions*. Variation in a strip is an example of a Weierstrass condition; it was discussed in Chapter 4, where we noted that the Weierstrass condition depends on the strip's orientation, that is, on the shape of the domain ω_ε.

The Formulas. To derive the Weierstrass-type condition we consider the increment of the functional caused by such a variation and round it up to $o(\varepsilon)$. The leading term in the increment is caused by the varied value $\mathbf{H}(\mathbf{x})$, $\mathbf{x} \in \omega_\varepsilon$, of ∇w. Rounding the leading term of the variation up to $o(\varepsilon)$, we replace $\mathbf{w}(\mathbf{x}) + \delta\mathbf{w}(\mathbf{x})$, $\mathbf{x} \in \omega_\varepsilon$ by $\mathbf{w}(\mathbf{x}_0)$ and $\nabla\mathbf{w}(\mathbf{x})$, $\mathbf{x} \in \omega_\varepsilon$ by $\nabla\mathbf{w}(\mathbf{x}_0)$:

$$\delta I = \int_{\omega_\varepsilon} [F(\mathbf{w} + \delta\mathbf{w}, \nabla(\mathbf{w} + \delta\mathbf{w}), \mathbf{x}) - F(\mathbf{w}, \nabla\mathbf{w}, \mathbf{x})]$$
$$= \varepsilon\left[F(\mathbf{w}(\mathbf{x}_0), \nabla\mathbf{w}(\mathbf{x}_0) + \mathbf{H}, \mathbf{x}_0) - F(\mathbf{w}(\mathbf{x}_0), \nabla\mathbf{w}(\mathbf{x}_0), \mathbf{x}_0)\right] + o(\varepsilon),$$

where $\mathbf{w}(\mathbf{x}_0)$ is the tested stationary solution at the point $\mathbf{x}_0$ and $\mathbf{H} = \nabla\delta\mathbf{w}$ is the variation of $\nabla\mathbf{w}$.

Example 6.2.1 As an example of a Weierstrass variation, consider the pyramidal variation (see Figure 6.1) in a domain shown in Figure 6.2. We assume that the variational problem has the form (6.2.1).

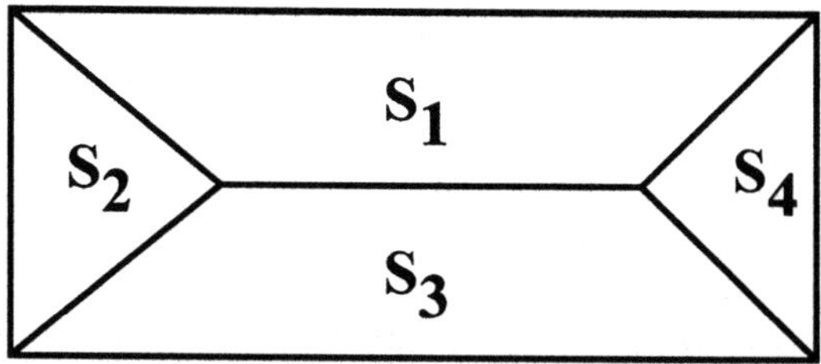

FIGURE 6.2. To the Weierstrass variation: The domains S_i.

Consider the function

$$H(\mathbf{x}) = \max\{0,\ \min\{\beta_1 - |\mathbf{i}_1 \cdot \mathbf{x}|,\ \beta_2 - |\mathbf{i}_2 \cdot \mathbf{x}|\}\} \tag{6.2.2}$$

shown in Figure 6.1. Here $\mathbf{i}_1$ and $\mathbf{i}_2$ are the orthogonal unit vectors, and β_1 and β_2 are two positive numbers; (6.2.2) says that H is finite:

$$H(\mathbf{x}) = 0 \ \text{ if } \ \mathbf{x} \notin \{[-\beta_1,\ \beta_1] \times \ [-\beta_2,\ \beta_2]\}.$$

Define the variation as

$$\delta\mathbf{w} = \varepsilon H\left(\frac{1}{\varepsilon}(\mathbf{x} - \mathbf{x}_0)\right)\boldsymbol{\alpha}$$

(see Figure 6.1), where $\boldsymbol{\alpha}$ is an arbitrary vector of magnitudes of the variations.

The variations of $\mathbf{w}$ are of order ε, and the variation of a component ∇w_i is (see Figure 6.1) a piecewise constant vector with components

$$\nabla w_i = \begin{cases} -\alpha_i \mathbf{i}_1 & \text{if } (\mathbf{x} - \mathbf{x}_0) \in S_1, \\ \alpha_i \mathbf{i}_1 & \text{if } (\mathbf{x} - \mathbf{x}_0) \in S_2, \\ -\alpha_i \mathbf{i}_2 & \text{if } (\mathbf{x} - \mathbf{x}_0) \in S_3, \\ \alpha_i \mathbf{i}_2 & \text{if } (\mathbf{x} - \mathbf{x}_0) \in S_4, \\ 0 & \text{if } (\mathbf{x} - \mathbf{x}_0) \notin \cup S_i. \end{cases}$$

Here S_i, $i = 1, \ldots, 4$ are the domains of constancy of the variations $\{\nabla w_1, \ldots, \nabla w_n\}$; see Figure 6.2.

Note that the areas $|S_1|$ and $|S_3|$ are equal: $|S_1| = |S_3|$. Likewise, $|S_2| = |S_4|$. Note also that the ratio of the areas $\frac{|S_1|}{|S_2|}$ is arbitrary.

The increment δI caused by this variation and rounded up to $o(\varepsilon)$ is composed of two terms, $\delta I = d(\mathbf{i}_1) + d(\mathbf{i}_2)$. The first term represents the variation of the Lagrangian integrated over the domains S_1 and S_3, and the second represents the variation of the Lagrangian integrated over the domains S_2 and S_4. Each term equals the second difference:

$$\begin{aligned} d(\mathbf{i}_k) = {}& 2|S_k|[F(\mathbf{w} + \delta\mathbf{w}, \nabla\mathbf{w} + \boldsymbol{\alpha} \otimes \mathbf{i}_k, \mathbf{x}_0) \\ & + F(\mathbf{w} + \delta\mathbf{w}, \nabla\mathbf{w} - \boldsymbol{\alpha} \otimes \mathbf{i}_k, \mathbf{x}_0) - 2F(\mathbf{w}, \nabla\mathbf{w}, \mathbf{x}_0)], \end{aligned} \tag{6.2.3}$$

where $|S_k|$ is the area of constancy of the gradient (see Figure 6.2). The nonnegativity of the increment implies that both second differences must be nonnegative: $d(\mathbf{i}_k) \geq 0$. Indeed, the ratio of the areas $\frac{|S_1|}{|S_2|}$ is not fixed, and the coefficient of each term $d(\mathbf{i}_1)$ and $d(\mathbf{i}_2)$ can be made arbitrarily large or small compared to the other.

Note that the domain ω_ε of variation can be turned through an arbitrary angle. The nonnegativity of the increment leads to the condition $d(\mathbf{n}) \geq 0$ (here, an arbitrary normal $\mathbf{n}$ replaces the unit vector $\mathbf{i}_1$ on the left-hand side of (6.2.3)). The last condition is a Weierstrass-type condition:

$$\mathcal{W}(F, \boldsymbol{\alpha} \otimes \mathbf{n}) \geq 0 \quad \forall \, \boldsymbol{\alpha}, \, \mathbf{n}, \tag{6.2.4}$$

where

$$\mathcal{W}(F(\mathbf{x}, \mathbf{w}, \nabla\mathbf{w} + \boldsymbol{\alpha} \otimes \mathbf{n}) = -F(\mathbf{x}, \mathbf{w}, \nabla\mathbf{w})$$
$$+ \tfrac{1}{2} F(\mathbf{x}, \mathbf{w}, \nabla\mathbf{w} + \boldsymbol{\alpha} \otimes \mathbf{n})$$
$$+ \tfrac{1}{2} F(\mathbf{x}, \mathbf{w}, \nabla\mathbf{w} - \boldsymbol{\alpha} \otimes \mathbf{n}),$$

and $\mathbf{w} = \mathbf{w}(\mathbf{x})$ is an optimal solution. A Weierstrass-type condition is the analogue of the requirement of convexity of the Lagrangian with respect to the derivative in the one-dimensional case.

Recall that the Weierstrass condition depends on the shape of a chosen variation. The question arises: What property of the Lagrangian is needed to make it stable against a variation in a domain of any given shape?

Rank-One Convexity

The condition (6.2.4) states that the Lagrangian $\alpha F(\mathbf{x}, \mathbf{w}, \mathbf{A})$ is convex with respect to some special trial matrices $\mathbf{R} = \boldsymbol{\alpha} \otimes \mathbf{n}$ of rank-one (but not with respect to arbitrary matrices). The corresponding property is called the rank-one convexity.

Definition 6.2.2 The scalar function F of an $n \times m$ matrix argument $\mathbf{A}$ is called *rank-one convex* at a point $\mathbf{A}_0$ if

$$F(\mathbf{A}_0) \leq \sum_{i=1}^{N} \alpha_i F\left(\mathbf{A}_0 + \alpha_i \xi_i \mathbf{R}\right)$$

for any α_i, ξ_i, $\mathbf{R}$, N that

$$\sum_{i=1}^{N} \alpha_i = 1, \quad \alpha_i \geq 0, \quad \sum_{i=1}^{N} \alpha_i \xi_i = 0, \quad \mathbf{R} = \mathbf{a} \otimes \mathbf{b}.$$

Here $\mathbf{a}$ and $\mathbf{b}$ are n-dimensional and m-dimensional vectors, respectively, and α_i are scalars.

Rank-one convexity requires convexity in some matrix "directions," namely, in the "directions" of the rank-one matrices. Obviously, the usual convexity implies rank-one convexity.

A detailed description of rank-one convexity can be found in many papers. We cite (Kohn and Strang, 1986a; Ball and Murat, 1991; Šverák, 1992b; Tartar, 1993; Rosakis and Simpson, 1994/95; Parry, 1995; Dacorogna and Haeberly, 1996; Pedregal, 1997; Pedregal and Šverák, 1998; Šilhavý, 1998).

There are two cases in which rank-one convexity coincides with convexity:

1. The Lagrangian depends on one independent variable: $\mathbf{x}$ is a scalar.

2. The Lagrangian depends on one dependent variable: $\mathbf{w}$ is a scalar.

In both cases, the matrix $\mathbf{A}_0 = \nabla \mathbf{w}$ degenerates into a rank-one matrix. These cases are studied in previous chapters: Chapter 1 deals with one-dimensional problems, and Chapters 4 and 5 discuss the scalar problem. The corresponding variational problems are discussed in (Ekeland and Temam, 1976).

The rank-one convexity of the Lagrangian is a necessary condition for the stability of the minimizer. If this condition is violated on a tested solution, then the special fine-scale perturbations (like the one described earlier) improve the cost; hence the solution is not optimal.

Remark 6.2.1 *Rank-one trial perturbation is consistent with the classical form* $L(\mathbf{x}, \mathbf{w}, \nabla \mathbf{w})$ *of Lagrangian. This form implies the special differential constraints* $\nabla \times (\mathbf{v}) = 0$ *that require the continuity of all but one component of the field* $\nabla \mathbf{w}$*. The definition of this necessary condition for the stability of the solution can be obviously generalized to the case where the differential constraints are given by the tensor* $\mathcal{A}$ *as in (6.1.7).*

6.2.2 Attainability of the Convex Envelope

In one-dimensional problems, the Weierstrass test requires convexity of the Lagrangian with respect to the derivative of a minimizer. The convexity of the Lagrangian F is sufficient to ensure stability against fine-scale perturbations also in multidimensional problems. This follows from considerations in Chapter 1. The multidimensional case differs from the one-dimensional case by additional constraints (6.1.7) of integrability that constrain the allowed fine-scale perturbations. The convexity of F is sufficient for the stability of the solution against unconstrained perturbations; certainly it is sufficient for the stability against constrained perturbations.

Therefore, we may want to obtain a solution without any short-term instabilities by convexification of the Lagrangian, i.e., by replacement of a nonconvex Lagrangian $F(\mathbf{v})$ with its convex envelope $CF(\mathbf{v})$. The convexity of $CF(\mathbf{v})$ makes the variational problem stable against any fine-scale oscillations to the solution.

The trouble is with another property of relaxation: It should preserve the cost of the problem. Generally, the cost of the convexified problem is less

than or equal to the cost of the initial problem because $CF(\mathbf{v})) \leq F(\mathbf{v})$.
Recall that preservation of the cost is achieved in the one-dimensional case
thanks to the explicit construction of a minimizing sequence that corresponds to the convex envelope (see Chapters 1 and 4).

The next example shows that the convex envelope of a multidimensional
Lagrangian may not correspond to any minimizing sequence.

Example 6.2.2 Consider the following nonconvex variational problem:

$$I = \min_{\mathbf{w}} \int_{\Omega} \Phi(\mathbf{V}), \quad \mathbf{V} = \nabla \mathbf{w} \text{ is periodic in } \Omega, \qquad (6.2.5)$$

where Ω is the unit square, $\mathbf{V}$ is the two-by-two matrix (6.1.8), and Φ is
the two-well Lagrangian:

$$\Phi = \begin{cases} 0 & \text{if } \mathbf{V} = 0, \\ 1 + \|\mathbf{V}\|^2 & \text{if } \mathbf{V} \neq 0. \end{cases}$$

The Lagrangian Φ is nonconvex. Therefore, the solution to (6.2.5) might
contain fine-scale oscillations. To relax the problem, we could try to replace
the Lagrangian Φ with its convex envelope $\mathcal{C}\Phi$, as in the one-dimensional
case.

Let us demonstrate that the convex envelope does not correspond to any
oscillating solution to the variational problem (6.2.5). (Recall that such
a sequence can be built for any one-dimensional variational problem; see
Chapter 1.)

We showed in Chapter 1 (Example 1.3.3) that the convex envelope $\mathcal{C}\Phi$
of the two-well Lagrangian Φ coincides with Φ if $\|\mathbf{V}\| \geq 1$, or it is a cone
$\|\mathbf{V}\| = 2$ with the vertex at $\mathbf{0}$ if $\|\mathbf{V}\| \geq 1$.

Consider a point $(\mathcal{C}\Phi(\mathbf{V}_0), \mathbf{V}_0)$ of the conical part of the convex envelope.
Here $\mathbf{V}_0$ is an arbitrary matrix, such that $\|\mathbf{V}_0\| < 1$. The point $\mathbf{V}_0$ lies on
a straight line that connects two points

$$(\Phi(\mathbf{V}_1) = 0, \ \mathbf{V}_1 = \mathbf{0}) \text{ and } (\Phi(\mathbf{V}_2) = 1 + \|\mathbf{V}_2\|^2, \ \mathbf{V}_2), \qquad (6.2.6)$$

where $\mathbf{V}_2$ is a supporting point in which the convex envelope touches the
graph of Φ. The first point $\mathbf{V}_1 = \mathbf{0}$ belongs to the degenerate well, and the
second point belongs to the well $1 + \|\mathbf{V}\|^2$. The matrix $\mathbf{V}_0$ is proportional
to $\mathbf{V}_2$:

$$\mathbf{V}_0 = c\mathbf{V}_2, \ c \in (0,1). \qquad (6.2.7)$$

Suppose that the point $\mathcal{C}\Phi(\mathbf{V}_0)$ is achievable by an oscillatory solution to the minimization problem (6.2.5). Hence, there exists a minimizing sequence of potentials $\{\nabla \mathbf{w}^k\}$ in (6.2.5) such that the sequence of
$\{\mathbf{V}^k\} = \{\nabla \mathbf{w}^k\}$ approximates $\mathcal{C}\Phi(\mathbf{V}_1)$ in the following sense:

$$\langle \Phi(\mathbf{V}^k) \rangle \to \mathcal{C}\Phi(\mathbf{V}_0) \quad k \to \infty$$

when

$$\langle \mathbf{V}^k \rangle = \mathbf{V}_0 \quad \forall k.$$

Consider a function $\mathbf{V}(\mathbf{x})$ from the approximating sequence $\{\mathbf{V}^k\}$ (we omit the index ^{k}). It is a piecewise constant matrix function with values $\mathbf{0}$ and $\mathbf{V}_2$:

$$\mathbf{V}(\mathbf{x}) = \begin{cases} \mathbf{V}_1 = \mathbf{0} & \text{if } \mathbf{x} \in \Omega_1 \\ \mathbf{V}_2 = \frac{1}{c}\mathbf{V}_0 & \text{if } \mathbf{x} \in \Omega_2 \end{cases} \quad \Omega_1 \cup \Omega_2 = \Omega_\varepsilon.$$

Here Ω_1 and Ω_2 are regions in the $\mathbf{x}$ plane, where $\{\mathbf{V}^k(\mathbf{x})\}$ takes values $\mathbf{V}_1 = \mathbf{0}$ and $\mathbf{V}_2$, respectively.

Let us demonstrate that the fields $\mathbf{V}_1$ and $\mathbf{V}_2$ cannot be neighbors in Ω. The representation $\mathbf{V}(\mathbf{x}) = \nabla \mathbf{w}(\mathbf{x})$ requires the continuity of the potential $\mathbf{w}(\mathbf{x})$ in Ω_ε. This continuity implies the continuity of tangent components $\mathbf{V} \cdot \mathbf{t}$ of $\mathbf{V}$ on the boundary between Ω_1 and Ω_2, where $\mathbf{t}$ is the tangent to this boundary. We have

$$(\mathbf{V}_1 - \mathbf{V}_2) \cdot \mathbf{t} = 0.$$

Because $\mathbf{V}_1 = \mathbf{0}$ (see (6.2.6)), then $\mathbf{V}_2 \cdot \mathbf{t} = \mathbf{0}$. Further, because $c \neq 0$, we obtain from (6.2.7)

$$\mathbf{V}_0 \cdot \mathbf{t} = \mathbf{0}. \tag{6.2.8}$$

This constraint (6.2.8) obviously contradicts the arbitrariness of the matrix $\mathbf{V}_0$. Hence, there is no sequence of solutions to the initial problem (6.2.5) that approximates its convex envelope.

This example shows that extension by the convex envelope generally does not correspond to any minimizing sequence. Replacement of the Lagrangian with its convex envelope does not preserve the cost of the original problem.

Remark 6.2.2 *The optimization problem considered in Chapter 4 dealt with the Lagrangian that depends on a scalar potential w. In this case, $\mathbf{v}_0$ is a rank-one matrix, and the discussed problem definitely has a solution: The proper tangent $\mathbf{t}$ can be chosen as a vector orthogonal to $\mathbf{v}_0$. The minimizing sequence is realized as a properly oriented laminate.*

In summary, the supporting points of the convex envelope are generally not compatible, so they cannot be neighbors in a microstructure. Hence, the points of the convex envelope are not attainable by any continuous solution w. The convexification of a Lagrangian provides a (may be rough) lower bound of the cost of the original problem, because $CF \leq \mathcal{F}$.

Attainability: Min-Max Variational Problems

Reformulation of the variational problem (6.1.6) to dual min-max form leads to the following problem:

$$J^{\text{dual}} = \max_\chi \min_{\mathbf{v}_*} \int_\Omega \left(\text{Tr}\, D^{\text{dual}}(\chi)(\mathbf{v}_* \otimes \mathbf{v}_*) + \gamma\chi \right),$$

where $\mathbf{v}_*$ is the variable dual to $\mathbf{v}$, and D^{dual} is the matrix inverse to D: $D^{\mathrm{dual}} = D^{-1}$.

We can proceed as in Chapter 4 and interchange the sequence of extremal operations $\max_\chi \min_{\mathbf{v}_*}$. We obtain the following well-posed variational problem with convex but nonsmooth Lagrangian

$$R(\mathbf{v}_*) = \max\{D^{\mathrm{dual}}(0)(\mathbf{v}_* \otimes \mathbf{v}_*), D^{\mathrm{dual}}(1)(\mathbf{v}_* \otimes \mathbf{v}_*) + \gamma\}.$$

By the maximin theorem (4.4.4), the new problem provides an upper bound for the original problem,

$$J^{\mathrm{dual}} \leq \min_{\mathbf{v}_*} \int_\Omega R(\mathbf{v}_*).$$

Generally, this bound is not achievable. If it were achievable by a minimizing sequence, then the same minimizing sequence would correspond to the convex envelope of the original problem (6.1.6). However, we have shown that there is no such sequence. For the discussion of relaxation of min-max problems, we refer to (Lurie, 1990b; Lurie, 1994).

6.3   Quasiconvexity

6.3.1   Definition of Quasiconvexity

Here we determine the requirements for the Lagrangian of a variational problem stable against fine-scale perturbations. An adequate property for stability is the quasiconvexity (Morrey, 1952) of the Lagrangian. There is an extended mathematical literature on quasiconvexity. We refer to (Reshetnyak, 1967; Ball and Murat, 1984; Ball and Murat, 1991; Dacorogna, 1982; Marcellini, 1985; Murat and Tartar, 1985b; Murat and Tartar, 1985a; Kohn and Strang, 1986a; Kohn, 1991; Šverák, 1992a; Pedregal, 1997; Müller and Fonseca, 1998) among others. Following the style of our exposition, we keep the explanation simple and omit some details that can be found in the mentioned papers.

The Definition

Consider the minimal problem

$$\min_{\mathbf{v} \text{ as in } (6.1.7)} \langle F(\mathbf{v}) \rangle,$$

where $\mathbf{v}$ are the discontinuous fields that are constrained by the differential conditions (6.1.7), which follow from the continuity of potentials and $\langle Z \rangle$ is the averaging operator (2.2.2).

Recall the definition of convexity of the functional (6.3.7) by Jensen's inequality (Chapter 1). Function $F(\mathbf{v})$ is convex at the point $\mathbf{v}$ if

$$F(\mathbf{v}) \leq \langle F(\mathbf{v} + \boldsymbol{\xi}) \rangle \quad \forall \boldsymbol{\xi} : \langle \boldsymbol{\xi} \rangle = 0. \tag{6.3.1}$$

It is assumed that the integral in (6.3.1) exists for all admissible perturbations $\boldsymbol{\xi}$. This definition compares the integrals of $F(\mathbf{v})$ with the constant argument $\mathbf{v} = \text{constant}(\mathbf{x})$ and with the perturbed argument $\mathbf{v} + \boldsymbol{\xi}$; the mean value of the perturbation $\boldsymbol{\xi}$ is zero.

To arrive at the definition of quasiconvexity, we impose the linear constraints (6.1.7) on a trial field $\boldsymbol{\xi}$.

Definition 6.3.1 The function $F(\mathbf{v})$ is called *quasiconvex* at the point $\mathbf{v}$ if

$$F(\mathbf{v}) \leq \langle F(\mathbf{v} + \boldsymbol{\xi}) \rangle, \quad \forall \boldsymbol{\xi} \in \Xi \tag{6.3.2}$$

where Ξ is:

$$\Xi = \left\{ \begin{array}{ll} \boldsymbol{\xi}: & \langle \boldsymbol{\xi} \rangle = 0, \quad \mathcal{A} : \nabla \boldsymbol{\xi} = \sum_{jk} a_{ijk} \frac{\partial \xi_j}{\partial x_k} = 0, \\ \boldsymbol{\xi} \text{ is } \varepsilon - \text{periodic}, & \xi_j \in L_\infty(\varepsilon) \end{array} \right\}. \tag{6.3.3}$$

Here ε is a cube in R_n. Note that the orientation and size of ε do not affect (6.3.2).

Remark 6.3.1 *The classical definition of quasiconvexity, given by Morrey (Morrey, 1952), dealt with the potential fields $\mathbf{V} = \nabla \mathbf{w}$. Here we prefer to deal with the more flexible form (6.1.7) of the constraints.*

A Finite-Dimensional Analogue

A finite-dimensional analogue of quasiconvexity is the convexity of the function of a linearly constrained argument. Consider the convexity of a twice-differentiable function $f(\mathbf{z})$ where $\mathbf{z}$ is an n-dimensional vector. The convexity requires the nonnegativity of the $n \times n$ Hessian,

$$H(f) = \left\{ \frac{\partial^2 f}{\partial z_i \partial z_j} \right\} \geq 0.$$

Suppose now that the vector $\mathbf{z}$ is subject to $n - m$ linear constrains, $L\mathbf{z} = 0$, where L is a matrix of the rank $n - m$, $1 \leq m < n$. Accordingly, $\mathbf{z}$ can be expressed through an m-dimensional vector $\mathbf{y}$,

$$\mathbf{z} = L_*\mathbf{y}. \tag{6.3.4}$$

Here the $m \times n$ matrix L_* is the projector to the subspace orthogonal to the subspace of the constraints: $L\mathbf{z} = L(L_*\mathbf{y}) = 0 \ \forall \mathbf{y}$. The convexity of $f(\mathbf{z})$, where $\mathbf{z}$ is subject to linear constraints (6.3.4), is defined as the convexity of the function $\phi(\mathbf{y}) = f(L_*\mathbf{y})$ of $\mathbf{y}$. Function $\phi(\mathbf{y})$ is convex if the $m \times m$ projection $L_*^T H L_*$ of the Hessian is nonnegative,

$$L_*^T H L_* \geq 0.$$

Obviously, the convexity of f implies the convexity of ϕ, but not vice versa.

Comparing the quasiconvexity with the convexity of the restricted finite-dimensional problem, we recognize that the linear restrictions $\mathcal{A} : \nabla \mathbf{v} = 0$ are imposed on the infinite-dimensional element $\mathbf{v}$ of the corresponding functional space L_p. The role of the orthogonal linear operator L_* is played by a potential. For example, the constraint $\nabla \cdot \mathbf{v} = 0$ corresponds to the vector potential $\boldsymbol{\kappa}$ through the representation $\mathbf{v} = \nabla \times \boldsymbol{\kappa}$.

Properties of Quasiconvex Functions

The following properties of quasiconvex functions are easy to check:

1. Convexity implies quasiconvexity. Any convex function is also quasiconvex, because the set of trial functions $\boldsymbol{\xi}$ in the definition is larger in the case of convexity than in the case of quasiconvexity.

2. If $F(\mathbf{v})$ is quasiconvex and $t > 0$, then $t\,F(\mathbf{v})$ is quasiconvex.

3. If $F(\mathbf{v})$ and $G(\mathbf{v})$ are quasiconvex, then $F(\mathbf{v}) + G(\mathbf{v})$ is quasiconvex.

Let us now demonstrate a quasiconvex but not convex function. We call such functions *translators*.

Example 6.3.1 Consider the function $\phi(\mathbf{V}) = \det \mathbf{V}$, where $\mathbf{V} = \nabla \mathbf{y}$ is as in Example 6.1.5; see (6.1.8):

$$\phi(\mathbf{V}) = \det \mathbf{V}, \quad \mathbf{V} = \nabla \mathbf{y}. \tag{6.3.5}$$

Obviously, $\phi(V_{ij})$ is not convex. For example, if $V_{11} = 0$, $V_{12} = V_{21}$, then $\phi(\mathbf{v})$ becomes concave: $\phi(\mathbf{V}) = -V_{12}^2$.

Let us prove, however, that $\phi(\mathbf{V})$ is quasiconvex. We rewrite $\phi(\mathbf{V})$ using (6.1.8) as the divergence of a vector Z:

$$\phi(\mathbf{V}) = V_{11}V_{22} - V_{12}V_{21} = \frac{\partial y_1}{\partial x_1}\frac{\partial y_2}{\partial x_2} - \frac{\partial y_1}{\partial x_2}\frac{\partial y_2}{\partial x_1}$$

$$= \frac{\partial}{\partial x_1}\left(y_1\frac{\partial y_2}{\partial x_2}\right) + \frac{\partial}{\partial x_2}\left(-y_1\frac{\partial y_2}{\partial x_1}\right) = \nabla \cdot \mathbf{Z},$$

where

$$\mathbf{Z} = \left(y_1\frac{\partial y_2}{\partial x_2}, \; -y_1\frac{\partial y_2}{\partial x_1}\right).$$

By Green's theorem, we have

$$\int_\omega \phi(\mathbf{V}) = \int_\omega \nabla \cdot \mathbf{Z} = \int_\Gamma \mathbf{Z} \cdot \mathbf{n}\, d\Gamma,$$

where $\mathbf{n} = [n_1, \; n_2]$ is the normal to $\Gamma = \partial\omega$. Next, we have

$$\mathbf{Z} \cdot \mathbf{n} = y_1\left(\frac{\partial y_2}{\partial x_2}n_1 - \frac{\partial y_2}{\partial x_1}n_2\right) = y_1\frac{\partial y_2}{\partial t},$$

where t is the direction along Γ.

Now substitute $\mathbf{V}+\boldsymbol{\xi}$ for $\mathbf{V}$ and take into account the periodicity of $\boldsymbol{\xi}$ and the condition $\langle\boldsymbol{\xi}\rangle = 0$. We observe that all terms that contain $\boldsymbol{\xi}$ disappear, which leads to the equality

$$\phi(\mathbf{V}) = \langle F(\mathbf{V}+\boldsymbol{\xi})\rangle \quad \forall\boldsymbol{\xi}\in\Xi. \tag{6.3.6}$$

We see that the nonconvex function $F(\mathbf{v}) = \det\mathbf{v}$ is quasiconvex; therefore ϕ is the translator. Moreover, we obtain an equality instead of inequality in the definition of quasiconvexity. Such quasiconvex functions are called *quasiaffine*.

Stability Against Fine-Scale Perturbations

Consider the variational problem (6.2.1),

$$\min_{\mathbf{w}\text{ as in }(6.1.7)} \int_\Omega L(\mathbf{x},\mathbf{w},\mathbf{v}), \tag{6.3.7}$$

where $\mathbf{w}$ are the continuous potentials and $\mathbf{v}$ are the discontinuous fields that are constrained by the differential conditions (6.1.7). Denote by $\mathbf{w}_0$ the solution to the Euler–Lagrange equation of (6.3.7) and let $\mathbf{v}_0$ be $\mathbf{v}_0 = \nabla\mathbf{w}_0$.

The solution $\mathbf{w}_0$ is stable against fine-scale perturbations if the Lagrangian $L(\mathbf{x},\mathbf{w},\mathbf{v})$ is quasiconvex with respect to its third argument, $\mathbf{v}$, at the point $\mathbf{x},\mathbf{w}_0,\mathbf{v}_0$. The fine-scale perturbations $\delta\mathbf{w}$ of the potential $\mathbf{w}$ have a form similar to the perturbation in the Weierstrass variations. Namely, the fine-scale perturbation is a function that

1. is located in the ε-neighborhood of a point $\mathbf{x}$,

2. has the magnitude $\|\delta\mathbf{w}\|$ of the order of ε,

3. has the finite magnitude of the components of the gradient $\delta\mathbf{v}$ (of course, $\delta\mathbf{v}$ is subject to the constraints (6.1.7)), and

4. is centered: $\langle\mathbf{v}\rangle = 0$.

If the Lagrangian L is a continuous function of $\mathbf{x}$ and $\mathbf{w}$, then the fine-scale perturbation $\delta\mathbf{w}$ changes it as follows:

$$L(\mathbf{x},\mathbf{w}+\delta\mathbf{w},\mathbf{v}+\delta\mathbf{v}) = L(\mathbf{x}_0,\mathbf{w},\mathbf{v}+\delta\mathbf{v}) + O(\varepsilon).$$

If, in addition, L is a quasiconvex function of $\mathbf{v}$, the definition of the quasiconvexity (6.3.2) gives

$$\langle L(\mathbf{x},\mathbf{w},\mathbf{v}+\delta\mathbf{v})\rangle \geq \langle L(\mathbf{x},\mathbf{w},\mathbf{v})\rangle.$$

Therefore, the fine-scale perturbation increases the cost of the problem (6.3.7).

The quasiconvexity of Lagrangians distinguishes the problems that can or cannot be improved by any strong local perturbation $\boldsymbol{\xi}$. If the variational problem has a minimizer that satisfies the Euler–Lagrange equation, then its Lagrangian is necessarily quasiconvex at the optimal solution. The lack of quasiconvexity means that a smooth solution is not optimal: there is a periodic perturbation of the solution that is consistent with the differential constraints and that decreases the cost.

Remark 6.3.2 *Note that quasiconvexity of the Lagrangian alone does not guarantee the existence of the solution to a variational problem; it only ensures that a solution, if it exists, does not have fine-scale spatial oscillations. The existence of the solution also requires growth conditions of the Lagrangian; see for example (Dacorogna, 1982; Zhikov, 1993).*

The Euler–Lagrange equations of the quasiconvex Lagrangian are elliptic. In turn, ellipticity is a necessary requirement for the existence of a solution to a boundary value problem.

Null-Lagrangians and Quasiconvexity

The example of the quasiaffine Lagrangian

$$L_0(\mathbf{V}) = \det \mathbf{V}; \quad \mathbf{V} = \nabla \mathbf{y}$$

is also interesting from the following viewpoint: The Euler–Lagrange equation $S(L_0)\mathbf{y} = 0$ of the corresponding variational problem vanishes identically. Indeed, we have

$$\delta I = \int_\omega \delta \mathbf{y} \cdot \left(\nabla \cdot \frac{\partial \det \nabla \mathbf{y}}{\partial \nabla \mathbf{y}} \right) = \int_\omega \sum_{j=1}^{2} \delta y_j \left(\frac{\partial^2 y_j}{\partial x_1 \partial x_2} - \frac{\partial^2 y_j}{\partial x_2 \partial x_1} \right) \equiv 0,$$

where $\mathbf{y} = [y_1, y_2]$. Recall that such functions are called null-Lagrangians (Chapter 1).

Adding a null-Lagrangian L_0 to a Lagrangian L does not affect the Euler–Lagrange equations $S(L) = 0$:

$$S(L) \equiv S(L + tL_0),$$

where t is an arbitrary parameter and L_0 is a null-Lagrangian. Therefore, the stability of the solution to a variational problem is independent of possibly adding a null-Lagrangian to its Lagrangian.

Recall that in the one-dimensional case, any null-Lagrangian is a linear function of the derivatives of the minimizers. Hence, adding a null-Lagrangian to the Lagrangian does not affect its convexity with respect to these variables (Chapter 1). In contrast, the multidimensional null-Lagrangians may be nonlinear and nonconvex functions of $\nabla \mathbf{w}$, as was demonstrated in Example 6.3.1.

Adding a nonlinear null-Lagrangian tL_0 to L affects the convexity of the Lagrangian $L + tL_0$. Particularly, $L + tL_0$ can be made nonconvex if t is large enough. However, the solution of the Euler–Lagrange equation remains stable or unstable regardless of null-Lagrangians and the convexity of the Lagrangian.

Unlike convexity, the quasiconvexity is insensitive to the null-Lagrangian terms because they are quasiaffine. Therefore, the quasiconvexity of a Lagrangian is the property needed to draw conclusions about the behavior of the solution to a variational problem.

6.3.2 Quasiconvex Envelope

Properties of Minimizers of Nonquasiconvex Variational Problems

Consider the variational problem (6.2.1). If the Lagrangian is not quasiconvex then the problem is unstable against fine-scale perturbation of the minimizer. Indeed, the definition (6.3.2) shows that there exists a value $\mathbf{v}_f$ and a discontinuous function $\boldsymbol{\xi} = \boldsymbol{\xi}(\mathbf{x})$ such that $\mathbf{v}_f + \boldsymbol{\xi}(\mathbf{x})$ corresponds to the smaller right-hand side of (6.3.2) than the constant $\mathbf{v}_f$:

$$\exists \boldsymbol{\xi} \in \Xi : \quad F(\mathbf{v}_f) > \langle F(\mathbf{v}_f + \boldsymbol{\xi}) \rangle$$

The minimizer $\mathbf{v}_f + \boldsymbol{\xi}$ is nonconstant in spite of the absence of external inhomogeneity, regularity of the domain, and periodicity of the boundary conditions.

The set of the values $\mathbf{v}_f$ that correspond to the previous inequality is called the *forbidden region* $\mathcal{V}_f$ or the region of nonquasiconvexity of the Lagrangian. The solution to the variational problem never belongs to the forbidden region. The simplest example of this behavior was discussed in Chapter 4.

Similarly to the problem in the Chapter 4, we expect that the minimizer $\mathbf{v}$ belongs to the boundary of the region $\mathcal{V}_f$ and oscillates along this boundary if its averaged value $\langle \mathbf{v} \rangle$ varies in $\mathcal{V}_f$. The average varies due to variation of positions of the supporting points and their proportions in the mixture. The Lagrangian of the relaxed nonquasiconvex problem corresponds to the minimum of the averaged Lagrangians, $\langle F(\mathbf{v}) \rangle$, among all positions of the supporting points. The computation of the relaxed Lagrangian is not easy because the differential constraints are assumed on $\mathbf{v}$.

Dealing with the nonquasiconvex problem, we want to find answers to the questions:

1. What is the region $\mathcal{V}_f$ of nonquasiconvexity?

2. What is the relaxed functional equal to?

3. What are pointwise values of the minimizers and their fractions when the average field, $\langle \mathbf{v} \rangle$, belongs to $\mathcal{V}_f$? In other terms, what are oscillating minimizing sequences?

Dealing with multiwell Lagrangians, we identify the optimal minimizer as a mixture of fields $\mathbf{v}_i$ that belong to different wells (materials). The minimizer corresponds to a mixture of the materials or to a composite. The structure of the optimal mixture is defined by the forbidden region $\mathcal{V}_f$: $\mathbf{v} \notin \mathcal{V}_f$ and by the jump conditions for $\mathbf{v}$ at the surface between domains $\mathcal{O}_i$ of different materials in the structure. For this problem, we pose the additional questions:

1. What are optimal geometrical patterns of the domains $\mathcal{O}_i$ (microstructures) formed by the materials in an optimal structure?

2. What *permitted region* $\mathcal{V}_i$ of the fields corresponds to the optimality of the ith well?

3. What function $\phi(\mathbf{v})$ of the field $\mathbf{v}$ stays constant when the field moves along the boundary of $\mathcal{V}_i$? In other terms, what function $\phi(\mathbf{v})$ stays constant inside the ith material in the optimal mixtures?

The constancy of a function $\phi(\mathbf{v}_i)$ and the differential constraints on $\mathbf{v}$ look contradictory. Indeed, the variation of the field $\mathbf{v}_i$ inside a solid domain $\mathcal{O}_i$ of the ith material is controlled by the differential constraints that are unrelated to $\phi(\mathbf{v})$ which is computed from the optimality requirements. To resolve this contradiction, we may assume that is the domains $\mathcal{O}_i$ are infinitesimally thin. This consideration again points out the fine-scale mixtures which make optimal solutions.

In the next chapters, we develop several techniques to answer these questions. Chapter 7 deals with special minimizing sequences, the laminates of kth rank, that approximate the relaxed problem, Chapter 8 describes the lower bounds on the relaxed Lagrangian, and Chapter 9 describes the fields in optimal structures. The following chapters illustrate these approaches on numerous examples.

Remark 6.3.3 *In this book, we consider minimizing sequences (fields) that appear in special laminate microstructures, and we derive the necessary conditions for the boundaries of forbidden regions, using the Weierstrass-type variations. However, we do not address a more general technique of the Young measures to describe the fast-oscillation solutions of variational problems. The reader is referred to (Young, 1942a; Ball, 1989; Kinderlehrer and Pedregal, 1991; Matos, 1992; Tartar, 1992; Kinderlehrer and Pedregal, 1994; Tartar, 1995; Pedregal, 1997) and references therein.*

Quasiconvex Envelope

The quasiconvexity can be used for relaxation of unstable multidimensional problems. To relax a nonquasiconvex variational problem, one needs to describe an infinitely often oscillating solution to the corresponding variational problem. Relaxation means the minimization of the average value of the Lagrangian among the set of arbitrary fine-scale oscillating trial fields $\boldsymbol{\xi}$ with zero mean. This procedure essentially repeats the relaxation procedure for one-dimensional problems (Chapter 1); however, the trial fields must be compatible with differential constraints (6.1.7). This requirement is adequate to the physical meaning of the problem: It states that trial fields are divergencefree, or curlfree, or satisfy similar physical requirements.

The necessary relaxation is given by the quasiconvex envelope of the Lagrangian, which is defined as follows.

Definition 6.3.2 The *quasiconvex envelope* QF of the Lagrangian $F(\mathbf{x}_0, \mathbf{w}_0, \mathbf{v})$ is a solution to the variational problem

$$QF(\mathbf{x}_0, \mathbf{w}_0, \mathbf{v}_0) = \inf_{\boldsymbol{\xi} \in \Xi} \langle F(\mathbf{x}_0, \mathbf{w}_0, \mathbf{v}_0 + \boldsymbol{\xi}) \rangle, \qquad (6.3.8)$$

where the set Ξ is defined in (6.3.3). Here, $\mathbf{w}_0$ is the differentiable variable.

Note that in (6.3.8) the smooth variables $\mathbf{w}_0$ and the coordinates $\mathbf{x}_0$ are fixed. We will not explicitly mention these fixed parameters; instead, we will write $F = F(\mathbf{v})$.

Like the convex envelope, the quasiconvex envelope QF of F is the largest quasiconvex function that does not exceed F at each point $\mathbf{v}$, i.e.,

$$QF(\mathbf{v}) = \max \left\{ G(\mathbf{v}) : \begin{array}{l} 1.\ G(\mathbf{v}) \text{ is quasiconvex,} \\ 2.\ G(\mathbf{v}) \leq F(\mathbf{v}) \end{array} \right\}.$$

The quasiconvex envelope provides the relaxation of a multidimensional variational problem in the same sense as the convex envelope provides the relaxation of a one-dimensional one: The new problem has a differentiable almost everywhere minimizer, and its cost is the same as that of the original problem:

$$\inf_{\mathbf{v} \text{ as in } (6.1.7)} I(F(\mathbf{v})) = \min_{\mathbf{v} \text{ as in } (6.1.7)} I(QF(\mathbf{v})).$$

For the detailed exposition of the relaxation by means of the quasiconvex envelope, we refer to (Kohn and Strang, 1986a; Pedregal, 1997).

Note that a quasiconvex envelope QL_0 of a null-Lagrangian L_0 coincides with the null-Lagrangian itself: $QL_0 = L_0$. Adding L_0 to a Lagrangian F does not affect the relaxation, as expected.

6.3.3 Bounds

The variational problem of the quasiconvex envelope looks similar to the original variational problem (6.2.1). However, there are two significant simplifications in the construction of the quasiconvex envelope:

1. The boundary conditions for $\boldsymbol{\xi}$ are periodic.

2. The differentiable variables $\mathbf{w}$ and the independent variables $\mathbf{x}$ are constant.

Both properties correspond to the fact that a quasiconvex envelope equals the Lagrangian averaged on an infinitesimally small neighborhood. They allow us to calculate the quasiconvex envelope for a number of problems discussed in the later chapters.

Still, calculation of the quasiconvex envelope is a complicated variational problem. We will construct upper and lower bounds of QF. For some Lagrangians corresponding to physically interesting problems, these bounds coincide and therefore determine the quasiconvex envelope.

The Simplest Bounds

No matter what the differential constraints (6.2.2) are, there are simple bounds for the quasiconvex envelope:

$$CF \leq QF, \quad QF \leq F. \tag{6.3.9}$$

The first inequality can be obtained if one enlarges the class of admissible trial functions $\boldsymbol{\xi}$ by neglecting all differential constraints in (6.3.3) or by setting

$$a_{ijk} = 0 \quad \forall i, j, k.$$

Clearly, this enlargement reduces the problem to the problem of the convex envelope. The inequality in (6.3.9) comes from the simple fact that the minimum over a larger set is less than or equal to the minimum over a smaller one.

The second inequality corresponds to the case where the additional constraints are posed and the class of trial functions $\boldsymbol{\xi}$ is abridged, even reduced to $\boldsymbol{\xi} = 0$. This is the case where every partial derivative of $\mathbf{V}$ is restricted

$$\frac{\partial \xi_j}{\partial x_i} = 0 \quad \forall i, j,$$

which implies that $\boldsymbol{\xi}_j = 0$. Generally, the quasiconvex envelope depends on the tensor $\mathcal{A}$.

The goal of the next three chapters is to tighten the bounds (6.3.9) of the quasiconvex envelope; we will construct better upper and lower bounds $\mathcal{L}F$ and $\mathcal{P}F$ that generally satisfy the inequalities

$$CF \leq \mathcal{L}F \leq QF \quad \text{and} \quad QF \leq \mathcal{P}F \leq F. \tag{6.3.10}$$

Moreover, we will see that in many cases these upper and lower bounds coincide, $\mathcal{L}F = \mathcal{P}F$ and their common value determines the quasiconvex envelope itself. This systematic approach was put forward in (Kohn and Strang, 1986a); it is based on the construction in (Morrey, 1952).

6.4   Piecewise Quadratic Lagrangians

Optimal Energy

Let us return to the problem of minimization of the energy stored in a composite. The energy $W(\chi, \mathbf{v})$ of an inhomogeneous medium (compare with (6.1.2)) is

$$W(\chi, \mathbf{v}) = \mathbf{v}^T D(\chi) \mathbf{v}, \quad D(\chi) = \sum_{i=1}^{N} \chi_i D_i,$$

where $D(\chi)$ represents the properties matrix of an inhomogeneous medium, and differential constraints (6.1.7) are imposed on the fields $\mathbf{v}$.

Assume that the volume fractions of the materials are fixed:

$$\langle \chi_i \rangle = m_i, \quad m_i \geq 0, \quad m_1 + \ldots + m_N = 1$$

and consider the problem of minimizing the energy and the cost of a composite:

$$\inf_{\chi_1, \ldots, \chi_N} \inf_{\boldsymbol{\xi} \in \Xi} \int_{\Omega} F(\chi, \mathbf{v} + \boldsymbol{\xi}), \quad F(\chi, \mathbf{v}) = \left(W(\chi, \mathbf{v}) + \sum_{i=1}^{N} \chi_i \gamma_i \right) \quad (6.4.1)$$

where γ_i is the cost of the ith material.

Relaxation of the problem (6.4.1) is achieved by the quasiconvex envelope QF of the Lagrangian F. The quasiconvex envelope can be written in the form

$$QF(m_i, \mathbf{v}) = \left\{ W_{\text{opt}}(m_i, \mathbf{v}) + \sum_{i=1}^{N} m_i \gamma_i \right\}$$

where $W_{\text{opt}}(m_i, \mathbf{v})$ is the energy stored in an optimal composite submerged into the field $\mathbf{v}$, and m_i denotes the fixed volume fractions of the components in the composite.

Relaxation physically means replacing the energy of a fine-scale inhomogeneous material with the energy $W_{\text{opt}}(m_i, \mathbf{v})$ of an optimal homogeneous effective material.

Quasiconvex Envelope and the Effective Properties

Here we establish some algebraic properties of $W_{\text{opt}}(m_i, \mathbf{v})$ that are valid independently of the differential constraints (6.1.7) or of the tensor $\mathcal{A}$. The relaxed Lagrangian is bounded by the inequalities (6.3.9). Let us compute the bounds.

The Upper Bound: Constant Trial Field. The upper bound of the relaxed Lagrangian $QF(m_i, \mathbf{v}_0)$ corresponds to the constant trial function $\boldsymbol{\xi} = $

0, $\mathbf{v}(\mathbf{x}) = \mathbf{v}_0$; the bound has the form

$$QF(m_i, \mathbf{v}_0) \leq F(m_i, \mathbf{v}_0),$$

$$F(m_i, \mathbf{v}_0) = \mathbf{v}_0^T \left(\sum_{i=1}^{N} m_i D_i \right) \mathbf{v}_0 + \sum_{i=1}^{N} \gamma_i m_i. \tag{6.4.2}$$

The Lower Bound: Convex Envelope. The lower bound (6.3.10) is given by the convex envelope $CF(m_i, \mathbf{v}_0)$ of the Lagrangian $F(m_i, \mathbf{v}_0)$. The convex envelope corresponds to the optimal choice of the fields $\mathbf{v}_i$ in the materials:

$$CF(m_i, \mathbf{v}_0) = \min_{\mathbf{v}_i \in \mathcal{V}_0} \sum_{i=1}^{N} m_i \mathbf{v}_i^T D_i \mathbf{v}_i + \sum_{i=1}^{N} \gamma_i m_i \tag{6.4.3}$$

where

$$\mathcal{V}_0 = \left\{ \mathbf{v}_i : \ \sum_{i=1}^{N} m_i \mathbf{v}_i = \mathbf{v}_0 \right\}.$$

Recall that the differential constraints on the field $\mathbf{v}$ do not appear in this construction. The fields $\mathbf{v}_i$ are constant, because the energy of each well is convex and the minimum corresponds to a constant solution inside each material.

Let us compute the minimum of the right-hand side of (6.4.3). The optimal fields $\mathbf{v}_i$ are found from the condition

$$\frac{\partial}{\partial \mathbf{v}_i} \left[\sum_{i=1}^{N} m_i \mathbf{v}_i^T D_i \mathbf{v}_i + \gamma \left(\sum_{i=1}^{N} m_i \mathbf{v}_i - \mathbf{v}_0 \right) \right] = 0,$$

where γ is the Lagrange multiplier. The calculation gives

$$\mathbf{v}_i = D_i \left(\sum_{i=1}^{N} m_i D_i^{-1} \right)^{-1} \mathbf{v}_0, \quad i = 1, \ldots, N \tag{6.4.4}$$

(here it is assumed that D_i are invertible). We substitute these fields $\mathbf{v}_i$ into (6.4.3) and obtain the lower bound $CF(m_i, \mathbf{v}_0)$ of the nonquasiconvex Lagrangian $F(m_i, \mathbf{v}_0)$:

$$CF(m_i, \mathbf{v}_0) = \mathbf{v}_0^T \left(\sum_{i=1}^{N} m_i D_i^{-1} \right)^{-1} \mathbf{v}_0 + \sum_{i=1}^{N} \gamma_i m_i. \tag{6.4.5}$$

Invariants of an Optimal Effective Tensor

Comparing the upper bound (6.4.2) and the lower bound (6.4.5) we may represent the quasiconvex envelope of F in the form

$$QF(m_i, \mathbf{v}_0) = W_{\mathrm{opt}} + \sum_{i=1}^{N} \gamma_i m_i, \quad W_{\mathrm{opt}} = \mathbf{v}_0^T D_{\mathrm{opt}}(\sigma_i, m_i, \mathbf{v}_0) \mathbf{v}_0$$

where D_{opt} is the effective properties tensor of an optimal composite.

The bounds (6.4.2) and (6.4.5) yield to the Wiener bounds for an effective tensor D_{opt} of an optimal composite:

$$\left(\sum_{i=1}^{N} m_i D_i^{-1}\right)^{-1} \geq D_{\text{opt}} \leq \sum_{i=1}^{N} m_i D_i.$$

Assume that each available material is either isotropic or an anisotropic material with arbitrary orientation of its main axes. The last condition says that if a material D is available, then all materials $\Phi^T D \Phi$ where Φ is a rotation matrix are also available. These assumptions make the energy W_i rotationally invariant.

The optimal tensor D_{opt} may depend on the field $\mathbf{v}$. Indeed, an optimal structure varies, together with the applied fields $\mathbf{v}$, and its effective tensor also varies. Let us analyze this dependency. Using bounds (6.4.2) and (6.4.5), we conclude that:

1. The quasiconvex envelope $QW(m_i, \mathbf{v})$ is rotationally invariant, that is,

$$QF(m_i, \Phi(\mathbf{v})) = QF(m_i, \mathbf{v}),$$

   where $\Phi(\mathbf{v})$ is the rotated field $\mathbf{v}$.

   Indeed, an optimal composite of isotropic material shows the same optimal properties for all orientations of the applied field $\mathbf{v}$, because the microstructures are free to turn as a whole.

2. The effective properties D_{opt} are independent of the magnitude $|\mathbf{v}|$. Hence, the optimal energy W_{opt} is a homogeneous function of the second degree of $\mathbf{v}$,

$$QW(m_i, \alpha\mathbf{v}) = \alpha^2 QW(m_i, \mathbf{v}),$$

   where α is an arbitrary scalar.

These two properties imply that the matrix D_{opt} (that corresponds to a composite of optimal properties) depends on the invariants $\text{Inv}(\mathbf{v})$ to a rotation of the applied field $\mathbf{v}$, but not on its magnitude:

$$D_{\text{opt}} = D_{\text{opt}}\left(\text{Inv}\left(\frac{\mathbf{v}}{|\mathbf{v}|}\right)\right).$$

For example, an optimal tensor D_{opt} of a conducting composite that minimizes the sum of energies caused by d orthogonal loadings depend only on the ratios between intensities of the loadings. An optimal tensor D_{opt} of a composite that minimizes the sum of energies of two applied fields of different types (say, electrical and thermal fields) depends only on the ratios between intensities of these loadings and the angle between them.

Finally, compare this consideration with the results of relaxation of the problem with a scalar potential w (Chapter 4) that is done by the convex envelope. In this case $\mathbf{v} = \nabla w$ is a vector that has only one invariant: its magnitude. Therefore, D_{opt} is independent of the applied fields and depends only on volume fractions m_i, which agrees with results of that chapter.

6.5   Problems

1. Formulate the problem of minimization of the energy of a structure if the external fields are time-dependent.

2. Formulate the problem of minimization of the energy of a structure for a periodic composite.

3. Suppose that the Lagrangian depends on a divergencefree current $\mathbf{j}$ and on a curlfree field $\mathbf{e}$. Find the tensor A and the differential constraints.

4. Prove the quasiconvexity of the function
$$\phi = -\det \mathbf{v}, \quad \mathbf{v} = \left(\nabla \mathbf{w} + (\nabla \mathbf{w})^T\right),$$
where vector $\mathbf{w}$ is a vector $\mathbf{w} = (w_1(x_1, x_2),\ w_2(x_1, x_2))$.
Hint: Represent ϕ as the sum of a convex function and a null-Lagrangian.

5. Derive the Euler–Lagrange equation for the Lagrangian ϕ.

6. Derive the Weierstrass condition for the Lagrangian ϕ.

7. Compute the convex envelope for problem (6.4.5).

8. The function $G(\mathbf{v})$ is defined as follows:
$$G(\mathbf{v}) = \begin{cases} \pi(\mathbf{v}) & \text{if } \mathbf{v} \in \mathcal{V}, \\ \infty & \text{if } \mathbf{v} \notin \mathcal{V}, \end{cases}$$
where π is a quasiconvex function and $\mathcal{V}$ is a convex set. Is $G(\mathbf{v})$ quasiconvex?

9. Consider the quasiconvex envelope of the Lagrangian equal to the energy of a linear composite that transports N different substances. On how many invariants does the optimal effective tensor depend?

7
Optimal Structures and Laminates

7.1 Laminate Bounds

In this book, two approaches will be examined to obtain the upper bounds of the quasiconvex envelope $QF(\mathbf{v})$. The first approach is discussed in this chapter. It is based on a straightforward construction of the minimizing sequences. The second approach is discussed in Chapter 9; it is based on necessary conditions of optimality.

For convenience, we rewrite here definition (6.3.8) of the quasiconvex envelope. It is the solution to the following extremal problem:

$$QF(\mathbf{x}_0, \mathbf{w}_0, \mathbf{v}_0) = \inf_{\boldsymbol{\xi} \in \Xi} \langle F(\mathbf{x}_0, \mathbf{w}_0, \mathbf{v}_0 + \boldsymbol{\xi}) \rangle,$$

where

$$\Xi = \{\boldsymbol{\xi} : \langle \boldsymbol{\xi} \rangle = 0, \ \mathcal{A} : \nabla \boldsymbol{\xi} = 0, \ \boldsymbol{\xi} \text{ is periodic}\}. \tag{7.1.1}$$

The minimizing sequences approach is a variant of the classical Ritz method. We choose a priori the family $\mathcal{V}_a$ of trial functions, where a is a parameter of this family, and we find the minimizer $\mathbf{v}$ in that family. All trial functions satisfy the differential constraints (7.1.1). The minimum over the family $\mathcal{V}_a$ gives an upper bound of QF, because the set $\mathcal{V}_a$ is smaller than Ξ.

The Simplest Bound. The simplest upper bound of a quasiconvex envelope QF corresponds to the trivial situation in which the set Ξ in (7.1.1) degenerates to one point $\boldsymbol{\xi} = 0$; this results in an inequality:

$$QF(\mathbf{v}) \leq F(\mathbf{v}). \tag{7.1.2}$$

This bound corresponds to the constant trial field $\mathbf{v}$. It is valid for all tensors $\mathcal{A}$ of constraints.

7.1.1 The Laminate Bound

Abridged Trial Fields

Consider one-dimensional trial fields: Choose a direction $\mathbf{n} = \sum \cos\phi_i \mathbf{e}_i$, and assume that $\boldsymbol{\xi}$ depends on one independent variable $\alpha = \mathbf{n}\cdot\mathbf{x}$ as follows: $\boldsymbol{\xi}^{\mathbf{n}}(\alpha) = \boldsymbol{\xi}(\alpha\mathbf{n})$. Here α measures the distance along $\mathbf{n}$. These functions form the set $\Xi^{\mathbf{n}} = \{\boldsymbol{\xi}^{\mathbf{n}}(\alpha)\}$. Obviously, $\Xi^{\mathbf{n}}$ is smaller than Ξ, or $\Xi^{\mathbf{n}} \subset \Xi$. Therefore, the minimum over $\Xi^{\mathbf{n}}$ is greater than the minimum over Ξ.

The differential constraints in (7.1.1), applied to $\boldsymbol{\xi}^{\mathbf{n}}(\alpha)$, become

$$B_{ij}\frac{d\xi_j^{\mathbf{n}}}{d\alpha} = 0, \quad i = 1,\ldots,r, \quad B_{ij} = \mathcal{A}\cdot\mathbf{n} = \sum_{k=1}^{d} a_{ijk}\cos\phi_k. \qquad (7.1.3)$$

These constraints allow us to express the components of $\boldsymbol{\xi}^{\mathbf{n}}(\alpha)$ through $n - r$ free parameters h_i and reduce the minimization problem to an unconstrained one. We have

$$\boldsymbol{\xi}^{\mathbf{n}}(\alpha) = \mathbf{qh}, \quad \mathbf{q} = \mathbf{q}(\mathbf{n}),$$

where $\mathbf{h}$ is the $(n-r)$-dimensional vector, $\mathbf{q}$ is an $(n-r)\times n$ matrix such that $B\mathbf{q} \equiv 0$, and $B = B(\mathbf{n})$ is the matrix of the elements B_{ij}.

Replacing Ξ with $\Xi^{\mathbf{n}}$ in definition (7.1.1) of a quasiconvex envelope, we arrive at its upper bound. Note that the new problem is algebraic:

$$\mathcal{Q}F(\mathbf{v}) \leq L_{(\mathbf{n})}(\mathbf{v}) = \min_{\mathbf{h}(\alpha)\in\mathcal{H}^{\mathbf{n}}} \int_{\Omega} F(\mathbf{v} + \mathbf{qh}(\alpha)), \qquad (7.1.4)$$

where

$$\mathcal{H}^{\mathbf{n}} = \{\mathbf{h}(\alpha), \ \langle\mathbf{h}\rangle = 0\}.$$

Minimizing the right-hand side of (7.1.4) over $\mathbf{h}$, we obtain the convex envelope of the Lagrangian $F(\mathbf{v} + \mathbf{qh}(\alpha))$.

The Carathéodory theorem says that the convex envelope of the Lagrangian is supported by $(n - r + 1)$ points $\boldsymbol{\xi}_i = \boldsymbol{\xi}^{\mathbf{n}}(\alpha_i)$:

$$L_{(\mathbf{n})}F(\mathbf{v}) = \min_{\alpha_k,\mathbf{h}_k,} \sum_{k=1}^{n-r+1} \alpha_k F(\mathbf{v} + \mathbf{q}(\mathbf{n})\mathbf{h}(\alpha_k)),$$

where

$$\sum_{k=1}^{n-r+1} \alpha_k\mathbf{h}(\alpha_k) = 0, \quad \sum_{k=1}^{n-r+1} \alpha_k = 1, \quad \alpha_k \geq 0, \ k = 1,\ldots,n-r+1.$$

Thus we have constructed an upper bound $L_{(\mathbf{n})}F(\mathbf{v})$,

$$QF \leq L_{(\mathbf{n})}F(\mathbf{v}) \leq F$$

that can be algebraically computed.

An Upper Bound $\mathcal{L}_1$

Next, we tighten the bound by eliminating the dependence of the normal $\mathbf{n}$, which so far has been chosen arbitrarily. Choose the direction $\mathbf{n}$ to minimize the value of $L_{(\mathbf{n})}F(\mathbf{v})$ over $\mathbf{n}$; we arrive at an upper bound called the $\mathcal{L}_1$-bound:

$$\mathcal{L}_1 F(\mathbf{v}) = \min_{\mathbf{n}} L_{(\mathbf{n})}F(\mathbf{v}).$$

The $\mathcal{L}_1$-bound improves the bound (7.1.2),

$$F(\mathbf{v}) \geq \mathcal{L}_1 F(\mathbf{v}) \geq QF(\mathbf{v}).$$

We note that $\mathcal{L}_1 F$ is rotationally invariant if $F(\mathbf{v})$ is isotropic, because the minimum has been taken over all orientations of $\mathbf{n}$.

When Does $\mathcal{L}_1 F = CF$ for Two-Well Lagrangians?

Here, we demonstrate that $\mathcal{L}_1 F$ coincides with the convex envelope CF of a two-well Lagrangian if the number of independent differential constraints r is less than the dimension d of the space of independent variables. Recall that a two-well Lagrangian is a minimum of two strongly convex functions called wells. Its convex envelope CF is either supported by two points $\mathbf{v}+\boldsymbol{\xi}_1$ and $\mathbf{v} + \boldsymbol{\xi}_2$ that belong to the first and second well or it coincides with F. The position of supporting points depends on the Lagrangian.

Let us determine when any pair of such points is compatible with differential constraints (7.1.3). Suppose that a laminate exists that corresponds to the fields $\mathbf{v} + \boldsymbol{\xi}_1$ and $\mathbf{v} + \boldsymbol{\xi}_2$ in the neighboring layers. The fields satisfy linear constraints (7.1.3). The constraints correspond to a homogeneous linear system for an unknown d-dimensional vector $\mathbf{n} = [\cos \phi_1, \ldots, \cos \phi_n]$:

$$\sum_{k=1}^{d} C_{ik} \cos \phi_k = 0, \ i = 1, \ldots, r; \quad C_{ik} = \sum_{j=1}^{n} a_{ijk}[\xi_j]. \tag{7.1.5}$$

This system is solvable if the rank of the matrix $C = \{C_{ik}\}$ is less than the dimension d no matter what the values of $[\xi_j]$ are. On the other hand, the rank of C is less than or equal to the number r of linearly independent constraints (7.1.1). Thus we end up with a sufficient condition

$$\mathcal{L}_1 F = CF \ \text{ if } \ r < d.$$

Notice that $\mathcal{L}_1 F$ is an upper bound for the quasiconvex envelope QF, and CF is a lower bound for QF. Whereas the upper and lower bounds coincide, the envelope itself is also defined:

$$QF = CF \ \text{ if } \ r < d.$$

A compatible microstructure is a properly oriented laminate.

Here, the differential constraints (7.1.1) do not forbid the neighboring of any two vectors $\boldsymbol{\xi}_1$ and $\boldsymbol{\xi}_2$. Generally, differential constraints always restrict the type of microstructures of an optimal composite, but they may not be restrictive enough to forbid the attainability of the convex envelope.

Example 7.1.1 Consider an upper bound for a two-well Lagrangian $W(\mathbf{e})$ that depends on a curlfree vector $\mathbf{e}$. The differential constraints $\nabla \times \mathbf{e} = 0$ require compatibility of the system (see (7.1.5))

$$[\mathbf{e}_1 - \mathbf{e}_2] \times \mathbf{n} = 0,$$

where $\mathbf{n}$ is a nonzero unknown vector that can be arbitrarily chosen. This system always has a solution: $\mathbf{n}$ is parallel to $\mathbf{e}_1 - \mathbf{e}_2$. Therefore, the convex envelope of a two-well Lagrangian $F(\nabla w)$ is attainable by a properly oriented laminate.

However, if $\mathbf{e}$ is a matrix gradient of a p-dimensional vector potential ($p > 1$), then the convex envelope is generally not attainable, because there is no vector $\mathbf{n}$ that is parallel to p differently oriented vectors.

Example 7.1.2 Consider a two-well Lagrangian $F(\mathbf{j})$ that depends on a divergencefree vector $\mathbf{j}$. The differential constraint $\nabla \cdot \mathbf{j}$ requires that $\mathbf{n} \cdot [\mathbf{j}_1 - \mathbf{j}_2] = 0$. This problem always has a solution, and therefore we have $\mathcal{L}_1 F(\mathbf{j}) = \mathcal{C}F(\mathbf{j})$. This time compatibility does not uniquely determine the orientation of laminates if $d > 2$.

Example 7.1.3 Consider a Lagrangian that depends on p divergencefree vectors $\mathbf{j}^1, \ldots, \mathbf{j}^p$

$$\nabla \cdot \mathbf{j}^k = 0, \quad k = 1, \ldots, p.$$

The differential constraint $\nabla \cdot \mathbf{j}$ requires that

$$[\mathbf{j}_1^k - \mathbf{j}_2^k] \cdot \mathbf{n} = 0, \quad k = 1, \ldots, p.$$

This linear system has a nontrivial solution if $p < d$. The normal to the optimal layers must be oriented perpendicular to all the vectors $[\mathbf{j}_i^k]$. This is possible if the number of vectors is less than the dimension of the space, $p < d$.

If $p \geq d$, then generally $\mathcal{L}_1 F$ is greater than CF, $\mathcal{L}_1 F \geq CF$. The convex envelope is generally not attainable, because there is no vector $\mathbf{n}$ that is perpendicular to p differently oriented vectors. However, we may improve the bound $\mathcal{L}_1 F$ by more complicated minimizing sequences.

7.1.2 Bounds of High Rank

Consider now the case where $r \geq d$ and, therefore, the convex envelope is not attainable, $\mathcal{L}_1 F > \mathcal{C}F$. The quasiconvex envelope is bounded from two

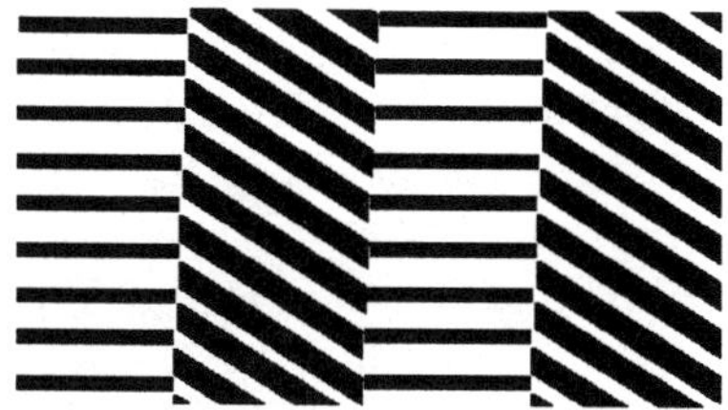

FIGURE 7.1. A second-rank laminate.

sides by two different functions $\mathcal{L}_1 F$ and $\mathcal{C}F$, $\mathcal{L}_1 F \geq \mathcal{C}F$, and there may be room for improving the upper bound $\mathcal{L}_1 F$.

To improve the bound $\mathcal{L}_1 F$, we repeat the preceding procedure and construct the bound

$$\mathcal{L}_2 F = \mathcal{L}_1 \left(\mathcal{L}_1 F \right).$$

The operator $\mathcal{L}_2$ geometrically corresponds to the iterative process of dividing the domain into layered subdomains: It corresponds to the structure of "laminates of second rank," that is, laminates assembled of smaller laminates. We treat a laminate composite as a new material with energy $L_{\mathbf{n}}$, and we laminate two such materials, assuming that the scale of the second step is larger than the scale of the first one. The scheme assumes the hierarchy of the scales: The thickness of inner laminates is much smaller than the thickness of the exterior laminates; see Figure 7.1. The simple laminates are included in the laminates of second rank; hence we obtain the inequalities

$$\mathcal{L}_2 F(\mathbf{v}) \leq \mathcal{L}_1 F(\mathbf{v}) \leq F(\mathbf{v}) \quad \forall \mathbf{v}.$$

Continuing the iterations, we construct the next bounds as follows:

$$\mathcal{L}_k F(\mathbf{v}) = \mathcal{L}_1 \left(\mathcal{L}_{k-1} F(\mathbf{v}) \right), \quad k = 2, \dots .$$

The procedure stops if two sequential bounds coincide, because the resulting bound $\mathcal{L}F$ cannot be further improved

$$\mathcal{L}_k F = \mathcal{L}_{k-1} F \quad \Rightarrow \quad \mathcal{L}_k = \mathcal{L}F.$$

Otherwise, we pass to the limit

$$\mathcal{L}F(\mathbf{v}) = \lim_{k \to \infty} \mathcal{L}_k F(\mathbf{v}).$$

The limit $\mathcal{L}F$ exists if the convex envelope $\mathcal{C}F$ of the energy is finite, $\mathcal{C}F(\mathbf{v}) \geq C > -\infty$ because the sequence $L_k F(\mathbf{v})$ monotonically decreases at each point $\mathbf{v}$ and is bounded from below by the quasiconvex envelope $\mathcal{Q}F$, which is greater than the convex envelope $\mathcal{C}F$,

$$\mathcal{L}F(\mathbf{v}) \geq \mathcal{Q}F(\mathbf{v}) \geq \mathcal{C}F \geq C \quad \forall \mathbf{v}.$$

If the Lagrangian depends on $\mathbf{v} = \nabla \mathbf{w}$, the fields in the laminates differ by a matrix of rank one. In this case, the bound $\mathcal{L}F$ is called the *rank-one bound*. We will keep this name also for the general case.

Relation between $\mathcal{L}_1 F$ and CF. For a two-well Lagrangian with two strongly convex wells, the upper bounds L_k, $k = 2, \ldots$ do not coincide with the convex envelope CF if the bound $\mathcal{L}_1 F$ does not coincide with it:

$$\mathcal{L}_1 F > CF \;\Rightarrow\; L_k F > CF.$$

Indeed, the convex envelope is realized by a field $\mathbf{v}(\mathbf{x})$ that has the following properties: (i) The field is constant within a material (well) and (ii) the fields $\mathbf{v}_1$ and $\mathbf{v}_2$ are compatible in two phases.

The inequality $\mathcal{L}_1 F > CF$ shows that $\mathbf{v}_1$ and $\mathbf{v}_2$ are generally not compatible. On the other hand, the L_k bound preserves the compatibility of the fields by using high-rank laminates. However, this construction increases the number of supporting points. Hence the field $\mathbf{v}$ takes more than one value in at least one well. Therefore, condition (i) is violated, and $\mathcal{L}_k F$ is different from CF.

Remark 7.1.1 *For multicomponent composites, the compatibility conditions could be more complicated. An example of the structures with constant fields in the material is given in Section 7.3 (the T-structure; Figure 7.4).*

Is Rank-One Bound Exact?

Generally, this bound is not exact. Counterexamples were presented starting from (Šverák, 1992b) (see (Bhattacharya, Firoozye, James, and Kohn, 1994; Milton, 2000)). However, the bound is exact for a large class of Lagrangians. Examples of such Lagrangians are given later. Note that the class of these Lagrangians is still not completely determined.

7.2 Effective Properties of Simple Laminates

Consider a variational problem for a multiwell piecewise quadratic Lagrangian that describes the energy of a composite of linear materials. The minimizing sequences for this problem correspond to micro-inhomogeneous composites. The quasiconvex envelope corresponds to the minimal energy among all structures.

Here we describe a large class of structures whose effective properties can be explicitly calculated. The minimum of the piecewise quadratic Lagrangian over the corresponding set of fields can be explicitly calculated. This minimum provides a tight upper bound for the quasiconvex envelope.

We generalize the description of effective properties of laminates given in Chapter 2 to media with arbitrary linear constitutive relations. We describe two approaches to derive the effective properties of laminates: the variational approach and the approach based on the constitutive relations.

7.2.1 Laminates from Two Materials

First, let us discuss the variational approach. Consider a two-component laminate. It is characterized by a normal $\mathbf{n}$ and by volume fractions m_1 and m_2 of subdomains Ω_1 and Ω_2 that are occupied with the materials D_1 and D_2. Suppose that this composite is submerged into a uniform field $\mathbf{v}_0$. The energy of a cell of periodicity Ω is characterized by the quadratic form

$$W = \mathbf{v}_0 \cdot D_* \mathbf{v}_0 = \min_{\mathbf{v}(\mathbf{x}) \in \mathcal{V}} \int_{\Omega} \mathbf{v}(\mathbf{x}) \cdot D(\mathbf{x})\mathbf{v}(\mathbf{x}), \quad \langle \mathbf{v} \rangle = \mathbf{v}_0,$$

where $\mathbf{v}(\mathbf{x})$ is an acting field (stress or strain in elasticity, current or gradient of potential in conductivity, etc.), $\mathcal{V}$ is the set of admissible fields $\mathbf{v}$, and $D(\mathbf{x})$ is a positive tensor of the material's properties, such as conductivity or resistance in a conductivity problem, compliance or stiffness in an elasticity problem, etc.

The tensor D is piecewise constant:

$$D(\mathbf{x}) = D_1 \chi_1(\mathbf{x}) + D_2 \chi_2(\mathbf{x}).$$

Fields in Laminates

Let us describe the set $\mathcal{V}$ of admissible fields $\mathbf{v}$:

1. The field $\mathbf{v}$ in the laminates is piecewise constant:

$$\mathbf{v}(\mathbf{x}) = \begin{cases} \mathbf{v}_1 & \text{if } \mathbf{x} \in \Omega_1, \\ \mathbf{v}_2 & \text{if } \mathbf{x} \in \Omega_2. \end{cases}$$

   Hence the variational problem of the optimal field $\mathbf{v}(\mathbf{x})$ is reduced to the finite-dimensional minimization problem for $\mathbf{v}_1$, $\mathbf{v}_2$:

$$W = \min_{\mathbf{v}_1, \mathbf{v}_2 \in \mathcal{V}} (m_1 \mathbf{v}_1 \cdot D_1 \mathbf{v}_1 + m_2 \mathbf{v}_2 \cdot D_2 \mathbf{v}_2). \tag{7.2.1}$$

2. The mean field $\mathbf{v}_0$ is fixed:

$$\langle \mathbf{v} \rangle = m_1 \mathbf{v}_1 + m_2 \mathbf{v}_2 = \mathbf{v}_0. \tag{7.2.2}$$

3. The differential constraints (7.1.1) on the field $\mathbf{v}$ imply that some linear combinations of its components are continuous on the boundary between layers,

$$\mathbf{p}[\mathbf{v}_1 - \mathbf{v}_2] = 0,$$

   which yields to

$$\mathbf{p}\mathbf{v}_1 = \mathbf{p}\mathbf{v}_2 = \mathbf{p}\mathbf{v}. \tag{7.2.3}$$

   Here $\mathbf{p}$ is a matrix projector on the subspace of continuous components. The matrix $\mathbf{p} = \{p_{ij}\}$ is equal to

$$\mathbf{p} = \mathcal{A} \cdot \mathbf{n} \quad \text{or } p_{ij} = \sum_{k=1}^{d} a_{ijk} n_k. \tag{7.2.4}$$

The components $\mathbf{pv}$ of $\mathbf{v}$ are constant in the laminate. According to the variational principle, the variable vector of orthogonal components $\mathbf{qv}$ of $\mathbf{v}$ should be chosen to minimize the energy of the structure.

Example 7.2.1 If $\mathbf{v} = \mathbf{j}$ is a divergencefree vector, then $\mathbf{p}$ is a (3×1) matrix $\mathbf{p} = \mathbf{n}$, due to the relation

$$\mathbf{n} \cdot [\mathbf{j}_1 - \mathbf{j}_2] = 0.$$

The orthogonal projector $\mathbf{q}$ is a 2×3 matrix $\mathbf{t}_1 \oplus \mathbf{t}_2$, where $\mathbf{t}_1, \mathbf{t}_2$ ($\mathbf{t}_1 \cdot \mathbf{n} = \mathbf{t}_2 \cdot \mathbf{n} = 0$) are tangents to layers:

$$\mathbf{q}^T \mathbf{j} = \begin{pmatrix} \mathbf{t}_1 \cdot \mathbf{j} \\ \mathbf{t}_2 \cdot \mathbf{j} \end{pmatrix}.$$

The notation $\oplus$ means the direct sum of the matrices: Matrix $C = A \oplus B$ consists of the rows of matrices A and B.

Example 7.2.2 If $\mathbf{v} = \mathbf{e}$ is a curlfree vector, then $\mathbf{p}$ is a (2×3) matrix $\mathbf{p} = \mathbf{t}_1 \oplus \mathbf{t}_2$, due to the relations

$$\begin{aligned} \mathbf{t}_1 \times [\mathbf{e}_1 - \mathbf{e}_2] &= 0, \\ \mathbf{t}_2 \times [\mathbf{e}_1 - \mathbf{e}_2] &= 0. \end{aligned}$$

The projector $\mathbf{q}$ is a 1×3 matrix

$$\mathbf{q} = \mathbf{n}^T, \quad \mathbf{qv} = (\mathbf{n}^T \mathbf{e})$$

or the normal vector $\mathbf{n}$.

Optimal Fields. Let us determine the fields $\mathbf{v}_1$ and $\mathbf{v}_2$ that minimize the energy of the cell. One can check that constraints (7.2.2) and (7.2.3) on the set $\mathcal{V}$ are satisfies if the following representation is used:

$$\mathbf{v}_1 = \mathbf{v}_0 - m_2 \mathbf{qh}, \quad \mathbf{v}_2 = \mathbf{v}_0 + m_1 \mathbf{qh}, \tag{7.2.5}$$

where $\mathbf{h}$ is an arbitrary vector in the space of discontinuous components and $\mathbf{p}\mathbf{q}^T = 0$.

To determine optimal fields, we substitute the fields (7.2.5) into the expression for the energy (7.2.1) and choose $\mathbf{h}$ to minimize the energy. The necessary condition of optimality $\frac{d\,W}{d\,\mathbf{h}} = 0$ yields to the representation $\mathbf{h}_0 = \mathcal{L}\mathbf{v}_0$, where

$$\mathcal{L} = [\mathbf{q}^T(m_1 D_2 + m_2 D_1)\mathbf{q}]^{-1}\mathbf{q}^T(D_2 - D_1).$$

Substitute $\mathbf{h}_0$ for $\mathbf{h}$ into the formulas for $\mathbf{v}_i$ and obtain

$$\mathbf{v}_1 = (I - m_2 \mathbf{q}\mathcal{L})\mathbf{v}_0, \quad \mathbf{v}_2 = (I + m_1 \mathbf{q}\mathcal{L})\mathbf{v}_0. \tag{7.2.6}$$

Finally, substitute (7.2.6) into (7.2.1) and obtain the energy W as a quadratic form $W = \mathbf{v}_0 \cdot D_{\mathrm{lam}} \mathbf{v}_0$, where D_{lam} is the effective properties tensor of a laminate, equal to

$$D_{\mathrm{lam}} = m_1 D_1 + m_2 D_2 - \mathcal{N}, \qquad (7.2.7)$$

and

$$\mathcal{N} = m_1 m_2 (D_2 - D_1)\mathbf{q}\left[\mathbf{q}^T(m_1 D_2 + m_2 D_1)\mathbf{q}\right]^{-1}\mathbf{q}^T(D_2 - D_1).$$

This formula for the effective properties of laminates can be used for any linear properties in media.

Example 7.2.3 Consider conducting laminate of two anisotropic materials with conductivities $D_i = \sigma_i$. The $(1 \times d)$ projector $\mathbf{q}$ is the normal $\mathbf{n}$ to the layers: $\mathbf{q} = \mathbf{n}$. The quantity

$$R = \mathbf{q}^T(m_1 D_2 + m_2 D_1)\mathbf{q}$$

is a scalar, equal to

$$R = m_1(\sigma_2)_{nn} + m_2(\sigma_1)_{nn}$$

where the subindex $_{nn}$ denotes the nn component of the tensor. The effective properties tensor σ_{lam} is

$$\begin{aligned}
\sigma_{\mathrm{lam}} &= m_1\sigma_1 + m_2\sigma_2 \\
&\quad - \tfrac{m_1 m_2}{R}(\sigma_2 - \sigma_1)\cdot\mathbf{n}\otimes\mathbf{n}\cdot(\sigma_2 - \sigma_1).
\end{aligned} \qquad (7.2.8)$$

In particular, consider a conducting laminate from two isotropic conductors: $\sigma_1 = s_1 I$, $\sigma_2 = s_2 I$. The effective tensor is

$$(m_1 s_1 + m_2 s_2)I - m_1 m_2 \frac{(s_1 - s_2)^2}{m_2 s_1 + m_1 s_2}\mathbf{n}\otimes\mathbf{n}.$$

Its eigenvectors correspond to the tangent(s) and normal to the layers. The $d - 1$ tangent eigenvalues are

$$\sigma_a = m_1 s_1 + m_2 s_2$$

and the last (normal) eigenvalue is

$$\sigma_h = m_1 s_1 + m_2 s_2 - m_1 m_2 \frac{(s_1 - s_2)^2}{m_2 s_1 + m_1 s_2} = \frac{s_1 s_2}{m_2 s_1 + m_1 s_2}.$$

(The reader will enjoy the last simplification.)

Example 7.2.4 Consider a two-dimensional laminate polycrystal formed from two orthogonally oriented fragments of an anisotropic material,

$$\boldsymbol{\sigma}_1 = \begin{pmatrix} s_1 & 0 \\ 0 & s_2 \end{pmatrix} \text{ and } \boldsymbol{\sigma}_2 = \begin{pmatrix} s_2 & 0 \\ 0 & s_1 \end{pmatrix}$$

with the normal $\mathbf{n} = [1, 0]$. The effective tensor $\boldsymbol{\sigma}_{\text{lam}}$, defined in (7.2.8), is equal to

$$\boldsymbol{\sigma}_{\text{lam}} = \begin{pmatrix} \frac{s_1 s_2}{s_p} & 0 \\ 0 & s_p \end{pmatrix}, \quad s_p = m_1 s_2 + m_2 s_1.$$

One can observe that $\det \boldsymbol{\sigma}_{\text{lam}}$ does not depend on volume fractions. Further, one can check that $\det \boldsymbol{\sigma}_{\text{lam}}$ remains constant when the tensors of properties of the mixing materials and the normal to the layers are arbitrarily oriented. This yields to a conjecture that the determinant of any two-dimensional conducting polycrystal must be equal to the determinant of the conductivity tensor of the initial material. In Chapter 11, we will prove this conjecture.

7.2.2 Laminate from a Family of Materials

Now, we find the effective properties of laminates from a family of materials with properties tensors $D(\phi)$ (ϕ is a parameter of the family). These materials are presented in laminates with measures (volume fractions) $\mu(\phi)$. This time we demonstrate a different technique of averaging the constitutive equations. This technique was suggested in (Backus, 1962).

The laminates are determined by the normal $\mathbf{n}$. The constitutive equations are written in the form

$$\mathbf{u} = D(\phi)\mathbf{v},$$

where $\mathbf{u}$ and $\mathbf{v}$ are the fields that are subject to linear differential constraints. Suppose that $\mathbf{p}$ is the projector to continuous components of $\mathbf{v}$ and $\mathbf{q}$ is the projector to continuous components of $\mathbf{u}$

$$\mathbf{p}[\mathbf{v}] = 0, \quad \mathbf{q}[\mathbf{u}] = 0;$$

$\mathbf{p}$ and $\mathbf{q}$ are orthogonal and complementary to each other.

Consider separately the continuous and discontinuous parts of the vectors $\mathbf{v}$ and $\mathbf{u}$:

$$\mathbf{u}' = \mathbf{p}\mathbf{u}, \quad \mathbf{v}' = \mathbf{p}\mathbf{v}, \quad \mathbf{u}'' = \mathbf{q}\mathbf{u}, \quad \mathbf{v}'' = \mathbf{q}\mathbf{v}.$$

The components $(\mathbf{u}', \mathbf{v}'')$ are continuous, and $(\mathbf{u}'', \mathbf{v}')$ are discontinuous.

The vector equations of the constitutive relations can be written in block form:

$$\begin{pmatrix} \mathbf{u}' \\ \mathbf{u}'' \end{pmatrix} = \mathbf{D} \begin{pmatrix} \mathbf{v}' \\ \mathbf{v}'' \end{pmatrix}, \quad \mathbf{D} = \begin{pmatrix} \mathcal{A} & \mathcal{B} \\ \mathcal{B}^T & \mathcal{C} \end{pmatrix},$$

where the matrices $\mathcal{A}$, $\mathcal{B}$, and $\mathcal{C}$ are as follows:

$$\mathcal{A} = \mathbf{p}^T D \mathbf{p}, \quad \mathcal{B} = \mathbf{p}^T D \mathbf{q}, \quad \mathcal{C} = \mathbf{q}^T D \mathbf{q}.$$

Let us determine the averaged constitutive equations. To average, we solve system (1.3.22) for the discontinuous components and obtain a representation for the effective properties tensor:

$$\begin{pmatrix} \mathbf{v}' \\ \mathbf{u}'' \end{pmatrix} = Z \begin{pmatrix} \mathbf{u}' \\ \mathbf{v}'' \end{pmatrix}, \tag{7.2.9}$$

where

$$Z = \begin{pmatrix} \mathcal{C}^{-1} & \mathcal{C}^{-1}\mathcal{B}^T \\ \mathcal{B}\mathcal{C}^{-1} & \mathcal{A} - \mathcal{B}\mathcal{C}^{-1}\mathcal{B}^T \end{pmatrix}. \tag{7.2.10}$$

(Note that the diagonal block $\mathcal{C}$ of the positive definite tensors $D(\phi)$ is positive definite; therefore, the inverse $\mathcal{C}^{-1}$ exists.)

The vector on the right-hand side of (7.2.9) is constant in the whole structure,

$$\mathbf{v}'' = \text{constant}(\mathbf{x}), \quad \mathbf{u}' = \text{constant}(\mathbf{x}),$$

and therefore the averaging of (7.2.9) requires only the averaging of the properties matrix Z. Here the averaging means

$$\langle Z \rangle = \int Z(\phi)\, d\mu(\phi),$$

where $\mu(\phi)$ is the fraction of the material $D(\phi)$ in the family.

These constraints yield to equalities $\mathbf{v}''(\mathbf{x}) = \langle \mathbf{v}'' \rangle$, $\mathbf{u}(\mathbf{x}) = \langle \mathbf{u}' \rangle$. The homogenized equation is

$$\begin{pmatrix} \langle \mathbf{v}' \rangle \\ \langle \mathbf{u}'' \rangle \end{pmatrix} = \langle Z \rangle \begin{pmatrix} \langle \mathbf{v}'' \rangle \\ \langle \mathbf{u}' \rangle \end{pmatrix}, \tag{7.2.11}$$

where

$$\langle Z \rangle = \begin{pmatrix} \langle \mathcal{C}^{-1} \rangle & \langle -\mathcal{C}^{-1}\mathcal{B}^T \rangle \\ \langle \mathcal{B}\mathcal{C}^{-1} \rangle & \langle \mathcal{A} - \mathcal{B}\mathcal{C}^{-1}\mathcal{B}^T \rangle \end{pmatrix}.$$

Finally, we rewrite the average constitutive relations in the initial form by solving (7.2.11) for the vector $\langle \mathbf{u} \rangle$,

$$\begin{pmatrix} \langle \mathbf{u}' \rangle \\ \langle \mathbf{u}'' \rangle \end{pmatrix} = D_{\text{lam}} \begin{pmatrix} \langle \mathbf{v}' \rangle \\ \langle \mathbf{v}'' \rangle \end{pmatrix},$$

where

$$D_{\text{lam}} = \begin{pmatrix} D_{11} & \langle \mathcal{B}^T \mathcal{C}^{-1} \rangle \langle \mathcal{C}^{-1} \rangle^{-1} \\ \langle \mathcal{C}^{-1} \rangle^{-1} \langle \mathcal{C}^{-1}\mathcal{B} \rangle & \langle \mathcal{C}^{-1} \rangle^{-1} \end{pmatrix} \tag{7.2.12}$$

and

$$D_{11} = \langle \mathcal{A} \rangle - \langle \mathcal{B}\mathcal{C}^{-1}\mathcal{B}^T \rangle + \langle \mathcal{B}^T \mathcal{C}^{-1} \rangle \langle \mathcal{C}^{-1} \rangle^{-1} \langle \mathcal{C}^{-1}\mathcal{B} \rangle.$$

We obtain an analytical expression for the effective tensor D_{lam} that depends on the initial set of materials $D(\omega)$, their measures $\mu(\omega)$, and the normal $\mathbf{n}$ to the layers:

$$D_{\text{lam}} = G(D(\omega), \mu(\omega), \mathbf{n}).$$

Remark 7.2.1 *If the matrices $D(\omega)$ are block diagonal, so that $\mathcal{B} = 0$, then D_{lam} becomes*

$$D_{lam} = \begin{pmatrix} \langle \mathcal{A} \rangle & 0 \\ 0 & \langle \mathcal{C}^{-1} \rangle^{-1} \end{pmatrix}.$$

It consists of arithmetic and harmonic averages.

7.3 Laminates of Higher Rank

Geometry

We now describe the laminates of high rank. They are defined by the following iterative process. We view a laminate structure as an anisotropic material with the effective properties tensor $\mathcal{L}(D_1, D_2, \mu, \mathbf{n})$ that depends on the initial material's properties D_1, D_2 and on structural parameters: the normal $\mathbf{n}$ and the volume fraction μ of the first material. Choosing two different sets

$$\mathbf{n}^{(11)}, \mu^{(11)} \quad \text{and} \quad \mathbf{n}^{(12)}, \mu^{(12)}$$

of values of the last two parameters, we determine two different laminates with the effective properties tensors

$$D^{(11)} = \mathcal{L}(D_1, D_2, \mu^{(11)}, \mathbf{n}^{(11)}) \text{ and } D^{(12)} = \mathcal{L}(D_1, D_2, \mu^{(12)}, \mathbf{n}^{(12)}).$$

The laminate of second rank is the laminate structure with normal $\mathbf{n}^{(2)}$ and fraction $\mu^{(2)}$ made of materials $D^{(11)}$ and $D^{(12)}$ (see Figure 7.1):

$$D_l^{(2)} = \mathcal{L}(D^{(11)}, D^{(12)}, \mu^{(2)}, \mathbf{n}^{(2)}).$$

This structure depends on the following structural parameters: normals $\mathbf{n}^{(11)}, \mathbf{n}^{(12)}, \mathbf{n}^{(2)}$ and concentrations $\mu^{(11)}, \mu^{(12)}, \mu^{(2)}$. It contains the materials D_1 and D_2 in the volume fractions

$$m_1 = \mu^{(11)} \mu^{(2)} + \mu^{(12)} \left(1 - \mu^{(2)}\right), \quad m_2 = 1 - m_1,$$

respectively.

By repeating this procedure one can obtain laminates of any rank. The procedure assumes separation of scales: The width of laminates of each succeeding rank is much greater than the width of the previous rank. The obtained composite is considered as a homogeneous effective material at each step. At the same time, all widths are much smaller than the characteristic length of the domain and of the scale of variation of exterior forces. Under these assumptions, it is possible to explicitly calculate the effective properties of the high-rank laminates. Namely, the laminates of the kth rank correspond to the tensors $D^{(k)}$ determined by the normal $\mathbf{n}^{(k)}$ and the concentration $\mu^{(k-1)}$:

$$D_l^{(k)} = \mathcal{L}(D^{(k-1,1)}, D^{(k-1,2)}, \mu^{(k-1)}, \mathbf{n}^{(k)}),$$

where $D^{(k-1,1)}$ and $D^{(k-1,2)}$ are two tensors of the $(k-1)$th rank. The hierarchical structures were first considered in (Bruggemann, 1930), where also their effective characteristics where introduced. They were first used in (Schulgasser, 1976; Schulgasser, 1977) to describe polycrystals and in (Milton, 1981a; Lurie et al., 1982) to optimize properties of composites.

In dealing with more than two mixing materials, one can either add them one by one or use the following procedure. Starting with the set of materials $D(\phi)$ and choosing a direction vector $\mathbf{n}$, we enlarge it to the set of laminates with properties tensors $\mathcal{L}^1$ described by (7.2.12).

Next we rotate the vector $\mathbf{n}$ by a rotation tensor $\Phi(\boldsymbol{\theta})$, where $\boldsymbol{\theta}$ is the angle(s) of rotation. A rotated tensor $R(\mathcal{L}^1)$ becomes $R(\mathcal{L}^1) = \Phi^T \mathcal{L}^1 \Phi$. The set of second-rank laminates corresponds to the extension of the set of tensors $R(\mathcal{L}^1)$ by the described procedure, and so on. The properties of the set of all laminates (the lamination closure (Milton, 1994)) are discussed later, in Chapter 10.

7.3.1 Differential Scheme

Here we describe a special class of laminates. It involves structures of infinite rank. We develop the following differential scheme: An infinitesimal portion of a pure material is added to the composite at each infinitesimal step. Such materials are described in many works, starting from (Bruggemann, 1935). They were rediscovered and systematically used in (Norris, 1985; Lurie and Cherkaev, 1985; Avellaneda, 1987a), among other papers.

Consider the process of formation of a laminate composite. Suppose that a portion $d\mu \ll 1$ of material D is added to the composite with the effective tensor $\Delta(\mu)$ and that the periodic cell has volume μ. The material is added in thin periodic layers with the normal $\mathbf{n}(\mu)$. The resulting composite cell has the volume $\mu + d\mu$ and an effective properties tensor denoted by $\Delta(\mu + d\mu)$.

Let us compute $\Delta(\mu + d\mu)$ by using (7.2.7), where we set

$$m_1 = \frac{d\mu}{\mu}, \quad m_2 = 1 - \frac{d\mu}{\mu},$$
$$D_1 = D, \quad D_2 = \Delta(\mu), \quad D_{\text{lam}} = \Delta(\mu + d\mu).$$

The relation (7.2.7) becomes

$$\Delta(\mu + d\mu) - \Delta(\mu) = \frac{d\mu}{\mu}\Psi(\Delta(\mu), D, \mathbf{n}) + o(d\mu)$$

where

$$\Psi(\Delta(\mu), D, \mathbf{n}) = -[(\Delta(\mu) - D) - (\Delta(\mu) - D)N(\Delta(\mu) - D)],$$
$$N = \mathbf{q}[\mathbf{q}^T D\mathbf{q}]^{-1}\mathbf{q}^T.$$

The function Ψ depends on $\mathbf{n}$ through $N = N(\mathbf{q})$, because $\mathbf{q}$ is determined by $\mathbf{n}$: $\mathbf{q} = \mathbf{q}(\mathbf{n})$.

As $d\mu$ tends to zero, we obtain the differential equation

$$\mu\frac{d}{d\mu}\Delta(\mu) = \Psi(\Delta(\mu), D, \mathbf{n}). \qquad (7.3.1)$$

This equation shows the rate of change of effective properties. It is integrated with respect to $\mu \in [0, 1]$. The functions $D = D(\mu)$ and $\mathbf{n} = \mathbf{n}(\mu)$ determine the structure of the composite. We assume that different materials with properties $D(\mu)$ are added to the composite in different "times" μ. We also assume that $\mathbf{n} = \mathbf{n}(\mu)$: The direction of laminates is generally changed during formation of the composite.

A Special Case. Suppose that $D = D_1 = \text{constant}(\mu)$ when μ varies in a certain interval. Accordingly, (7.3.1) can be transferred to the linear equation in that interval as follows. We introduce the tensor

$$Z(\mu) = (\Delta(\mu) - D)^{-1}$$

and we compute

$$\frac{d}{d\mu}\Delta = -Z^{-1}\left(\frac{d}{d\mu}Z\right)Z^{-1}, \quad \Psi = -Z^{-1} - Z^{-1}NZ^{-1}.$$

Remark 7.3.1 *We assume here that the tensor Z^{-1} exists or, equivalently, that $\det(\Delta(\mu) - D) \neq 0$.*

Dealing with the degenerate matrix $\Delta(\mu) - D$, we project the matrix equation (7.3.1) to the nonzero space of $\Delta(\mu) - D$ and apply the procedure to the projection. The components of Z that belong to the zero subspace of $\Delta(\mu) - D$ remain constant, because in that subspace the properties of the added material D are equal to the properties of the composite Δ. The corresponding example is discussed in Chapter 16.

Equation (7.3.1) takes the form

$$\mu^2\frac{d}{d\mu}\left(\frac{Z(\mu)}{\mu}\right) + N = 0, \quad N = \mathbf{q}[\mathbf{q}^T D\mathbf{q}]^{-1}\mathbf{q}^T. \qquad (7.3.2)$$

Equation (7.3.2) is a matrix linear differential equation for Z; Matrix $N = N(\mathbf{q}(\mu))$ controls the direction of lamination.

Example 7.3.1 Let us obtain properties of a simple two-component laminate by integrating (7.3.2). We set $D = D_1 = \text{constant}(\mu))$ and $N = \text{constant}(\mu)$. The corresponding initial condition takes the form

$$Z(m_2) = (D_2 - D_1)^{-1}, \tag{7.3.3}$$

because the process of formation of the composite starts with adding D_1 to the pure material D_2, which has the volume fraction m_2.

Under the above assumptions, equation (7.3.2) becomes

$$\frac{d}{d\mu}\left(\frac{Z(\mu) - N}{\mu}\right) = 0$$

and we have

$$Z(\mu) = -N + \mu C$$

where C is a constant matrix. The constant C is defined from (7.3.3); solution becomes

$$Z(\mu) = \frac{\mu}{m_2}(D_2 - D_1)^{-1} + \frac{\mu - m_2}{m_2}N$$

The effective property of the laminate corresponds to the final point $\mu = 1$ of the process of formation: $Z(1) = (D_{\text{lam}} - D_1)^{-1}$. We compute

$$Z(1) = \frac{1}{m_2}\left[(D_2 - D_1)^{-1} + m_1 N\right] \tag{7.3.4}$$

(recall that N is defined by (7.3.2)).

This example introduces another useful representation for the effective properties of a laminate:

$$m_2(D_{\text{lam}} - D_1)^{-1} = (D_2 - D_1)^{-1} + m_1\mathbf{q}\left(\mathbf{q}^T D\mathbf{q}\right)^{-1}\mathbf{q}^T, \tag{7.3.5}$$

which is obtained from (7.3.4) by substituting the values of Z and N. One can algebraically show that (7.3.5) is equivalent to (7.2.7). This representation is used in the next subsection when we build the effective properties of "matrix laminates." Note that (7.3.5) is not symmetric with respect to D_1 and D_2.

Remark 7.3.2 *Formula (7.3.5) was derived in (Gibiansky and Cherkaev, 1987); a close form was suggested in (Francfort and Murat, 1986). Both papers deal with optimization of three-dimensional elastic structures.*

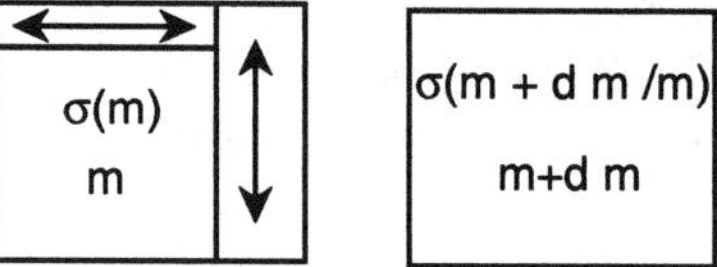

FIGURE 7.2. An infinitesimal step of constructing an isotropic laminate poly-crystal of infinite rank. On the left is the scheme of adding an anisotropic material (arrow shows the direction of larger conductivity); on the right is the homogenized composite. The step is repeated infinitely many times.

Isotropic Layered Polycrystal

The differential scheme, slightly modified, is used to obtain effective properties of an isotropic polycrystal. Consider a three-dimensional inhomogeneous anisotropic material σ characterized by the conductivity tensor

$$\sigma(\mathbf{x}) = \Phi(\mathbf{x})\sigma_0\Phi^T(\mathbf{x}),$$

where $\Phi(\mathbf{x})$ is a rotation matrix that varies from point to point and

$$\sigma_0 = \begin{pmatrix} s_1 & 0 & 0 \\ 0 & s_2 & 0 \\ 0 & 0 & s_2 \end{pmatrix}$$

is an anisotropic material (crystallite) with two equal eigenvalues s_2.

Let us find an isotropic polycrystal that is the composite of differently oriented fragments of a crystallite. Here we describe a layered isotropic composite. In Chapter 11, we will prove that it has extremal conductivity.

We use the following variant of the differential scheme. At each step we add to the polycrystal three orthogonal infinitely small layers with equal thickness $\frac{d\mu}{3}$, and we orient the crystallite σ_0 so that the normal to the layer coincides with the eigenvector of the single eigenvalue s_1 in all three added layers (see Figure 7.2).

The conductivity of the described structure is found from the differential equation (compare with (7.3.1))

$$\mu\frac{d}{d\mu}\Delta(\mu) = \sum_{i=1}^{3}\frac{1}{3}\Psi(\Delta(\mu), D_i, \mathbf{n}_i), \qquad (7.3.6)$$

where $\mathbf{n}_i$ are three orthogonal directions and D_i are the three tensors of properties rotated through $90°$ to each other.

Computing Ψ we obtain the equation for the matrix $\Delta(\mu) = \text{diag}(\delta_1, \delta_2, \delta_3)$ with eigenvalues $\delta_i(\mu)$:

$$3\mu\frac{d}{d\mu}\begin{pmatrix} \delta_1 & 0 & 0 \\ 0 & \delta_2 & 0 \\ 0 & 0 & \delta_3 \end{pmatrix} = \begin{pmatrix} \delta_1\frac{s_1-\delta_1}{s_1} & 0 & 0 \\ 0 & s_2-\delta_2 & 0 \\ 0 & 0 & s_2-\delta_3 \end{pmatrix}$$

$$+ \begin{pmatrix} s_2 - \delta_1 & 0 & 0 \\ 0 & \delta_2 \frac{s_1 - \delta_2}{s_1} & 0 \\ 0 & 0 & s_2 - \delta_3 \end{pmatrix}$$

$$+ \begin{pmatrix} s_2 - \delta_1 & 0 & 0 \\ 0 & s_1 - \delta_2 & 0 \\ 0 & 0 & \delta_2 \frac{s_2 - \delta_3}{s_1} \end{pmatrix}.$$

All the eigenvalues $\delta_i(\mu)$ of the "growing crystal" satisfy the same differential equation

$$3\mu \frac{d\delta_i}{d\mu} = \delta_i(\mu) \frac{s_1 - \delta_i(\mu)}{s_1} + 2(s_2 - \delta_2(\mu)), \quad i = 1, 2, 3.$$

The effective conductivity of the polycrystal corresponds to the stationary solution $\frac{d\delta_i}{d\mu} = 0$. The stationarity expresses the fact that the properties of the polycrystal are stable to adding new portions of the crystallite. Notice that there is no dependence on the initial conditions, which is physically obvious.

All stationary eigenvalues δ_i^s share the same value, which is also the effective isotropic conductivity σ_* of the polycrystal

$$\delta_i^s = \sigma_* \quad i = 1, 2, 3.$$

Hence $\sigma_* = \delta_i^s$ satisfies the algebraic equation,

$$\sigma_*(s_1 - \sigma_*) + 2s_1(s_2 - \sigma_*) = 0,$$

which follows from the stationarity conditions $\frac{d}{d\mu} \delta_i^s = 0$, $i = 1, 2, 3$.

The conductivity σ_* of an isotropic three-dimensional polycrystal is equal to

$$\sigma_* = \frac{1}{2} \left(\sqrt{s_1^2 + 8s_1 s_2} \right). \tag{7.3.7}$$

This solution demonstrates that the resulting conductivity of laminates of infinite rank has irrational dependence of the parameters on the initial materials, which cannot be achieved by finite-rank laminates. We will show in Chapter 11 that the conductivity σ_* is in fact the minimal conductivity of a polycrystal of an arbitrary microstructure.

Asymptotically, when $s_1 \ll s_2$, the conductivity σ_* becomes

$$\sigma_* \simeq \sqrt{2s_1 s_2}.$$

Laminates of Infinite Rank with Controllable Properties

This differential scheme enables us to formulate the problem of optimal microstructures as a standard problem of optimal control. This approach was introduced in (Lurie and Cherkaev, 1985) for a special example.

The sequence of added materials $D(\mu)$ can be treated as the control. One chooses $D(\mu)$ to minimize the conductivity of the composite, $\Delta(1)$.

Equation (7.3.1) describes the differential constraints. We can generalize the scheme if we enlarge the set of controls, allowing the addition not only of pure materials but also of composites.

The following scheme considers adding the laminates to the composite at each infinitesimal step. The laminate are made from N initially given isotropic materials with properties $D_i, i = 1, \ldots, N$. The properties of the added laminate are treated as the additional controls. Those are the normal $\tilde{\mathbf{n}}$ and the volume fractions m_i of the materials in the laminate. The controls can vary with μ:

$$D(\mu) = D_{\mathrm{lam}}(\tilde{\mathbf{n}}(\mu), m_i(\mu)).$$

The minimizing quantity is the function $f(\Delta(1))$ of the elements of the resulting effective tensor $D_* = \Delta(1)$. The additional integral constraints may specify the available volume fractions M_i of the materials.

The problem of the optimal control becomes

$$\min_{\mathbf{n}(\mu), \tilde{\mathbf{n}}(\mu), m_i(\mu)} f(\Delta(1)). \tag{7.3.8}$$

The optimization is performed over the following controls: $m_i(\mu)$ $\mathbf{n}(\mu)$ is the normal to the added layer; $\tilde{\mathbf{n}}(\mu)$ is the normal to the laminates inside the added layer.

The minimum in (7.3.8) is subject to the following constraints:

1. the matrix differential equation (7.3.1)

$$\mu \frac{d}{d\mu} \Delta(\mu) = \Psi\left(\Delta(\mu), D_{\mathrm{lam}}(\tilde{\mathbf{n}}, m_i), \mathbf{n}\right),$$

where $m_i(\mu)$ is the volume fraction of the ith material in a laminate with the normal $\mathbf{n}$ added to the composite at the "time" μ;

2. the integral constraints

$$\int_0^1 m_i(\mu)d\mu = M_i,$$

which specify the available amounts of materials; and

3. the geometrical piecewise constraints,

$$m_i \geq 0, \quad \sum_i m_i(\mu) \equiv 1.$$

Remark 7.3.3 *An obvious generalization of this scheme is the following. One adds n differently oriented infinitesimal layers with relative concentration $\alpha_1, \ldots \alpha_n$ at each "moment" μ, as in the preceding problem of a* *polycrystal. The matrix differential equation (7.3.1) takes a form similar to (7.3.6),*

$$\mu\frac{d}{d\mu}\Delta(\mu) = \sum_{i=k}^{n}\alpha_k\Psi\left(\Delta(\mu), D_{lam}(\tilde{\mathbf{n}}_k, m_{ik}), \mathbf{n}_k\right),$$

where $m_{ik}(\mu)$ is the volume fraction of the ith material in a laminate with the normal $\mathbf{n}_k$ added to the composite at the "time" μ. In that scheme the set of controls is enlarged. The constraints must be correspondingly modified.

This approach enables us to minimize the properties of a broad range of structures, including structures topologically equivalent to coated spheres or circles, and to the optimal structures described in the example in Chapter 4 (Figure 4.7). The structures may consist of laminates that form spirals with controllable angle, or multicoated spheres, discussed later in this chapter.

In spite of this generality, these structures are still constructed under some a priori assumptions on the type of the added composite; hence there is no guarantee that they describe all of the G-closure.

7.3.2 Matrix Laminates

Let us describe a special class of laminates called matrix laminates. They have been introduced and studied in (Lurie and Cherkaev, 1982; Lurie et al., 1982; Kohn and Strang, 1986b; Milton, 1986; Milton, 1991b; Francfort and Murat, 1986; Kohn and Lipton, 1988; Strang and Kohn, 1988; Lipton, 1992; Bendsøe, Díaz, and Kikuchi, 1993; Milton, 1994; Díaz, Lipton, and Soto, 1994; Lipton and Díaz, 1995) and other papers, and they represent the most investigated class of high-rank laminates.

Matrix laminates are obtained by iterative lamination of an already built composite with the same initial materials at each step of the procedure. After several steps, we end up with a structure in which the disconnected inclusions of one of the materials is wrapped into another; the first material forms the envelope (matrix), and the second forms the nuclei.

Matrix Laminates of the Second Rank

The form (7.3.5) is convenient for finding the effective moduli of the matrix laminates. Let us calculate effective properties of a second-rank structure that is a layered composite with the normal $\mathbf{n}$, assembled from the material D_1 and the laminate D_{lam}. The concentrations of materials D_1 and D_{lam} are denoted by μ_{21} and μ_{22}. The laminate D_{lam} has a different normal $\mathbf{n}'$ and is assembled from the materials D_1 and D_2, taken in the volume fractions μ_{11}, μ_{12}, respectively. The total amount of the material D_2 is fixed equal to m_2. Therefore, the parameters μ_i satisfy the constraints

$$m_2 = \mu_{12}\,\mu_{22}, \quad \mu_{11} + \mu_{12} = 1, \quad \mu_{21} + \mu_{22} = 1. \tag{7.3.9}$$

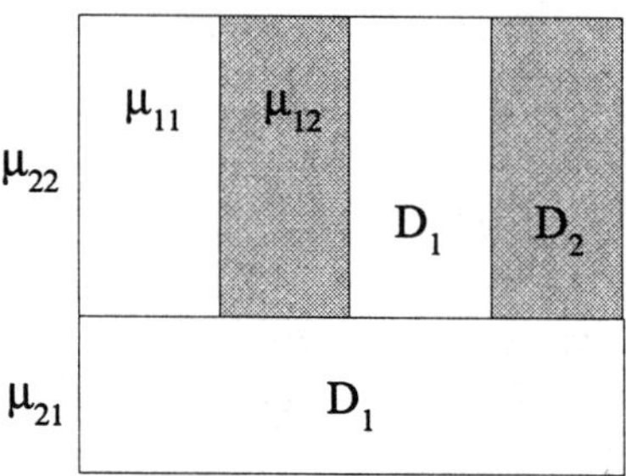

FIGURE 7.3. The element of periodicity of a matrix laminate of the second rank.

That leaves one degree of freedom to fix them.

We call this structure the *matrix laminate of second rank*, and denote its tensor of effective properties by $D_{\text{ml-2}}$. The tensor $D_{\text{ml-2}}$ can be calculated from (7.3.5) taking the form:

$$(D_{\text{ml-2}} - D_1)^{-1} = \frac{1}{\mu_{22}}(D_{\text{lam}} - D_1)^{-1} + \frac{\mu_{21}}{\mu_{22}}\mathbf{q}_1(\mathbf{q}_1^T D_1 \mathbf{q}_1)^{-1}\mathbf{q}_1^T. \quad (7.3.10)$$

where $\mathbf{q}_1 = \mathbf{q}(\mathbf{n}_1)$ and $\mathbf{n}_1$ is the normal to the first rank layers. Tensor D_{lam} represents the tensor of properties of the laminates made in the first step of the procedure. This tensor is again found from (7.3.5), where one replaces $\mathbf{q} = \mathbf{q}(\mathbf{n})$ by $\mathbf{q}(\mathbf{n}_2)$, m_1 by μ_{11}, and m_2 by μ_{12}:

$$(D_{\text{lam}} - D_1)^{-1} = \frac{1}{\mu_{12}}(D_2 - D_1)^{-1} + \frac{\mu_{11}}{\mu_{12}}\mathbf{q}_2(\mathbf{q}_2^T D_1 \mathbf{q}_2)^{-1}\mathbf{q}_2^T.$$

We substitute $(D_{\text{lam}} - D_1)^{-1}$ in (7.3.10) and obtain

$$\begin{aligned}
(D_{\text{ml-2}} - D_1)^{-1} = {} & \frac{1}{\mu_{22}\,\mu_{12}}(D_2 - D_1)^{-1} \\
& + \frac{\mu_{11}}{\mu_{22}\,\mu_{12}}\mathbf{q}_2(\mathbf{q}_2^T D_1 \mathbf{q}_2)^{-1}\mathbf{q}_2^T \\
& + \frac{\mu_{21}}{\mu_{22}}\mathbf{q}_1(\mathbf{q}_1^T D_1 \mathbf{q}_1)^{-1}\mathbf{q}_1^T.
\end{aligned}$$

Now we use (7.3.9), and we observe the following relations between the volume fractions:

$$\frac{1}{\mu_{22}\,\mu_{12}} = \frac{1}{m_2}, \qquad \frac{\mu_{11}}{\mu_{22}\,\mu_{12}} + \frac{\mu_{21}}{\mu_{22}} = \frac{m_1}{m_2}.$$

Finally, we obtain the formula

$$(D_{\text{ml-2}} - D_1)^{-1} = \frac{1}{m_2}(D_2 - D_1)^{-1} + \frac{m_1}{m_2}G, \qquad (7.3.11)$$

where

$$G = \sum_{i=1}^{2} \alpha_i \mathbf{q}_i(\mathbf{q}_i^T D_1 \mathbf{q}_i)^{-1}\mathbf{q}_i^T.$$

Here the parameters α_1 and α_2 are

$$\alpha_1 = \left(\frac{m_2}{m_1}\right)\frac{\mu_{11}}{\mu_{22}\mu_{12}} \geq 0, \quad \alpha_2 = \left(\frac{m_2}{m_1}\right)\frac{\mu_{21}}{\mu_{22}} \geq 0.$$

They are chosen so that

$$\alpha_1 + \alpha_2 = 1, \quad \alpha_1 \geq 0, \quad \alpha_2 \geq 0.$$

Matrix Laminates of an Arbitrary Rank

The effective properties of matrix laminates of an arbitrary rank are derived in a similar way. By repeating the outlined procedure k times, we derive a formula for the effective properties $D_{\mathrm{ml}-k}$ of a matrix laminate the of kth rank:

$$D_{\mathrm{ml}-k} = D_1 + m_2\left[(D_2 - D_1)^{-1} + m_1 G\right]^{-1}, \tag{7.3.12}$$

where

$$G = \sum_{i=1}^{k} \alpha_i \mathbf{q}(\mathbf{n}_i) H_i^{-1} \mathbf{q}^T(\mathbf{n}_i)$$

and

$$\alpha_i \geq 0, \quad \sum_{i=1}^{k} \alpha_i = 1, \quad H_i = \mathbf{q}^T(\mathbf{n}_i) D_1 \mathbf{q}(\mathbf{n}_i).$$

Notice that the last term on the right-hand side of (7.3.12) is a convex envelope of the set of $m \times m$ matrices, stretched on the matrices $N_i = \mathbf{q}(\mathbf{n}_i) H_i^{-1} \mathbf{q}^T(\mathbf{n}_i)$.

If the enveloping material D_1 is isotropic, then H_i is independent of the direction $\mathbf{n}_i$, $H_i = \mathrm{constant}(i) = H$ is an isotropic positive definite matrix, and G becomes

$$G = \sum_{i=1}^{k} \alpha_i \mathbf{q}(\mathbf{n}_i) H^{-1} \mathbf{q}^T(\mathbf{n}_i).$$

Remark 7.3.4 *This formula was obtained independently in (Francfort and Murat, 1986; Gibiansky and Cherkaev, 1987) for three-dimensional elastic composites.*

Example 7.3.2 Consider the matrix laminates of two isotropic conducting materials $\boldsymbol{\sigma}_1 = \sigma_1 I$ and $\boldsymbol{\sigma}_2 = \sigma_2 I$. We set $\mathbf{q} = \mathbf{n}$ and $\mathbf{q}^T \boldsymbol{\sigma}_1 \mathbf{q} = (\sigma_1)$. Then the formula (7.3.12) takes the form

$$(\boldsymbol{\sigma}_* - \sigma_1 I)^{-1} = \frac{1}{m_2}\left[\frac{1}{\sigma_2 - \sigma_1}I + m_1\frac{1}{\sigma_1}G\right], \tag{7.3.13}$$

where

$$G = \frac{1}{\sigma_1}\sum_{i=1}^{p}\alpha_i \mathbf{n}_i \otimes \mathbf{n}_i, \quad \sum_{i=1}^{p}\alpha_i = 1.$$

Effective tensors of matrix composites have several remarkable properties. The trace of the tensor $(\boldsymbol{\sigma}_* - \sigma_1 I)^{-1}$ is constant. Indeed,

$$\mathrm{Tr}(\boldsymbol{\sigma}_* - \sigma_1 I)^{-1} = \frac{1}{m_2}\,\mathrm{Tr}(\boldsymbol{\sigma}_2 - \boldsymbol{\sigma}_1)^{-1} + \frac{m_1}{m_2}\,\mathrm{Tr}\,G,$$

where

$$\mathrm{Tr}\,G = \frac{1}{\sigma_1}\,\mathrm{Tr}\sum_{i=1}^{p}\alpha_i \mathbf{n}_i \otimes \mathbf{n}_i = \frac{1}{\sigma_1}\sum_{i=1}^{p}\alpha_i\,\mathrm{Tr}(\mathbf{n}_i \otimes \mathbf{n}_i) = \frac{1}{\sigma_1}. \qquad (7.3.14)$$

Here we use the equalities $\mathrm{Tr}(\mathbf{n}_i \otimes \mathbf{n}_i) = 1$ and $\sum_{i=1}^{p}\alpha_i = 1$. We have from (7.3.14), (7.3.15),

$$\mathrm{Tr}(\boldsymbol{\sigma}_* - \sigma_1 I)^{-1} = \frac{1}{m_2}\left(\frac{d}{\sigma_2 - \sigma_1} + m_1\frac{1}{\sigma_1}\right). \qquad (7.3.15)$$

Every effective tensor of a matrix laminate lies on the surface (7.3.15) independently of the inner geometrical parameters of these structures.

Particularly, the effective conductivity of the isotropic matrix laminates is uniquely determined as

$$\sigma_{\mathrm{isotr}}^* = \sigma_1 + m_2\left(\frac{1}{\sigma_2 - \sigma_1} + \frac{m_1}{d\sigma_1}\right)^{-1}.$$

It coincides with the effective conductivity of the coated circles (Chapter 2). It is remarkable that the last conductivity coincides with the maximal possible conductivity of any isotropic composite if $\sigma_1 \geq \sigma_2$, and with minimal possible conductivity of any isotropic composite, if $\sigma_1 \leq \sigma_2$, as it follows from the Hashin–Shtrikman bounds (Hashin and Shtrikman, 1962a) for isotropic composites. It turns out that anisotropic matrix composites of both types form components of the boundary of the G_m-closure (Lurie and Cherkaev, 1982; Lurie and Cherkaev, 1984a; Murat and Tartar, 1985a); see Chapter 11.

The other constraints on the class of effective tensors are inequalities that follow from the nonnegativity of N; see (Lurie and Cherkaev, 1984a). We have

$$(\boldsymbol{\sigma}_* - \sigma_1 I)^{-1} - \left(\frac{1}{m_2}(\sigma_2 I - \sigma_1 I)^{-1}\right) \geq 0,$$

or, assuming that $\sigma_2 > \sigma_1$,

$$(m_1\sigma_1 + m_2\sigma_2)I - \boldsymbol{\sigma}_* \geq 0. \qquad (7.3.16)$$

We recognize the arithmetic mean bound in the last inequality.

If the rank k of the matrix laminates is less than d, then $d - k$ eigenvalues of the matrix on the left-hand side of the last expression are equal to zero.

This demonstrates that a matrix laminate of rank k, which is less than the dimension of the space d, has $d - k$ eigenvalues equal to the arithmetic mean bound.

One can also prove the following inverse statements (see (Lurie and Cherkaev, 1984a)):

1. Any matrix $\boldsymbol{\sigma}_*$ that satisfies the equality (7.3.15) and the inequalities (7.3.16) is equal to the tensor of effective properties of some matrix laminate structure of rank not more than d, with mutual orthogonal layers.

2. Any matrix σ_* that satisfies the equality (7.3.15) and the inequalities (7.3.16) is equal to the tensor of effective properties of some matrix laminate structure of rank $k < d$ with mutual orthogonal layers if the rank of the matrix on the left-hand side of (7.3.16) is equal to k.

It follows that the class of dth-rank matrix laminates with orthogonal layers contains elements that are equivalent to any other matrix laminate.

7.3.3 Y-Transform

The Wiener bounds for effective tensors are valid independent of the differential properties of the field. Any effective tensor D_* lies in the Wiener box

$$D_* \leq m_1 D_1 + m_2 D_2, \quad D_*^{-1} \leq m_1 D_1^{-1} + m_2 D_2^{-1}. \tag{7.3.17}$$

To account for these bounds, it is convenient to introduce a linear-fractional transform of D_* that maps the Wiener box into the positive cone

$$Y(D_*) \geq 0 \text{ if and only if } D_* \text{ satisfies (7.3.17)}.$$

This representation, suggested and developed in (Milton and Golden, 1985; Milton, 1991a; Cherkaev and Gibiansky, 1992) is called the Y-transform.

Y-Transform of a Scalar. Let us introduce the Y-transform of a scalar parameter d_*:

$$Y(d_*) = \frac{d_*(m_1 d_2 + m_2 d_1) - d_1 d_2}{m_1 d_1 + m_2 d_2 - d_*} \tag{7.3.18}$$

or

$$Y(d_*) = (m_1 d_2 + m_2 d_1)\frac{d_* - d_h}{d_a - d_*}.$$

One can see and that the identity

$$\frac{1}{Y(d_*) + d_*} = \frac{m_1}{Y(d_*) + d_1} + \frac{m_2}{Y(d_*) + d_2}$$

holds and that

$$Y(d_*) \in [0, \infty) \quad \text{if and only if} \quad d_* \in [d_h, d_a),$$
$$d_h = \lim_{Y \to 0} d_*(Y), \quad d_a = \lim_{Y \to \infty} d_*(Y).$$

Y-Transform of a Tensor. The Y-transform of a tensor D_* is defined by the analogue of the previous equality

$$(D_* + Y)^{-1} = m_1(D_1 + Y)^{-1} + m_2(D_2 + Y)^{-1}. \tag{7.3.19}$$

It represents the shifted tensor $(D_* + Y)$ as the harmonic mean of the shifted tensors $(D_1 + Y)$, $(D_2 + Y)$.

Generally, the shift $Y \geq 0$ determines the position of D_* within the Wiener box. In other terms, Y maps the domain $D_h \leq D_* \leq D_a$ onto the cone $Y(D_*) \geq 0$.

Let us determine properties of $Y(D_*)$ for specific geometries.

Example 7.3.3 Consider again the formula for laminates (7.2.7). The lamination formula can be rewritten as

$$\begin{aligned} D_* = {}&m_1 D_1 + m_2 D_2 \\ &- m_1 m_2 (D_2 - D_1)(m_1 D_2 + m_2 D_1 + Y)^{-1}(D_2 - D_1), \end{aligned} \tag{7.3.20}$$

where Y is the improper tensor $Y = Y_{\text{lam}}$ of the form

$$Y_{\text{lam}} = \lim_{c \to \infty} c\mathbf{p}\mathbf{p}^T. \tag{7.3.21}$$

To derive this representation, we use the identity:

$$\lim_{c \to \infty}(Z + c\mathbf{p}\mathbf{p}^T)^{-1} = \mathbf{q}(\mathbf{q}^T Z \mathbf{q})^{-1}\mathbf{q}^T$$

where $\mathbf{q}$ is the supplement to $\mathbf{p}$: $\mathbf{q} \oplus \mathbf{p} = I$.

In the space of eigenvalues of the Y tensor, composites correspond to the improper corner: the eigenvalues corresponding to the harmonic mean of properties are zero, and the those corresponding to the arithmetic mean tend to infinity. Note that (7.3.20) can be also rewritten in the form (7.3.19).

Example 7.3.4 The matrix laminates of conducting materials correspond to the Y-tensors

$$\text{Tr}(Y^{\text{ml}} + \sigma_1 I)^{-1} = \frac{1}{\sigma_1} \tag{7.3.22}$$

(see (Cherkaev and Gibiansky, 1992)). To derive this representation, we substitute the expression (7.3.19) for Y into the formula for matrix laminates (7.3.12) and do the necessary simplifications. The asymmetry of the first and second materials is due to the matrix laminate structure.

Note that both representations (7.3.21) and (7.3.22) are independent of the volume fractions. Tensor Y measures how close the property tensor is to the boundary of the Wiener bounds. It defines the topological properties of the structure and its anisotropy.

Properties of Y-Tensors

We will show that the use of Y-tensors is convenient to describe G_m-closures. Here we mention several useful equalities for that tensor. We assume here that the matrices D_1 and D_2 are commutative: $D_1 D_2 = D_2 D_1$. The noncommutative case is considered in (Milton, 1990b).

The proofs of the following equalities are straightforward; they are left to the reader:

1. When the matrices D_1, D_2, D_* commute, $Y(D_*, D_1, D_2)$ is equal to

$$Y(D_*) = -(D_* - D_h)(D_* - D_a)^{-1} D_1 D_2 D_h^{-1}. \qquad (7.3.23)$$

2. If the components D_1 and D_2 are isotropic, then the tensor $Y(D_*)$ has the same eigenvectors as D_*. The isotropy of D_* leads to the isotropy of $Y(D_*)$.

3. Consider the Y-transform as a function of the effective tensor D_* and the tensors D_1 and D_2 of materials' properties. The equality

$$Y(D_*^{-1}, D_1^{-1}, D_2^{-1}) = Y^{-1}(D_*, D_1, D_2) \qquad (7.3.24)$$

holds; it states that the substitution of the inverse values of the tensors D_1, D_2, D_* in (7.3.20) is equivalent to the inversion of Y.

4. $Y(D_*, D_1, D_2)$ satisfies the formal equalities

$$\begin{aligned} Y(D_1, D_1, D_2) &= -D_1, \\ Y(D_2, D_1, D_2) &= -D_2. \end{aligned} \qquad (7.3.25)$$

The equalities (7.3.25) are formal in the sense that the tensor D_* of a composite with nonzero volume fraction is never equal to the tensors D_1 or D_2 of the components; the formal nature of these inequalities corresponds to the negativity of the right-hand side of (7.3.25)

Remark 7.3.5 *The Y-transform was also used in the analytical method (Milton and Golden, 1990) where this transform was introduced as the solution to the variational problem of polarization.*

7.3.4 Calculation of the Fields Inside the Laminates

The Algorithm. The following is a procedure for calculating the fields in a laminate of a high rank viewed as laminates from substructures that could be either laminates or substructures of a deeper level. An iterative scheme is needed to compute the fields in the pure materials that form the deepest level of the structure.

The algorithm is as follows: First we calculate the effective properties of a laminate as a function of the known properties of its components. Then we calculate the fields inside the layers of the laminate using the known average field and the effective properties. We use formula (7.2.6) to compute the fields inside the layers in an arbitrary laminate structure. In this formula, D_1 and D_2 are the effective tensors of substructures in layers. The fields in deeper layers are found by iterations of the procedure.

To illustrate the algorithm, consider two-dimensional conductivity. Suppose that laminates are assembled from two anisotropic materials with the codirected conductivity tensors σ_1 and σ_2 mixed in proportions c_1 and c_2 $(c_1 + c_2 = 1)$. The normal $\mathbf{n}$ to the laminate and the matrices of materials' properties are

$$\mathbf{n} = \begin{pmatrix} 1 \\ 0 \end{pmatrix}, \quad \sigma_i = \begin{pmatrix} \mathbf{s}_A^{(i)} & 0 \\ 0 & \mathbf{s}_B^{(i)} \end{pmatrix}, \; i = 1, 2. \tag{7.3.26}$$

The effective properties tensor σ_* of the laminate is

$$\sigma_* = \begin{pmatrix} \dfrac{\mathbf{s}_A^{(1)}\mathbf{s}_A^{(2)}}{c_1 \mathbf{s}_A^{(2)} + c_2 \mathbf{s}_A^{(1)}} & 0 \\ 0 & c_1 \mathbf{s}_B^{(1)} + c_2 \mathbf{s}_B^{(2)} \end{pmatrix}; \tag{7.3.27}$$

the normal component is given by the harmonic mean of the materials' properties, and the tangent component is given by the arithmetic mean.

Suppose that the laminate is submerged into two mutual orthogonal fields described by the symmetric matrix $\mathbf{e}$ that are codirected with the eigenvectors of σ_*. The field $\mathbf{e}_1$ in the material σ_1 is computed from the equality $c_1 \mathbf{e}_1 + c_2 \mathbf{e}_2 = \mathbf{e}$ and from the jump condition. It is equal to

$$\mathbf{e}_1 = \mathbf{K}_1 \mathbf{e}, \quad \mathbf{K}_1 = \begin{pmatrix} \dfrac{\mathbf{s}_A^{(2)}}{c_1 \mathbf{s}_A^{(2)} + c_2 \mathbf{s}_A^{(1)}} & 0 \\ 0 & 1 \end{pmatrix}. \tag{7.3.28}$$

The field $\mathbf{e}_2$ in σ_2 is computed similarly.

To calculate the fields in a laminate of a high rank one must first compute the effective properties of the substructures that form the composite and find the matrices $\mathbf{K}_i$. Then one computes the fields using (7.3.28).

Example 7.3.5 As an example let us compute the effective properties of the "T-structure" shown in Figure 7.4. First, we compute the properties of laminate substructure $R_{(13)}$ of σ_1 and σ_3. The relative fraction of the first and third materials in the substructure are $\dfrac{m_1}{m_1+m_3}$ and $\dfrac{m_3}{m_1+m_3}$. Equations (7.3.26) and (7.3.27) give

$$\mathbf{n} = \begin{pmatrix} 0 \\ 1 \end{pmatrix}, \quad \sigma_{(13)} = \begin{pmatrix} \mathbf{s}_A^{(13)} & 0 \\ 0 & \mathbf{s}_B^{(13)} \end{pmatrix},$$

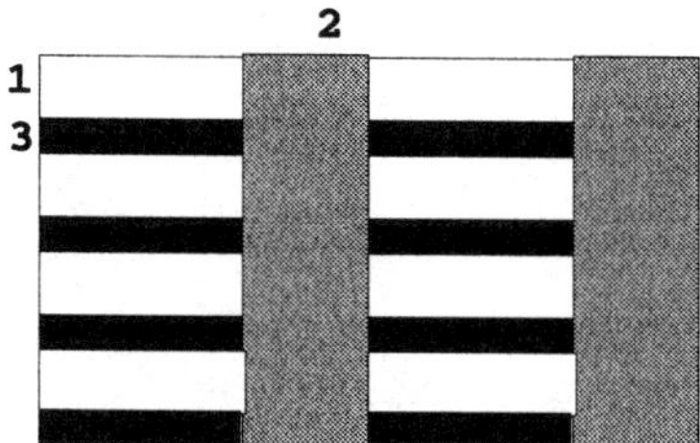

FIGURE 7.4. T-structure. Three materials with conductivities σ_1, σ_2, σ_3 and the volume fractions m_1, m_2, m_3 are combined as follows: The first and the third materials form a laminate $R_{(13)}$; then the second material and $R_{(13)}$ form a laminate of second rank. The normal to the last laminate is orthogonal to the normal of the first-rank laminate.

where (see (7.3.27))

$$s_A^{(13)} = \frac{m_1\sigma_1 + m_3\sigma_3}{m_1 + m_3}, \quad s_B^{(13)} = \frac{(m_1 + m_3)\sigma_1\sigma_3}{m_1\sigma_3 + m_3\sigma_1}.$$

Now we can compute the fields in domain $\mathcal{O}_2$ and the average field in the composed domain $\mathcal{O}_{13}$

$$\mathbf{e}_2 = \mathbf{K}_2\,\mathbf{e}, \qquad \mathbf{K}_2 = \begin{pmatrix} \dfrac{s_A^{(13)}}{(m_2+m_3)s_A^{(13)}+m_2\sigma_2} & 0 \\ 0 & 1 \end{pmatrix},$$

$$\mathbf{e}_{(13)} = \mathbf{K}_{(13)}\,\mathbf{e}, \qquad \mathbf{K}_{(13)} = \begin{pmatrix} \dfrac{\sigma_2}{(m_2+m_3)s_A^{(13)}+m_2\sigma_2} & 0 \\ 0 & 1 \end{pmatrix}.$$

Let us compute the fields in the domains $\mathcal{O}_1$ and $\mathcal{O}_3$ that compose the first rank laminate $\mathcal{O}_{13}$:

$$\mathbf{e}_1 = \mathbf{K}_{(1)}\mathbf{e}_{(13)}, \quad \mathbf{K}_{(1)} = \begin{pmatrix} 1 & 0 \\ 0 & \dfrac{\sigma_3(m_1+m_3)}{m_1\sigma_3+m_3\sigma_1} \end{pmatrix},$$

$$\mathbf{e}_{(3)} = \mathbf{K}_{(3)}\mathbf{e}_{(13)}, \quad \mathbf{K}_{(3)} = \begin{pmatrix} 1 & 0 \\ 0 & \dfrac{\sigma_1(m_1+m_3)}{m_1\sigma_3+m_3\sigma_1} \end{pmatrix}.$$

Finally, we obtain the dependence of the fields in the materials on the geometric parameters of the structure (the tensors $\mathbf{K}_i$):

$$\mathbf{e}_1 = \mathbf{K}_{(1)}\,\mathbf{K}_{(13)}\,\mathbf{e}, \quad \mathbf{e}_2 = \mathbf{K}_2\,\mathbf{e}, \quad \mathbf{e}_3 = \mathbf{K}_{(3)}\,\mathbf{K}_{(13)}\,\mathbf{e}.$$

Controlling the coefficients $\mathbf{K}_{(i)}$ one varies the fields $\mathbf{e}_i$.

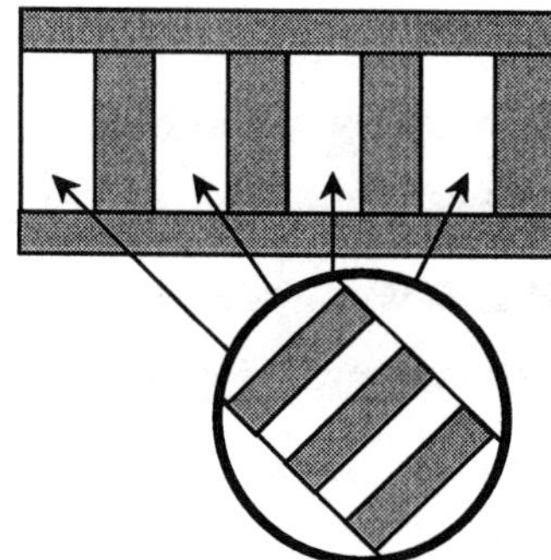

FIGURE 7.5. Scheme of multicoated matrix laminates.

7.4   Properties of Complicated Structures

7.4.1   Multicoated and Self-Repeating Structures

Multicoated Structures MC

Note that a matrix laminate in two dimensions has only one free parameter: the relative width of the first-rank layers. This parameter completely determines the degree of anisotropy of a homogenized composite. Matrix laminates correspond to the boundary of the G_m-closure of two-phase conducting materials, that is, a set of the pairs of eigenvalues of effective tensors σ_*.

However, matrix laminates cannot describe the boundary of the G_m-closure of materials that simultaneously conducts several substances, like heat, electricity, diffusive liquids, or of electromagnetic composites. For these problems, we must consider a type of microstructure with more structural parameters.

The bounds for an isotropic composite (two-dimensional problem) with an arbitrary number of properties such as dielectric, magnetic, and thermal were obtained in (Milton, 1981a; Milton, 1981c). The optimal microstructures that realize the isotropic bounds are the sequences of "coated spheres" called *multicoated spheres*. These structures consist of sequences of concentric spherical layers that are filled with two materials in alternating order. The properties of these isotropic structures are controlled by the relative thickness of the alternating layers.

The construction of "multicoated matrices" (see Figure 7.5) combines the anisotropy of matrix laminates and the topological complexity of multicoated spheres. These structures have been described in (Milton, 1991b; Cherkaev and Gibiansky, 1992), where they are used to build the G_m-closure for coupling conductivities (see Chapter 11). We follow (Cherkaev and Gibiansky, 1992).

Let us demonstrate the multicoated matrices using the simplest example of a two-dimensional conducting composite (see Figure 7.5) that is made of materials with isotropic conductivities $\boldsymbol{\sigma}_1 = \sigma_1 I$ and $\boldsymbol{\sigma}_2 = \sigma_2 I$.

The following iterative procedure is considered. At the first step we assemble a matrix laminate composite (MLC) from the initial materials taken in the volume fractions c_{11} and c_{12}, respectively. The second material $\boldsymbol{\sigma}_2$ is wrapped around the first one. The effective properties tensor $\boldsymbol{\sigma}^{(2,1)}$ of such a structure (see (7.3.13)) is

$$\boldsymbol{\sigma}^{(2,1)} = \boldsymbol{\sigma}_2 + c_{12}[(\boldsymbol{\sigma}_1 - \boldsymbol{\sigma}_2)^{-1} + c_{11}\boldsymbol{\sigma}_2^{-1}(\alpha \mathbf{n}_\alpha \otimes \mathbf{n}_\alpha + (1 - \alpha)\tilde{\mathbf{n}}_\alpha \otimes \tilde{\mathbf{n}}_\alpha)]^{-1}.$$

Here α and $(1-\alpha)$ are the relative concentrations of the enveloping material $\boldsymbol{\sigma}_2$ in the layers of the first and second rank. We also assume that the normals $\mathbf{n}_\alpha$ and $\tilde{\mathbf{n}}_\alpha$ are orthogonal.

At the second step we assemble a matrix laminate in which the homogenized composite obtained in the first step is wrapped by the first material (see Figure 7.5). Assume that volume fractions of the components in this step are c_{21} (inclusions) and c_{22} (envelope). The effective properties tensor $\boldsymbol{\sigma}^{(1,2,1)}$ of such a structure is computed by (7.3.13), which has the form

$$\begin{aligned}
\boldsymbol{\sigma}^{(1,2,1)} = \ &\boldsymbol{\sigma}_1 + c_{21}[(\boldsymbol{\sigma}^{(2,1)} - \boldsymbol{\sigma}_1)^{-1} \\
&+ c_{22}\boldsymbol{\sigma}_1^{-1}(\beta \mathbf{n}_\beta \otimes \mathbf{n}_\beta + (1 - \beta)\tilde{\mathbf{n}}_\beta \otimes \tilde{\mathbf{n}}_\beta)]^{-1}
\end{aligned} \tag{7.4.1}$$

where β is the parameter defined analogous to α; the normals $\mathbf{n}_\beta$ and $\tilde{\mathbf{n}}_\beta$ are orthogonal. The total amounts of initial materials are fixed:

$$m_1 = c_{11}c_{22} + c_{21}, \quad m_2 = c_{12}c_{22}.$$

We will call these structures *"multicoated matrices of type 1–2–1"* or MCM(1,2,1). The sequence of indices shows the order of enveloping of materials in the microstructure; see Figure 7.5.

The structures of MCM are more general than the matrix layered composites; these last composites correspond to MCM(1,2) (inclusions of the first material) or MCM(2,1) (inclusions of the second material).

Using the tensor $Y(\boldsymbol{\sigma}^{(1,2,1)})$, one can transform (see (Cherkaev and Gibiansky, 1992)) the relationship (7.4.1) to the form

$$\begin{aligned}
(Y(\boldsymbol{\sigma}^{(1,2,1)}) + \boldsymbol{\sigma}_1)^{-1} = \ &\gamma(Y(\boldsymbol{\sigma}^{(2,1)}) + \boldsymbol{\sigma}_1)^{-1} \\
&+ (1 - \gamma)(Y(\boldsymbol{\sigma}^{(1,2)}) + \boldsymbol{\sigma}_1)^{-1},
\end{aligned} \tag{7.4.2}$$

where the parameter γ is equal to

$$\gamma = \frac{c_{11}}{m_1}, \quad (1 - \gamma) = c_{12}\frac{c_{21}}{m_1}, \quad \gamma \in [0, 1].$$

Notice that (7.4.2) defines $\boldsymbol{\sigma}^{(1,2,1)}$ through the convex envelope supported by tensors $Y(\boldsymbol{\sigma}^{(2,1)})$ and $Y(\boldsymbol{\sigma}^{(1,2)})$.

The effective tensor depends on four rotationally invariant parameters: the scalar product $\mathbf{n}_\alpha \cdot \mathbf{n}_\beta$ (recall that the layers in matrices are orthogonal: $\mathbf{n}_\alpha \cdot \tilde{\mathbf{n}}_\alpha = 0$ and $\mathbf{n}_\beta \cdot \tilde{\mathbf{n}}_\beta = 0$) and the three parameters α, β, and γ. The parameters α and β determine the degree of anisotropy of the first and second envelopes, respectively, and γ controls the layout of the first material between the inclusions and the external envelope. The orientation of $\mathbf{n}_\alpha$ and $\mathbf{n}_\beta$ determines the orientation of the principal axis of the effective properties tensor, and the scalar parameter $\mathbf{n}_\alpha \cdot \mathbf{n}_\beta$ affects its invariant characteristic.

These MCM structures can degenerate into the simpler types described earlier. Indeed, if $\alpha = \beta = \frac{1}{2}$, the structure becomes isotropic and its effective properties coincide with the properties of the multicoated spheres introduced in (Milton, 1981c; Milton, 1981a).

It is possible to construct a more complex MCM by enveloping the structure MCM(1,2,1) with the material σ_2. Such a structure is denoted by MCM(2,1,2,1); its properties tensor $\sigma^{(2,1,2,1)}$ can be obtained from the next iteration of the described procedure.

Remark 7.4.1 *In Chapter 11, we will demonstrate that MCM(1,2,1) and MCM(2,1,2) form G_m-closure of materials that have two permeabilities, like porosity and thermal conductivity. In these problems, each effective property is described by the formula (7.4.2). The scalar product $\mathbf{n}_\alpha \cdot \mathbf{n}_\beta$ is responsible for the disorientation of two effective properties tensors that generally are anisotropic, and γ is responsible for coupling of properties even if the structure is isotropic.*

Self-Repeating Structures

In dealing with multiphase composites and polycrystals one needs to consider a much greater variety of topologically different microstructures. Some results are discussed in Chapter 12, but the whole picture is not clear yet. The multicomponent optimal structures often have fractal geometries. The same geometric construction is repeated again and again as in an Escher picture. Here we describe an elegant yet unexpected approach to the calculation of polycrystal structures suggested in (Nesi and Milton, 1991).

Consider an anisotropic material D. Let us laminate this material with some anisotropic material A in a laminate with a normal $\mathbf{n}$ and a volume fraction m_D of D. Denote the resulting composite by $\mathcal{L}(D, A, m_D, \mathbf{n})$. Suppose that it is possible to choose the parameters $m_D \in (0,1)$, $\mathbf{n}$, and the unknown material A so that

$$\mathcal{L}(D, A, m_D, \mathbf{n}) = \Phi A \Phi^T, \tag{7.4.3}$$

where Φ is the tensor of rotation through an angle. The formula (7.4.3) states that lamination of the given anisotropic material D with a material with properties A leads to the composite with the properties equal to A rotated on some angle.

If the relation (7.4.3) is satisfied, then A is a polycrystal assembled of variously oriented fragments of D.

To show this, we again use the process with an infinite number of steps. At the first step, we obtain the composite (7.4.3). At the second step, we laminate the composite $\mathcal{L}(D, A, m_D, \mathbf{n}) = \Phi A \Phi^T$ with the rotated material $\Phi D \Phi^T$ in the same way as in the first step. Obviously, the result of this lamination is $\Phi^2 A (\Phi^2)^T$, i.e., the material A rotated on the double angle. This step can be repeated infinitely many times; still, the resulting material remains equal to the rotated material A.

Let us calculate the volume fractions of materials in a composite obtained at the ith step. The material A is used only on the first step of the procedure. Its volume fraction is equal to $1 - m_D$ in the composite obtained after the first step. After the second step this fraction becomes $(1 - m_D)^2$, because the material D has been added to the mixture in the volume fraction m_D. Similarly, after the ith iteration the fraction of the initial "seed" A is equal to $(1 - m_D)^i$. When i increases, the volume fraction of the "seed" A becomes arbitrarily small, and almost all the volume is occupied with differently oriented fragments of material D. In the limit $i \to \infty$, the described composite becomes a polycrystal of D.

To find this polycrystal, we solve equation (7.4.3) for the unknown tensor A. The parameter m_D and the rotation tensor Φ can be arbitrarily assigned. Any solution represents a polycrystal of D. The set of solutions $A(\Phi, m_D, \mathbf{n})$ represents a class of polycrystals of D that can be obtained by the described method.

7.4.2 Structures of Contrast Properties

Consider a composite of two materials with contrast properties,

$$\|D_1^0\| \leq \varepsilon \quad \|(D_2^0)^{-1}\| \leq \varepsilon, \tag{7.4.4}$$

and suppose that $\varepsilon \to 0$. The question is, what effective tensors can be obtained by laminating these two extremes?

We solve the problem in two steps. First, we find the "extremal materials" that are characterized by the tensors D_{extr} of either zero or infinite eigenvalues. We want to find extremal materials with a maximally large space of eigenvectors.

Generally, a composite of two extremal materials is also an extremal material; the effective tensor of a composite has zero of infinite eigenvalues. The values "zero" and "infinity" must be understood according to condition (7.4.4).

The extremal materials in a sense separate the material's parameters from the structural parameters, because the properties of the composite entirely depend on the geometry of the structure. These structures of extremal materials were introduced in (Milton, 1992) where they were used to obtain an elastic composite with Poisson coefficient asymptotically close to -1. In (Cherkaev, 1994) they were used for optimization of elastic composites, and in (Milton and Cherkaev, 1995) they were used to prove that all elastic materials are achievable by laminates of extremal materials.

Algebra of Extremal Materials

Each extremal material can be characterized by a subspace $\mathcal{X}$ of eigenvectors that correspond to zero eigenvalues or by an orthogonal subspace $I - \mathcal{X}$ of eigenvectors that correspond to infinite eigenvalues. Its effective tensor D_{extr} has the form

$$D_{\text{extr}} = \left(\sum_{i=1}^{r} \lambda_i \mathbf{e}_i \otimes \mathbf{e}_i \right) + \left(\sum_{i=r+1}^{n} \frac{1}{\lambda_i} \mathbf{e}_i \otimes \mathbf{e}_i \right).$$

Here $\lambda_i \to \infty$, $i = 1, \ldots, n$, r is the rank of the subspace $\mathcal{X}$ of improper eigenvalues λ_i and $\mathbf{e}_i$ are the eigenvectors that correspond to these eigenvalues.

Consider the constitutive equation $\mathbf{u} = D_{\text{extr}} \mathbf{v}$, where D_{extr} is an extremal material. Some projections of $\mathbf{v}$ and $\mathbf{u}$ are equal to zero; the others are undefined. Namely, the projection of $\mathbf{v}$ onto the subspace $\mathcal{X}$ is zero, and the projection of $\mathbf{u}$ onto the subspace $I - \mathcal{X}$ is zero.

The subspace of zeros $\mathcal{X}$ is scratched on the eigenvectors $\mathbf{e}_i, i = 1, \ldots, r$,

$$\mathcal{X} = \bigoplus_{i=1}^{r} \mathbf{e}_i. \tag{7.4.5}$$

Particularly, the subspace $\mathcal{X}$ equals the entire space for the material D_2^0 with infinite properties, $\mathcal{X} = I$, and it equals the zero subspace, $\mathcal{X} = 0$, for the material D_1^0 with zero properties.

Subspace of Zero Fields in a Laminate. Consider laminates of two extremal materials D_1 and D_2 characterized by two subspaces $\mathcal{X}_1$ and $\mathcal{X}_2$. The effective material D_0 of the laminate is also an extremal material, and therefore it is determined by its subspace of zeros $\mathcal{X}_0$. The subspace of zeros $\mathcal{X}_0$ of the laminate is determined by the subspaces $\mathcal{X}_1$, $\mathcal{X}_2$ and by the normal $\mathbf{n}$ to the layers. Let us find $\mathcal{X}_0$.

The laminate structure yields to the split of the space of vectors $\mathbf{u}$ into the subspace $\mathcal{Z}$ of components continuous everywhere in the structure and the subspace of the discontinuous components. The subspace $\mathcal{Z}$ is determined by the normal $\mathbf{n}$ to the layer $\mathcal{Z} = \mathcal{Z}(\mathbf{n})$ and by the tensor A of differential constraints $A : \nabla \mathbf{v} = \mathbf{g}$ as in (7.2.4). For example, the conducting material corresponds to the subspace $\mathcal{Z}$ of the tangent components, because the tangent fields are continuous. The projections of $\mathbf{v}$ on the subspace $\mathcal{Z}$ are equal in both laminated materials; $\mathcal{Z}\mathbf{v}_1 = \mathcal{Z}\mathbf{v}_2$. Therefore, if $\mathcal{Z}\mathbf{v}_1 = 0$ then $\mathcal{Z}\mathbf{v}_2 = 0$ and vice versa.

Now we can describe the subspace $\mathcal{X}_0$ of zeros. We mention the following:

1. The projection of the field $\mathbf{v}$ in a direction $\mathbf{p}$ is zero in the entire structure if this projection of $\mathbf{v}$ is equal to zero in both components. These directions form the subspace

$$\mathcal{X}_1 \cap \mathcal{X}_2 \subset \mathcal{X}_0. \tag{7.4.6}$$

2. The projection of the field $\mathbf{v}$ in a direction $\mathbf{p}$ is zero in the entire structure if this direction corresponds to zero field $\mathbf{v}_1$ in the first material D_1, $\mathbf{p} \in \mathcal{X}_1$ and if this field stays continuous in the laminate, $\mathbf{p} \in \mathcal{Z}$. Indeed, the continuity implies that the projection of the field $\mathbf{u}_2$ in the second material into the direction $\mathbf{p}$ is also zero.

   These directions belong to the intersection of the subspaces $\mathcal{X}_1$ and $\mathcal{Z}$:

$$\mathcal{X}_1 \cap \mathcal{Z} \subset \mathcal{X}_0.$$

3. By symmetry, the supplementary subspace of zeros is:

$$\mathcal{X}_2 \cap \mathcal{Z} \subset \mathcal{X}_0. \tag{7.4.7}$$

In summary, the subspace of zeros in the laminate is the linear envelope scratched on the three subspaces (7.4.6), (7.4.7), and (7.4.7):

$$\mathcal{X}_0 = (\mathcal{X}_1 \cap \mathcal{X}_2) \oplus (\mathcal{X}_1 \cap \mathcal{Z}) \oplus (\mathcal{X}_2 \cap \mathcal{Z}),$$

or

$$\mathcal{X}_0 = (\mathcal{X}_1 \cap \mathcal{X}_2) \oplus [(\mathcal{X}_1 \oplus \mathcal{X}_2) \cap \mathcal{Z}]. \tag{7.4.8}$$

Any projections of $\mathbf{v}$ onto the subspace $\mathcal{X}_0$ are equal to zero, and the orthogonal projections of $\mathbf{v}$ are not defined. On the contrary, the corresponding projections of the dual vector $\mathbf{u}$ onto the subspace $\mathcal{X}_0$ are not defined, and the orthogonal projections are equal to zero.

The laminates of the extremal materials can be viewed as a mapping $\mathcal{L}$ of the subspaces $\mathcal{X}_1, \mathcal{X}_2$ and $\mathcal{Z}(\mathbf{n})$ to the subspace $\mathcal{X}_0$,

$$\mathcal{L}[\mathcal{Z}(\mathbf{n}), \mathcal{X}_1, \mathcal{X}_2] = \mathcal{X}_0.$$

Examples. The simple algebra of the extremal materials allows us to build the "skeleton" of the laminate structures and to describe various constructions of laminates of high rank. Several examples of extreme laminates follow.

Example 7.4.1 Let us compute the $\mathcal{X}_0$-space for the laminates of several materials. First, we laminate three extremal materials in a simple laminate: We substitute $\mathcal{L}[\mathcal{Z}, \mathcal{X}_1, \mathcal{X}_2]$ for $\mathcal{X}_1$, and $\mathcal{X}_3$ for $\mathcal{X}_2$ into (7.4.8) and we obtain the representation

$$\mathcal{L}[\mathcal{Z}, \mathcal{L}[\mathcal{Z}, \mathcal{X}_1, \mathcal{X}_2], \mathcal{X}_3] = (\mathcal{X}_1 \cap \mathcal{X}_2 \cap \mathcal{X}_3) \oplus [(\mathcal{X}_1 \cap \mathcal{Z}) \oplus (\mathcal{X}_2 \cap \mathcal{Z}) \oplus (\mathcal{X}_3 \cap \mathcal{Z})]$$

for the zero subspace of the laminate.

For an N-material laminate, we obtain, by induction,

$$\mathcal{X}_N = (\mathcal{Z} \cap \bigoplus_{i=1}^{N} \mathcal{X}_i) \oplus (\bigcap_{i=1}^{N} \mathcal{X}_i).$$

Example 7.4.2 Let us compute the $\mathcal{X}_0$-space for the matrix laminates. Suppose that the matrix laminate is assembled from the "weak" material $\mathcal{X}_1 = 0$ that forms the nucleus and the "rigid" material $\mathcal{X}_2 = I$ that forms the envelope. The laminate corresponds to the subspace $\mathcal{Z}_1 = \mathcal{Z}(\mathbf{n}_1)$. The subspace of zeros for the laminate of the first rank, $\mathcal{X}^1$, is defined from the basic formula (7.4.8)

$$\mathcal{X}^1 = \mathcal{L}[\mathcal{Z}_1, 0, I] = \mathcal{X}_l^1 = \mathcal{Z}_1.$$

A second-rank laminate is a composite of the "rigid" material ($\mathcal{X} = I$) and the material $\mathcal{X}^1 = \mathcal{Z}_1$; the normal $\mathbf{n}_2$ corresponds to the subspace $\mathcal{Z}_2 = \mathcal{Z}(\mathbf{n}_2)$. The subspace of zeros, $\mathcal{X}^2$ (compare with (7.3.13)) is equal to

$$\mathcal{X}^2 = \mathcal{L}[\mathcal{Z}_2, I, \mathcal{Z}_1] = \mathcal{Z}_1 \oplus \mathcal{Z}_2.$$

Similarly, the Nth-rank matrix laminate has a subspace of zeros, $\mathcal{X}^N$, equal to

$$\mathcal{X}^N = \mathcal{L}[\mathcal{Z}_N, I, \mathcal{L}[\mathcal{Z}_{N-1}, I, \ldots, \mathcal{L}[\mathcal{Z}_1, I, 0]] \ldots] = \mathcal{Z}_1 \oplus \cdots \oplus \mathcal{Z}_N.$$

Example 7.4.3 The matrix laminate of second rank with the zero material in the envelope and the rigid material in the nucleus is considered similarly. This structure is characterized by the subspace of zeros, $\mathcal{X}_w^1$,

$$\mathcal{X}_w^1 = \mathcal{L}[\mathcal{Z}_2, 0, \mathcal{Z}_1] = \mathcal{Z}_1 \cap \mathcal{Z}_2.$$

By induction, we obtain the formula for the subspace of zeros, $\mathcal{X}_w^N$, of the matrix composite of Nth-rank,

$$\mathcal{X}_w^N = \mathcal{Z}_1 \cap \ldots \cap \mathcal{Z}_N.$$

Structures of Extremal Materials with Arbitrary Properties

Having all possible extremal materials, one can build a laminate composite that "mimics" material with prescribed finite effective properties. This problem was considered in (Milton and Cherkaev, 1995) for three-dimensional elastic structures.

Suppose that we want to mimic a material

$$D_{\text{target}} = \sum_{i=1}^{n} \mu_i \mathbf{e}_i \otimes \mathbf{e}_i,$$

where $\mu_i > 0$ are the arbitrary eigenvalues and $\mathbf{e}_i$ are the eigenvectors. Suppose also that we have all extremal materials available.

Attainability of One Finite Eigenvalue. As we mentioned, a composite of extremal materials is generally an extremal material, too. The exception is a mixture of two extremal materials

$$D_{\text{extr}}^{(1,1)} = \left(\sum_{i=1}^{r} \lambda_i \mathbf{e}_i \otimes \mathbf{e}_i \right) + \left(\sum_{i=r+1}^{n} \frac{1}{\lambda_i} \mathbf{e}_i \otimes \mathbf{e}_i \right)$$

and

$$D_{\text{extr}}^{(1,2)} = \left(\sum_{i=1}^{r-1} \lambda_i \mathbf{e}_i \otimes \mathbf{e}_i \right) + \left(\sum_{i=r}^{n} \frac{1}{\lambda_i} \mathbf{e}_i \otimes \mathbf{e}_i \right),$$

where $\lambda_i \to \infty$, $i = 1, \ldots n$. These materials differ only by one eigenvalue, λ_r, that goes to infinity in the first component and to zero in the second one.

We argue that in this case a composite has the rth eigenvalue equal to an arbitrary number between the two extremes. By using the conservation property of G-closures (see Chapter 3), we find that the structure of the effective tensor of any composite from these materials is

$$D_{\text{extr}}^{(2,*)} = \left(\sum_{i=1}^{r-1} \lambda_i \mathbf{e}_i \otimes \mathbf{e}_i \right) + \left(\sum_{i=r+1}^{n} \frac{1}{\lambda_i} \mathbf{e}_i \otimes \mathbf{e}_i \right) + \mu_r (\mathbf{e}_r \otimes \mathbf{e}_r),$$

where the eigenvalue $\mu_r \in [0, \infty]$ has an intermediate value. This value depends on the composite structure and can be made equal to any number between the extremes. The other eigenvalues and eigenvectors of the composite are equal to the corresponding eigenvalues of the components.

Scheme of Attainability of an Arbitrary Tensor. Repeating this procedure one more time, we consider two materials

$$D_{\text{extr}}^{(2,1)} = \left(\sum_{i=1}^{r-2} \lambda_i \mathbf{e}_i \otimes \mathbf{e}_i \right) + \left(\sum_{i=r+1}^{n} \frac{1}{\lambda_i} \mathbf{e}_i \otimes \mathbf{e}_i \right)$$
$$+ \frac{1}{\lambda_{r-1}} (\mathbf{e}_{r-1} \otimes \mathbf{e}_{r-1}) + \mu_r (\mathbf{e}_r \otimes \mathbf{e}_r)$$

and

$$D_{\text{extr}}^{(2,2)} = \left(\sum_{i=1}^{r-2} \lambda_i \mathbf{e}_i \otimes \mathbf{e}_i \right) + \left(\sum_{i=r+1}^{n} \frac{1}{\lambda_i} \mathbf{e}_i \otimes \mathbf{e}_i \right)$$
$$+ \lambda_{r-1} (\mathbf{e}_{r-1} \otimes \mathbf{e}_{r-1}) + \mu_r (\mathbf{e}_r \otimes \mathbf{e}_r),$$

where $\lambda_i \to \infty$, $i = 1, \ldots, n$.

These materials (i) have one finite common eigenvalue μ_r and the corresponding common eigenvector $\mathbf{e}_r$; (ii) have the $n - 2$ common eigenvalues equal to either zero or infinity, and the corresponding common eigenvectors;

and (iii) have different eigenvalues (equal to zero and infinity, respectively) corresponding to the eigenvector $\mathbf{e}_{r-1}$.

We mix these obtained materials $D_{\text{extr}}^{(2,1)}$ and $D_{\text{extr}}^{(2,2)}$ and obtain the new composite

$$D_{\text{extr}}^{(3,*)} = \left(\sum_{i=1}^{r-2} \lambda_i \mathbf{e}_i \otimes \mathbf{e}_i\right) + \left(\sum_{i=r+1}^{n} \frac{1}{\lambda_i} \mathbf{e}_i \otimes \mathbf{e}_i\right)$$
$$+\mu_{r-1}(\mathbf{e}_{r-1} \otimes \mathbf{e}_{r-1}) + \mu_r(\mathbf{e}_r \otimes \mathbf{e}_r)$$

with two finite eigenvalues, μ_{r-1} and μ_r, in the directions of $\mathbf{e}_r$ and $\mathbf{e}_{r-1}$. There eigenvalues can be arbitrary assigned.

Finally, we repeat the procedure until all eigenvalues are equal to the prescribed values $\mu_1, \ldots, \mu_n$.

Remark 7.4.2 *In order to apply this method, one should check that all extremal materials are available. The availability depends on the set of initially given materials and on the $\mathcal{Z}(\mathbf{n})$, i.e., on the differential restrictions on the field $\mathbf{v}$. The corresponding example is discussed in Chapter 14.*

7.5 Optimization in the Class of Matrix Composites

Upper Bound by Laminates

We return to the discussion of the upper bound for the quasiconvex envelope. Recall that an upper bound corresponds to special minimizing sequences. The previously described laminate structures are examples of the oscillating sequences that may be used to minimize the functional.

To find an upper bound for the quasiconvex envelope one chooses a class of laminate composites like the matrix laminates. The necessity to choose a suitable topology is a weak point of the approach: It is not formalized and requires considerable intuition. On the other hand, the freedom of choice provides the needed flexibility to the method. One uses common sense and physical analogies to justify the choice, but doubt remains and is eliminated only at the last step when (and if!) the solution coincides with the lower bound obtained independently by the sufficient condition method. After the class of laminates is chosen, we minimize the energy by the parameters of laminate structures. This minimization is a finite-dimensional optimization problem.

With this in mind, it is truly amazing that a large number of quasiconvex envelopes have been built by using a rule of thumb justified mainly by intuition. Of course, some clue is provided by the necessary and sufficient conditions that we discuss in the next chapters.

Minimum of the Sum of Energies: Setting

Consider the Lagrangian (see (6.1.3))

$$W(\mathbf{U}) = \min_{\chi} \; (\chi(\sigma_1 \operatorname{Tr} \mathbf{U} + \gamma) + (1 - \chi)\sigma_2 \operatorname{Tr} \mathbf{U})$$

equal to the sum of the mean energy of two loadings and the cost of the materials. Here χ is the characteristic function of the first material, γ is the cost of that material, and

$$\mathbf{U} = \sum_{i=1}^{p} \mathbf{e}_i \otimes \mathbf{e}_i, \quad \nabla \times \mathbf{e}_i = 0 \tag{7.5.1}$$

is the matrix of loadings (see (6.1.3)). Note that $\mathbf{U}$ is symmetric and non-negative definite: $\mathbf{U} = \mathbf{U}^T$, $\mathbf{U} \geq 0$.

Let us consider the energy of an optimal matrix laminate structure in which the poor conductor σ_1 forms an envelope and the good conductor σ_2 is placed in the inclusion. This topology corresponds to the intuitive idea of a medium with poor conductivity in all directions. The minimum on these structures gives the upper bound $\mathcal{L}_2 W(\mathbf{U})$ of the quasiconvex envelope of $W(\mathbf{U})$.

The problem for $\mathcal{L}_2 W(\mathbf{U})$ is

$$\mathcal{L}_2 W(\mathbf{U}) = \min_{\boldsymbol{\sigma}_{\mathrm{ml}}} \; (\operatorname{Tr} \boldsymbol{\sigma}_{\mathrm{ml}} \mathbf{U}) + \gamma m, \tag{7.5.2}$$

where $m = \langle \chi \rangle$ is the volume fraction of the first material in the composite.

The effective tensor of a matrix laminate (7.3.13) depends on a number of inner parameters: the orientation $\mathbf{n}_i$ of laminates and their relative concentration α_i. These parameters must be chosen to minimize the total energy.

However, it is technically easier to use the constraints (7.3.15), (7.3.16) on an arbitrary effective tensor $\boldsymbol{\sigma}_{\mathrm{ml}}$ of matrix laminates to describe the class of these composites. The constraints consist of the inequalities

$$\boldsymbol{\sigma}_{\mathrm{ml}} \leq (m_1 \sigma_1 + m_2 \sigma_2)I, \tag{7.5.3}$$

and the equality

$$\boldsymbol{\sigma}_{\mathrm{ml}} = S + \sigma_1 I, \quad \operatorname{Tr} S^{-1} = c(m_1), \tag{7.5.4}$$

where (see (7.3.15)) the constant c is

$$c(m_1) = \frac{dm_2}{\sigma_2 - \sigma_1} + \frac{m_2}{m_1}\left(\frac{1}{\sigma_1}\right)$$

and d is dimensionality of the space. Any effective tensor of matrix laminates $\boldsymbol{\sigma}_{\mathrm{ml}}$ satisfies the constraints.

Optimal Structures (Fixed Volume Fractions)

Consider problem (7.5.2). Let us specify the dependence of the functional on m. Rewrite (7.5.2) as

$$\mathcal{L}_2 W = \min_{m \in [0,\, 1]} \left(J(m) + \gamma m \right),$$

where

$$J(m) = \min_{\boldsymbol{\sigma}_{\mathrm{ml}}} \left(\operatorname{Tr} \boldsymbol{\sigma}_{\mathrm{ml}} \mathbf{U} \right).$$

First, assume that the volume fraction m is fixed, and calculate $J(m)$. We use the Lagrange method to find the optimal parameters of a structure. Namely, we add the constraint (7.5.4) with Lagrange multiplier λ to (7.5.2). We also suppose here that the restrictions (7.5.3) are fulfilled as strong inequalities. This assumption physically means that the optimal structure is a matrix laminate of the rank not less than d (it does not degenerate into laminates of lower rank).

The problem becomes

$$J(m) = \min_{S} \max_{\lambda} \left\{ \sigma_1 \operatorname{Tr} \mathbf{U} + \operatorname{Tr}(\mathbf{U}\, S) + \lambda(\operatorname{Tr} S^{-1} - c) \right\}. \tag{7.5.5}$$

The minimum value of J is taken with respect to an unknown matrix S that represents the composite structure.

Remark 7.5.1 *Recall that the derivative of a scalar function $\phi(a_{ij})$ with respect to the matrix $A = \{a_{ij}\}$ is defined as*

$$D = \frac{d\phi}{dA} = \{d_{ij}\}, \quad d_{ij} = \frac{d\phi}{da_{ji}}.$$

This implies the immediately verifiable formulas for symmetric matrices S:

$$\frac{d}{dS} \operatorname{Tr} S = I, \quad \frac{d}{dS} \operatorname{Tr} S^{-1} = S^{-2}, \quad \frac{d}{dS} \operatorname{Tr}(\mathbf{U}\, S) = \mathbf{U}.$$

Using these formulas, we compute the stationary conditions,

$$\frac{\partial}{\partial S} [\operatorname{Tr}(\mathbf{U}\, S) + \lambda(\operatorname{Tr} S^{-1} - c)] = \mathbf{U} + \lambda S_{\mathrm{opt}}^{-2} = 0.$$

This gives

$$S^{-1} = (\sqrt{-\lambda})\mathbf{U}^{\frac{1}{2}}. \tag{7.5.6}$$

Generally, $\mathbf{U}^{\frac{1}{2}}$ is not uniquely defined. However, equation (7.5.6) requires the symmetry of $\mathbf{U}^{\frac{1}{2}}$ and fixes the uncertainty in its definition as follows. If we denote the eigenvalues of the nonnegative symmetric matrix $\mathbf{U}$ by $e_1^2, \ldots e_d^2$, $(e_i \geq 0, i = 1, \ldots, d)$, then the matrix $\mathbf{U}^{\frac{1}{2}}$ has the eigenvalues $e_1, \ldots e_d$. The eigenvectors of $\mathbf{U}$ and $\mathbf{U}^{\frac{1}{2}}$ coincide.

To find λ we compute the trace of (7.5.6) and then use (7.5.4). We have

$$\sqrt{-\lambda}\,\mathrm{Tr}\,\mathbf{U}^{\frac{1}{2}} = \mathrm{Tr}\,S^{-1} = c,$$

which gives

$$\sqrt{-\lambda} = \frac{c}{\mathrm{Tr}\,\mathbf{U}^{\frac{1}{2}}}.$$

Now we find the optimal matrix S_{opt} from (7.5.6):

$$S_{\mathrm{opt}} = \frac{c}{\mathrm{Tr}\,\mathbf{U}^{\frac{1}{2}}}\mathbf{U}^{-\frac{1}{2}}$$

and, finally, we compute the optimal Lagrangian (7.5.5)

$$J(m) = \sigma_1\,\mathrm{Tr}\,\mathbf{U} + c(m)\left(\mathrm{Tr}(\mathbf{U}^{\frac{1}{2}})\right)^2.$$

In notations $e_1,\dots e_d$, the last equality takes an especially symmetric form:

$$J(m) = \sigma_1\sum_{i=1}^{d}(e_i)^2 + c(m)\left(\sum_{i=1}^{d}e_i\right)^2.$$

The functional $J(m)$ represents the energy stored in the best-adapted matrix laminate that contains the materials in the given proportion m and $1-m$. Notice that $J(m)$ is an isotropic function of $\mathbf{U}$.

Optimal Volume Fractions

The upper bound of the functional (7.5.2) is computed when the volume fractions are chosen in the best way. We have

$$\mathcal{L}_2 W = \sigma_1\sum_{i=1}^{d}e_i^2 + \min_{m\in[0,1]}\left\{c(m)\left(\sum_{i=1}^{d}e_i\right)^2 + \gamma m\right\}.$$

The optimal value m_0 of m has the form

$$m_0 = A\sum_{i=1}^{d}e_i + B, \tag{7.5.7}$$

where

$$A = \frac{\sqrt{\sigma_1}\,\sqrt{(2\,d-1)\,\sigma_1 + \sigma_2}}{\sqrt{\gamma}\,\sqrt{\sigma_2 - \sigma_1}}, \qquad B = \frac{2\,d\,\sigma_1}{\sigma_1 - \sigma_2},$$

and it is assumed that the right-hand side of (7.5.7) lies between zero and one. Substituting this value into the problem, we obtain

$$J(m_0) + \gamma m_0 = \sigma_1\sum_{i=1}^{d}e_i^2 - \sigma_1\left(\sum_{i=1}^{d}e_i\right)^2$$

$$+2\gamma A \sum_{i=1}^{d} e_i + \gamma B$$

$$= -2\sigma_1 \sum_{i,j=1,\ i\neq j}^{d} e_i e_j + 2\gamma A \sum_{i=1}^{d} e_i + \gamma B.$$

This expression is a component the upper bound for the quasiconvex envelope. Note that it is a polylinear function off e_i

Finally, we find the bound $\mathcal{L}_2 W$ in the form

$$\mathcal{L}_2 W(\mathbf{U}) = \begin{cases} \sigma_1 \sum_{i=1}^{d} e_i^2 + \gamma & \text{if } m_0 \leq 0, \\ J(m_0) + \gamma m_0 & \text{if } m_0 \in [0, 1], \\ \sigma_2 \sum_{i=1}^{d} e_i^2 & \text{if } m_0 \leq 0. \end{cases} \qquad (7.5.8)$$

Recall again that the constraints (7.5.3) are supposed to be satisfied in this calculation as strong inequalities. Therefore, the solution must be supplemented by these inequalities. They require that

$$\min_{k} \left\{ \frac{e_k}{\sum_{i=1}^{d} e_i} \right\} \geq \frac{m_2}{\sigma_2 - \sigma_1} + \frac{m_1}{m_2 \sigma_1 d}. \qquad (7.5.9)$$

This inequality expresses a requirement that magnitudes of loadings from different directions are close to one another. If the mean field $\mathbf{U}$ does not satisfy the inequalities (7.5.9), then one must examine the matrix laminates of lower rank.

Example 7.5.1 Consider the "hydrostatic" loading; $e_k = e\ \forall k$, where e is the intensity. We have $\mathbf{U} = e^2 I$, $m_0(e) = dAe$, and

$$\mathcal{L}_2 W = \begin{cases} \sigma_1 d e^2 + \gamma & \text{if } m_0 \geq 1, \\ -(d-1)\sigma_1 e^2 + 2d\gamma A e + \gamma B & \text{if } m_0 \in [0, 1], \\ \sigma_2 d e^2 & \text{if } m_0 \leq 0. \end{cases}$$

Observe that if $d > 1$, then $\mathcal{L}_2 W(eI)$ is not convex if it does not coincide with $W(eI)$, $\mathcal{L}_2 W(eI) < W(eI)$.

Other Components of the Upper Bound

The remaining parts of the components of $\mathcal{L}_2 W$ are found in the same fashion. These components correspond to the range of $\mathbf{U}$ that violates (7.5.9). The optimal structures degenerate into matrix laminates of lower rank. The calculation of their properties is done similarly to the previous calculations.

In the two-dimensional case, for example, the optimal structures are either matrix laminates or simple laminates. The energy of the matrix laminates is given by (7.5.8), where $d = 2$. The energy of the simple laminates

$(e_2 \leq e_1)$ is:

$$W_{\mathrm{lam}}(e_1, e_2) = \min_{m \in [0, 1]} \left\{ (m\sigma_1 + (1 - m)\sigma_2)e_1^2 + \frac{e_2^2}{\frac{m}{\sigma_1} + \frac{1-m}{\sigma_2}} + \gamma m \right\}$$

$$= 2e_1 \sqrt{\frac{\sigma_1\sigma_2(\gamma - e_2^2(\sigma_2 - \sigma_1))}{\sigma_2 - \sigma_1}} + \frac{-\gamma\sigma_1 - e_2^2(\sigma_2^2 - \sigma_1^2)}{\sigma_2 - \sigma_1},$$

if $m = m_{\mathrm{opt}} \in [0, 1]$. Here, the stationary value of m that minimizes the functional is

$$m_{\mathrm{opt}} = \left(\frac{\sigma_1}{\sigma_2 - \sigma_1} + e_1 \sqrt{\frac{\sigma_1\sigma_2}{(\sigma_2 - \sigma_1)(\gamma - e_2^2(\sigma_2 - \sigma_1))}} \right).$$

One can see that the energy W_{lam} is nonconvex. It becomes convex only when $e_2 = 0$; then it degenerates into the convex envelope of the Lagrangian that corresponds to one loading (Chapter 4):

$$W_{\mathrm{lam}}(e_1, 0) = 2e_1 \sqrt{\frac{\sigma_1\sigma_2\gamma}{\sigma_2 - \sigma_1}} - \frac{\gamma\sigma_1}{\sigma_2 - \sigma_1},$$

$$m_{\mathrm{opt}} = \left(\frac{\sigma_1}{\sigma_2 - \sigma_1} + e_1 \sqrt{\frac{\sigma_1\sigma_2}{\gamma(\sigma_2 - \sigma_1)}} \right) \in [0, 1].$$

Observe that this formula coincides with (4.2.7).

To summarize, we observe that the upper bound $\mathcal{L}_2 W(\mathbf{U})$ given by the energy on matrix laminates is an isotropic nonconvex surface. It consists of several analytical components, and it degenerates into a convex envelope when $\mathbf{U}$ becomes a rank-one matrix. The energy $\mathcal{L}_2 W(\mathbf{U})$ obviously degenerates into the energy of a pure material, if the optimal volume fraction is one or zero.

7.6   Discussion and Problems

Computation of the Upper Bound

Constructing $\mathcal{L}_2 F$ requires the solution to a finite-dimensional optimal problem in each step. However, even for piecewise quadratic two-well energy this program can be effectively realized analytically only under additional restrictions on the matrix of properties or under some a priori assumptions on the minimization procedure at each step. The point is that the number of minimizing parameters, although finite, grows exponentially with the rank of the laminate. The effective properties tensors are expressed as matrix continuous fractions, and formulas become messy.

Additional consideration based on necessary and sufficient conditions of optimality described in Chapters 8 and 9, asymptotic behavior, and physical intuition a priori determine the topology of the optimal structures and thus reduce the number of minimizing parameters. The laminates represent optimal structures known for the majority of the already solved problems of structural optimization (the exceptional example (Sigmund, 1998) is shown in Chapter 16). On the other hand, it is fair to say that the problems solved so far are those that lead to laminate solutions. The laminate structures are well investigated partly because they are simple enough.

Problems

1. Find the formula for lamination of a process described by a system of elliptic equations

$$\nabla \cdot \sum_{j=1}^{n} \mathbf{A}_{ij} \nabla w_j = f_i, \quad i = 1, \ldots, n,$$

   where $\mathbf{A}_{ij}$ are symmetric $d \times d$ matrices. Introduce variables and define the constitutive relations in the standard form. Determine the differential constraints and construct the projectors $\mathbf{p}$ and $\mathbf{q}$.

2. Prove the equivalence of the representations (7.2.7) and (7.3.4) for the effective tensors of laminates.

3. Prove formulas (7.3.23) and (7.3.24), using the definition of the Y-tensor.

4. Consider two extreme isotropic materials with the following properties: One material is an ideal conductor of heat and an ideal resistor of electricity, and the second material is an ideal conductor of electricity and an ideal resistor of heat. Suggest an isotropic structure (in three-dimensions) that ideally conducts both the electricity and heat. Can an ideal conductor be built in two dimensions? Can an ideal resistant structure be built?

5. Suppose that A is an $n \times m$ matrix with elements a_{ij}, and $f(A)$ is a scalar function of these elements. Justify the formula of matrix differentiation

$$\frac{\partial f}{\partial A} = D = \{d_{ij}\}, \quad \text{where } d_{ij} = \frac{\partial f}{\partial a_{ji}}.$$

   Check the following formulas for symmetric matrices S:

$$\frac{\partial}{\partial S} \operatorname{Tr} S = I, \quad \frac{\partial}{\partial S} \operatorname{Tr} S^{-1} = S^{-2}, \quad \frac{\partial}{\partial S} \det S = (\det S)S^{-1}.$$

8
Lower Bound: Translation Method

8.1   Translation Bound

In this chapter we construct a lower bound $\mathcal{P}F$ for the quasiconvex envelope $\mathcal{Q}F$ that is more restrictive than the convex bound $\mathcal{C}F$:

$$\mathcal{Q}F \geq \mathcal{P}F \geq \mathcal{C}F.$$

The idea of the bound is similar to the idea of the Lyapunov functions: The solution to the variational problem for the quasiconvex envelope is bounded from below by a special quasiconvex but not convex function $\mathcal{P}F$ that depends on several free parameters. We choose the values of these parameters to obtain a bound.

Return to definition (6.3.8) of a quasiconvex envelope. Let us enlarge the set Ξ of trial functions $\boldsymbol{\xi} = (\xi_1, \ldots, \xi_m)$ to make the computations of the infimum in the right-hand side explicit. The scheme is the following: we neglect the differential constraints in the definition of Ξ (6.3.3) as it has been done in the derivation of the convex bound, but this time we take into account integral equalities that follow from those constraints.

The idea of the translation method was developed in the early eighties by several groups.

In (Lurie et al., 1980a) the method was suggested to prove the optimality of the laminate structure in a problem of plane elasticity. In (Lurie and Cherkaev, 1981c) it was used to bound the bulk and shear moduli of a two-dimensional elastic body. In (Lurie and Cherkaev, 1982; Lurie and Cherkaev, 1984a) it was used to find the G_m-closure of two conducting materials. In (Gibiansky and Cherkaev, 1984) we used the method to solve the variational problem of the minimal energy of a composite made from elastic materials. In these papers, the translation method was treated as the Lagrange method for the minimization of the Lagrangian with a periodic minimizer or as a variant of the Lyapunov method.

Simultaneously and independently, the method was developed by Tartar and Murat (Murat and Tartar, 1985a; Tartar, 1985) as an application of the earlier developed theory of compensated compactness (see below). The construction of the translation bound was used to find the G_m-closure of two conducting materials. The earlier papers (Tartar, 1975; Murat, 1977; Tartar, 1979b) discuss the application of weak convergence as a tool for bounding the effective properties.

In the same time, the series of papers (Kohn and Strang, 1982; Kohn and Strang, 1983; Kohn and Strang, 1986a), and (Strang and Kohn, 1988) suggested the method from the general perspective of the calculus of variations. The authors derived the translation method in the general context of the theory of quasiconvexity. The earlier construction of polyconvexity by Morrey (Morrey, 1952) that had been used to test quasiconvexity of a Lagrangian became the method to construct the lower bound for non-quasiconvex Lagrangians. This idea was realized in the paper (Kohn and Strang, 1986a) and further developed in (Firoozye, 1991).

The next steps included development, simplification and modification of the method and applying it to various problems of physics and mechanics.

In (Lurie and Cherkaev, 1986a) the known examples were summarized and the method was presented in the complete form. In (Gibiansky and Cherkaev, 1987) we found the exact lower bound for the energy of a three-dimensional elastic structure. In (Lurie and Cherkaev, 1988) the method was used to study the structures of phase equilibrium. In (Avellaneda, Cherkaev, Lurie, and Milton, 1988) it was applied to find the best lower bound for the conductivity of an isotropic polycrystal (see Chapter 11).

Milton developed the method in (Milton, 1990a; Milton, 1990b), and he suggested the name for it in (Milton, 1990b). Particularly, the method was viewed as a projection of the convex envelope of a vector-valued function onto the subspace of gradients. The connections between the Hashin-Shtrikman method and the translation method were explained in (Milton, 1990a). In (Zhikov, 1986; Zhikov, 1991a) some translation bounds were obtained using classical integral inequalities.

Many papers on the development of the translation method and its applications were published in the 1990s. We refer to some of them when we discuss related topics.

Translators

Functions $\phi(\mathbf{v})$ that are quasiconvex but not convex are called *translators*. An example of a translator was demonstrated in Chapter 6. The function $\phi = \det \mathbf{v}$, $\mathbf{v} = \nabla \mathbf{y}$ (see (6.3.5)) is a translator.

Definition 8.1.1 The translator $\phi(\mathbf{v})$ is a function with the following properties:
Function $\phi(\mathbf{v})$ is not convex:

$$\exists \boldsymbol{\xi} : \quad \phi(\mathbf{v}) > \langle \phi(\mathbf{v} + \boldsymbol{\xi}(\mathbf{x})) \rangle, \quad \langle \boldsymbol{\xi} \rangle = 0. \tag{8.1.1}$$

But $\phi(\mathbf{v})$ is quasiconvex:

$$\phi(\mathbf{v}) \leq \langle \phi(\mathbf{v} + \boldsymbol{\xi}(\mathbf{x})) \rangle \quad \forall \boldsymbol{\xi} \in \Xi,$$

where

$$\Xi = \{ \boldsymbol{\xi} : \ \langle \boldsymbol{\xi} \rangle = 0, \quad \mathcal{A}(\boldsymbol{\xi}) = \mathcal{A} : \nabla \boldsymbol{\xi} = 0, \ i = 1, \ldots, r \}, \tag{8.1.2}$$

vector $\mathcal{A} : \nabla \boldsymbol{\xi}$ is r dimensional vector of differential constraints

$$\mathcal{A} : \nabla \boldsymbol{\xi} == \sum_{j=1}^{m} \sum_{k=1}^{d} a_{ijk} \frac{\partial \xi_j}{\partial x_k} \ i = 1, \ldots, r$$

and $\boldsymbol{\xi}$ are periodic.

These properties indicate that the translator satisfies Jensen inequality for all admissible trial functions $\boldsymbol{\xi} \in \Xi$ that is subject to differential constraints but fails to satisfy this inequality for arbitrary trial functions $\boldsymbol{\xi} \notin \Xi$. In the next section we discuss the systematic method to determine quadratic translators.

The Translation Bound

Translators are used to bound the functional $QF(\mathbf{v})$ as follows. Suppose several translators $\phi_1(\mathbf{v}), \ldots, \ \phi_l(\mathbf{v})$ are determined. The convex combination of translators is a quasiconvex function, because each translator is quasiconvex. In other words, if translators are multiplied by some positive numbers $t_1, \ldots, t_l$ and then summed, the result is a new translator, which we call $\Phi(\mathbf{t}, \mathbf{v})$

$$\Phi(\mathbf{t}, \mathbf{v}) = \sum_{i=1}^{l} t_i \phi_i(\mathbf{v}), \quad \mathbf{t} = [t_1, \ldots, t_l]; \tag{8.1.3}$$

it depends on a positive parameter vector $\mathbf{t} = [t_1, \ldots, t_l]$, $t_i \geq 0$.

To obtain a lower bound of the quasiconvex envelope $\mathcal{Q}F(\mathbf{v})$ of $F(\mathbf{v})$, we start with the identity

$$F(\mathbf{v}) = [F(\mathbf{v}) - \Phi(\mathbf{t}, \mathbf{v})] + \Phi(\mathbf{t}, \mathbf{v}).$$

Estimate the first term $F(\mathbf{v}) - \Phi(\mathbf{t}, \mathbf{v})$ on the right-hand side by its convex envelope $\mathcal{C}_{\mathbf{v}}(F - \Phi)$ with respect to $\mathbf{v}$ ($\mathbf{t}$ is considered a vector-valued parameter). We get an inequality:

$$F(\mathbf{v}) \geq \mathcal{C}_{\mathbf{v}}(F(\mathbf{v}) - \Phi(\mathbf{t}, \mathbf{v})) + \Phi(\mathbf{t}, \mathbf{v}) \ \forall \ t_i \geq 0. \tag{8.1.4}$$

To make the bound (8.1.4) maximally effective, we maximize the right-hand-side term of (8.1.4) over the parameters $t_i \geq 0$ and we call the resulting lower bound $\mathcal{P}F(\mathbf{v})$:

$$F(\mathbf{v}) \geq \mathcal{P}F(\mathbf{v}) \ \forall \mathbf{v},$$

where
$$\mathcal{P}F(\mathbf{v}) = \max_{\mathbf{t} \geq 0} \left\{ \mathcal{C}(F(\mathbf{v}) - \phi(\mathbf{t}, \mathbf{v})) + \phi(\mathbf{t}, \mathbf{v}) \right\}. \tag{8.1.5}$$

The bounding function $\mathcal{P}F(\mathbf{v})$ (8.1.5) is the sum of a convex function $\mathcal{C}(F(\mathbf{v}) - \Phi(\mathbf{v}, \mathbf{t}))$ and a translator $\Phi(\mathbf{v}, \mathbf{t})$, and therefore $\mathcal{P}F(\mathbf{v})$ is a quasiconvex, but not necessarily convex function of $\mathbf{v}$. The quasiconvexity of $\mathcal{P}F$ implies the inequality

$$\mathcal{P}F(\mathbf{v}) \leq \mathcal{Q}F(\mathbf{v})$$

because the quasiconvex envelope $\mathcal{Q}F$ is by definition the maximum of all quasiconvex functions that are less than or equal to F.

Properties of the Translation Bound

1. If $\mathbf{t} = 0$, the right-hand side of (8.1.5) coincides with the convex envelope $\mathcal{C}_{\mathbf{v}}F(\mathbf{v})$ of F; therefore, its maximal with respect to $\mathbf{t}$ value $\mathcal{P}F$ is greater than or equal to $\mathcal{C}F$:

$$\mathcal{P}F(\mathbf{v}) \geq \mathcal{C}F(\mathbf{v}).$$

   Therefore the translation bound (8.1.5) is generally better than the bound by convex envelope.

2. The definition (8.1.5) is an algebraic minimization problem of a constrained minimum. The constraints are the inequalities of quasiconvexity

$$\langle \phi(\mathbf{v} + \boldsymbol{\xi}) \rangle \geq \phi(\langle \mathbf{v} + \boldsymbol{\xi} \rangle) \quad \forall \boldsymbol{\xi} \in \Xi;$$

   they are added with the Lagrange multipliers t_i to the finite-dimensional minimal problem (the Carathéodory problem) that defines the convex envelope $\mathcal{C}F$.

3. If both $\phi(\mathbf{v})$ and its negative $-\phi(\mathbf{v})$ are quasiconvex, we call the function $\phi(\mathbf{v})$ quasiaffine. All null-Lagrangians are quasiaffine. The corresponding multiplier t by a quasiaffine function in (8.1.5) can be arbitrary and not necessarily nonnegative.

4. Generally, the translators ϕ may nonlinearly depend on additional parameters r:

$$\phi = \phi(\mathbf{v}, r). \tag{8.1.6}$$

In that case, the lower bound is obviously found as the maximum over the extended set of parameters $[\mathbf{t},\ r]$:

$$\mathcal{P}F(\mathbf{v}) = \max_{\mathbf{t}\geq 0} \max_{r} \mathcal{C}(F(\mathbf{v}) - \Phi(\mathbf{t}, r, \mathbf{v})) + \Phi, (\mathbf{t}, r, \mathbf{v})$$

where Φ is determined as in (8.1.3).

Is the Translation Bound Exact?

Whether the translation bound is exact depends on the set of translators ϕ used. Few types of translators are described, and each translator gives a new finite-dimensional approximation of the variational problem of the quasiconvex envelope.

Generally, we cannot expect exactness of the finite-dimensional approximation of an infinite-dimensional variational problem for an arbitrary Lagrangian F. However, we will demonstrate that for some Lagrangians F it is enough to use several known translators to obtain the exact bound. The general problem of explicit definition of all required translators is equivalent to the original problem of a quasiconvex envelope. Notwithstanding, we still may use only special translators to obtain bounds that are not necessarily exact but that are better than the bounds obtained by a convex envelope.

The hunt for new translators is an exciting problem. Generally, one should prove the quasiconvexity of the guessed nonconvex function. Each new class of translators can potentially be used to tighten the bounds for a new class of variational problems. We refer to (Šverák, 1992a; Zhang, 1997; Pedregal and Šverák, 1998).

Translations of Polynomial Growth

We have seen already that the two-by-two Jacobian is quasiaffine (6.3.6). The result can be extended to an arbitrary Jacobian. Consider an $(m \times d)$-matrix M whose columns are gradients of potentials $\mathbf{w} = [w_1, \ldots, w_m]$,

$$M = \nabla\mathbf{w},$$

where $\mathbf{w}$ is an m-dimensional vector of potentials.

A remarkable property of that matrix has been established in (Morrey, 1966; Reshetnyak, 1967)): The determinant of each minor of M is quasi-affine,

$$\langle \mathrm{subdet} M \rangle = \mathrm{subdet}\, \langle M \rangle .$$

The proof is analogous to the one for the two-by-two Jacobian (see Chapter 6): Jacobians can be represented as the divergence of a vector.

The subdeterminants are examples of quasiaffine functions of polynomial growth. They were studied in (Ball and Murat, 1984).

Remark 8.1.1 *The construction of the lower bound suggested in (Kohn and Strang, 1982; Kohn and Strang, 1983; Kohn and Strang, 1986a) explicitly uses this property of subdeterminants. It was suggested to represent the function $F(\frac{\partial w_j}{\partial x_k})$ as a function of an enlarged set of arguments: $G(\frac{\partial w_j}{\partial x_k}, \mathrm{subdet} M)$. Obviously, this can be done in infinitely many ways. A lower bound of the average G corresponds to the convex envelope of G with respect to all its arguments. The best lower bound corresponds to the best choice of this representation. The construction is similar to (8.1.4).*

How does the Translation Method Work?

Here we discuss another explanation of the translation method based on the physical properties of the translations.

Consider the system of equations

$$\nabla \cdot \mathbf{j}_1 = f_1, \quad \mathbf{j}_1 = \boldsymbol{\sigma}(\chi)\nabla w_1$$
$$\nabla \cdot \mathbf{j}_2 = f_2, \quad \mathbf{j}_2 = \boldsymbol{\sigma}(\chi)\nabla w_2 \tag{8.1.7}$$

and let us minimize the energy of that system. The corresponding Lagrangian is

$$W(\chi, w_1, w_2) = \begin{pmatrix} \nabla w_1 \\ \nabla w_2 \end{pmatrix}^T A(\boldsymbol{\sigma}(\chi)) \begin{pmatrix} \nabla w_1 \\ \nabla w_2 \end{pmatrix} \\ + 2f_1 w_1 + 2f_2 w_2 + \gamma(\chi),$$

where

$$A(\boldsymbol{\sigma}) = \begin{pmatrix} \sigma & 0 \\ 0 & \sigma \end{pmatrix} .$$

Consider the variational problem

$$\min_{w_1, w_2, \chi} \int_\Omega W(\chi, w_1, w_2),$$

where ∇w_1 and ∇w_2 are periodic in Ω functions with fixed mean values. The problem needs relaxation due to the minimization over χ. The relaxed Lagrangian keeps the same form, but the oscillating properties tensor $\boldsymbol{\sigma}(\chi)$ is replaced by tensor $\boldsymbol{\sigma}_*$ of the homogenized properties.

To estimate the relaxed functional from below, we may use the harmonic mean bound. For the Lagrangian $W(\chi, w_1, w_2)$, this bound is

$$\begin{pmatrix} \nabla w_1 \\ \nabla w_2 \end{pmatrix}^T A(\boldsymbol{\sigma}_*) \begin{pmatrix} \nabla w_1 \\ \nabla w_2 \end{pmatrix} \geq \begin{pmatrix} \nabla w_1 \\ \nabla w_2 \end{pmatrix}^T A_h \begin{pmatrix} \nabla w_1 \\ \nabla w_2 \end{pmatrix},$$

where A_h is the harmonic mean,

$$A_h \leq \left\langle A(\boldsymbol{\sigma}(\chi))^{-1} \right\rangle^{-1}.$$

We notice that the coefficients of the properties matrix A are not uniquely defined by the relations between f and w. One can add identically zero terms to (8.1.7) that correspond to the Euler–Lagrange equation or a null-Lagrangian. In other words, we can add null-Lagrangian to the Lagrangian $W(\chi, w_1, w_2)$ without affecting f and w.

Here we use the identity

$$t\nabla \cdot R\nabla w = t\left(\frac{\partial^2 w}{\partial x_1 \partial x_2} - \frac{\partial^2 w}{\partial x_2 \partial x_1} \right) \equiv 0 \quad \forall w$$

to represent the integrability conditions. Here t is a real parameter and R is the tensor of rotation through a right angle (see (4.4.8)). Let us add this identity to both equations of (8.1.7) as follows:

$$\begin{aligned} \nabla \cdot [\boldsymbol{\sigma}(\chi)\nabla w_1 + tR\nabla w_2] &= f_1, \\ \nabla \cdot [\boldsymbol{\sigma}(\chi)\nabla w_2 - tR\nabla w_1] &= f_2. \end{aligned} \tag{8.1.8}$$

The new system possesses the same solutions w_1 and w_2, but it corresponds to different constitutive relations,

$$\begin{aligned} \nabla \cdot \mathbf{j}_1' = f_1, \quad &\mathbf{j}_1' = \boldsymbol{\sigma}(\chi)\nabla w_1 + tR\nabla w_2, \\ \nabla \cdot \mathbf{j}_2' = f_2, \quad &\mathbf{j}_2' = \boldsymbol{\sigma}(\chi)\nabla w_2 - tR\nabla w_1 \end{aligned}$$

and to different Lagrangian

$$\begin{aligned} W_T(\chi, w_1, w_2) = &\begin{pmatrix} \nabla w_1 \\ \nabla w_2 \end{pmatrix}^T A_T(\boldsymbol{\sigma}(\chi)) \begin{pmatrix} \nabla w_1 \\ \nabla w_2 \end{pmatrix} \\ &+ 2f_1 w_1 + 2f_2 w_2 + \gamma(\chi). \end{aligned} \tag{8.1.9}$$

The matrix of properties $A(\boldsymbol{\sigma})$ is replaced by the translated matrix

$$A_T = \begin{pmatrix} \boldsymbol{\sigma} & tR \\ -tR & \boldsymbol{\sigma} \end{pmatrix};$$

if t is small enough, matrix symmetric A_T remains positive.

System (8.1.8) describes the same conducting process as (8.1.7) in the same medium with conductivity $\sigma(\mathbf{x})$. The vectors $\mathbf{j}_i'$ are divergencefree and can be considered the currents in that medium.

Obviously, the effective properties tensor $\boldsymbol{\sigma}_*$ does not depend on the added identities, particularly on the coefficient t. However, the bound depends on this coefficient. Indeed, the Wiener bound for $\boldsymbol{\sigma}_*$, based on the system (8.1.9), is expressed through the harmonic mean of A_T as

$$\begin{pmatrix} \boldsymbol{\sigma}_* & 0 \\ 0 & \boldsymbol{\sigma}_* \end{pmatrix} + \begin{pmatrix} 0 & tR \\ -tR & 0 \end{pmatrix} \geq \left\langle \left(\begin{pmatrix} \boldsymbol{\sigma}(\chi) & tR \\ -tR & \boldsymbol{\sigma}(\chi) \end{pmatrix} \right)^{-1} \right\rangle^{-1}.$$

This bound for $\boldsymbol{\sigma}_*$ depends on the parameter t due to the nonlinearity of the harmonic averaging. This dependence entirely comes from the method of estimating the energy by its convex envelope. Therefore, one may consider the family of harmonic mean bounds that correspond to different values of t and choose the best bound from this family; this is what the translation method does.

Remark 8.1.2 *This consideration motivated the introduction of the translation method in early papers (Lurie et al., 1982); it was used in (Cherkaev, Lurie, and Milton, 1992) for the problem of plane elasticity, and it was developed in a general setting in (Milton, 2000). In the papers (Milton and Movchan, 1995; Milton and Movchan, 1998), the translations are used to transform the system of equations of the plane elasticity to a system of two equations of conductivity.*

8.2   Quadratic Translators

Here we present a method to find quadratic translators of the form $\phi(\mathbf{v}) = \mathbf{v} \cdot T\mathbf{v}$, where T is the matrix of the translator. The number of such translators is finite, because the number of linearly independent matrices is finite. The matrix T is defined by the differential constraints on the field $\mathbf{v}$. In fact, there are few practically interesting cases corresponding to constraints of the form $\nabla \mathbf{v} = 0$, $\nabla \times \mathbf{v} = 0$ and their combinations.

8.2.1   Compensated Compactness

Quadratic translators of the type $\phi(\mathbf{v}) = \mathbf{v} \cdot T\mathbf{v}$ can be found, as in Chapter 6, by constructing a finding a vector-function $\mathbf{k} : \nabla \cdot \mathbf{k} = \phi(\mathbf{v})$ However, the more direct approach is to use the compensated compactness method suggested in (Tartar, 1979a) and (Murat, 1981). The elegant idea of the approach is based on the Fourier transform.

Let $\hat{\mathbf{v}}$ be the Fourier transform of $\mathbf{v}$:

$$\hat{\mathbf{v}}(\boldsymbol{\omega}) = (2\pi)^{-d/2} \int_{\Omega} v(x_1, \ldots, x_d) e^{i[x_1\omega_1 + ,\ldots, + x_d\omega_d]} dx_1 \cdots dx_d,$$

where $\boldsymbol{\omega} = (\omega_1, \ldots, \omega_d)$ is the complex-valued frequencies vector.

The compensated compactness method directly applies Plancherel's formula:

$$\int_\Omega \mathbf{v} \cdot T\mathbf{v}\, d\mathbf{x} = \int_{\boldsymbol{\omega}} \hat{\mathbf{v}}^*(\boldsymbol{\omega}) \cdot T\hat{\mathbf{v}}(\boldsymbol{\omega}) d\boldsymbol{\omega},$$

where $\hat{\mathbf{v}}^*$ denotes the complex conjugate to $\hat{\mathbf{v}}$.

This equality can be used for determination of translators, because the linear differential constraints (8.1.2) imposed on the perturbation $\boldsymbol{\xi}$ (see definition of the quasiconvex envelope) become linear algebraic constraints on its Fourier transform $\hat{\boldsymbol{\xi}}$:

$$\mathcal{A} : \hat{\boldsymbol{\xi}} \otimes \boldsymbol{\omega} = \left(\sum_{k=1}^{d} \sum_{j=1}^{n} a_{ijk}\omega_k \right) \hat{\boldsymbol{\xi}}_j = 0, \ i = 1, \ldots, r.$$

Let us return to the definition of translators (8.1.1)–(8.1.2). Consider a quadratic function $\mathbf{v} \cdot T\mathbf{v}$, where T is a constant real matrix. The definition of quasiconvexity implies that its Fourier transform satisfies the equation

$$\int_{\boldsymbol{\omega}} \left[(\hat{\mathbf{v}} + \hat{\boldsymbol{\xi}}(\boldsymbol{\omega}))^* \cdot T(\hat{\mathbf{v}} + \hat{\boldsymbol{\xi}}(\boldsymbol{\omega})) - \hat{\mathbf{v}} \cdot T\hat{\mathbf{v}} \right] d\boldsymbol{\omega}$$

$$= \int_{\boldsymbol{\omega}} \hat{\boldsymbol{\xi}}^*(\boldsymbol{\omega}) \cdot T\hat{\boldsymbol{\xi}}(\boldsymbol{\omega})\, d\boldsymbol{\omega} \geq 0 \quad \forall \hat{\boldsymbol{\xi}} \in \hat{\Xi},$$

where $\hat{\Xi}$ is the linear subspace of Fourier images of perturbations $\hat{\boldsymbol{\xi}}(\boldsymbol{\omega})$:

$$\hat{\Xi} = \left\{ \hat{\boldsymbol{\xi}} : \ \int_{\boldsymbol{\omega}} \hat{\boldsymbol{\xi}}\, d\boldsymbol{\omega} = 0, \quad \mathcal{A} : \hat{\boldsymbol{\xi}} \otimes \boldsymbol{\omega} = 0 \right\}. \tag{8.2.1}$$

The last equality allows us to check the quasiconvexity of quadratic forms. Indeed, any quadratic form of $\mathbf{v}$ is a translator if (i) it is a convex function of the vectors $\hat{\mathbf{v}}$ that belong to the linear subspace (8.2.1) but (ii) it is not a convex function of the arbitrary vectors $\hat{\mathbf{v}}$. The subspace (8.2.1) must be independent of the frequencies vector $\boldsymbol{\omega}$.

This simple principle enables us to immediately find several translators. We discuss following examples following (Dacorogna, 1982).

Example 8.2.1 Quasiconvexity of the Energy. Suppose that the $2d$-dimensional vector $\mathbf{v} = (\mathbf{v}_1, \mathbf{v}_2)$ consists of a d-dimensional divergencefree part $\mathbf{v}_1$ and a d-dimensional curlfree part $\mathbf{v}_2$:

$$\mathbf{v} = (\mathbf{v}_1, \mathbf{v}_2), \quad \nabla \cdot \mathbf{v}_1 = 0, \quad \nabla \times \mathbf{v}_2 = 0.$$

Consider the quadratic form $W(\mathbf{v}) = \mathbf{v}_1 \cdot \mathbf{v}_2$ (the energy). We represent it as

$$W(\mathbf{v}) = \mathbf{v} \cdot T_1\mathbf{v}, \quad T_1 = \frac{1}{2} \begin{pmatrix} 0 & I \\ I & 0 \end{pmatrix}.$$

Here I is a d-dimensional unit matrix.

The Fourier transforms of the differential constraints have the form (8.2.1)

$$\boldsymbol{\omega} \cdot \hat{\boldsymbol{\xi}}_1(\boldsymbol{\omega}) = 0, \quad \boldsymbol{\omega} \times \hat{\boldsymbol{\xi}}_2(\boldsymbol{\omega}) = 0, \tag{8.2.2}$$

where $\hat{\boldsymbol{\xi}}_1(\boldsymbol{\omega})$ and $\hat{\boldsymbol{\xi}}_2(\boldsymbol{\omega})$ are Fourier images of centered perturbations: $\hat{\boldsymbol{\xi}}_1(0) = \hat{\boldsymbol{\xi}}_2(0) = 0$. Equations (8.2.2) say that the vectors $\hat{\boldsymbol{\xi}}_1(\boldsymbol{\omega})$ and $\hat{\boldsymbol{\xi}}_2(\boldsymbol{\omega})$ are orthogonal to each other: $\hat{\boldsymbol{\xi}}_1(\boldsymbol{\omega}) \cdot \hat{\boldsymbol{\xi}}_2(\boldsymbol{\omega}) = 0$. The orthogonality implies that $\hat{\boldsymbol{\xi}}_1^* T \cdot \hat{\boldsymbol{\xi}}_2 = 0 \ \forall \boldsymbol{\omega} \neq 0$.

The Fourier prototype of $\hat{\mathbf{v}}_1(0)$ and $\hat{\mathbf{v}}_2(0)$ are the average fields $\langle \mathbf{v}_1 \rangle$ and $\langle \mathbf{v}_2 \rangle$ respectively. Using Plancherel's formula, we obtain

$$\langle \mathbf{v} \cdot T_1 \mathbf{v} \rangle = \langle \mathbf{v}_1 \rangle \cdot T_1 \langle \mathbf{v}_2 \rangle.$$

We conclude that the averaged energy $\langle W(\mathbf{v}) \rangle$ is a quasiaffine function of the currents $\mathbf{v}_1$ and the fields $\mathbf{v}_2$,

$$\langle W(\mathbf{v}) \rangle = W(\langle \mathbf{v} \rangle).$$

Let us give a physical illustration of the quasiaffineness of $W(\mathbf{v})$. Suppose that the vectors $\mathbf{v}_1$ and $\mathbf{v}_2$ are continuous in the domains Ω_1 and Ω_2 and are discontinuous along some plane Γ that divides these domains. Recall that some components of $\mathbf{v}_1$ and $\mathbf{v}_2$ are still continuous on Γ:

$$[\mathbf{v}_1] \cdot \mathbf{n} = 0, \quad [\mathbf{v}_2] \cdot \mathbf{t} = 0,$$

where $\mathbf{n}$, $\mathbf{t}$ are the normal and tangent to Γ. The averaged values $\langle W(\mathbf{v}) \rangle$ of W are equal:

$$\langle W(\mathbf{v}) \rangle = \langle \mathbf{v}_1 \cdot \mathbf{v}_2 \rangle = \langle (\mathbf{v}_1 \cdot \mathbf{n})(\mathbf{v}_2 \cdot \mathbf{n}) \rangle + \langle (\mathbf{v}_1 \cdot \mathbf{t})(\mathbf{v}_2 \cdot \mathbf{t}) \rangle.$$

Observe that one of the multipliers in each term on the right-hand side is constant and therefore equal to its averaged value. We have

$$\langle W(\mathbf{v}) \rangle = E(\langle \mathbf{v}_1 \rangle, \langle \mathbf{v}_2 \rangle).$$

Any discontinuous component of one vector is multiplied by a continuous component of the other one. In other words, the discontinuity of a component of one vector is "compensated" by the continuity of the corresponding component of the second vector. This explains the name *compensated compactness* (Tartar, 1983; Murat and Tartar, 1985a; Tartar, 1989; Tartar, 1997).

The compensated compactness of the energy naturally follows from the corresponding variational principle. Indeed, the discontinuity of a component of the gradient is matched by the continuity of the component of the vector of dual variables. These boundary conditions are called *natural* or *variational boundary conditions* (Gelfand and Fomin, 1963), since they correspond to minimum of the energy.

Example 8.2.2 The quasiaffineness of the quadratic form

$$\phi_2 = \mathbf{v}_1 \cdot \Phi_{12}\,\mathbf{v}_2 \tag{8.2.3}$$

can be demonstrated in the same way. Here $\mathbf{v}_1$ and $\mathbf{v}_2$ are two curlfree vectors,

$$\mathbf{v} = (\mathbf{v}_1,\ \mathbf{v}_2),\quad \nabla \times \mathbf{v}_1 = 0,\quad \nabla \times \mathbf{v}_2 = 0,$$

and Φ_{12} is an antisymmetric $d \times d$ matrix: $\Phi_{12}^T = -\Phi_{12}$. The bilinear form (8.2.3) can be rewritten as a quadratic form:

$$\phi_2 = \mathbf{v} \cdot \Phi\mathbf{v},\quad \Phi = \frac{1}{2}\begin{pmatrix} 0 & \Phi_{12} \\ -\Phi_{12} & 0 \end{pmatrix}. \tag{8.2.4}$$

Note that Φ is the symmetric $2d \times 2d$ matrix.

We observe that the Fourier image of each curlfree vector $\hat{\mathbf{v}}_i$, $i = 1,2$ is parallel to the frequency vector $\boldsymbol{\omega}$,

$$\boldsymbol{\omega} = \alpha_1\hat{\mathbf{v}}_1,\quad \boldsymbol{\omega} = \alpha_2\hat{\mathbf{v}}_2,$$

where α_1 and α_2 are real constants. Therefore, they are parallel to each other: $\hat{\mathbf{v}}_1 = \alpha\hat{\mathbf{v}}_2$. We obtain

$$\hat{\mathbf{v}}_1^* \cdot \Phi_{12}\,\hat{\mathbf{v}}_2 = 0 \tag{8.2.5}$$

for any antisymmetric matrix Φ_{12}. Equation (8.2.5) states the quasiaffineness of ϕ_2, i.e., the quasiconvexity of both ϕ_2 and $-\phi_2$.

Example 8.2.3 In particular, the quadratic form of a pair of two-dimensional gradients $\mathbf{v}_1$ and $\mathbf{v}_2$,

$$\phi_2 = \mathbf{v}_1 \cdot R\mathbf{v}_2, \tag{8.2.6}$$

is quasiaffine. Here R is a matrix of rotation through a right angle (see (4.4.8)).

Remark 8.2.1 *Note that the function (8.2.6) is equal to the determinant of the matrix* $(\mathbf{v}_1, \mathbf{v}_2) = \nabla[w_1, w_2]$:

$$\mathbf{v}_1 \cdot R\mathbf{v}_2 = \det(\mathbf{v}_1, \mathbf{v}_2).$$

We established the quasiaffineness of ϕ_2 earlier (in Section 6.3) by a straight calculation.

Example 8.2.4 Consider the function

$$\phi_3 = \mathbf{j}_1 \cdot R\mathbf{j}_2, \tag{8.2.7}$$

where $\mathbf{j}_1$, $\mathbf{j}_2$ are divergencefree two-dimensional vectors. This function is quasiaffine:

$$\langle \phi_3 \rangle = \langle \mathbf{j}_1 \cdot R\mathbf{j}_2 \rangle = \langle \mathbf{j}_1 \rangle \cdot R \langle \mathbf{j}_2 \rangle.$$

Indeed, each two-dimensional vector $\hat{\mathbf{j}}_1$, $\hat{\mathbf{j}}_2$ is perpendicular to $\boldsymbol{\omega}$, and therefore they are parallel to each other:

$$\boldsymbol{\omega} \perp \hat{\mathbf{j}}_1, \quad \boldsymbol{\omega} \perp \hat{\mathbf{j}}_2 \quad \rightarrow \quad \hat{\mathbf{j}}_1 = \alpha \hat{\mathbf{j}}_2.$$

This equality demonstrates that

$$\hat{\mathbf{j}}_1^{\,*} \cdot R\hat{\mathbf{j}}_2 = 0, \tag{8.2.8}$$

which implies the quasiconvexity of ϕ_3.

Remark 8.2.2 *One could expect this result from the previous example, because each two-dimensional divergencefree vector rotated through a right angle becomes curlfree and vice versa.*

Notice that (8.2.8) is not valid for three-dimensional divergencefree vectors: In this case, the perpendicularity of $\hat{\mathbf{j}}_1$, $\hat{\mathbf{j}}_2$ to $\boldsymbol{\omega}$ does not imply their parallelism.

8.2.2 Determination of Quadratic Translators

So far, the compensated compactness was used to check the quasiconvexity of a given quadratic form. Here we describe a method that allows us to find quadratic translators systematically. We will construct quasiconvex functions by analysing the differential constraints.

Consider (8.2.2) as a linear homogeneous system for the d-dimensional vector $\boldsymbol{\omega} = (\omega_1, \dots \omega_d)$,

$$C\boldsymbol{\omega} = 0, \tag{8.2.9}$$

where the $(r \times d)$ matrix $C = \{c_{ij}\}$ linearly depends on $\hat{\mathbf{v}}$,

$$C = \{c_{ik}\}, \quad c_{ik} = \sum_{j=1}^{m} a_{ijk}\hat{v}_j, \quad i = 1,\dots,r, \quad k = 1,\dots,d. \tag{8.2.10}$$

For all admissible vectors $\mathbf{v} \in \Xi$, system (8.2.10) has a nontrivial solution $\boldsymbol{\omega}$.

This solvability condition could imply some constraints on the coefficients $c_{ij}(\hat{\mathbf{v}})$ and consequently on the admissible vector $\hat{\mathbf{v}}$. If we expressed those constraints in the form of positiveness of a quadratic form of $\hat{\mathbf{v}}$, then this form would define a quadratic translator.

Case Where No Translators Exist

If the number of linearly independent differential constraints r is less than the dimension of the space, $r < d$, the situation is trivial: The rank of the matrix C is less than d. Therefore, the system (8.2.9) has a nontrivial solution ω for all fields $\hat{\mathbf{v}}$. The differential constraints do not require additional constraints on $\hat{\mathbf{v}}$; hence there are no translators. Recall that in this case the quasiconvex envelope coincide with the convex envelope (Section 7.1). The problems in Chapter 4 are examples of this situation.

Finding Quadratic Translators

Suppose now that the number of linearly independent constraints r is greater than or equal to d. We have

$$\operatorname{rank} C = d.$$

In this case, (8.2.9) does not has a solution for an arbitrary matrix $C(\hat{\mathbf{v}})$; it has a solution only for those $\hat{\mathbf{v}}$ that make the matrix C degenerate:

$$\operatorname{rank} C(\hat{\mathbf{v}}) < d \quad \text{if } \hat{\mathbf{v}} \in \Xi. \tag{8.2.11}$$

This observation gives an idea of the method for finding quadratic translators. The number of linearly independent rows of C (8.2.10) is equal to the number of linearly independent differential restrictions r. If $r \geq d$, then the requirement (8.2.11) implies a linear dependence among the rows $\mathbf{c}_i$ of C:

$$\sum_{i=1}^{r} \alpha_i \mathbf{c}_i = 0,$$

where $\mathbf{c}_i = (c_{i1}, \ldots c_{id})$, and not all α_i are equal to zero.

The elements c_{ik} of C are linear functions of $\hat{\mathbf{v}}$. Therefore, the quadratic translator is also a quadratic function of c_{ik}.

The problem becomes the following: Construct a real-valued quadratic function of the elements c_{ik}; this function must (i) be positive definite if the rows of C are linearly dependent and (ii) be not positive definite if the rows of C are arbitrary. The linear dependence of rows states that equation (8.2.9) has a solution and the differential constraints are satisfied.

First, we transform C to a quadratic matrix. Multiply the equation $C \cdot \omega = 0$ on the left by an arbitrary $r \times d$ matrix K of rank d. The equation becomes

$$X \cdot \omega = 0. \tag{8.2.12}$$

X is a $d \times d$ matrix:

$$X = KC = K \cdot \mathcal{A} \cdot \hat{\mathbf{v}}, \quad X_{mn} = \sum_{i,j} K_{mi} a_{ijn} . \hat{v}_j, \tag{8.2.13}$$

and K_{mi} are elements of K. The elements of the square matrix X are linear combinations of $\hat{\mathbf{v}}$,

$$X_{mn} = N_{mnj}\hat{v}_j \quad N_{mnj} = \sum_i K_{mi}a_{ijn};$$

they depend on the elements of K as on parameters. For admissible fields $\hat{\mathbf{v}}$, the rank of X is $d-1$, due to condition (8.2.12).

There is a difference between the two practically interesting cases $d = 2$ and $d = 3$.

Two-Dimensional Quadratic Translators

This case is especially simple. The linear dependence of the rows is naturally expressed as $\det X = 0$. On the other hand, the determinant of a 2×2 matrix is the quadratic function of its elements. Hence, a quadratic translator is

$$\phi(K, \hat{\mathbf{v}}) = \det X = \hat{\mathbf{v}}^* \cdot T(K)\hat{\mathbf{v}}. \tag{8.2.14}$$

To compute $T(K)$ we set

$$\phi(K) = X_{11}X_{22} - X_{12}X_{21} \quad X_{mn} = N_{mn}\hat{\mathbf{v}};$$

$$N_{mn} = \left(\sum_{m,j} K_{im}a_{ijn}\right)\mathbf{e}_j,$$

where $\mathbf{e}_j$ are the basis vectors for $\mathbf{v}$, $\hat{\mathbf{v}} = \sum \hat{v}_j\mathbf{e}_j$.

Accordingly, we have, from (8.2.14),

$$T(K) = N_{11} \otimes N_{22} - N_{12} \otimes N_{21}.$$

Remark 8.2.3 *The equality* $\phi(K, \hat{\mathbf{v}}) = 0$ *holds. Therefore* $\phi(K, \mathbf{v})$ *is a quasiaffine function of* $\mathbf{v}$.

One can check that the previous examples correspond to the scheme.

Three-Dimensional Quadratic Translators

Consider now the three-dimensional case, $d = 3$. Now X is a degenerate 3×3 matrix. Let λ_i be the eigenvalues of X. The solvability condition $\det X = 0$ implies that $\lambda_3 = 0$. We represent the eigenvalues of X as

$$\lambda_1 = a_1 + i\, b_1, \quad \lambda_2 = a_2 + i\, b_2, \quad \lambda_3 = 0.$$

This time, the condition $\det X = 0$ cannot be used to define the quadratic translator, because $\det X$ is a cubic function of $\mathbf{v}$. Instead, we analyze the positivity of a real quadratic form of elements of X. The only special property of X is its degeneracy: $\lambda_3 = 0$. Therefore, we consider only functions of the eigenvalues λ_i. The translators should also be real quadratic functions of the elements of the complex-valued matrix X; therefore, we look on symmetric quadratic functions of the eigenvalues (these are the only functions of eigenvalues that are expressed as polynomial functions of the coefficients).

Consider the two real-valued symmetric polynomials of the elements of X (recall that $\lambda_3 = 0$)

$$\phi_1 = \mathrm{Tr}(X\,X^*) = (|\lambda_1|^2 + |\lambda_2|^2) = (a_1^2 + a_2^2 + b_1^2 + b_2^2),$$
$$\phi_2 = (\mathrm{Tr}\,X)(\mathrm{Tr}\,X^*) = |\lambda_1 + \lambda_2|^2 = (a_1 + a_2)^2 + (b_1 + b_2)^2,$$

and their linear combination

$$\Phi(X, \alpha, \beta) = \alpha\phi_1 + \beta\phi_2.$$

Let us choose parameters α, β such that Φ is nonnegative for the degenerate matrix X.

The function $\Phi(X, \alpha, \beta)$ can be represented as

$$\Phi = \frac{2\alpha + \beta}{2}\left((a_1 + a_2)^2 + (b_1 + b_2)^2\right) + \frac{\beta}{2}\left((a_1 - a_2)^2 + (b_1 - b_2)^2\right).$$

The conditions of nonnegativity of $\Phi(X, \alpha, \beta)$ are

$$2\alpha + \beta \geq 0, \quad \beta \geq 0. \tag{8.2.15}$$

The critical value of the parameters corresponds to the strongest results. Assigning $\beta = 2$, $\alpha = -1$ we obtain the translator

$$\phi(X) = 2\ \mathrm{Tr}(X\,X^T) - (\mathrm{Tr}\,X)^2 \ \geq \ 0$$

from (8.2.15). Notice that $\phi(X)$ is not positive if X is an arbitrary 3×3 matrix, but iis positive if the rank of X equals two.

Substituting the expression (8.2.13) of X, we find the sought quadratic form of $\hat{\mathbf{v}}$:

$$\phi_K(\hat{\mathbf{v}}) = 2(\mathrm{Tr}(K \cdot \mathcal{A} \cdot \hat{\mathbf{v}}))^2 - \mathrm{Tr}[(K \cdot \mathcal{A} \cdot \hat{\mathbf{v}})^T \cdot (K \cdot \mathcal{A} \cdot \hat{\mathbf{v}})] \geq 0. \tag{8.2.16}$$

Here K is an arbitrary matrix of parameters of the rank d. The nonnegativity of $\phi_K(\hat{\mathbf{v}})$ implies the quasiconvexity of $\phi_K(\mathbf{v})$.

Remark 8.2.4 *Note that $\phi(\hat{\mathbf{v}})$ is quasiconvex but not quasiaffine.*

Remark 8.2.5 *The reader can check that the d-dimensional case corresponds to the translator similarly to (8.2.16), where the coefficient 2 on the right-hand side is replaced by $d - 1$.*

The function $\phi_K(\mathbf{v})$ depends on an arbitrary matrix K, and this dependence is bilinear. The coefficients of K play the role of additional parameters r, which we introduced at the beginning of this chapter (see (8.1.6)).

Example 8.2.5 Consider a $3 \times k$ matrix V built of k divergencefree vectors:

$$V = [\mathbf{v}(1), \ldots, \mathbf{v}(k)], \quad \nabla \cdot \mathbf{v}(i) = 0, \ i = 1, \ldots, k.$$

Let us find a quadratic translator for this matrix. Here $\mathcal{A} \cdot V = V$.

If $k < 3$, then the rank of V is less than d, and there are no translators, because the equation $V\omega = \mathbf{0}$ always has a nontrivial solution.

If $k \geq 3$, then a translator is found from (8.2.16) as

$$\phi_K(V) = 2 \operatorname{Tr}(KVV^T K^T) - \operatorname{Tr}(KV)^2, \tag{8.2.17}$$

where K is an arbitrary $k \times d$ matrix of rank d.

This example (for k=d, K=I) was discovered by Tartar in the early 1980s (Tartar, 1985). The inequality (8.2.17) where K is an arbitrary diagonal matrix, has been obtained and used in (Gibiansky and Cherkaev, 1987) for bounding the elastic energy (see below, Chapter 15). In that problem, an optimal choice of the elements of K makes the bound exact.

8.3   Translation Bounds for Two-Well Lagrangians

8.3.1   Basic Formulas

Let us apply the obtained bounds for the problem of optimal composites. In this section we work out the algebraic form of the bound of two-well quadratic Lagrangians. Recall that a two-well piecewise quadratic Lagrangian(see (6.1.4)) is

$$F(\mathbf{v}) = \min_m (W(m, \mathbf{v}) + \gamma m), \tag{8.3.1}$$

where

$$W(m, \mathbf{v}) = \min_{\mathbf{v}} \ \min_{\chi : \langle \chi \rangle = m} \langle \chi \mathbf{v}^T D_1 \mathbf{v} + (1 - \chi) \mathbf{v}^T D_2 \mathbf{v} \rangle.$$

Let us determine the bound for the energy $W(m, \mathbf{v})$. Assuming that the volume fraction $m = \langle \chi \rangle$ of the first phase is fixed we construct the translation bounds (8.1.5) for $W(\chi, \mathbf{v})$. We use a quadratic translator $\Phi(\mathbf{v}, \mathbf{t})$,

$$\Phi(\mathbf{v}, \mathbf{t}) = \mathbf{v}^T T(\mathbf{t}) \mathbf{v},$$

where $T(\mathbf{t})$ is a constant translation matrix depending on a parameter vector $\mathbf{t}$. The translator depends only on the fields $\mathbf{v}$ and is independent of χ.

The translation bound (see (6.1.4)) is

$$W \geq \mathcal{P}W = \max_{\mathbf{t}} \mathcal{C}\left(W(\chi, \mathbf{v}) - \Phi(\mathbf{v}, \mathbf{t}) \right) + \Phi(\mathbf{v}, \mathbf{t}).$$

Here the convex envelope $CV = \mathcal{C}_{\mathbf{v}}\left(W(\chi, \mathbf{v}) - \Phi(\mathbf{v}, \mathbf{t})\right)$ is computed as

$$CV = \begin{cases} \mathbf{v}^T(D - T(\mathbf{t}))_h\mathbf{v} & \text{if } D(\chi(\mathbf{x})) - T(\mathbf{t}) \geq 0 \quad \forall \mathbf{x}, \\ -\infty & \text{otherwise,} \end{cases}$$

where $(\ )_h$ is the harmonic mean: $X_h = (m_1 X_1^{-1} + m_2 X_2^{-1})^{-1}$. To make the bound nontrivial, we choose $\mathbf{t} \in \mathcal{T}$ such that

$$\mathcal{T} = \{\mathbf{t} : D_1 - T(\mathbf{t}) \geq 0, \quad D_2 - T(\mathbf{t}) \geq 0\}. \tag{8.3.2}$$

The set $\mathcal{T}$ is not empty because (8.3.2) are satisfied if $\mathbf{t} = 0$, $T(\mathbf{0}) = 0$.

The bound $\mathcal{P}W$ is given by:

$$\mathcal{P}W = \max_{T \in \mathcal{T}} \mathbf{v}_0^T[(D - T(\mathbf{t}))_h + T(\mathbf{t})]\mathbf{v}_0 = \max_{T \in \mathcal{T}} \mathbf{v}_0^T D_p(\mathbf{t})\mathbf{v}_0, \tag{8.3.3}$$

where

$$D_p(\mathbf{t}) = (D - T(\mathbf{t}))_h + T(\mathbf{t})$$

or

$$D_p = [m_1(D_1 - T)^{-1} + m_2(D_2 - T)^{-1}]^{-1} + T. \tag{8.3.4}$$

This bound looks similar to the harmonic mean bound. The difference is that the matrices D_1, D_2, and D_p are translated (shifted) by the matrix T (the last property explains the name "translation method" suggested in (Milton, 1990b)). The optimal translation depends on optimal values $\mathbf{t}_0$ of $\mathbf{t}$ that are determined by the vector $\mathbf{v}$ in the maximization operation in (8.3.3).

If D_1, D_2 are isotropic then the value $\mathbf{t}_0$ depends only on rotation invariants of $\mathbf{v}$, but not on its magnitude. Particularly, there are no translation bounds different from the harmonic bound for the problem of the minimization of energy of conducting composites because a vector field $\mathbf{v}$ does not have any rotational invariants other than the magnitude. This again suggests why the harmonic mean bound is exact for that case. On the other hand, the sum of energies of several gradient fields has a nontrivial translation bound: The translation parameters depend on the ratios of magnitude of acting fields and on their mutual orientation.

8.3.2 Extremal Translations

The next feature of the translation bound makes it different from the harmonic bound. It is always possible to choose parameters of translations $\mathbf{t}$ to make at least one of the matrices $\mathcal{D}_1 - T(\mathbf{t})$ or $\mathcal{D}_2 - T(\mathbf{t})$ degenerate:

$$\exists \mathbf{t}: \ \det(D_1 - T(\mathbf{t}))\det(D_2 - T(\mathbf{t})) = 0.$$

Indeed, the translation matrices T have eigenvalues of different signs, because T is nonpositive by the definition of translators. Both matrices $D_1 -$

$T(\mathbf{t})$ and $D_2 - T(\mathbf{t})$ are positive if $\mathbf{t}$ is close to zero and they are not positive for some large $\mathbf{t}$. Therefore, values of $\mathbf{t}$ exist for which one or more of the eigenvalues of $\mathcal{D}_1 - T(\mathbf{t})$ or $\mathcal{D}_2 - T(\mathbf{t})$ are zero and the other eigenvalues are positive. The degeneration leads to a special algebraic form of the bounds and influences the attainability of the bounds.

Example 8.3.1 Suppose that $\det(D_2 - T(\mathbf{t}_0)) > 0$, but $\det(D_1 - T(\mathbf{t}_0)) = 0$. The degenerate matrix $D_1 - T(\mathbf{t}_0)$ admits the representation $D_1 - T(\mathbf{t}_0) = \mathbf{p}\mathcal{K}\mathbf{p}^T$, where $\mathcal{K}$ is the positive $k \times k$ matrix, $k < n$, and $\mathbf{p}$ is the $n \times k$ projector.

In this case it is convenient to express the translation $T(\mathbf{t}_0)$ in the form

$$T = D_1 - \mathbf{p}\mathcal{K}\mathbf{p}^T,$$

substitute it into the expression (8.3.4) for D_p, and simplify the expression. After calculations, we obtain the formula

$$D_p = D_1 + m_2 \mathbf{p}[\mathbf{p}^T(D_2 - D_1)^{-1}\mathbf{p} + m_1\mathcal{K}^{-1}]^{-1}\mathbf{p}^T,$$

which remains the expression for the effective properties of a matrix laminate (7.3.12).

The case where both matrices $D_1 - T$ and $D_2 - T$ degenerate is considered similarly. We will analyse this case in Section 11.3.

Asymptotics

Consider the case where the nonconvex energy has an especially simple form:

$$W(\mathbf{v}) = \begin{cases} 0 & \text{if } \mathbf{v} = \mathbf{0}, \\ \gamma + \mathbf{v}^T D \mathbf{v} & \text{if } \mathbf{v} \neq \mathbf{0}. \end{cases} \tag{8.3.5}$$

In the context of optimal design, this case corresponds to an optimal composite of two materials if one of them (the void) has ideal properties, $D_1 \geq \beta I, \beta \to \infty$ and the zero cost, and the second material has the finite properties D and the cost γ. We minimize the sum of the total energy and cost of the composite.

The estimating matrix $\mathcal{D}_p(\mathbf{t})$ (the limit of (8.3.4) when $D_1 \to \infty$) has an extremely simple form:

$$D_p(\mathbf{t}) = \frac{1}{m_2}(D - T(\mathbf{t})) + T(\mathbf{t}) = \frac{1}{m_2}D - \frac{m_1}{m_2}T(\mathbf{t}).$$

The finiteness of the matrix D ensures the finiteness of the set $\mathcal{T}$ of possible values of the translator. The translator satisfies the condition $D - T \geq 0$. The translation bound

$$W(\mathbf{v}) \geq \max_{\mathbf{t} \in \mathcal{T}} \mathbf{v}^T D_p(\mathbf{t})\mathbf{v} + m\gamma \tag{8.3.6}$$

necessarily corresponds to the optimal value t_0 that belongs to the boundary ∂T of the domain of its definition, because the bound depends monotonically on T. The bound becomes

$$W(\mathbf{v}) \geq \max_{t \in \partial T} \left\{ \mathbf{v}^T T \mathbf{v} + \frac{1}{m_2} \mathbf{v}^T (D - T) \mathbf{v} \right\} + m\gamma, \qquad (8.3.7)$$

where

$$\partial T = \{t : \ \det(D - T(t)) = 0, \quad D - T(t) \geq 0\}.$$

A Lower Bound for the Quasiconvex Envelope

To find the lower bound (8.3.1) for the quasiconvex envelope, we find the optimal volume fractions. Let us comment on the structure of the bound. For simplicity, consider the asymptotic case (8.3.5). Suppose that $T = T_0$ is the optimal translator found as in (8.3.6). The translation bound (8.3.7) for fixed m (we omit the subindex $_2$) is

$$W_p(m, \mathbf{v}) = \mathbf{v}^T T_0 \mathbf{v} + \frac{1}{m} \mathbf{v}^T (D - T_0) \mathbf{v} + m\gamma,$$

where T_0 is the optimal translator. The matrix $(D - T_0)$ is degenerate $(\det(D - T_0) = 0)$.

The optimal volume fraction m_0 is determined from (8.3.1) as

$$m_0 = \sqrt{\frac{\mathbf{v}^T (D - T_0) \mathbf{v}}{\gamma}}.$$

The optimal energy (the quasiconvex envelope) is bounded from below by

$$W_p = \mathbf{v}^T T_0 \mathbf{v} + 2\sqrt{\gamma \mathbf{v}^T (D - T_0) \mathbf{v}}. \qquad (8.3.8)$$

If $T_0 = 0$, equation (8.3.8) becomes the convex envelope. Geometrically, the convex envelope is an elliptical cone with the vertex at the point $(0, 0)$, which is tangent to the paraboloid $1 + \mathbf{v}^T D \mathbf{v}$, see Example 1.3.3.

If $T_0 \neq 0$, the nonconvex quadratic term $\mathbf{v}^T T_0 \mathbf{v}$ (the hyperboloid) is added to the envelope. Also, the cone $\sqrt{\gamma \mathbf{v}^T (D - T_0) \mathbf{v}}$ degenerates because $\det(D - T_0) = 0$.

Remark 8.3.1 $\mathbf{v}^T (D - T_0) \mathbf{v}$ *is a degenerate quadratic form. Particularly,* $D - T_0$ *may be a rank-one matrix,* $D - T_0 = \mathbf{d}\mathbf{d}^T$. *(We observe this case in elasticity problems considered in Chapter 15) In this case, the Lagrangian* W_q *is a rational function of* $\mathbf{v}$:

$$W_q = \mathbf{v}^T T_0 \mathbf{v} + \sqrt{\gamma} |\mathbf{v}^T \mathbf{d}|.$$

8.3.3 Example: Lower Bound for the Sum of Energies

Consider the minimization of the sum of energies of two two-dimensional conducting equilibria (Problem (6.1.1)). Denote by $\mathbf{w} = (w_1, w_2)$ the vector of potentials. Let $\mathbf{E}$ be the matrix $\mathbf{E} = \nabla \mathbf{w}$, and let the properties tensor be an isotropic tensor $D = \sigma I$; the isotropic conductivity takes the values σ_1 and σ_2 in the first and second materials. The energy of the equilibrium W is the sum of two energies corresponding to two conductivities. It is equal to

$$W = \mathrm{Tr}\left[(\chi D_1 + (1 - \chi)D_2)\mathbf{E}\mathbf{E}^T\right]$$

where χ is the characteristic function of the first material in the composite.

Construct the lower bound $\mathcal{P}W$ of the energy $\mathcal{Q}W$ of an optimal composite assuming that the volume fractions are fixed: $\langle \chi \rangle = m_1$, $\langle 1 - \chi \rangle = m_2$. The energy $\mathcal{Q}W$ has the form

$$\mathcal{Q}W(m, \mathbf{E}) = \mathrm{Tr}\left(\boldsymbol{\sigma}_* \mathbf{E}^T \mathbf{E}\right).$$

Here, $\boldsymbol{\sigma}_*$ is the effective tensor of the optimal composite. The tensor $\boldsymbol{\sigma}_*$ depends on the volume fraction $m \in [0, 1]$ and invariants of the field matrix $\mathbf{E}^T\mathbf{E}$ but not on its magnitude.

The symmetric 2×2 matrix has one invariant to rotation different from its magnitude. One can choose this invariant as $\kappa = 2\frac{\det \mathbf{E}}{\|\mathbf{E}\|^2}$, $|\kappa| \leq 1$, where $\|\mathbf{E}\| = \sqrt{\sum_{ij} e_{ij}^2}$. We have $\boldsymbol{\sigma}_* = \boldsymbol{\sigma}_*(m, \kappa)$.

Structure of the Translation Bound

We introduce a four-dimensional vector $\tilde{\mathbf{v}} = [\tilde{v}_1, \tilde{v}_2, \tilde{v}_3, \tilde{v}_4]$ simply by rewriting the elements of the matrix $\mathbf{E}$ as a vector:

$$\tilde{v}_1 = \mathbf{E}_{11} = \frac{\partial w_1}{\partial x_1}, \quad \tilde{v}_2 = \mathbf{E}_{12} = \frac{\partial w_1}{\partial x_2},$$

$$\tilde{v}_3 = \mathbf{E}_{21} = \frac{\partial w_2}{\partial x_1}, \quad \tilde{v}_4 = \mathbf{E}_{22} = \frac{\partial w_2}{\partial x_2}.$$

The energy W can be rewriten as

$$W = \tilde{\mathbf{v}}^T[\chi A(\sigma_1) + (1 - \chi)A(\sigma_2)]\tilde{\mathbf{v}}$$

where $A(\sigma_i) = \sigma_i I$ is proportional to 4×4 unit matrix I.

The quasiaffine translator $\phi(\mathbf{v}) = 2t \det \mathbf{E}$ is represented as

$$\phi(\tilde{\mathbf{v}}) = \tilde{\mathbf{v}}^T T \tilde{\mathbf{v}}, \quad T(t) = \begin{pmatrix} 0 & 0 & 0 & -t \\ 0 & 0 & t & 0 \\ 0 & t & 0 & 0 \\ -t & 0 & 0 & 0 \end{pmatrix}.$$

The translation bound (8.3.3) becomes

$$\mathcal{Q}W(m, \tilde{\mathbf{v}}) \geq \max_{t \in \mathcal{T}}\left\{\tilde{\mathbf{v}}^T A_p(\sigma_1, \sigma_2, t)\tilde{\mathbf{v}}\right\}. \tag{8.3.9}$$

Here

$$A_p = \left(m_1 \left(A(\sigma_1) - T(t) \right)^{-1} + m_2 \left(A(\sigma_2) - T(t) \right)^{-1} \right)^{-1} + T(t)$$

and

$$\mathcal{T} = \{ t : \; A(\sigma_1) - T(t) \geq 0, \quad A(\sigma_1) - T(t) \geq 0 \} .$$

Let us analyze the expression (8.3.9). Each matrix

$$A(\sigma_i) - T(t) = \begin{pmatrix} \sigma_i & 0 & 0 & -t \\ 0 & \sigma_i & t & 0 \\ 0 & t & \sigma_i & 0 \\ -t & 0 & 0 & \sigma_i \end{pmatrix}, \quad i = 1, 2,$$

has two pairs of equal eigenvalues,

$$\lambda_1 = \lambda_2 = \sigma_i - t, \quad \lambda_3 = \lambda_4 = \sigma_i + t. \tag{8.3.10}$$

The set $\mathcal{T}$ of admissible t corresponds to the nonnegativity of the eigenvalues (8.3.10):

$$\mathcal{T} = \{ t : \; |t| \leq \sigma_1 \} .$$

(Recall that we agree that $\sigma_1 < \sigma_2$.)

The eigenvectors

$$\mathbf{f}_1 = \frac{1}{\sqrt{2}} \begin{pmatrix} 1 \\ 0 \\ 0 \\ -1 \end{pmatrix}, \quad \mathbf{f}_2 = \frac{1}{\sqrt{2}} \begin{pmatrix} 0 \\ 1 \\ 1 \\ 0 \end{pmatrix},$$

$$\mathbf{f}_3 = \frac{1}{\sqrt{2}} \begin{pmatrix} 1 \\ 0 \\ 0 \\ 1 \end{pmatrix}, \quad \mathbf{f}_4 = \frac{1}{\sqrt{2}} \begin{pmatrix} 0 \\ 1 \\ -1 \\ 0 \end{pmatrix}$$

are the same for both matrices $A(\sigma_1) - T(t)$ and $A(\sigma_2) - T(t)$.

The translation bound has the form

$$PW(m, \tilde{\mathbf{v}}_0) = \max_{t \in \mathcal{T}} q(m, \tilde{\mathbf{v}}_0, t),$$

where

$$q(m, \tilde{\mathbf{v}}_0, t) = \left\{ \mu_1(t) c_1^2 + \mu_2(t) c_2^2 \right\}$$

and c_i are

$$c_1 = \sqrt{ \left(\tilde{\mathbf{v}} \cdot \mathbf{f}_1 \right)^2 + \left(\tilde{\mathbf{v}} \cdot \mathbf{f}_2 \right)^2 } = \| \tilde{\mathbf{v}} \| \sqrt{1 - \kappa},$$

$$c_2 = \sqrt{ \left(\tilde{\mathbf{v}} \cdot \mathbf{f}_3 \right)^2 + \left(\tilde{\mathbf{v}} \cdot \mathbf{f}_4 \right)^2 } = \| \tilde{\mathbf{v}} \| \sqrt{1 + \kappa}.$$

Parameters c_1, c_2 represent the applied external fields $\mathbf{E}$. They depend on the invariant parameter κ of $\mathbf{E}$, and they satisfy the equality

$$c_1^2 + c_2^2 = 2\| \tilde{\mathbf{v}} \| = \| \mathbf{E} \|.$$

If the two applied fields are orthogonal, one can choose the Cartesian axes in the reference basis so that $\tilde{\mathbf{v}} \cdot \mathbf{f}_2 = \tilde{\mathbf{v}} \cdot \mathbf{f}_3 = 0$; then c_1, c_2 become the sum and the difference of magnitudes of the applied fields, respectively.

Coefficients μ_i depend only on the translation parameter t:

$$\mu_1 = \left(\frac{m_1}{\sigma_1 - t} + \frac{m_2}{\sigma_2 - t} \right)^{-1} + t, \quad \mu_2 = \left(\frac{m_1}{\sigma_1 + t} + \frac{m_2}{\sigma_2 + t} \right)^{-1} - t.$$

Calculation of the Constants

The stationary values of t either coincide with the boundaries of the set $\mathcal{T}$ of the admissible values of t

$$t_1 = \sigma_1, \quad t_2 = -\sigma_1,$$

or correspond to the stationary points t_3, t_4, where $\frac{\partial q}{\partial t} = 0$. A straightforward calculation gives

$$t_3 = (m_1\sigma_2 + m_2\sigma_1)\frac{-c_1 + c_2}{c_1 + c_2}, \quad t_4 = (m_1\sigma_2 + m_2\sigma_1)\frac{c_1 + c_2}{-c_1 + c_2}.$$

The stationary values t_3 or t_4 are in the set $\mathcal{T}$, $(|t_3| \leq \sigma_1, |t_3| \leq \sigma_1)$, which implies the inequalities:

$$t_3 \in \mathcal{T} \quad \text{if} \quad \frac{\sigma_1}{m_1\sigma_2 + m_2\sigma_1} \leq \left| \frac{-c_1 + c_2}{c_1 + c_2} \right|,$$

$$t_4 \in \mathcal{T} \quad \text{if} \quad \frac{\sigma_1}{m_1\sigma_2 + m_2\sigma_1} \leq \left| \frac{c_1 + c_2}{-c_1 + c_2} \right|.$$

Observe that t depends only on κ and the conductivities of the phases. If the two applied fields are orthogonal, then $|-c_1 + c_2|$ and $|c_1 + c_2|$ are the intensities of these fields.

The bound $q(m, \tilde{\mathbf{v}}_0, t_s)$ that corresponds to the stationary values t_3, t_4 of t is

$$q_{3,4} = \sigma_a \frac{(|c_1| - |c_2|)^2}{2} + \sigma_h \frac{(|c_1| + |c_2|)^2}{2};$$

where σ_a and σ_h are the arithmetic and harmonic means of the conductivities. One can see that $q_{3,4}$ is equal to the energy in the laminate structure.

The limiting values t_1, t_2 lead to the bound

$$q_{1,2} = \sigma_1(c_1^2 + c_2^2) + \min\{c_1^2, c_2^2\} m_2 \left(\frac{m_1}{2\sigma_1} + \frac{m_2}{\sigma_2 - \sigma_1} \right)^{-1}.$$

One can check that $q_{1,2}$ is equal to the energy in the optimal second-rank matrix laminate. The upper and lower bounds coincide; therefore, the quasiconvex envelope is determined. The calculations are led for the reader.

8.3.4 Translation Bounds and Laminate Structures

Here we discuss attainability of translation bounds. The bound is exact if there exists a minimizing sequence (the structure) that realizes it. As with the attainability of the convex envelope, the attainability of the translation bounds is determined by the conditions on the fields (Milton, 1990b). Namely, the bound is exact if the fields inside the structure satisfy the condition

$$(D_i - T(\mathbf{t}))\mathbf{v}_i = \text{constant}(\mathbf{x}) \qquad (8.3.11)$$

everywhere. This condition is similar to (6.4.4).

Nonextremal Translations and the Fields on Optimal Structures

If the optimal value $\mathbf{t}_0$ of the translation parameter $\mathbf{t}$ corresponds to nondegenerate matrices $D_i - T(\mathbf{t}_0)$, then the fields inside each of the materials in the composite must be constant due to (8.3.11). The translator affects only the jump conditions; for a two-component composite, (8.3.11) becomes

$$D_1\mathbf{v}_1 - D_2\mathbf{v}_2 = T(\mathbf{t})(\mathbf{v}_1 - \mathbf{v}_2). \qquad (8.3.12)$$

where $\mathbf{v}_1$ and $\mathbf{v}_2$ are the fields in the first and second materials, respectively.

The following example shows how these conditions are satisfied on an appropriately oriented laminate structure. To satisfy them, we choose parameters of the structure (the normal to the laminates) and the parameter $\mathbf{t}$ of the translation. The last control allows us to satisfy (8.3.12) for a number of cases.

Example 8.3.2 We saw in Chapter 4 that a laminate structure minimizes the energy of a composite made from two isotropic conductors with conductivities σ_1 and σ_2. Now we investigate the minimum of the sum of the energies, considered in Example 8.3.3.

Consider a laminate with normal $\mathbf{n}$ and tangent $\mathbf{t}$. Suppose that the structure is submerged into two external orthogonal fields directed along and across the layers; they are described by a diagonal matrix $\mathbf{E}_0$.

The fields in the laminates satisfy the continuity conditions (see Chapter 2)

$$\mathbf{t} \cdot (\mathbf{E}_1 - \mathbf{E}_2)\mathbf{t} = 0, \quad \mathbf{n} \cdot (\sigma_1\mathbf{E}_1 - \sigma_2\mathbf{E}_2)\mathbf{n} = 0, \quad \mathbf{t} \cdot \mathbf{E}_1\mathbf{n} = \mathbf{t} \cdot \mathbf{E}_2\mathbf{n} = 0.$$

Here $\mathbf{E}$ is the matrix $\mathbf{E} = \nabla \mathbf{w} = (\nabla w_1, \nabla w_2)$ of the fields. The fields in the first and second phases are denoted by $\mathbf{E}_1$ and $\mathbf{E}_2$, respectively. Both matrices are diagonal.

Suppose that the translation bound is exact and the laminate structure realizes it. The matrix of the translator $T(\mathbf{t})$ is proportional to the matrix (4.4.8) of rotation through a right angle R, $T(\mathbf{t}) = \theta R$, where θ is the parameter of the translation. Condition (8.3.12) links the currents and the fields in the orthogonal directions.

$$\sigma_1 \mathbf{E}_1 - \sigma_2 \mathbf{E}_2 = \theta R^T (\mathbf{E}_1 - \mathbf{E}_2) R. \qquad (8.3.13)$$

The projection of the equality (8.3.13) on the normal $\mathbf{n}$ is

$$[\mathbf{n} \cdot (\sigma_1 \mathbf{E}_1 - \sigma_2 \mathbf{E}_2)]\mathbf{n} = \theta \, \mathbf{t} \cdot [\mathbf{E}_1 - \mathbf{E}_2]\mathbf{t}.$$

(We have used the relation $\mathbf{n} \cdot R = \mathbf{t}$, where $\mathbf{t}$ is a unit tangent vector.) This condition is satisfied independently of θ, because the differences in the brackets on the left- and right-hand sides equal zeros, due to the continuity conditions (the normal currents and the tangent fields are continuous).

The projection of (8.3.13) on the tangent $\mathbf{t}$ links the nonzero differences:

$$\mathbf{t} \cdot (\sigma_1 \mathbf{E}_1 - \sigma_2 \mathbf{E}_2)\mathbf{t} = -\theta \, \mathbf{n} \cdot (\mathbf{E}_1 - \mathbf{E}_2)\mathbf{n}.$$

This equality may be satisfied by the choice of the translation parameter θ:

$$\theta = \theta_0 = \frac{\mathbf{t} \cdot (\sigma_1 \mathbf{E}_1 - \sigma_2 \mathbf{E}_2)\mathbf{t}}{\mathbf{n} \cdot (\mathbf{E}_1 - \mathbf{E}_2)\mathbf{n}}, \quad \theta_0 \in [-\sigma_1, \, \sigma_1]. \qquad (8.3.14)$$

If this parameter is chosen as in (8.3.14), the translation lower bound coincides with the energy of a laminate. This proves that the bound is exact, and the laminate is optimal.

Note that the condition $\theta_0 \in [-\sigma_1, \sigma_1]$ that follows from the positiveness of $D_i - T(\theta_0)$ restricts the applicability of the construction. It is valid in a certain range of parameters. Outside of this range the laminate structure and the bound do not coincide. This case corresponds to the degenerate matrix $D_1 - T$ in the translation bound and to a second rank laminate structure.

Extremal Translations and Fields in Optimal Structures

The important difference between the translation bound and the convex envelope bound is that the matrices $D_i - T(\mathbf{t})$ can degenerate. They have the form $D_i - T(\mathbf{t}) = \mathbf{p}_i \mathbf{p}_i^T$, where $\mathbf{p}_i$ is the projector to the nondegenerate subspace of $D_i - T(\mathbf{t})$. In this case, the fields $\mathbf{v}_i$ in subdomains Ω_i are not necessarily constant. Indeed, the constancy of $(D_i - T(\mathbf{t}))\mathbf{v}_i$ (see the condition (8.3.11)) in the subdomain Ω_i implies only the constancy of a projection

$$\mathbf{p}^T \mathbf{v}_i = \text{constant}(\mathbf{x}) \quad \forall \mathbf{x} \in \Omega_i, \qquad (8.3.15)$$

but not the constancy of all components of $\mathbf{v}_i$.

Particularly, degeneration of only one matrix $D_1 - T$ suggests the search within matrix laminates because they possess one constant field of the second phase (the nucleus) and several different fields of the first phase (envelope). In this case, the rank of a matrix laminate corresponds to the defect of the matrices $D_i - T$ in the translation bound. Indeed, consider a matrix laminate of the rank k. The field in the enveloping material D_1 is piecewise constant. It takes k different values, $\mathbf{v}_1, \ldots, \mathbf{v}_k$.

The degeneracy of the matrix $D_1 - T = \mathbf{p}^T \mathcal{K} \mathbf{p}$ matches the multiplicity of the field in the envelope in the matrix laminate. If the matrix laminate matches the translation bound, the fields $\mathbf{v}_i$ in different layers $i = 1, \ldots, k$ filled with the same material D_1 satisfy (8.3.15).

Generally, (8.3.15) expresses conditions for the fields in optimal structures and provides the hint for searching them. The defect of matrices $D_i - T$ in the translation bound corresponds to the number of pieces in a piecewise constant field that can realize the bound. This number suggests how many differently oriented layers of this material are contained in a high-rank laminate structure if this structure agrees with the bounds. The rank of "suspicious" matrix laminates is greater than or equal to the defect of $D_1 - T$.

8.4 Problems

1. Prove the quasiaffineness of $\det \nabla \mathbf{w}$, where $\mathbf{w}$ is a three-component vector in the three-dimensional space.

   Hint: Find a divergence form of $\det \nabla \mathbf{w}$.

2. Prove the quasiconvexity of $\phi = -\det(\nabla \mathbf{w} + \nabla \mathbf{w}^T)$, where $\mathbf{w}$ is a two-component vector in the two-dimensional space.

   Hint: Represent ϕ as the sum of a convex and a quasiaffine function.

3. Draw a graph of the lower bound $\mathcal{P}W(c_1, c_2)$ in Example (8.3.3) and the convexity bound $\mathcal{C}W(c_1, c_2)$. Where do they coincide?

4. Show the coincidence of the upper and lower bounds in Example (8.3.3). Consider a pair of orthogonal external fields.

5. Derive the formula for the translation bound of a piecewise quadratic two-well Lagrangian if the matrices $D_1 - T$ and $D_2 - T$ both degenerate.

9
Necessary Conditions and Minimal Extensions

9.1 Variational Methods for Nonquasiconvex Lagrangians

Necessary Conditions of Weierstrass Type and an Extension

We are considering a technique of necessary conditions for nonconvex variational problems. Based on these conditions, we also suggest *a minimal extension* of unstable variational problems that makes the problem stable against these variations. The extended Lagrangian gives an upper bound for the quasiconvex envelope. We follow (Cherkaev, 1999). The method is based on the classical variational technique that requires a comparison of nearby configurations, rather than a comparison of all configurations, as sufficient methods do. In contrast to the laminates technique, which deals with an a priori chosen set of effective properties and geometric configurations, the necessary conditions analyze the fields in an optimal structure.

Recall the arguments of the method of necessary conditions (Chapters 1 and 4). We consider an unstable variational problem,

$$\min_{\mathbf{v}} J(F(\mathbf{v})), \quad J(F(\mathbf{v})) = \int_{\omega} F(\mathbf{v}), \; A : \mathbf{v} = \mathbf{G},$$

and we introduce the variations $\delta\mathbf{v}$ compatible with the differential constraints. Then we detect the forbidden region $\mathcal{V}_{\mathrm{f}}$ where the condition

$$\delta J(F(\mathbf{v})) = \min_{\delta\mathbf{v}}[J(F(\mathbf{v} + \delta\mathbf{v})) - J(F(\mathbf{v}))] \geq 0$$

is violated. In that region, we define an extended Lagrangian $\mathcal{S}F(\mathbf{v})$ from the condition

$$\delta J(\mathcal{S}F(\mathbf{v})) = 0 \quad \forall \mathbf{v} \in \mathcal{V}_{\mathrm{f}}.$$

The choice of the class of variations is nontrivial. The variations must be sufficiently simple to permit calculation of the increment of the functional. At the same time, the variations must be sensitive enough to detect various types of instabilities. The technique is problem-oriented. We demonstrate the technique on an example of the problem of the minimization of the sum of the energies of a conducting medium.

Remark 9.1.1 *This type of optimality condition (strong variations of the material's properties) was suggested in (Lurie, 1963; Lurie, 1967; Lurie, 1970b) where the variation in the strip was used (see Chapter 4) and the forbidden regions were detected. This way, the problem was proven to be ill-posed. In the papers (Lurie, 1975; Lurie et al., 1982; Lurie, 1993), these variations were used for analysis of several problems of optimal design of conducting and elastic media. The paper (Petukhov, 1995) considers more complicated variations for problems in elasticity.*

The method is a multidimensional analogue of the Weierstrass variations in classical calculus of variations and Pontryagin's maximum principle (Pontryagin et al., 1964; Rozonoèr, 1959) in the theory of control. Here we generalize the method using more sophisticated variations, supplement it with the method of minimal extension, and compare the extensions with those obtained by sufficient conditions, described in Chapter 8.

Formulation of the Problem

Again, we consider the problem of optimizing the structure of a composite of two conducting materials in a two-dimensional domain, clearly the simplest problem of structural optimization.

Recall the notation. A periodic cell Ω is parted into two subdomains Ω_i, $i = 1, 2$. The subdomains Ω_i are filled with isotropic materials with conductivities σ_1 and $\sigma_2 > \sigma_1$ so that a material σ_i occupies a domain Ω_i.

Suppose that a pair of uniform external electrical fields $\mathbf{E}_1\ \mathbf{E}_2$ are applied to the structure. They cause the fields $\tilde{\mathbf{e}}_1 = \nabla w_1$ and $\tilde{\mathbf{e}}_2 = \nabla w_2$, respectively, at each point of Ω. The field $\tilde{\mathbf{e}}_i(\mathbf{x})$ is a solution to the variational problem (6.1.1).

We minimize the functional J equal to the sum of energies $J(\mathbf{E}_1, \chi)$ $+ J(\mathbf{E}_2, \chi)$ plus the cost of materials. The minimization is performed by choosing the characteristic functions χ_i and the local fields $\tilde{\mathbf{e}}_1, \tilde{\mathbf{e}}_2$.

We introduce, as in Chapter 6, the 2×2 symmetric positive definite matrix $\mathbf{e}$:

$$\mathbf{e} = \left(\tilde{\mathbf{e}}^T \cdot \tilde{\mathbf{e}}\right)^{\frac{1}{2}} = \begin{pmatrix} e_A & 0 \\ 0 & e_B \end{pmatrix}. \tag{9.1.1}$$

Its eigenvalues e_A and e_B are nonnegative.[1]

The symmetric matrix of the external loadings is defined similarly: $\mathbf{E} = (\tilde{\mathbf{E}}\tilde{\mathbf{E}}^T)^{1/2}$. Here $\tilde{\mathbf{E}} = (\mathbf{E}_1\mathbf{E}_2)$ is the nonsymmetric matrix composed of the vector fields $\mathbf{E}_1$ and $\mathbf{E}_2$.

In the introduced notation, the optimization problem becomes

$$J(\mathbf{E}) = \min_{\chi_i} \ \min_{\mathbf{e}:\langle\mathbf{e}\rangle=\mathbf{E}} \int_\Omega W(\chi_i, \mathbf{e}),$$

where

$$W(\chi_i, \mathbf{e}) = \sum_{i=1}^{N} \left(\mathrm{Tr}(\sigma_i \mathbf{e}^2) + \gamma_i \right) \chi_i. \tag{9.1.2}$$

9.2 Variations

9.2.1 Variation of Properties

Here we describe the Weierstrass-type "structural variation" that is used to check the optimality of a structure. To perform this variation we implant an infinitesimal inclusion of an admissible material σ' at a point $\mathbf{x}$ in the domain Ω occupied by a host material σ and compute the difference in energies and cost. If the examined structure is optimal, then the increment of the cost is nonnegative. The increment of energy depends on a shape of the region of variation; this shape must be adjusted to the field so that the increment reaches its minimal value (which must remain nonnegative).

Here we consider the following variation. We replace material σ in a neighborhood of a point $\mathbf{x}$ by a quasiperiodic dilute composite of second-rank laminates. In these laminates, the envelope is made of the host material σ, and the inclusions are made of material σ'. In other words, we construct a second-rank laminate from the isotropic host medium in the envelope. This matrix laminate composite of second rank is characterized by its effective tensor $\boldsymbol{\sigma}_*$ (see (7.3.13)),

$$\boldsymbol{\sigma}_* = \sigma I + m \left((\sigma' I - \sigma I)^{-1} + (1 - m)G(\sigma) \right)^{-1}, \tag{9.2.1}$$

where m is the volume fraction of the inclusions and matrix G determines the degree of anisotropy,

$$G = \frac{1}{\sigma} \begin{pmatrix} \alpha & 0 \\ 0 & 1 - \alpha \end{pmatrix}, \quad 0 \leq \alpha \leq 1.$$

The eigenvectors of G coincide with the directions of lamination, α is an inner parameter of the structures that determine the relative elongation (intensities in orthogonal directions) of the inclusions.

---

[1] Here we use the indices $(\ )_A$ and $(\ )_B$ to indicate the axes.

Let us compute the increment $\delta\boldsymbol{\sigma}$ of the tensor properties caused by the array of infinitely dilute nuclei with conductivity $\sigma' = \sigma'I$ and infinitesimal volume fraction δm. We replace m by δm and $\boldsymbol{\sigma}_*$ by $\boldsymbol{\sigma} + \delta\boldsymbol{\sigma}$ in (9.2.1) and obtain

$$\delta\boldsymbol{\sigma} = \Lambda\delta m + o(\delta m), \quad \Lambda = \left((\boldsymbol{\sigma}' - \boldsymbol{\sigma})^{-1} + G(\sigma)\right)^{-1}. \tag{9.2.2}$$

Denote the eigenvalues Λ by λ_A and λ_B. They are

$$\lambda_A = \frac{\sigma(\sigma' - \sigma)}{\sigma + \alpha(\sigma' - \sigma)}, \quad \lambda_B = \frac{\sigma(\sigma' - \sigma)}{\sigma + (1 - \alpha)(\sigma' - \sigma)}. \tag{9.2.3}$$

We call Λ the matrix of variation. Note that Λ depends on the parameter α and the orientation of laminates.

Remark 9.2.1 *Using technique of (Eshelby, 1957; Eshelby, 1961), one can check that the form of the increment (9.2.2) coincides with the increment caused by a single elliptical inclusion of equal area.*

Remark 9.2.2 *The discussing approach is applicable to the corresponding elasticity problem; the increments caused by a single elliptical inclusion be calculated using (Eshelby, 1975; Movchan and Movchan, 1995). However, the array of inclusions generally leads to more sensitive variation than a single inclusion of the same area; see (Cherkaev, Grabovsky, Movchan, and Serkov, 1998) and Chapter 15.*

9.2.2   Increment

Let us compute the variation of energy caused by the Weierstrass-type variation (9.2.2) of properties $\delta\boldsymbol{\sigma}$. For simplicity, we assume that the main axes of $\delta\boldsymbol{\sigma}$ are codirected with the principal axes of the matrix $\mathbf{e}$ (later we will see that the obtained necessary conditions are the strongest ones, which justifies this assumption). The cost consists of the increment of energy due to the variation of conductivity $\delta\boldsymbol{\sigma}$ and the "direct cost" of the variation, that is, the change in the total cost due to change of quantities of the materials used. The direct cost of the variation is determined by its type: We replace the material σ (with the specific cost γ) by the material σ' (with the specific cost γ'). The change in the total cost is

$$(\gamma' - \gamma)\,\delta m.$$

The increment of the energy δW is equal to

$$\delta W = \mathrm{Tr}\left(\mathbf{e}^2\,\delta\boldsymbol{\sigma}\right) = \left(\lambda_A e_A^2 + \lambda_B e_B^2\right)\delta m$$

where e_A and e_B are defined by (9.1.1). Using (9.2.3), we transform the increment to the form

$$\delta W(\alpha) = \sigma(\sigma' - \sigma)\left(\frac{e_A^2}{\sigma + \alpha(\sigma' - \sigma)} + \frac{e_B^2}{\sigma' - \alpha(\sigma' - \sigma)}\right)\delta m. \tag{9.2.4}$$

The total cost of the variation, $\Delta J \delta m$, is

$$\Delta J \delta m = \left(\gamma' - \gamma + \delta W(\alpha)\right) \delta m. \tag{9.2.5}$$

It depends on the shape of the variation, specifically, on the parameter α of anisotropy (elongation) of the inclusions.

If the structure is optimal, then all variations (including the most sensitive ones) lead to the nonnegative increment

$$\Delta J_0 = \gamma' - \gamma + \min_{\alpha} \delta W(\alpha) \geq 0. \tag{9.2.6}$$

Otherwise, the cost will be reduced by this variation, and the structure fails the test. Condition (9.2.6) is called the Weierstrass-type condition.

The Most Sensitive Variations

Let us compute the most sensitive variations. One can check that the second derivative of the increment with respect to α has the sign of $\sigma' - \sigma$. Therefore, the expression for an optimal value α_0 of α depends on the sign of $\sigma' - \sigma$.

Case $\sigma' - \sigma < 0$. In this case, α_0 always belongs to the boundary of the interval $[0, 1]$: $\alpha = 0$ or $\alpha = 1$. The minimal increment ΔJ_0 is

$$\Delta J_0 = \min_{\alpha} \Delta J(\alpha) = \gamma' - \gamma + \sigma(\sigma' - \sigma) F_2(\sigma, \sigma', E), \tag{9.2.7}$$

where

$$F_2(\sigma', \sigma, E) = \begin{cases} \frac{e_A^2}{\sigma'} + \frac{e_B^2}{\sigma} & \text{if} \quad \frac{e_A}{e_B} \geq 1, \\ \frac{e_A^2}{\sigma} + \frac{e_B^2}{\sigma'} & \text{if} \quad \frac{e_A}{e_B} \leq 1. \end{cases} \tag{9.2.8}$$

In the plane of e_A, e_B, the permitted region is the interior of the intersection of two mutual orthogonal ellipses. The elongation of these ellipses is determined by the ratio $\frac{\sigma'}{\sigma}$, and their scale is defined by the difference in costs $\gamma' - \gamma$. Note that the boundary of the permitted set has the corner point $e_A = e_B$.

The result can be interpreted in the following way: If an inclusion is less conducting than the host material, then the best shape of the inclusion is a strip, elongated along the direction of the maximal field. Physically speaking, the less-conducting material "tries to expose" itself by creating an elongated obstacle across the direction of the maximal current to maximally reduce the conductivity of the structure.

Case $\sigma' - \sigma > 0$. In this case, the optimal value α_0 varies in $[0, 1]$ depending on the field $\mathbf{e}$. We find the stationary value A of α from the equation $\frac{\partial \delta J(\alpha)}{\partial \alpha} = 0$ as:

$$A = \frac{e_B \sigma' - e_A \sigma}{(e_A + e_B)(\sigma' - \sigma)}.$$

The optimal elongation rate α_0 is

$$
\alpha_0 = \begin{cases} A & \text{if } A \in (0,1), \\ 1 & \text{if } A \geq 1, \\ 0 & \text{if } A \leq 0. \end{cases}
$$

We substitute this value of α in (9.2.4) and find that the minimal increment ΔJ_0 is

$$
\Delta J_0 = \min_\alpha \delta J(\alpha) = \gamma' - \gamma + \sigma(\sigma' - \sigma) F_1(\sigma, \sigma', E), \tag{9.2.9}
$$

where

$$
F_1(\sigma', \sigma, E) = \begin{cases} \frac{e_A^2}{\sigma'} + \frac{e_B^2}{\sigma} & \text{if } \frac{e_A}{e_B} < \frac{\sigma}{\sigma'}, \\ \frac{(e_A + e_B)^2}{\sigma + \sigma'} & \text{if } \frac{e_A}{e_B} \in \left[\frac{\sigma}{\sigma'}, \frac{\sigma'}{\sigma}\right], \\ \frac{e_A^2}{\sigma} + \frac{e_B^2}{\sigma'} & \text{if } \frac{e_A}{e_B} > \frac{\sigma'}{\sigma}. \end{cases} \tag{9.2.10}
$$

In the plane of e_A, e_B, the permitted region is the convex envelope supported by two orthogonal ellipses; note that the boundary of the set has a straight component. The elliptical parts of the boundary of the permitted region correspond to the strip-like inclusions (the case where $\alpha_0 = 0$ or $\alpha_0 = 1$). The straight part corresponds to the dilute second-order laminates or the elliptical inclusion (the case where $\alpha_0 \in (0,1)$).

Physically, we interpret the result using the image of an equivalent elliptical trial inclusion. If the inclusion has a higher conductivity σ' than the host medium, its best shape is a circle (if the field $\mathbf{e}$ is isotropic), an ellipse (if the eigenvalues of the field $\mathbf{e}$ are close to each other), or a strip elongated across the direction of the minimal field (if the ratio of the eigenvalues of the field $\mathbf{e}$ is large enough). The highly conducting inclusions "want to hide" in the domain to minimize the decrease of the total conductivity.

Remark 9.2.3 *Interestingly, the variational technique is connected with the problem of "the best hidden" inclusion. This problem of detectability has application in nondestructive testing. It was introduced and investigated in (Cherkaeva and Cherkaev, 1995).*

Let us look on these results from a different viewpoint. First, we fix the shape of the inserted inclusion, that is the parameter α of the variation. The necessary condition $\delta W \geq 0$ (see (9.2.5)) says that the magnitudes e_A, e_B of the field in the tested (host) material belong to interior of the ellipse

$$
\frac{e_A^2}{\sigma + \alpha(\sigma' - \sigma)} + \frac{e_B^2}{\sigma' - \alpha(\sigma' - \sigma)} + \frac{\gamma' - \gamma}{\sigma(\sigma' - \sigma)} = 0
$$

if $\sigma' < \sigma$; or to exterior of that ellipse if $\sigma' > \sigma$.

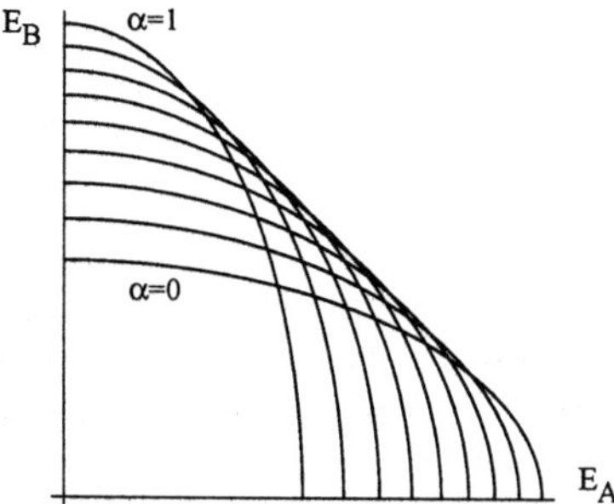

FIGURE 9.1. Family of optimality conditions obtained by inserting of various second-rank inclusions with variable shape (parameter α).

Varying the shape of the inclusion: $\alpha \in [0, 1]$, we obtain a family of necessary conditions that corresponds to the family of ellipses; see Figure 9.1. All these tests must be satisfied, therefore the field in the tested material belongs to intersection of all ellipses when $\sigma' < \sigma$ and it belongs to exterior of their union when $\sigma' > \sigma$. Figure 9.1 shows that the intersection of ellipses is bounded by a curved quadrangle formed by the outer ellipses that correspond to $\alpha = 0$ and $\alpha = 1$; the corresponding increment is given by (9.2.7). The union of ellipses is bounded by the envelope of the family which correspond to $\alpha = \alpha_0$; here, it coincides with the convex envelope supported by the outer ellipses, as shown in Figure 9.1. This case corresponds to the increment (9.2.9).

Necessary Conditions

The obtained inequalities allow us to find the range of admissible fields for each material. In an optimal structure, the field $\mathbf{e}_i$ in the material σ_i leads to a nonnegative increment $\Delta J(\sigma_j, \sigma_i, \mathbf{e}_i)$ if the host material σ_i is replaced by an inclusion of any admissible material σ_j. The field $\mathbf{e}_i$ remains optimal in the domain $\mathcal{V}_i$ where all increments are nonnegative:

$$\mathcal{V}_i = \{\mathbf{e}_i : \ \Delta J(\sigma_j, \sigma_i, \mathbf{e}_i) \geq 0 \ \ \forall j = 1, \ldots, N, \ j \neq i\}.$$

The union $\mathcal{V} = \cup \mathcal{V}_i$ of these sets does not coincide with the whole space $\mathcal{E}$ of $\mathbf{e}$. The remaining part $\mathcal{V}_f$ is forbidden; none of the materials is optimal in $\mathcal{V}_f$ and the fields in an optimal structure never belong to that region.

Suppose that an optimal periodic structure is submerged into a constant field $\mathbf{E}$. If the external field $\mathbf{E}$ lies in the forbidden region, it is split into several parts in an optimal structure, so that each part belongs to a region $\mathcal{V}_i$ and the mean field is equal to $\mathbf{E}$. This phase separation corresponds to the optimality of a composite structure.

If the external field $\mathbf{E}$ belongs to $\mathcal{V}_i$, then the solution to the variational problem (9.1.2) is homogeneous and corresponds to optimality of one of the pure materials: $\sigma_{\mathrm{opt}} = \sigma_i$.

Generalization: Other Variations. These results depend on the type of variations one uses. For multicomponent composites, one can introduce more sophisticated variations. For example, one can consider a variation such as (9.2.2), but the inclusions in the dilute matrix composites are filled not with a given material but with a laminate or other composite of several available materials. The requirement is that the effective properties of this composite must be an explicitly computable function of the conductivities of components and volume fractions. Accordingly, one can optimize the variations by choosing the most suitable material in the inclusion and the shape of the inclusions. The example of such a variation is discussed in Section 12.2.

9.2.3 Minimal Extension

The necessary conditions allow us to determine an extended Lagrangian in the forbidden region. Physically, the extended Lagrangian describes the minimal energy stored in a composite medium, assembled from the given materials. The extended Lagrangian $\mathcal{S}W(\mathbf{E})$ has the following properties:

1. It preserves the cost of the variational problem (9.1.2).

2. It leads to a classical solution defined for all mean fields (including those in the forbidden region), which cannot be improved by the class of variations considered.

If the mean field $\mathbf{E}$ belongs to $\mathcal{V}_i$, then the relaxed Lagrangian $\mathcal{S}W(\mathbf{E})$ coincides with the original Lagrangian:

$$\mathcal{S}W(\mathbf{E}) = W(\mathbf{E}) \quad \forall \mathbf{E} \in \mathcal{V}_i, \ i = 1, \ldots, N.$$

However, if the mean field $\mathbf{E}$ belongs to the forbidden region $\mathcal{V}_{\mathrm{f}}$, no homogeneous solutions exist. The pointwise field $\mathbf{e} = \mathbf{e}(\mathbf{x})$ never belongs to the forbidden region, $\mathbf{e}(\mathbf{x}) \notin \mathcal{V}_i \ \forall \mathbf{x}$.

The problem needs a relaxation. It is convenient to define the *extended Lagrangian $\mathcal{S}W(\mathbf{e})$* in the form

$$\mathcal{S}W(\mathbf{e}) = \mathrm{Tr}(\mathbf{e}\,\boldsymbol{\sigma}_*) + \gamma_*, \quad \boldsymbol{\sigma}_* = \boldsymbol{\sigma}_*(\mathbf{e}). \tag{9.2.11}$$

Here $\boldsymbol{\sigma}_*$ is a tensor that depends on $\mathbf{e}$. The tensor $\boldsymbol{\sigma}_*$ can be interpreted as an anisotropic effective tensor of composite, made of initially given materials. The structure of the optimal composite varies together with the external field $\mathbf{e}$.

Denote by m_i the volume fractions of the materials in the composite and by γ_i their cost; the cost γ_* of the composite is:

$$\gamma_* = \sum_{i=1}^{N} m_i \gamma_i.$$

To determine the extended Lagrangian we perform the following Weierstrass variation: Insert one of the original materials σ_i in the field $\mathbf{e} = \in \mathcal{V}_\mathrm{f}$ instead of the unknown optimal material $\boldsymbol{\sigma}_*$. We call the extension $\mathcal{S}W(\boldsymbol{\sigma}_*, \mathbf{e})$ neutral with respect to the variation if

$$\Delta_* = \min_i \{\Delta J(\sigma_i, \boldsymbol{\sigma}_*, \mathbf{e}) - \gamma_i + \gamma_*\} = 0. \tag{9.2.12}$$

This *condition of neutrality* states that the introduced material with the properties $\boldsymbol{\sigma}_*$ and the cost γ_* is neutral with respect to the most "dangerous" variations. Equality (9.2.12) determines the optimal tensor $\boldsymbol{\sigma}_*(\mathbf{e})$ and the extended Lagrangian.

This analysis shows that the minimal extension $\mathcal{S}W$ of the Lagrangian W is defined by a variational inequality:

$$\begin{aligned}
\mathcal{S}W(\mathbf{e}) &= W_i(\mathbf{e}), & \Delta_*(\mathbf{e}) &\geq 0, & (\mathbf{e} &\in \mathcal{V}_i), \\
\mathcal{S}W(\mathbf{e}) &\leq W_i(\mathbf{e}), & \Delta_*(\mathbf{e}) &= 0, & (\mathbf{e} &\notin \cup \mathcal{V}_i)
\end{aligned}$$

(compare with the minimal extension of the one-dimensional variational problems, Chapter 1).

This extension gives an upper boundary of the "final" extension, the quasiconvex envelope of the functional. Indeed, one could think of a wider class of variations that would lead to another extension with smaller minimal value for the variation. However, we will demonstrate that the extension is final for the problem under consideration.

Remark 9.2.4 *Equation (9.2.12) determines the effective tensor $\boldsymbol{\sigma}_*$ that corresponds to a composite with extremal conductivity. The set $\mathcal{W}(\boldsymbol{\sigma}_*)$ of these tensors estimates G_m-closure (the set of effective properties of all possible composites) from inside. The boundary of $\mathcal{W}(\boldsymbol{\sigma}_*)$ corresponds to those composites that are stable against the chosen type of variation.*

Another Scheme of Extension. It is often more convenient to use another way to determine the extended Lagrangian. We perform the inverse variation: Replace the material σ_i at a point of the allowed region $\mathbf{e}_i \in \mathcal{V}_i$ with the unknown composite material $\boldsymbol{\sigma}_*$, and choose the best geometry of that inclusion. The resulting variation has the form

$$\delta J(\boldsymbol{\sigma}_*, \sigma_i, \mathbf{e}) = \delta W(\boldsymbol{\sigma}_*, \sigma_i, \mathbf{e}) - \gamma_* + \gamma_i. \tag{9.2.13}$$

The minimal extension corresponds to the equality

$$\delta_0 J(\boldsymbol{\sigma}_*, \sigma_i) = \min_i \ \min_{\mathbf{e}_i \in \mathcal{V}_i} J(\boldsymbol{\sigma}_*, \sigma_i, \mathbf{e}) = 0,$$

which determines the extended Lagrangian. This condition again says that the introduced material with the properties $\boldsymbol{\sigma}_*$ and the cost γ_* are neutral under the most "dangerous" variations. The varying parameters include the shape of the inclusion, the field $\mathbf{e}_i \in \mathcal{V}_i$ in the phase σ_i, and the number i of the phase. Again, the Lagrangian extended in that way preserves the cost of the variational problem.

9.3 Necessary Conditions for Two-Phase Composites

Let us apply the technique to minimize the sum of two energies in a conducting composite and to find optimal two-phase composites. This example was considered in the two previous chapters.

9.3.1 Regions of Stable Solutions

Consider an optimal composite made of two materials σ_1 and $\sigma_2 > \sigma_1$ and find the range of fields permitted by the described variations. Applying formulas (9.2.8) and (9.2.10) where the material constants σ and σ' are properly chosen, we obtain the following inequalities.

1. The increment of the functional due to the inclusion of material σ_2 inserted into the domain Ω_1 is

$$\delta_{12} J = (\sigma_2 - \sigma_1)\sigma_1 F_1(\sigma_1, \sigma_2, \mathbf{e}_1) + \gamma_2 - \gamma_1 \geq 0, \qquad (9.3.1)$$

where $\mathbf{e}_1$ is the field in the domain Ω_1 of the first material.

2. The increment of the functional due to the inclusion of material σ_1 inserted into the domain Ω_2 is

$$\delta_{21} J = (\sigma_1 - \sigma_2)\sigma_2 F_2(\sigma_2, \sigma_1, \mathbf{e}_2) - \gamma_2 + \gamma_1 \geq 0, \qquad (9.3.2)$$

where $\mathbf{e}_2$ is the field at a point of the domain Ω_2 of the second material.

Let the set of permitted values of the field in the first material be $\mathcal{V}_1$ and the set of permitted values of the field in the second material be $\mathcal{V}_2$:

$$\mathcal{V}_1 = \{\mathbf{e} : \ \delta_{12} J \geq 0\}, \quad \mathcal{V}_2 = \{\mathbf{e} : \ \delta_{21} J \geq 0\}.$$

Assume that the eigenvalues of $\mathbf{e}$ are ordered as

$$0 \leq e_A \leq e_B. \qquad (9.3.3)$$

Computing the sets $\mathcal{V}_1, \mathcal{V}_2$ from (9.2.8) and (9.2.10), we have

$$\left.\begin{array}{ll} \sigma_1(\sigma_2 - \sigma_1)\left(\dfrac{e_A^2}{\sigma_2} + \dfrac{e_B^2}{\sigma_1}\right) + \gamma_2 - \gamma_1 \geq 0 & \text{if } \dfrac{e_A}{e_B} \leq \dfrac{\sigma_1}{\sigma_2}, \\[2ex] \dfrac{\sigma_1(\sigma_2 - \sigma_1)}{\sigma_1 + \sigma_2}\left(|e_A| + |e_B|\right)^2 + \gamma_2 - \gamma_1 \geq 0 & \text{if } \dfrac{e_A}{e_B} \geq \dfrac{\sigma_1}{\sigma_2}, \end{array}\right\} \text{ in } \mathcal{V}_1$$

and

$$\frac{e_B^2}{\sigma_1} + \frac{e_A^2}{\sigma_2} \leq (\gamma_1 - \gamma_2)\frac{1}{\sigma_2(\sigma_2 - \sigma_1)} \quad \text{in } \mathcal{V}_2.$$

The corresponding graph of the permitted fields is presented in Figure 9.2.

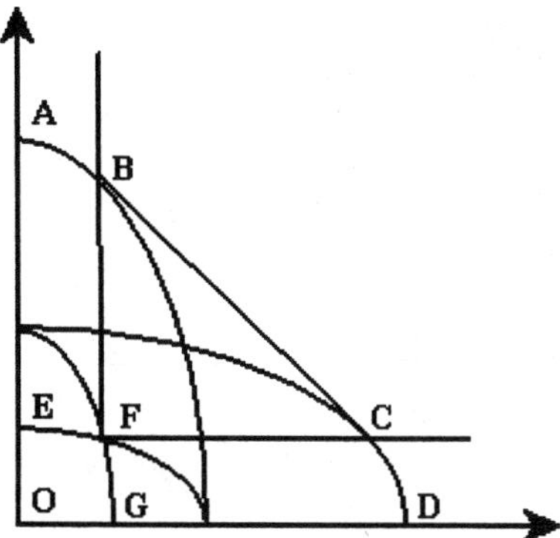

FIGURE 9.2. Permitted fields e_A, e_B in an optimal two-component composite. Region $\mathcal{V}_1$ lies outside ABCD, region $\mathcal{V}_2$ lies inside EFG, and in between lies the forbidden region $\mathcal{V}_f$.

Note that the forbidden region $\mathcal{V}_f$ in which none of the given materials is optimal lies between regions $\mathcal{V}_1$ and $\mathcal{V}_2$.

Let us analyze the fields in an optimal structure. Assume that an external field $\mathbf{E}$ is given. The field $\mathbf{e}$ in an optimal structure depends on the Lagrange multipliers or, equivalently, on the volume fraction of the materials in the composite. If the Lagrange multipliers are chosen so that the external field belongs to $\mathcal{V}_1$ or $\mathcal{V}_2$, then the composite consists of one material σ_1 or σ_2 only; the volume fractions are zero and one, respectively; and the field is constant everywhere in the cell. The nontrivial case occurs when the mean field belongs to the forbidden region $\mathcal{V}_f$. For any point $\mathbf{x}$ in the cell Ω, the field $\mathbf{e}(\mathbf{x})$ cannot belong to this region. Therefore, $\mathbf{e}$ alternates values in $\mathcal{V}_1$ and $\mathcal{V}_2$. In this situation we are dealing with a true composite, and the solution to the variational problem is given by a nonsmooth minimizer, i.e., the field $\mathbf{e}$ jumps on the boundary between these two regions.

9.3.2 Minimal Extension

Let us perform the minimal extension of the Lagrangian in the forbidden region. We demonstrate that the minimal extension leads to determination of effective properties of optimal composites without our having to guess optimal microstructures.

We assume that the extended Lagrangian $\mathcal{S}W(\mathbf{e})$ is given by the formula (9.2.11) and compute the tensor σ_* that determines $\mathcal{S}W(\mathbf{e})$. Let $\delta J_{*,1}(\mathbf{e}, \alpha)$ be the increment caused by replacing an isotropic material σ_1 with an anisotropic inclusion made of material σ_* with eigenvalues λ_A and λ_B:

$$\sigma_* = \begin{pmatrix} \lambda_A & 0 \\ 0 & \lambda_B \end{pmatrix}.$$

Here α is the parameter of elongation of the inclusion.

The calculation of the increment $\delta J_{*,1}(\mathbf{e}, \alpha)$ due to an anisotropic inclusion is similar to the calculation described earlier. Similarly to that calculation, we determine the increment caused by inserting an unknown *anisotropic* material $\boldsymbol{\sigma}_*$ into the domain $\mathcal{O}_1$ of the material σ_1, where $\mathbf{e} \in \mathcal{V}_1$. The increment $\delta J_{*,1}(\mathbf{e}, \alpha)$ is

$$\delta J_{*,1}(\mathbf{e}, \alpha) = \frac{\sigma_1(\lambda_A - \sigma_1)e_A^2}{\alpha\lambda_A + (1 - \alpha)\sigma_1} + \frac{\sigma_1(\lambda_B - \sigma_1)e_B^2}{\alpha\sigma_1 + (1 - \alpha)\lambda_B} + \gamma_* - \gamma_1. \quad (9.3.4)$$

The difference $\gamma_* - \gamma_1$ of the cost of materials σ_* and σ_1 is proportional to the amount of the second material used in the formation of the composite σ_*:

$$\gamma_* - \gamma_1 = (\gamma_2 - \gamma_1)m_2.$$

The increment depends on the parameter α that shows the elongation of the inserted inclusion. As before, we determine the minimal increment by calculating the optimal value of α and substituting it into (9.3.4). The resulting increment (compare with (9.2.13)) is

$$\delta J(\boldsymbol{\sigma}_*, \sigma_i, \mathbf{e}) = \sigma_1 \left(e_A + e_B\right)^2 \frac{(\lambda_A - \sigma_1)(\lambda_B - \sigma_1)}{\lambda_A\lambda_B - \sigma_1^2} + (\gamma_2 - \gamma_1)m_2. \quad (9.3.5)$$

(It is assumed in this calculation that the optimal value of α lies in $(0, 1)$ and does not takes boundary values.)

The minimal increment (9.3.5) is zero if the extension is neutral to the considered variations:

$$\delta_0 J(\boldsymbol{\sigma}_*, \sigma_i) = \min_{\mathbf{e}\in\partial\mathcal{V}_1} \delta J(\boldsymbol{\sigma}_*, \sigma_i, \mathbf{e}) = 0. \quad (9.3.6)$$

This equation serves to restrict the eigenvalues λ_A and λ_B of the tensor $\boldsymbol{\sigma}_*$ associated with the minimal extension.

First, assume that the optimal fields e_A and e_B in (9.3.6) belong to the straight segment of the boundary $\partial\mathcal{V}_1$. One can check that the the fields e_A, e_B belong to the straight segment of $\partial\mathcal{V}_1$, if $\mathbf{E}$ is in the triangle BFC, Figure 9.2. In this case, e_A, e_B (see (9.2.10)) satisfy the equality

$$(e_A + e_B)^2 = (\gamma_1 - \gamma_2)\frac{\sigma_1 + \sigma_2}{\sigma_1(\sigma_2 - \sigma_1)} \quad \text{if} \quad \frac{e_A}{e_B} \geq \frac{\sigma_1}{\sigma_2}.$$

We substitute this value into (9.3.6) to obtain the equality for the eigenvalues of the effective tensor:

$$\frac{(\lambda_A - \sigma_1)(\lambda_B - \sigma_1)}{\lambda_A\lambda_B - \sigma_1^2} = m_2 \frac{\sigma_2 - \sigma_1}{\sigma_2 + \sigma_1}.$$

After obvious manipulations, this equality is transferred to the familiar form

$$\frac{1}{\lambda_A - \sigma_1} + \frac{1}{\lambda_B - \sigma_1} = \frac{1}{m_2}\left(\frac{2}{\sigma_2 - \sigma_1} + \frac{m_1}{\sigma_1}\right).$$

One can see that λ_A and λ_B are the eigenvalues of the second-rank laminates. On the other hand, this energy is equal to the bound obtained by the translation method; therefore, it is optimal.

In a similar way, one can obtain the minimal extension for the case where $\mathbf{E} \in \mathcal{V}_{\mathrm{f}}$ and $\mathbf{E} \notin BFC$. The optimal value of α in (9.3.5) is either zero or one. One can check that the extension corresponds to the effective tensor with eigenvalues $\lambda_A = \frac{m_1}{\sigma_1} + \frac{m_2}{\sigma_2}$ and $\lambda_B = m_1\sigma_1 + m_2\sigma_2$ that also match the sufficient conditions (Wiener bounds).

The Extension and the Regions $\mathcal{V}_i$. The picture of the regions $\mathcal{V}_i$ (Figure 9.2) corresponds to the extension and the quasiconvex envelope in the following way: The extended Lagrangian $\mathcal{S}W$ coincides with the ith well of the Lagrangian W when $\mathbf{e} \in \mathcal{V}_i$. In the forbidden region $\mathcal{V}_{\mathrm{f}}$ we have $\mathcal{S}W < W$.

Recall that the extended Lagrangian is the upper bound for the quasiconvex envelope, $\mathcal{S}W \geq \mathcal{Q}W$. Therefore,

$$\mathcal{Q}W(\mathbf{e}) < W(\mathbf{e}) \quad \forall \mathbf{e} \in \mathcal{V}_{\mathrm{f}}.$$

However, the coincidence of W and $\mathcal{S}W$ does not prove that $\mathcal{Q}W = W$; other variations might enlarge the forbidden region $\mathcal{V}_{\mathrm{f}}$.

9.3.3 Necessary Conditions and Compatibility

The conditions of optimality of the variational problem and the extensions point to the appearance of the microstructures, but the geometry of the optimal microstructure remains unknown. Let us link optimality of the fields to optimal microstructures.

Compatibility

To begin, we find a form for the necessary conditions that does not include the Lagrange multipliers γ_1 and γ_2. Consider the sum of the variations (9.2.8) and (9.2.10): We interchange two small inclusions of equal volume, placing the inclusion of σ_1 into the domain Ω_2 occupied by σ_2 and vice versa. Note that optimal inclusions may be different shapes, only their volumes must be equal. This variation does not change the total amounts of materials σ_1 and σ_2; it only replaces them. Therefore, the cost of materials is irrelevant.

The total change in the functional due to the interchange of the materials is found from (9.3.1) and (9.3.2):

$$\delta_{12}J + \delta_{21}J = (\sigma_2 - \sigma_1)\left(\sigma_2 F_1(\sigma_1, \sigma_2, \mathbf{e}_1) - \sigma_1 F_2(\sigma_2, \sigma_1, \mathbf{e}_2)\right).$$

The variation cannot decrease the energy of the optimal structure (which has the minimal energy). Therefore, the following inequality must hold

$$\delta_{12}I + \delta_{21}I \geq 0 \quad \forall \mathbf{e}_1 \in \mathcal{V}_1, \quad \forall \mathbf{e}_2 \in \mathcal{V}_2,$$

or
$$\sigma_2 F_1(\sigma_1, \sigma_2, \mathbf{e}_1) - \sigma_1 F_2(\sigma_2, \sigma_1, \mathbf{e}_2) \geq 0.$$

Consider the question: How do these optimality conditions match the jump conditions on the phase boundaries?

The Jump Conditions. Let us compute the jump of the fields $\mathbf{e}$ on the boundary line between zones Ω_1 and Ω_2. Denote the normal to the boundary by $\mathbf{n} = (\cos\theta,\ \sin\theta)$. The elements of the matrix $\tilde{\mathbf{e}} = \nabla\mathbf{w}$ are discontinuous along the boundary. The continuity conditions that follow from the curlfree nature of the fields $\tilde{\mathbf{e}}_i$ are

$$[\sigma\tilde{\mathbf{e}}_i \cdot \mathbf{n}]_-^+ = 0 \text{ and } [\tilde{\mathbf{e}}_i \cdot \mathbf{t}]_-^+ = 0, \ i = 1, 2, \tag{9.3.7}$$

where $[.]_-^+$ denotes the jump: $[z]_-^+ = z_+ - z_-$.

The conditions (9.3.7) can be rewritten in the matrix form as

$$[H(\phi)\tilde{\mathbf{e}}_i]_-^+ = 0 \tag{9.3.8}$$

where
$$H(\phi) = \begin{pmatrix} \sigma & 0 \\ 0 & 1 \end{pmatrix} \begin{pmatrix} \cos\phi & \sin\phi \\ -\sin\phi & \cos\phi \end{pmatrix}$$

and ϕ in the angle of the inclination of the normal to the boundary.

We want to compare the jump conditions with the necessary optimality conditions. The technical difficulty is that the jump conditions are expressed in terms of the matrix $\tilde{\mathbf{e}}_i = \nabla\mathbf{w}$ of gradients, while the optimality conditions are written in terms of the symmetric matrix $\mathbf{e} = \left(\tilde{\mathbf{e}}\tilde{\mathbf{e}}^T\right)^{-1}$. Therefore, we have to express the jump conditions in terms of the matrix $\mathbf{e}$.

The conditions (9.3.8) imply in particular that the determinants of the matrices $\tilde{\mathbf{e}}$ and $\mathbf{e}$ on the opposite sides of the boundary are connected by an equality,

$$[\sigma \det \tilde{\mathbf{e}}]_-^+ = [\sigma \det \mathbf{e}]_-^+ = 0,$$

applied to both matrices $\tilde{\mathbf{e}}$ and $\mathbf{e}$.

The calculation of the jump of the trace of the matrix $\tilde{\mathbf{e}}$ is more complicated. It depends on the normal $\mathbf{n}$ to the boundary. Using (9.3.7), we can relate the jumps of $\mathrm{Tr}\,\tilde{\mathbf{e}}$ and $\mathrm{Tr}\,\mathbf{e}$. The relation depends on the angle ϕ and the antisymmetric part of $\tilde{\mathbf{e}}$.

We do not show here the details of this rather technical calculation performed by Maple; we simply review the results instead. The jump condition leads to the inequality

$$\mathrm{Tr}(\mathbf{e}_1) \leq \frac{\sigma_1}{\sigma_2} \min\{(e_2)_A,\ (e_2)_B\} + \max\{(e_2)_A,\ (e_2)_B\}. \tag{9.3.9}$$

where indices $_1$ and $_2$ denote the number of the material and the field in that material. The maximum value of the jump of the trace of $\mathbf{e}$ (9.3.9) corresponds to the equality sign in (9.3.9) and the following *requirements on the fields* in the neighborhood of the dividing line:

1. The matrices $\tilde{\mathbf{e}}$ are symmetric ($\tilde{\mathbf{e}} = \mathbf{e}$) and their eigenvectors are codirected at both sides of the boundary.

2. The normal $\mathbf{n}$ is codirected with the eigenvectors of $\tilde{\mathbf{e}}_1, \tilde{\mathbf{e}}_2$ that corresponds to thier minimal eigenvalue.

3. In neighborhood of the dividing line, the field in phase σ_1 belongs to the boundary of the set of permitted fields $\mathcal{V}_1$ and the field on the other side of it belongs to the boundary of $\mathcal{V}_2$.

If these conditions are satisfied, the field jumps over the forbidden region, i.e., the jump conditions become compatible with the necessary conditions. If one of these conditions is violated, the magnitude of the jump is insufficient to pass from a point in $\mathcal{V}_1$ to a point in $\mathcal{V}_2$. For example, if the field on one side of the dividing line belongs to interior of the set $\mathcal{V}_1$ then the field on the other side of the boundary lies in the forbidden region.

Hence, the derived necessary condition is the strongest possible or the considered variation is the most sensitive. Indeed, if there were a variation that would lead to larger forbidden region, then it would be impossible to jump across this region. The system of such hypothetical necessary conditions would be inconsistent with the jump on the dividing line.

From the preceding considerations, we conclude that the normal to the boundary between $\mathcal{V}_1$ and $\mathcal{V}_2$ is not uniquely determined only if the field $\mathbf{e}$ in $\mathcal{V}_2$ is "isotropic": $\mathbf{e}_2 = \beta I$. Either the field in the second phase must be isotropic, or the optimal structure is a simple laminate.

If the field $\mathbf{e}$ in $\mathcal{V}_2$ is isotropic, then the field on the boundary of $\mathcal{V}_1$ satisfies the equality

$$\frac{e_A}{e_B} = \frac{\sigma_1}{\sigma_2}. \tag{9.3.10}$$

The field (see Figure 9.2) belongs to points B or C of the boundary of $\mathcal{V}_1$ where the elliptical segment meets the straight segment.

9.3.4 Necessary Conditions and Optimal Structures

To illustrate the realization of necessary conditions, we examine two optimal structures.

Example 9.3.1 First, consider the construction of coated spheres (circles) shown in Figure 9.3 and described in Chapter 2. This geometry provides the best isotropic effective modulus (Hashin and Shtrikman, 1962a). If it is submerged into two equal orthogonal fields of the same magnitude ($\mathbf{E} = I$), it delivers the minimum of the sum of the stored energy. The properties of the structure were described in Chapter 2. Recall that the coated spheres(circles), submerged into homogeneous medium with effective isotropic conductivity σ_*, leave the outside field uniform. The effective conductivity of this structure, obtained in (Hashin and Shtrikman, 1962a),

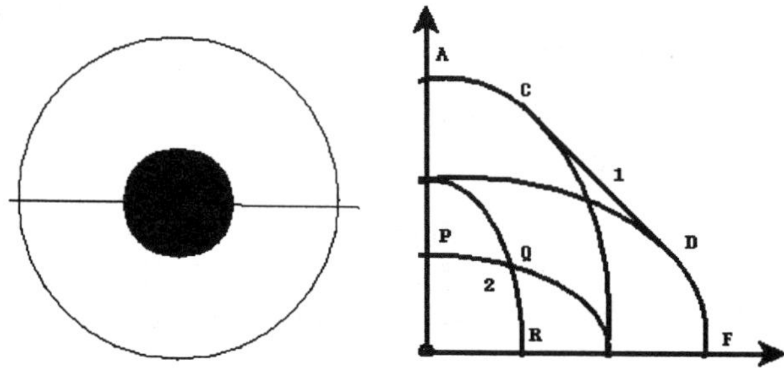

FIGURE 9.3. The structure of coated circles. The field inside the second material is isotropic (point Q), the field in the core varies along the straight component CD, and the boundary between regions corresponds to the point C.

is described by (2.2.19). Suppose that the field $\mathbf{e}$ outside the coated circles is equal to I, $\mathbf{e} = I$. The potentials are

$$u_1 = R(r)\cos\theta, \quad u_2 = R(r)\sin\theta,$$

where r is the radial coordinate and $R(r)$ is defined by the equations (2.2.17), (2.2.18).

Let us compute the field $\mathbf{e} = \nabla w$. The eigenvalues of $\mathbf{e}$ are

$$e_r = R', \quad e_\theta = \frac{R}{r},$$

and the eigenvectors coincide with the coordinate lines in polar coordinates.

1. From the solution (2.2.17), (2.2.18), one sees that the field in the nucleus is isotropic:

$$e_r^1 = e_\theta^1 = a_0 = \frac{2\,\sigma_2}{m_2\,\sigma_1 + (1 + m_1)\,\sigma_2}.$$

2. The jump conditions (9.3.10) on the boundary between the nucleus and the envelope are satisfied:

$$e_r = a_1 - \frac{b_1}{r^2} = \left(\frac{\sigma_2}{\sigma_1}\right) a_0, \quad e_\theta = a_1 + \frac{b_1}{r^2} = a_0.$$

3. The requirements on the fields on the dividing line are also satisfied. The field on one side of the boundary is isotropic ($\tilde{\mathbf{e}} = I$); on the other side the matrix $\tilde{\mathbf{e}}$ is symmetric, and the eigenvectors are codirected at each point with the radial direction, that is, with the normal.

4. The field in the envelope has the constant trace:

$$e_r^1 + e_\theta^1 = 2a_1,$$

meaning that it belongs to the straight component of the boundary of $\mathcal{V}_1$ everywhere.

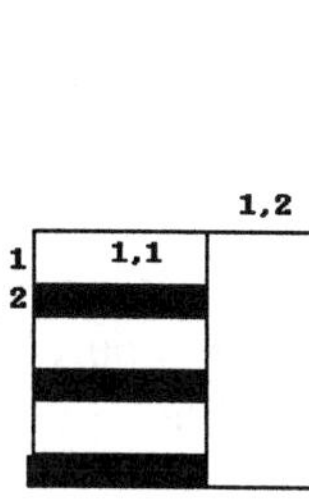 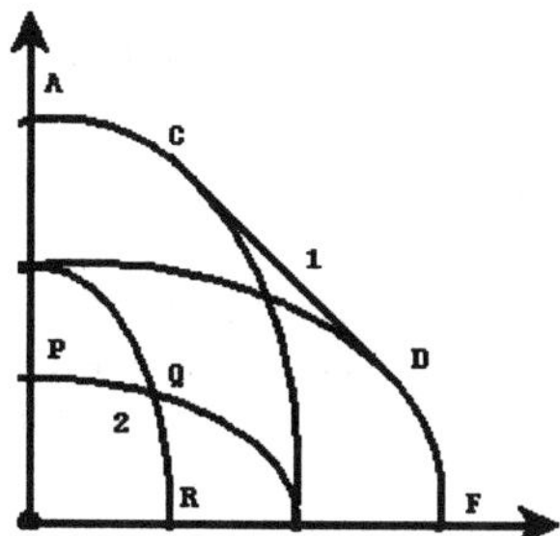

FIGURE 9.4. The fields in matrix laminates. The field on the second material is isotropic (point Q in $\mathcal{V}_2$). The field in region (1,1) is in the rank-one connection with the field in $\mathcal{V}_2$ (point C). The field in region (1,2) varies with the external field, and it belongs to the interval CD.

Let us also check the optimality conditions at the boundary point $r = 1$ where the material σ_1 and the effective material σ_{HS} (see Chapter 2) meet. We have

$$e_r = a_1 - b_1 = \frac{(1 + m_1)\,\sigma_1 + m_2\,\sigma_2}{m_2\,\sigma_1 + (1 + m_1)\,\sigma_2} = \frac{\sigma_{HS}}{\sigma_1}$$

and

$$e_\theta = a_1 + b_1 = 1.$$

This says that the necessary conditions are also satisfied on the boundary between the material σ_1 and the effective medium $\sigma_* = \sigma_{HS}$. The structure satisfies the necessary conditions as equalities everywhere.

This result is anticipated because the optimality of this structure has been established by the sufficient Hashin–Shtrikman bound (Hashin and Shtrikman, 1962a). This calculation rather checks the limiting character of the suggested necessary conditions.

Example 9.3.2 The conditions can be checked on the other known optimal structure, the second-rank matrix laminates (see Figure 9.4).

The technique for calculation of the fields in a laminate structure is described in Chapter 7. The field $\mathbf{e}_{1,1}$ is in the rank-one connection with the field $\mathbf{e}_2$ in the second material,

$$(\mathbf{e}_{1,1} - \mathbf{e}_2) \cdot \mathbf{i}_A = 0,$$

where $\mathbf{i}_A$ is the vector of the coordinate axis. The field $\mathbf{e}_{1,2}$ is in the rank-one connection with the composite $\mathbf{e}_\mu = \mu \mathbf{e}_{1,1} + (1 - \mu)\mathbf{e}_2$ of the fields $\mathbf{e}_{1,1}$ and $\mathbf{e}_2$,

$$(\mathbf{e}_{1,2} - \mathbf{e}_\mu) \cdot \mathbf{i}_B = 0.$$

Notice that $\mathbf{e}_{1,2}$ is not in the rank-one connection with $\mathbf{e}_{1,1}$ or $\mathbf{e}_2$ separately.

One can check that optimization of the sum of the energies with respect to μ leads to the following results:

1. The field in the nucleus is isotropic and belongs to the corner point of $\mathcal{V}_2$ (see Figure 9.4) even if the exterior field $\mathbf{E}$ is anisotropic.

2. In an optimal structure, $\mathrm{Tr}(\mathbf{e}_{1,1}) = \mathrm{Tr}(\mathbf{e}_{1,2})$. Both fields $\mathbf{e}_{1,1}$ and $\mathbf{e}_{1,2}$ belong to the straight component of the boundary of $\mathcal{V}_1$; see Figure 9.4. The mean field in Ω_1 belongs to that boundary as well.

3. If the external field is outside of the triangle CQD (see Figure 9.4), the structure degenerates into a simple laminate structure. In that case, the fields in both materials belong on elliptical parts of the boundaries of Ω_1 and Ω_2, Figure 9.4.

This shows that all necessary conditions are satisfied as equalities.

Nonuniqueness of Optimal Structures

The preceding two examples demonstrate that the fields are different in the coated spheres and isotropic matrix laminates. However, both structures have the same extremal effective properties, which also can be obtained by the formal procedure of minimal extension. Why does this happen?

The minimal extension procedure treats the fields in each phase as single tensors, not as functions of the point. These fields can be considered as the mean fields within the corresponding subdomains Ω_i; they belong to the boundary of the permitted domains $\partial \mathcal{V}_i$. On the other hand, the field $\mathbf{e} = \mathbf{e}(\mathbf{x})$ satisfies the Weierstrass conditions, therefore it varies in the region $\mathcal{V}_i$. It is geometrically clear that the mean field $\mathbf{e}_i$ belongs to the boundary $\partial \mathcal{V}_i$ of $\mathcal{V}_i$ only if the field $\mathbf{e}(\mathbf{x})$, $\mathbf{x} \in \Omega_i$ varies along a component $\partial_1 \mathcal{V}_i$ of the boundary $\partial \mathcal{V}_i$ and his component is linear.

In the considered structures, the trace $e_A + e_B$ of the field in the first material is constant everywhere in both coated spheres and matrix laminates. Obviously, the mean field has the same trace. Therefore, the necessary conditions (the constancy of the trace of $\mathbf{e}_1(\mathbf{x})$) are satisfied at every point $\mathbf{x}$ and also for the mean field.

One could expect a nonunique microstructure of the optimal composite if a permitted region $\mathcal{V}_i$ has a linear component of the boundary; the field in the ith material can vary along this component.

To the contrary, the strong convexity of the permitted region $\mathcal{V}_i$ (like the region $\mathcal{V}_2$ in the checked examples) suggests that the field in the ith material is constant at every point of Ω_i. Otherwise the mean field $\mathbf{e}_i$ cannot belong to the boundary $\partial \mathcal{V}_i$ of $\mathcal{V}_i$. These fields may correspond to either an inclusion (like the nucleus in the checked geometries) or a layer in a laminate.

9.4   Discussion and Problems

Discussion

We have demonstrated a variety of properties of the necessary condition method:

1. The necessary conditions naturally express the optimality of the variational functional in terms of the fields at every point of the structure. In contrast, the lamination technique deals with the algebraic transformation of the effective properties. The lamination technique is simpler, but it requires more guessing.

2. The problem of an optimal structure has multiple solutions that are geometrically different, such as second-rank laminates and coated spheres. These structures correspond to different optimal fields. However, the sets of permitted values of the field are the same because they are defined independently of the geometry of the structure.

3. The minimal extension replaces the original problem with a problem that is stable against an explicitly described class of fine-scale perturbations.

4. The extensions can be not final. On the other hand, if a counterexample emerges, the corresponding structure should be included in the set of tests. Then the enlarged set will guarantee stability to the new type of perturbations.

The necessary conditions are effective in problems of optimization of multiphase structures (see Chapter 12).

Problems

1. Derive the optimal variation for an anisotropic inclusion and for an anisotropic matrix.

2. In what range of external fields are the laminate structures optimal? Compare the optimality conditions and the jump conditions for the case where $\mathbf{e}$ has very different eigenvalues and the laminate structure is optimal.

3. Calculate the increment in the three-dimensional case. What do the domains $\mathcal{V}_1$ and $\mathcal{V}_2$ look like?

4. Consider the problem of a polycrystal: the composite of the materials

$$\sigma_1 = \begin{pmatrix} s_1 & 0, \\ 0 & s_2 \end{pmatrix} \quad \text{and} \quad \sigma_2 = \begin{pmatrix} s_2 & 0, \\ 0 & s_1 \end{pmatrix}$$

that has the minimal energy. What do the domains $\mathcal{V}_1$ and $\mathcal{V}_2$ look like?

5. Calculate the minimal extension for the problem of a polycrystal.

Part IV

G-Closures

10
Obtaining G-Closures

10.1 Variational Formulation

G_m-Closures and Structural Optimization

We pass to methods of description of G_m-closures. Recall that a G_m-closure is the set of effective properties of all possible composites that are assembled from materials with known properties mixed in certain proportions. The G_m-closure problem describes effective properties of all composites that correspond to different microstructures.

The properties of a structure are determined up to a rotation of the structure as a whole. Hence, the G_m-closures are defined as the sets of rotationally invariant characteristics of the effective properties tensor. For example, the effective conductivity tensor $\boldsymbol{\sigma}_*$ is characterized by three eigenvalues $\lambda_1, \lambda_2, \lambda_3$. The G_m-closure of conductivity tensors is a domain in the space of eigenvalues λ_i. It consists of all possible triplets of eigenvalues that describe an effective tensor of a composite.

The notion of G_m-closures allows us to solve a large class of optimal design problems. The explicit description of all G-limits (the G_m-closure) makes an optimal control problem well-posed (see Chapter 5). The G-closures are also used in the theory of composites because they produce a complete set of a priori bounds for effective tensors.

10.1.1   Variational Problem for G_m-Closure

Variational Principles

Consider a periodic composite made of N materials with properties D_i, $i = 1, \ldots, N$. Assume that ith material occupies the subdomain Ω_i of the cell of periodicity Ω. The structure is defined by the Ω-periodic vector-valued characteristic function $\chi = [\chi_1, \ldots, \chi_N]$ of the subdomains Ω_i. The structure is submerged into uniform external field $\mathbf{v}_0$ that causes the Ω-periodic field $\mathbf{v} = \mathbf{v}(\mathbf{x})$ inside the structure.

Assume that the equilibrium corresponds to the minimum of the energy:

$$W_0(\chi, \mathbf{v}_0) = \mathbf{v}_0 \cdot D_*(\chi)\mathbf{v}_0 = \min_{\mathbf{v}} \langle \mathbf{v} \cdot D(\chi_i)\mathbf{v} \rangle,$$

where the field $\mathbf{v}$ is subject to differential constraints (see Chapter 6) and its mean value is equal to the external field,

$$\langle \mathbf{v} \rangle = \mathbf{v}_0, \quad \sum_{k=1}^{d}\sum_{j=1}^{m} a_{ijk}\frac{\partial v_j}{\partial x_k} = 0, \ i = 1, 2, \ldots, r; \tag{10.1.1}$$

the characteristic function χ_i of ith material is subject to constraints:

$$\langle \chi_i \rangle = m_i, \quad D(\chi) = \sum_{i=1}^{N} \chi_i D_i, \quad \sum_{i=1}^{N} m_i = 1$$

and $\mathbf{v}$ and χ_i are Ω-periodic.

This variational principle is accompanied by the dual principle that requires minimization of the energy W_0^{dual},

$$W_0^{\text{dual}}(\chi, \mathbf{v}_0^{\text{dual}}) = \mathbf{v}_0^{\text{dual}} \cdot D_*^{-1}(\chi)\mathbf{v}_0^{\text{dual}} = \min_{\mathbf{v}^{\text{dual}}} \left\langle \mathbf{v}^{\text{dual}} \cdot D^{-1}(\chi_i)\mathbf{v}^{\text{dual}} \right\rangle,$$

which depends on the dual variable $\mathbf{v}^{\text{dual}}(\mathbf{x}) = D(\mathbf{x})\mathbf{v}(\mathbf{x})$. The variable $\mathbf{v}^{\text{dual}}$ is subject to the differential restrictions similar to those imposed on $\mathbf{v}$ (see Chapter 6). These two variational problems allow us to bound the set of effective tensors.

Wiener-Type Bounds

The convex envelopes of $W_0(\chi, \mathbf{v})$ and $W_0^{\text{dual}}(\chi, \mathbf{v}^{\text{dual}})$ lead to Wiener-type bounds for $D_*(\chi)$; the bounds provide simple geometrically independent inequalities of G_m-closure. The Wiener bounds are

$$(D_*)^{-1} \leq \langle (D)^{-1} \rangle = \sum_{i=1}^{N} m_i D_i^{-1}$$

and

$$D_* \leq \langle D \rangle = \sum_{i=1}^{N} m_i D_i.$$

These bounds constrain from below the minimum of eigenvalues $\lambda_1 = \min\{\lambda_i\}$ of the effective tensors and constrain from above the maximum of its eigenvalues $\lambda_3 = \max\{\lambda_i\}$. They determine the minimal cube (Wiener box) that contains the G_m-closure:

$$\|D_*\| \leq \sum_{i=1}^{N} m_i D_i, \quad \|D_*^{-1}\| \leq \sum_{i=1}^{N} m_i D_i.$$

Remark 10.1.1 *The Wiener-type bounds can be tightened by using the quasiconvex envelopes of $W_0(\chi, \mathbf{v})$ and $W_0^{dual}(\chi, \mathbf{v}^{dual})$ instead of the convex envelopes. In this case, we take into account the translations that may exist due to differential constrains $A : \nabla \mathbf{v} = 0$ on the set of $\mathbf{v}$ (see Chapter 8). We will demonstrate an example of the translation bounds for the elastic energy in Chapter 15.*

Sum of Energies

Energy $W_0(\chi, \mathbf{v})$ estimates the composite's resistance only in the direction of the applied field $\mathbf{v}_0$ and is not sensitive to the properties of the composite in the orthogonal directions. The Wiener-type bound corresponds to media with minimal resistance in the direction of $\mathbf{v}_0$. We may expect that an optimal medium is anisotropic and its resistance in other directions is greater that the minimized resistance in a chosen direction. Hence, obtaining complete bounds for the G_m-closure requires consideration of the coupled bounds on resistivity in different directions.

On the other hand, the energy and the complementary energy are the functionals that determine the tensor of the effective properties D_* via the variational principle. These functionals should be used for bounding the set of D_*. A natural idea is to derive the bound for the sum of the values of specific energy stored in a structure and caused by a number of external fields, $\mathbf{v}_0(1), \ldots \mathbf{v}_0(n)$:

$$J_1 = \sum_{i=1}^{n} \mathbf{v}_0(i) \cdot D_* \mathbf{v}_0(i) = \left\langle \sum_{i=1}^{n} \mathbf{v}(i) \cdot D(\chi) \mathbf{v}(i) \right\rangle,$$

where

$$\langle \mathbf{v}(i) \rangle = \mathbf{v}_0(i), \quad A : \nabla \mathbf{v}(i, \mathbf{x}) = 0, \quad \mathbf{v}(i, \mathbf{x}) \text{ are } \Omega\text{-periodic.}$$

We assume that the constant external fields $\mathbf{v}_0(1), \ldots, \mathbf{v}_0(n)$ are linearly independent; therefore, the number of them is not greater than the dimension n of $\mathbf{v}$.

A Coupled Bound. Consider, as an example, a conducting medium. The number of linearly independent external fields is equal to the dimensionality d of the space. We may also assume that the external fields are orthogonal.

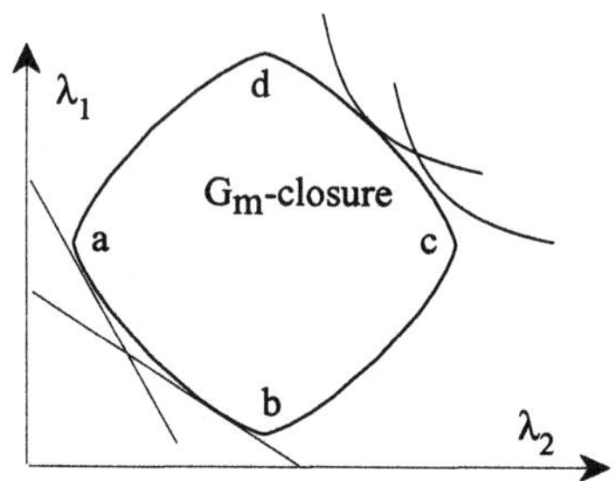

FIGURE 10.1. The scheme of bounding of a G_m-closure.

Functional J_1 is equal to a convex combination of eigenvalues λ_i of the effective tensor $\boldsymbol{\sigma}_*$,

$$J_1 = \sum_{i=1}^{d} \lambda_i \alpha_i^2,$$

where α_i are coefficients that depend on the magnitudes and directions of the applied fields. Various applied fields $\mathbf{v}_0(i)$ lead to bounds on various linear combinations of λ_i. The union of these inequalities forms the bound for the component of the G-closure boundary closest to the origin (see curve a b, Figure 10.1).

Complementary Bound. Another component of the G_m-closure boundary is obtained by a bound on the sum of complementary energies:

$$J_d = \left\langle \sum_{i=1}^{n} \mathbf{v}^{\mathrm{dual}}(i, \mathbf{x}) \cdot D^{-1}(\chi) \mathbf{v}^{\mathrm{dual}}(i, \mathbf{x}) \right\rangle.$$

In the conducting medium, J_d is equal to a convex combination of $\frac{1}{\lambda_1}, \ldots, \frac{1}{\lambda_d}$:

$$J_d = \sum_{i=1}^{d} \frac{1}{\lambda_i} (\alpha_i^*)^2,$$

or a convex combination of the eigenvalues of the inverse tensor D_*^{-1}. An estimate of J_d gives a bound for the supplementary component of the G-closure that is most distant from the origin (the component c d, Figure 10.1).

Other Bounds. We may consider functionals that correspond to linear combinations of "direct" and "dual" loadings. This way we bound other components of the boundary of the G_m-closure (Figure 10.1) that correspond to the minimization of the combination

$$J_k = \sum_{i=1}^{k} \lambda_i \alpha_i^2 + \sum_{i=k+1}^{d} \frac{1}{\lambda_i} (\alpha_i^*)^2 \tag{10.1.2}$$

of the eigenvalues of $\boldsymbol{\sigma}_*$, and they correspond to the most anisotropic components of the G_m-closure (the components b c and a d, Figure 10.1).

The Bounds from Outside by the Translation Method

Translation bounds for the functionals J_k, $k = 1, \ldots, n$ lead to inequalities (analogues of the Wiener bounds) that are valid for any tensor $D_* \in G_m$-closure. We call $P_m U$ the set of tensors satisfying these inequalities. Clearly,

$$P_m U \supseteq G_m U.$$

The formal scheme of the translation bounds for the G_m-closure is as follows (see (Lurie and Cherkaev, 1986a)).

Consider the quadratic form J_k

$$J_k = \mathbf{Z}^T A(D_*) \mathbf{Z},$$

where $A(D)$ is the $n^2 \times n^2$ block diagonal matrix of the structure

$$A(D) = \mathrm{Diag}\left(\underbrace{D, \ldots, D}_{k \text{ times}}, \underbrace{D^{-1}, \ldots, D^{-1}}_{n-k \text{ times}}\right); \qquad (10.1.3)$$

it consists of k identical blocks D and $n - k$ blocks D^{-1}, $0 \le k \le n$. Here $\mathbf{Z}$ is a n^2-dimensional vector that consists of the k "direct" constituent vectors $\mathbf{v}_i$ and $(n - k)$ "dual" vectors $\mathbf{v}_i^{\text{dual}}$

$$\mathbf{Z} = \left[\mathbf{v}_1, \ldots \mathbf{v}_k, \mathbf{v}_{k+1}^{\text{dual}}, \ldots, \mathbf{v}_n^{\text{dual}}\right].$$

Each vector $\mathbf{v}_i$, $i = 1, \ldots, k$ satisfies differential constraints (10.1.1); the vectors $\mathbf{v}_i^{\text{dual}}$, $i = k + 1, \ldots, n$ satisfy the dual constraints. To obtain the translation bounds, we take into account the constraints $A_z : \nabla \mathbf{z} = 0$ for the whole vector $\mathbf{z}$.

Notice that the number of constraints is greater than the number of n linearly independent fields $\mathbf{v}_i$, $i = 1, \ldots, n$ and therefore greater than the dimension d, because each vector $\mathbf{v}_i$ corresponds to at least one constraint. Therefore, these constraints necessarily define a quadratic translator (Chapter 8),

$$\phi(\mathbf{Z}) = \mathbf{Z}^T T(t) \mathbf{Z},$$

where T is an $n^2 \times n^2$ symmetric matrix, dependent on the translation parameters $t_1, \ldots, t_m$.

Example 10.1.1 Consider the two fields $\mathbf{v}_1 = \nabla w_1$ and $\mathbf{v}_2 = \nabla w_2$, where ∇w_i, $i = 1, 2$ are the two-dimensional vectors. They correspond to the following three quasiaffine bilinear forms (Chapter 8),

$$t_1 (\nabla w_1)^T R(\nabla w_1), \quad t_2 (\nabla w_2)^T R(\nabla w_2), \quad \text{and } t_3 (\nabla w_1)^T R(\nabla w_2)$$

where R is the matrix of rotation through a right angle and t_i, $i = 1, 2, 3$ are real parameters. The first two quadratic forms are identically zero, because R is antisymmetric, and the third one leads to the translator matrix T:

$$T(t_3) = \begin{pmatrix} \mathbf{0} & t_3 R \\ t_3 R^T & \mathbf{0} \end{pmatrix}.$$

Notice that, unlike the matrix A, the matrix T of the translator is not block-diagonal. Translators bond the fields that correspond to different external sources. Matrix T is independent of the minimizing Lagrangian but only depends on differential constraints.

Bounds. We use the translation method to estimate J_k. The corresponding inequality has the form (see (8.3.3))

$$Z^T(A_* - T)Z \geq Z^T \left(m_1(A_1 - T)^{-1} + m_2(A_2 - T)^{-1} \right)^{-1} Z \qquad (10.1.4)$$

where $A_* = A(D_*)$, $A_1 = A(D_1)$, $A_2 = A(D_2)$; see (10.1.3). The translation parameter t must be chosen to make the matrices $(A_1 - T)$ and $(A_2 - T)$ positive:

$$t \in \mathcal{T} = T = \{t : A_i - T(t) \geq 0, \ i = 1, 2\}. \qquad (10.1.5)$$

The bound (10.1.4) is valid for all Z. It implies the matrix inequality

$$(A_* - T) \geq \left(m_1(A_1 - T)^{-1} + m_2(A_2 - T)^{-1} \right)^{-1} \qquad (10.1.6)$$

or

$$U(D_*) = (A_* - T)^{-1} - m_1(A_1 - T)^{-1} - m_2(A_2 - T)^{-1} \geq 0. \qquad (10.1.7)$$

The matrix inequality (10.1.7) reaches its limit if $U(D_*)$ has at least one zero eigenvalue, and the other eigenvalues are nonnegative. The equations for the boundary of that set are

$$\det U(D_*) = 0, \quad U(D_*) \geq O.$$

The corresponding matrix $U(D_*)$ implicitly defines the boundary of the permitted set of effective tensors D_* that contains the G_m-closure or coincides with it. The obtained bound is called the *translation bound* for G_m-closure.

Other Bounds. The calculation must be repeated for different settings (10.1.3) of the matrix A that consists of k terms with the matrix D and $n - k$ terms with the matrix D^{-1}. The resulting inequalities restrict the G-closure from different sides. The G-closure lies inside the intersection of these inequalities.

Y-Transform

The translation bounds for two-component composites take especially elegant form in the Y-tensor notation introduced in Chapter 7. Recall that every structure can be characterized by a tensor $Y(D_*)$ defined by the equality (7.3.19),

$$(D_* + Y)^{-1} = m_1(D_1 + Y)^{-1} + m_2(D_2 + Y)^{-1}, \qquad (10.1.8)$$

and that the Wiener inequality corresponds to the positiveness of Y. As with $Y(D)$, we can define $Y(A_*)$, where the matrix A_* has the structure (10.1.3).

Comparing the formula (10.1.8) with the translation bound (10.1.6), we transform it to the especially simple form (Cherkaev and Gibiansky, 1992):

$$Y(A_*) + T \geq 0 \quad \text{if} \quad -Y(A_i) + T \geq 0, \; i = 1, 2. \qquad (10.1.9)$$

The last inequality is obtained by using (10.1.5) and the property (7.3.25): $Y(A_i) = -A_i, \; i = 1, 2$.

The boundary of the closure corresponds to the vanishing of one of the eigenvalues of (10.1.9). The equation for the boundary is

$$Y(A_*) + T \geq 0, \quad \det(Y(A_*) + T) = 0, \qquad (10.1.10)$$

where the extremal translator T satisfies the relations

$$\det(-Y(A_1) + T)\det(-Y(A_2) + T) = 0, \quad -Y(A_i) + T \geq 0.$$

The simplicity of the last representation makes it convenient to use. The representation has several remarkable properties.

1. The bound has the same form for all volume fractions.

2. If T is isotropic, then $\det(Y(A) - T)$ is an isotropic function of $Y(A)$.

3. The G_m-closures may correspond to unbounded sets in the Y notations, because the simple laminate corresponds to an improper point.

Notice that the tensor $Y(A)$ represents the structure. It is block-diagonal, similarly to A. On the other hand, the translator T is not block-diagonal because it represents differential properties of fields, including the coupling of them.

Remark 10.1.2 *The Y-transform of $A(D_*)$ cannot be defined by (10.1.8) if A_1 and A_2 have the common eigenvalue and eigenvector, because this definition leads to uncertainty. In this case, the translation bound must be modified. The modification is done in Section 16.2.*

Bounds for Volume Fractions

The translation method can be inverted to bound unknown volume fractions of materials in a composite if its effective tensor is known. Suppose that $A_2 - A_1$ is nonnegative. Rewrite the translation bound (10.1.6) in the form

$$(A_* - T)^{-1} - (A_1 - T)^{-1} \leq m_2 \left((A_2 - T)^{-1} - (A_1 - T)^{-1} \right)$$

or

$$S(T) \leq m_2 I, \tag{10.1.11}$$

where

$$S(T) = \left((A_* - T)^{-1} - (A_1 - T)^{-1} \right) \left((A_2 - T)^{-1} - (A_1 - T)^{-1} \right)^{-1}$$

(here, the equality $m_1 = 1 - m_2$ is used). The matrix inequality (10.1.11) produces the bound for m_2. Namely, m_2 is bounded from below by the spectral norm $\|S\| = \lambda_{\max}(S)$ of the matrix on the left-hand side of (10.1.11):

$$m_2 \geq \max_{T \in \mathcal{T}} \|S(T)\|.$$

Exactness of Translation Bounds and Multicomponent Composites

As we have mentioned, the translation bounds generally restrict a bigger than G_m-closure set. We cannot conclude a priori when these bounds are exact. However, we can point out problems where the translation bounds are definitely too broad; these cases are the G_m-closures for multicomponent composites.

Assume that the materials $D_1, \ldots, D_k$ are combined with the volume fractions $m_1, \ldots, m_k$, respectively, in a structure. The bounds for this multicomponent composite can be formally found by the previous scheme. The bounds take the form

$$(A_* - T)^{-1} \geq \sum_{i=1}^{N} m_i (A_i - T)^{-1}, \tag{10.1.12}$$

where

$$T(t) \in \mathcal{T}, \quad \mathcal{T} = \{ T(t): \ A_i - T(t) \geq 0, \ i = 1, \ldots, n \}.$$

However, these bounds are not exact for all values of the volume fractions of phases. Indeed, the set $\mathcal{T}$ of translation matrices depends on the materials' properties D_i, but not on their volume fraction. Suppose that the material D_1 limits this set so that the translators are constrained by D_1:

$$A_1 - T_{\mathrm{opt}} \geq 0. \tag{10.1.13}$$

Consider composites with an infinitely small amount m_1 of D_1. Clearly, the G_m-closure of such composites is arbitrary close to the G_m -closure of the set $D_2, \ldots, D_k$ if $m_1 \to 0$. However, the translation bound still depends on D_1 through (10.1.13). This observation was made in (Kohn and Milton, 1986).

The translation bounds are still valid, but they fail to be exact, unless the optimal translation T_{opt} belongs to the interior of the permitted interval. This would mean that T_{opt} is independent of the materials' properties, and that the fields in the matching structures are constant in each material. Wiener bounds are an example of this situation: They have the expected asymptotic property. For multicomponent composites, they could be more restrictive than the more sophisticated translation bounds. Moreover, Wiener bounds can be sometimes exact; see Chapter 12.

Of course, the general ideas of translations are still applicable for multicomponent composites. The critical question is the type of translators used. The quasiconvexity of quadratic functions of the fields $\mathbf{Z}$ is not sufficient to deal with multicomponent composites. Some other information must be incorporated.

10.1.2 G-Closures

Here we discuss use of the translation method to describe G-closures of materials, that is, the set of all possible effective tensors corresponding to fixed-volume fractions.

Straight Calculation. An obvious way to describe the G-closure of two materials is to directly calculate

$$G(U) = \bigcup_{m \in [0,\ 1]} G_m(U). \tag{10.1.14}$$

The corresponding calculation is straightforward, but often inconvenient. It requires unnecessary information about the G_m-closures. Therefore, we prefer to derive an inequality from (10.1.6) that does not depend on the volume fractions m_1, m_2.

The First Case. This method is applicable to the problem of polycrystals; see (Avellaneda et al., 1988). Suppose that we can find a translator T_0 such that the matrix on the right-hand side of inequality (10.1.12),

$$(A_* - T_0)^{-1} \leq \sum_{i=1}^{N} m_i (A_i - T_0)^{-1}, \tag{10.1.15}$$

is singular in all directions. This happens, for example, if the A_i differ by a rotation and T is isotropic. The translation inequality (10.1.6) becomes

$$(A_* - T_0) \geq 0, \tag{10.1.16}$$

where T_0 is determined from the condition $\det(A_i - T_0) = 0$. The component of the boundary of the G-closure is defined by the equality

$$\det(A_* - T_0) = 0. \qquad (10.1.17)$$

The Second Case. This method is applicable to any two-phase composite; it was suggested by Gibiansky in 1990 (private communication).

Consider again inequality (10.1.6) for a two-material composite. Recall that A_1, A_2 are $n^2 \times n^2$ matrices of the type (10.1.3). From the variety of these matrices, choose a matrix A that corresponds to the eigenvalues of $A_1 - A_2$ of different signs.

We find an $n^2 \times m$, $(m < n^2)$ matrix $\mathbf{L}$ that corresponds to the equality

$$\mathbf{L}^T (A_1 - T)^{-1} \mathbf{L} = \mathbf{L}^T (A_2 - T)^{-1} \mathbf{L} = \mathbf{K}.$$

Here K is the $m \times m$ matrix.

The matrix $\mathbf{L}$ projects $(A_1 - T)^{-1}$ and $(A_2 - T)^{-1}$ to a common subspace. Such a projector $\mathbf{L}$ necessarily exists, because $(A_1 - T)^{-1} - (A_2 - T)^{-1}$ is not positive.

From (10.1.7) we have

$$\mathbf{L}^T (A_* - T)^{-1} \mathbf{L} \geq m_1 \mathbf{K} + m_2 \mathbf{K} = \mathbf{K}. \qquad (10.1.18)$$

This inequality is valid for all effective tensors A_* and does not depend on the volume fractions of the materials.

It describes a boundary component for the G-closure:

$$\mathbf{L}^T (A_* - T)^{-1} \mathbf{L} = \mathbf{K}. \qquad (10.1.19)$$

Note that the inequality (10.1.19) depends on the translation matrix T, both directly and via the projector $\mathbf{L} = \mathbf{L}(T)$. In the next chapter, we will give examples of these approaches.

10.2   The Bounds from Inside by Laminations

The L-Closure

The effective properties of a specific class of microstructures obviously form a subset of the G-closure set. It is convenient to consider the laminates of some rank because their effective properties can be explicitly found.

The effective tensors of all high-rank laminate structures determine the set $L_m U$ that bounds the G_m-closure from inside:

$$L_m U \subset G_m U.$$

When the rank of laminates increases, the described sets tend to a set of all tensor properties of all possible laminates. In (Lurie and Cherkaev, 1986a), it was suggested to describe this subset of G_m-closure to approximate this set. Following (Milton, 1994), where this problem was studied, we call this set the *lamination closure* and denote it L-closure.

An Algorithm

Properties of All Simple Laminates. Consider the procedure described in Chapter 7 for calculation of properties of laminates. According to this procedure, we split the vectors $\mathbf{u}$ and $\mathbf{v}$ in the constitutive relations $\mathbf{u} = D\mathbf{v}$ into continuous and discontinuous parts by the orthogonal projectors $\mathbf{p}$ and $\mathbf{q}$. These projectors depend on the normal $\mathbf{n}$ to the laminates: $\mathbf{p} = \mathbf{p}(\mathbf{n})$, $\mathbf{q} = \mathbf{q}(\mathbf{n})$. Using these projectors, we solve the linear constitutive relations for the discontinuous components of $\mathbf{u}$ and $\mathbf{v}$. The matrix of properties $Z(D)$ of this form of the relation is introduced in (7.2.10) as

$$Z(D) = \begin{pmatrix} D_{11}^{-1} & -D_{11}^{-1}D_{12} \\ -D_{12}^T D_{11}^{-1} & D_{22} - D_{12}D_{11}^{-1}D_{12}^T \end{pmatrix}, \qquad (10.2.1)$$

where $D_{11} = \mathbf{p}^T D\mathbf{p}$, $D_{22} = \mathbf{p}^T D\mathbf{q}$, $D_{12} = \mathbf{q}^T D\mathbf{q}$.

The laminates with normal $\mathbf{n}$ have the properties matrix $Z(D_{\mathrm{lam}})$, which is the convex envelope of the matrices $Z(D_i)$ of the laminated materials:

$$Z(D_{\mathrm{lam}}) = \sum_i^N m_i Z(D_i), \quad \sum_i m_i = 1, \quad m_i \geq 0, \ i = 1, \ldots N.$$

or

$$Z(D_{\mathrm{lam}}) = \mathcal{C}Z(\mathcal{D})$$

where $\mathcal{D} = \{D_1, \ldots, D_N\}$ and $\mathcal{C}(Z)$ is the convex envelope of Z.

The set of tensors $\mathcal{D}_1 = \mathcal{D}_{\mathrm{lam}}$ of all the laminates made of the original set $\mathcal{D}_0$ is defined by the mapping

$$\mathcal{D}_{\mathrm{lam}} = \mathcal{L}(\mathbf{n}, \mathcal{D}). \qquad (10.2.2)$$

Here

$$\mathcal{L}(\mathbf{n}, \mathcal{D}) = Z^{-1}(\mathbf{n}, \ \mathcal{C}(Z(\mathbf{n}, \mathcal{D}))),$$

matrix $Z(\mathbf{n}, D)$ is defined by transform (10.2.1), and $Z^{-1}(\mathbf{n}, A)$ is the inverse transform.

Iterations. Consider a set $\mathcal{D}_0$ of the initially given materials. We call $\mathcal{D}_1 = \mathcal{D}_{\mathrm{lam}}$ the set of all laminates from $\mathcal{D}_0$; according to (10.2.2) it is equal to

$$\mathcal{D}_1 = \mathcal{L}(\mathbf{n}_1, \mathcal{D}_0).$$

The set of second-rank laminates is obtained similarly, starting with the set $\mathcal{D}_1$ instead of $\mathcal{D}_0$ and choosing a different normal $\mathbf{n}_2$ of laminates: $\mathcal{D}_2 = \mathcal{L}(\mathbf{n}_2, \mathcal{D}_1)$. Then the iterations continue,

$$\mathcal{D}_k = \mathcal{L}(\mathbf{n}_k, \mathcal{D}_{k-1}),$$

The iterative procedure for constructing the *L*-closure is illustrated in Figure 10.2.

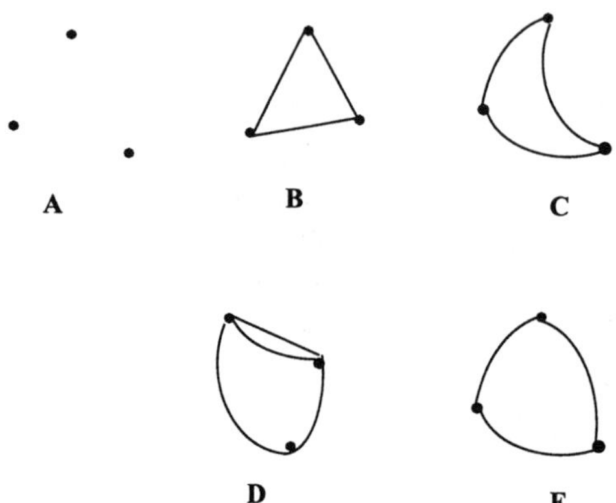

FIGURE 10.2. Construction of L-closure: (A) initial set $\mathcal{D}_0$; (B) transform $Z(\mathbf{n}_1, \mathcal{D}_0)$; laminates of the first rank (convexification); (C) set $\mathcal{D}_1 = \mathcal{L}_1(\mathbf{n}_1, \mathcal{D}_0)$; (D) transform $Z(\mathbf{n}_2, \mathcal{D}_1)$, laminates of the second rank (convexification); (E) set $\mathcal{D}_2 = \mathcal{L}_1(\mathbf{n}_2, \mathcal{D}_1)$.

The procedure monotonically increases the set $\mathcal{L}_k$ of all tensors of laminates of rank k:

$$\mathcal{L}^k D \subseteq \mathcal{L}^{k+1} D \quad \forall k.$$

On the other hand, each effective tensor has a finite number of components, and they are obviously bounded by the Wiener-type bounds. Therefore the described sequence of the sets $\mathcal{L}^k D$ has a limit: the L-closure.

Properties. The L-closure has several remarkable properties similar to the properties of G-closures. Let us describe this set. We use here the results of (Lurie and Cherkaev, 1981a; Lurie and Cherkaev, 1981b; Lurie and Cherkaev, 1986a; Milton, 1994; Grabovsky and Milton, 1998).

We suppose that the original set $\mathcal{D}_0$ is isotropic: The set $\mathcal{D}_0$ contains either isotropic materials or it contains, together with an anisotropic material, all materials obtained by an arbitrary rotation of it:

$$\text{If } D \in \mathcal{D}_0 \quad \text{then } \Phi D \Phi^T \in \mathcal{D}_0,$$

where Φ is an arbitrary rotation tensor.

Let us denote the lamination closure as the set $\mathcal{D}_\infty$ of tensors D_∞ of effective properties of laminates of infinite rank; its boundary is denoted by $\partial \mathcal{D}_\infty$. The boundary is described as

$$\phi(D_\infty) = 0 \quad \text{iff } D \in \partial \mathcal{D}_\infty.$$

where ϕ is a function of the tensor D.

The set $\mathcal{D}_\infty$ is the result of enlarging the original set $\mathcal{D}_0$ by adding to it a sequence of laminates of high rank. The sets $\mathcal{D}k$ depend on the normals $\mathbf{n}_k, \ldots \mathbf{n}_1$ to laminates. However, the limiting set is invariant to rotation and it does not depend on these normals.

The L-closure $\mathcal{D}_\infty$ has the following properties:

(i) Set $\mathcal{D}_\infty$ is isotropic:

$$\phi(\mathcal{D}_\infty) = 0 \;\Rightarrow\; \phi(\Phi^T \mathcal{D}_\infty \Phi) = 0,$$

The boundary of the lamination closure depends only on the invariants of tensors $D \in \mathcal{D}_\infty$.

(ii) The initial set $\mathcal{D}_0$ belongs to $\mathcal{L}D$:

$$\mathcal{D}_0 \subset \mathcal{D}_\infty.$$

Equations for L-Closure. Let us find the equation for the L-closure. Consider the set $\mathcal{Z}_k = \mathcal{C}Z(\mathbf{n}_k, \mathcal{D}_{k-1})$ of laminates of the kth rank, transformed by (10.2.1). Let us denote the boundary surface of this set by $\partial \mathcal{Z}_k$. The boundary corresponds to an equation

$$\phi_k(D_k) = 0 \quad \forall D \in \partial \mathcal{D}_k.$$

The set $\mathcal{Z}_k$ is convex. Its boundary $\partial \mathcal{Z}_k$ consists of three components,

$$\partial \mathcal{Z}_k = \partial_1 \mathcal{Z}_k \cup \partial_2 \mathcal{Z}_k \cup \partial_3 \mathcal{Z}_k.$$

The component $\partial_1 \mathcal{Z}_k$ is composed of the points that are added on the kth step in the process of convexification of the set $Z(\mathbf{n}_k, \mathcal{D}_{k-1})$. The component $\partial_2 \mathcal{Z}_k$ is left unchanged by this procedure, but it was added to the boundary on a previous step m, $m < k$. The component $\partial_3 \mathcal{Z}_k$ corresponds to the points of the original set $\mathcal{D}_0$.

The surface $\partial_1 \mathcal{Z}_k$ is a part of the convex envelope of the surface $\partial \mathcal{Z}_k$ that does not coincide with it. Therefore, the surface $\partial_1 \mathcal{Z}_k$ is a convex, but not strongly convex; as such, it satisfies the relations

$$\begin{aligned}
H(\phi_k, Z_k) &\geq 0 && \text{(convexity)}, \\
\det H(\phi_k, Z_k) &= 0 && \text{(not strong convexity)}, \\
\forall Z_k &= Z(D_k), \quad D_k \in \partial_1 \mathcal{Z}_k,
\end{aligned} \qquad (10.2.3)$$

at each analytical point. Here $H(\phi, Z)$ is the Hessian of $\phi(Z)$ with respect to Z:

$$H(\phi, Z) = \frac{\partial^2 \phi}{\partial Z^2}.$$

The boundary component $\partial_2 \mathcal{Z}_k(\mathcal{D}_0)$ is not changed during the convexification on the kth step but is added on some previous step m by similar convexification. Hence, $\partial_2 \mathcal{Z}_k(\mathcal{D}_0)$ satisfies the same relation (10.2.3), but for the function $\phi_m(Z_m(\mathcal{D}_{m-1}))$, that is, for a different normal $\mathbf{n}_m$ of a previous step.

The component $\partial_3 \mathcal{Z}_k(\mathcal{D}_k) = \partial_3 \mathcal{Z}_k(\mathcal{D}_0)$ is determined by the boundary of $\mathcal{D}_0$.

We have obtained the following: All points of the boundary of $\partial \mathcal{Z}_k$ satisfy at least one equation from the set

$$\det H(\phi_n(Z_n(\mathcal{D}_{n-1}))) = 0, \ H(\phi_n(Z_n(\mathcal{D}_{n-1}))) \geq 0, \ n = 1, \ldots, k$$

everywhere it does not coincide with $\partial Z(\mathcal{D}_0)$.

The same property is obviously valid for the final laminate envelope. Namely, the surface $f(\mathbf{z}) = \partial \mathcal{L}(\mathcal{D})$ either coincides with $\partial Z(\mathcal{D})$ or satisfies the nonlinear differential equation

$$\min_{\mathbf{n}} \det \ (H(\phi(\mathcal{Z}(\mathbf{n}, D_\infty)))) = 0 \tag{10.2.4}$$

and the inequalities

$$H(f(\mathcal{Z}_\infty)) \geq 0 \quad \forall \mathbf{n}. \tag{10.2.5}$$

The analytical components of the envelope meet at some manifolds of lower dimensions where more than one eigenvalue of $H(f(\mathcal{Z}_\infty))$ is zero.

To determine the L-closure, one has to solve the boundary value problem (10.2.4) (10.2.5) with the boundary conditions that specify the original set of materials. The problem can be effectively used to check the L-closure. Notice also that the isotropy of the L-closure simplifies the problem, as shown by the following example.

10.2.1   The L-Closure in Two Dimensions

Consider an L-closure of a set of anisotropic conducting materials $\boldsymbol{\sigma}_1, \ldots,$ $\boldsymbol{\sigma}_n$, and suppose that each material can be arbitrarily rotated in the microstructure. The problem was investigated in (Lurie and Cherkaev, 1981a; Lurie and Cherkaev, 1981b; Francfort and Milton, 1987; Lurie and Cherkaev, 1986a). It was proved in these papers that the G-closure coincides with the L-closure.

Suppose that all materials in $\mathcal{D}_0$ shear the same eigenvectors. Each anisotropic material $\boldsymbol{\sigma}_k$ is represented by two matrices,

$$\boldsymbol{\sigma}_i(k, 1) = \begin{pmatrix} \lambda_A^{(k)} & 0 \\ 0 & \lambda_B^{(k)} \end{pmatrix}, \quad \boldsymbol{\sigma}_i(k, 2) = \begin{pmatrix} \lambda_B^{(k)} & 0 \\ 0 & \lambda_A^{(k)} \end{pmatrix} \tag{10.2.6}$$

differing by a rotation through a right angle. These matrices form the set $\mathcal{D}_0$. The set $\mathcal{D}_0$ of matrices is characterized by the set $S(\mathcal{D}_0)$ of its eigenvalues $[\lambda_A, \lambda_B]^1$.

---

[1]Here we use the subindexes $_A$ and $_B$ to show the Cartesian axes x_A and x_B

Consider laminates with the normal codirected with an eigenvector of materials. The eigenvalues of such laminates are the arithmetic and harmonic means of corresponding eigenvalues of initial materials. The transformation $Z_1(D)$ transforms the matrices in (10.2.6) to the matrices

$$Z_1(\sigma_i(k,1)) = \begin{pmatrix} \frac{1}{\lambda_A^{(k)}} & 0 \\ 0 & \lambda_B^{(k)} \end{pmatrix}, \quad Z_1(\sigma_i(k,2)) = \begin{pmatrix} \frac{1}{\lambda_B^{(k)}} & 0 \\ 0 & \lambda_A^{(k)} \end{pmatrix}, \quad (10.2.7)$$

respectively. At the first step, the laminates have the normal $\mathbf{n}_1 = [1,0]$.

The effective tensors of laminates corresponds to the convexification of the set $Z(\sigma_i)$

$$\mathcal{C}Z_1(\mathcal{D}_0) = \begin{pmatrix} \frac{1}{\lambda_A^*} & 0 \\ 0 & \lambda_B^* \end{pmatrix},$$

where the pairs $\left[\frac{1}{\lambda_A^*}, \lambda_B^*\right]$ lie on the straight segments of the lines that join all elements of the set $Z_1(\mathcal{D}_0)$:

$$\left[\frac{1}{\lambda_A^*}, \lambda_B^*\right] = \bigcup_{\lambda_A^i, \lambda_B^i, \lambda_A^j, \lambda_B^j \in S(\mathcal{D}_0)} \left\{ \eta \left[\frac{1}{\lambda_A^i}, \lambda_B^i\right] + (1-\eta) \left[\frac{1}{\lambda_A^j}, \lambda_B^j\right] \right\}.$$

Here $\eta \in [0, 1]$ is a parameter of the position of a point in the segment. We rewrite this representation in the form

$$(\lambda_A^*)^{-1} = \langle \lambda_A^{-1} \rangle_1, \quad \lambda_B^* = \langle \lambda_B \rangle_1$$

where $\langle \lambda_B \rangle_1$ denotes the averaging at the first step.

The effective properties of laminates σ_*^1 becomes

$$\mathcal{L}_1(\mathcal{D}_0) = Z_1^{-1} \mathcal{C} Z_1(\mathcal{D}_0) = \begin{pmatrix} \langle \lambda_A^{-1} \rangle_1^{-1} & 0 \\ 0 & \langle \lambda_B \rangle_1 \end{pmatrix},$$

On the second step, consider laminates of the second rank and suppose that the normal to the second-rank layers is orthogonal to the normal to the first-rank layers. The second-rank layers are made of materials from the set σ_*^1. The Z_2-transform is

$$Z_2(\mathcal{D}) = \begin{pmatrix} \lambda_A & 0 \\ 0 & \frac{1}{\lambda_B} \end{pmatrix}.$$

The set of properties of the second-rank laminates corresponds to the convexification

$$\left[\lambda_A^*, \frac{1}{\lambda_B^*}\right] = \bigcup_{\lambda_A^i, \lambda_B^i : \lambda_A^j, \lambda_B^j \in S(\mathcal{L}_1(\mathcal{D}_0)),} \left\{ \mu \left[\lambda_A^i, \frac{1}{\lambda_B^i}\right] + (1-\mu) \left[\lambda_A^j, \frac{1}{\lambda_B^j}\right] \right\}.$$

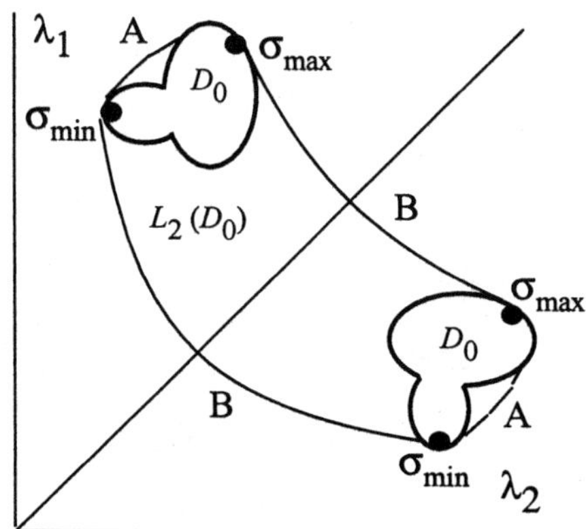

FIGURE 10.3. The L-closure of the set D_0 of two-dimensional conducting materials: The A-curves correspond to laminates and the B-curves to polycrystals.

where $\mu \in [0, 1]$, or

$$\lambda_A^* = \langle \lambda_A \rangle_2, \quad (\lambda_B^*)^{-1} = \langle \lambda_B^{-1} \rangle_2,$$

where $\langle \, \rangle_2$ denotes the averaging at the second step. The effective properties of second-rank laminates σ_*^2 are

$$L_2(D_0) = Z_2^{-1} C Z_2(L_1(D_0)) = \begin{pmatrix} \left\langle \left(\langle \lambda_A^{-1} \rangle_1 \right)^{-1} \right\rangle_2 & 0 \\ 0 & \left\langle \langle \lambda_B \rangle_1^{-1} \right\rangle_2^{-1} \end{pmatrix}.$$

Convexity Properties. We obtain the set S of pairs $[\lambda_A, \lambda_B]$ of the eigenvalues of $L_2 D_0$. The described procedure shows that S is the minimal set that: (i) contains eigenvalues of D_0: $S(L_2(D_0)) \supset S(D_0)$ and (ii) is convex in the coordinates $\frac{1}{\lambda_A}, \lambda_B$ and $\lambda_A, \frac{1}{\lambda_B}$.

The components of the boundary of the L-closure corresponds to either the boundary of the original set $S(D_0)$ or the linear functions of the type $C_1(\lambda_A^{-1}) + C_2 \lambda_B + C_3 = 0$ and $C_1(\lambda_B^{-1}) + C_2 \lambda_B + C_3 = 0$. The constants C_1, C_2 and C_3 are determined from the condition that the corresponding curves are tangents to the original set D_0 in the coordinates $\lambda_A, \frac{1}{\lambda_B}$.

Finally, the boundary of the envelope is the curve of type

$$C_3 \lambda_i + C_2 \lambda_A \lambda_B + C_1 = 0, \quad i = A, B,$$

which is attached to two points of the boundary of D_0. Particularly, the laminated polycrystal corresponds to the curve

$$C_2 \lambda_A \lambda_B + C_1 = 0.$$

that passes through the pair (10.2.7)

The L-closure corresponds to the convexity in the coordinates ($\det\boldsymbol{\sigma}$, an eigenvalue of $s(\boldsymbol{\sigma})$). An example of the L-closure is shown in Figure 10.3. Each point of the L-closure is achieved by the second-rank laminates. Its boundary consists either of laminates or of polycrystals of the extremal materials $\boldsymbol{\sigma}_{\min}$ and $\boldsymbol{\sigma}_{\max}$ (we assume that the materials are ordered as follows: $\det\boldsymbol{\sigma}_{\min} \leq, \ldots, \leq \det\boldsymbol{\sigma}_{\max}$).

One can check that no further steps are needed for the G-closure (Lurie and Cherkaev, 1986a; Francfort and Milton, 1987). The consideration is based on the fact that the boundary of the G-closure consists of the composites (i) with extremal value of $\det\boldsymbol{\sigma}_*$ and (ii) with extremal degree of anisotropy, $\max|\lambda_A - \lambda_B|$, and fixed values of $\det\boldsymbol{\sigma}_*$. These structures are the described laminates.

Remarks: Inner Bounds of G-Closure

Relations with the G-Closure. Of course, we cannot conclude that a G-closure and the L-closure coincide for any set of initial materials. However, such coincidence has been observed for many problems; this suggests the proposition that these two sets coincide at least for a large class of linear operators; see (Lurie and Cherkaev, 1986a; Milton, 1986). On the other hand, a counterexample has been recently built by Milton (private communication), see also (Milton, 2000), who demonstrated that the L-closure is smaller than G-closure for a problem of an optimal composite of seven elastic materials.

The Minimal Extensions by Necessary Conditions. An alternative way to produce the inside bounds for G-closures is to investigate the necessary conditions of optimality of a composite and to build the minimal extensions based on these conditions. This approach is discussed in Chapters 9 and 12. Necessary conditions produce an extension that gives an inside bound for the G_m-closure that cannot be improved by inserting any inclusion of an explicitly described shape.

The minimal extension is a variational technique based on calculus of infinitesimals, while the lamination closure technique deals with integral quantities, groups of rotation, symmetries, and similar algebraic features. Both techniques have unique features. The minimal extension deals with the fields in the structures, while the L-closure approach operates with geometrical characteristics of layouts. The minimal extension deals with infinitesimal changing of the fields due to small inclusions, while the L-closure deals with the final integral properties of the composites. The minimal extension a priori restricts the class of tests and the L-closure a priori restricts the geometry of would-be optimal layouts.

11
Examples of G-Closures

11.1 The G_m-Closure of Two Conducting Materials

In this chapter we find the G_m-closures and G-closures of various composites of conducting materials. We start with G_m-closures of two isotropic materials.

11.1.1 The Variational Problem

Denote again the isotropic conductivities of components by σ_1 and σ_2 ($\sigma_1 \leq \sigma_2$) and the conductivity tensors by $\boldsymbol{\sigma}_1$ and $\boldsymbol{\sigma}_2$: $\boldsymbol{\sigma}_i = \sigma_i I$.

The tensor of anisotropic conductivity of a composite made of σ_1 and σ_2 is denoted by $\boldsymbol{\sigma}_*$ and its eigenvalues are denoted by λ_i, $i = 1, \ldots, d$. The volume fractions of the materials in the composite are denoted by m_1 and m_2 ($m_1 + m_2 = 1$, $m_i \geq 0$).

To obtain the bounds for $\boldsymbol{\sigma}_*$, we minimize the sum J_1 of energies:

$$J_1 = \left\langle \sum_{i=1}^{d} \sigma \mathbf{e}_i^2 \right\rangle,$$

caused by d fixed orthogonal fields $\langle \mathbf{e}_1 \rangle, \ldots, \langle \mathbf{e}_d \rangle$. Here $\sigma(\mathbf{x}) = \sigma_1 \chi(\mathbf{x}) + \sigma_2(1 - \chi(\mathbf{x}))$. The functional J_1 is a convex combination of the eigenvalues λ_i of the tensor $\boldsymbol{\sigma}_*$

$$J_1 = \sum_{i=1}^{d} \langle \mathbf{e}_i \rangle \cdot \boldsymbol{\sigma}_* \langle \mathbf{e}_i \rangle = \sum_{i=1}^{d} \lambda_i \alpha_i^2; \tag{11.1.1}$$

real parameters α_i depend on $\langle \mathbf{e}_i \rangle$. A bound of J_1 corresponds to the component $a\,b$) of the boundary of the G_m-closure (see Figure 10.1).

The complementary component of the G_m-closure boundary can be obtained by estimating the sum of complementary energies, equal to a convex combination of the inverse values of the eigenvalues of the tensor σ_* (see Figure 10.1),

$$J_d = \sum_{i=1}^{d} \left\langle \mathbf{j}_i \cdot (\sigma_*)^{-1} \mathbf{j}_i \right\rangle = \sum_{i=1}^{d} \frac{1}{\lambda_i} \beta_i^2,$$

where β_i are real parameters that depend on $\langle \mathbf{j}_i \rangle$.

The simplest bound of the stored energy is associated with a single applied field and is given by the convex envelope. This bound leads to Wiener bounds for effective tensors. The Wiener bounds for the G_m-closure are

$$(\sigma_*)^{-1} \leq \left(\frac{m_1}{\sigma_1} + \frac{m_2}{\sigma_2} \right) I, \quad \sigma_* \leq (m_1 \sigma_1 + m_2 \sigma_2)\, I; \qquad (11.1.2)$$

they determine the minimal cube (Wiener box) that contains the G_m-closure. These bounds constrain from below the minimal eigenvalues of the effective tensors σ_* and from above the maximal of them.

11.1.2   The G_m-Closure in Two Dimensions

Let us calculate the G_m-closure for a composite of two isotropic conductors in two dimensions. This problem was solved in (Lurie and Cherkaev, 1982) using the technique presented here. Tartar and Murat independently solved the problem at approximately the same time (Tartar, 1985; Murat and Tartar, 1985a). Two decades earlier, Hashin and Shtrikman obtained the isotropic points of the G_m-closure using their variational method (Hashin and Shtrikman, 1962a).

Functionals

The bounds of the G_m-closure are obtained by estimating the functional (11.1.1). We introduce the four-dimensional vector $Z = [\mathbf{e}_1, \mathbf{e}_2]$ of the fields. The vector Z satisfies the differential constraints

$$\nabla \times \mathbf{e}_1 = \nabla \times \mathbf{e}_2 = 0,$$

which imply the existence of the translator $\mathbf{e}_1 \times \mathbf{e}_2 = Z^T T Z$. The translator matrix is

$$T(t) = \begin{pmatrix} 0 & 0 & 0 & t \\ 0 & 0 & -t & 0 \\ 0 & -t & 0 & 0 \\ t & 0 & 0 & 0 \end{pmatrix}.$$

The material properties $A_i = A(\boldsymbol{\sigma}_i)$ correspond to the block-diagonal (4×4) matrices. The block form of them is

$$A_i = A(\boldsymbol{\sigma}_i) = \begin{pmatrix} \sigma_i & \mathbf{O} \\ \mathbf{O} & \sigma_i \end{pmatrix}, \quad i = 1, 2.$$

Bounds

The bound (10.1.7) has the form

$$(A_* - T)^{-1} \leq m_1(A_1 - T)^{-1} + m_2(A_2 - T)^{-1}, \tag{11.1.3}$$

where $A_* = A(\boldsymbol{\sigma}_*)$.

Let us analyze the bound (11.1.3). We choose a Cartesian basis in which $\boldsymbol{\sigma}_*$ is diagonal. The matrices $(A_* - T)$ and $(A_i - T)$, $i = 1, 2$, become

$$A_* - T = \begin{pmatrix} \lambda_1 & 0 & 0 & t \\ 0 & \lambda_2 & -t & 0 \\ 0 & -t & \lambda_1 & 0 \\ t & 0 & 0 & \lambda_2 \end{pmatrix}, \quad A_i - T = \begin{pmatrix} \sigma_i & 0 & 0 & t \\ 0 & \sigma_i & -t & 0 \\ 0 & -t & \sigma_i & 0 \\ t & 0 & 0 & \sigma_i \end{pmatrix}.$$

The matrices $(A_i - T)$, $i = 1, 2$, have the following eigenvalues λ_i:

$$\lambda_1 = \lambda_2 = \sigma_i + t, \quad \lambda_3 = \lambda_4 = \sigma_i - t, \quad i = 1, 2.$$

These matrices are nonnegative if $t \in \mathcal{T}$, where

$$\mathcal{T} = \{t : \; |t| \leq \sigma_1\}.$$

Both matrices $(A_i - T)$, $i = 1, 2$, have the same eigenvectors $\mathbf{f}_i$:

$$\mathbf{f}_1 = \frac{1}{\sqrt{2}} \begin{pmatrix} 1 \\ 0 \\ 0 \\ 1 \end{pmatrix}, \quad \mathbf{f}_2 = \frac{1}{\sqrt{2}} \begin{pmatrix} 0 \\ 1 \\ -1 \\ 0 \end{pmatrix}, \quad \mathbf{f}_3 = \frac{1}{\sqrt{2}} \begin{pmatrix} 1 \\ 0 \\ 0 \\ -1 \end{pmatrix}, \quad \mathbf{f}_4 = \frac{1}{\sqrt{2}} \begin{pmatrix} 0 \\ 1 \\ 1 \\ 0 \end{pmatrix}.$$

The projection of (11.1.3) to the eigenvector $\mathbf{f}_1$ gives the scalar inequality

$$\mathbf{f}_1^T (A_* - T)^{-1} \mathbf{f}_1 \leq m_1 \mathbf{f}_1^T (A_1 - T)^{-1} \mathbf{f}_1 + m_2 \mathbf{f}_1^T (A_2 - T)^{-1} \mathbf{f}_1.$$

We calculate the projections and obtain

$$f_t(\lambda_1, \lambda_2) = \frac{\lambda_1 + \lambda_2 - 2t}{\lambda_1 \lambda_2 - t^2} - \frac{m_1}{\sigma_1 + t} - \frac{m_2}{\sigma_2 + t} \geq 0 \quad \forall t \in [-\sigma_1, \sigma_1].$$

The last inequality is valid for all $t \in \mathcal{T}$. The corresponding linear-fractional curves $f_t(\lambda_1, \lambda_2) = 0$ all pass through the corners (σ_h, σ_a) and (σ_a, σ_h) of the Wiener box. The strongest inequality corresponds to the equality $t = \sigma_1$;

$$\frac{\lambda_1 + \lambda_2 - 2\sigma_1}{\lambda_1 \lambda_2 - \sigma_1^2} \geq \frac{m_1}{2\sigma_1} + \frac{m_2}{\sigma_1 + \sigma_2}. \tag{11.1.4}$$

Remark 11.1.1 *The consideration of projections on eigenvectors $\mathbf{f}_2 - \mathbf{f}_4$ leads to identical results.*

In particular, the isotropic effective conductivity $\sigma_* = \lambda_1 = \lambda_2$ coincides with the Hashin-Shtrikman bound

$$\frac{1}{\sigma_1 + \sigma_*} \geq \frac{m_1}{2\sigma_1} + \frac{m_2}{\sigma_1 + \sigma_2}.$$

A straightforward calculation brings (11.1.4) to the form

$$\frac{1}{\lambda_1 - \sigma_1} + \frac{1}{\lambda_2 - \sigma_1} + \frac{1}{\sigma_1} \leq \frac{1}{m_2}\left(\frac{2}{\sigma_2 - \sigma_1} + \frac{1}{\sigma_1}\right). \tag{11.1.5}$$

The tensor form of the last inequality is especially symmetric

$$\mathrm{Tr}\left(\boldsymbol{\sigma}_1^{-1}\boldsymbol{\sigma}_* - I\right)^{-1} + 1 \leq \frac{1}{m_2}\left(\mathrm{Tr}\,\boldsymbol{\sigma}_1^{-1}(\boldsymbol{\sigma}_2 - I)^{-1} + 1\right). \tag{11.1.6}$$

We will show that (11.1.6) is valid for any dimension of the space. Moreover, it also is valid for anisotropic materials $\boldsymbol{\sigma}_1 \leq \boldsymbol{\sigma}_2$, if their eigenvectors are parallel (Kohn and Milton, 1986; Zhikov, 1986; Grabovsky, 1993).

The Complementary Bound

The complementary bound is obtained from the dual problem. The derivation is parallel to the previous case. We put

$$Z = [\mathbf{j}_1,\ \mathbf{j}_2], \quad A_i = \begin{pmatrix} \boldsymbol{\sigma}_i^{-1} & 0 \\ 0 & \boldsymbol{\sigma}_i^{-1} \end{pmatrix}.$$

Translator T has the same form as before; it corresponds to the quasiaffine form $\det(\mathbf{j}_1,\mathbf{j}_2)$ (see (8.2.7)).

After manipulations, we obtain the bound

$$\frac{1}{\lambda_1 - \sigma_2} + \frac{1}{\lambda_2 - \sigma_2} \leq \frac{1}{m_1}\left(\frac{2}{\sigma_1 - \sigma_2} + \frac{m_2}{\sigma_2}\right) \tag{11.1.7}$$

or, in tensor form,

$$\mathrm{Tr}\left(\boldsymbol{\sigma}_2^{-1}\boldsymbol{\sigma}_* - I\right)^{-1} + 1 \leq \frac{1}{m_2}\left(\mathrm{Tr}(\boldsymbol{\sigma}_2^{-1}\boldsymbol{\sigma}_1 - I)^{-1} + 1\right)$$

Remark 11.1.2 *One expects symmetry in the complementary bound, because any two-dimensional curlfree vector* $\mathbf{e}$ *becomes the divergencefree vector* $\mathbf{j}$ *when rotated through $90°$, and vice versa.*

The obtained bounds completely describe the G_m-closure. One can check that the curves (11.1.5) and (11.1.7) meet at the points $(\lambda_1 = \sigma_h, \lambda_2 = \sigma_a)$

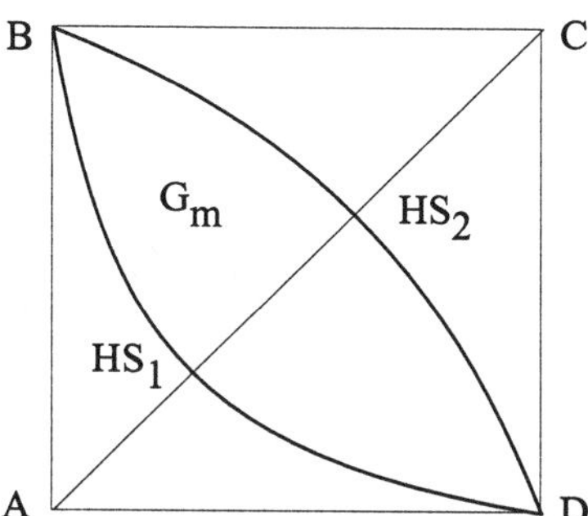

FIGURE 11.1. The G_m-closure $\mathbf{G}_m$ of two conducting materials in two dimensions. The square A B C D shows the Wiener box. The points HS_1 and HS_2 show the isotropic Hashin–Shtrikman bounds.

and $(\lambda_1 = \sigma_a, \lambda_2 = \sigma_h)$, which are also corner points of the Wiener bounds (see Figure 11.1).

We mention that (i) The bounds are stronger than the Wiener bound; (ii) consideration of other functionals (10.1.2) is unnecessary, because the G_m-closure is shaped like a lens with two sharp edges; and (iii) the isotropic points HS_1 and HS_2 (see Figure 11.1) correspond to the Hashin–Shtrikman bounds.

The Optimal Structures

It is easy to verify that the matrix laminates of second rank correspond to both components of the G_m-closure. One component corresponds to the envelope of the first material, and the second to the envelope of the second one. Indeed, the formulas for matrix laminates (7.3.13) produce equality in inequalities (11.1.5) and (11.1.7).

Recall that the properties of matrix laminates depend on an inner parameter. When this parameter varies, the corresponding point slides along the boundary curve. The components of the G_m-closure meet at the points that correspond to the degeneration of the structure to simple laminates.

Remark 11.1.3 *The second-rank laminates are not the only structures that realize the bounds of the G_m-closure. Other known structures are the isotropic coated spheres (Hashin and Shtrikman, 1962a), the coated ellipses (Milton, 1981b), and "truly periodic" structures (Vigdergauz, 1988). As we argued in Chapter 9, the only feature of optimality is the limiting character of the fields; the geometry of the optimal structures is not unique.*

Remark 11.1.4 *The boundary of the G_m-closure corresponds to a straight line in the plane of the main invariants $\operatorname{Tr} \sigma_*$ and $\det \sigma_*$. This line passes through the points corresponding to the laminate and coated circles.*

Y-Transform

The bounds are especially elegant in the Y-notation (Chapter 7). The matrix $Y(A_*)$ has the form

$$Y(A_*) = \begin{pmatrix} y_1 & 0 & 0 & 0 \\ 0 & y_2 & 0 & 0 \\ 0 & 0 & y_1 & 0 \\ 0 & 0 & 0 & y_2 \end{pmatrix}$$

where

$$y_i = \frac{(m_1\sigma_2 + m_2\sigma_1)\lambda_i - \sigma_1\sigma_2}{m_1\sigma_1 + m_2\sigma_2 - \lambda_i}.$$

The bounds (10.1.9) become

$$\sigma_1^2 \leq \det Y(\boldsymbol{\sigma}_*) \leq \sigma_2^2.$$

In the Y-plane, the G_m-closure set is bounded by two symmetric parabolas and is independent of the volume fractions. Isotropic composites correspond to the isotropic tensor $Y(\boldsymbol{\sigma}_*)$; the isotropic bounds are $\sigma_1 \leq Y(\boldsymbol{\sigma}_*) \leq \sigma_2$. The laminates correspond to the improper points $(0, \infty)$ and $(\infty, 0)$.

11.1.3 Three-Dimensional Problem

Now we calculate the G_m-closure of two isotropic conductors in the three-dimensional case. This problem was independently solved in (Lurie and Cherkaev, 1984a) and (Tartar, 1985).

First, we compute the lower bound for the eigenvalues of $\boldsymbol{\sigma}_*$. We bound the sum of three energies stored in three external orthogonal fields. The block-diagonal (9×9) matrices A_i have the form

$$A_i = \begin{pmatrix} \sigma_i & \mathbf{O} & \mathbf{O} \\ \mathbf{O} & \sigma_i & \mathbf{O} \\ \mathbf{O} & \mathbf{O} & \sigma_i \end{pmatrix}, \quad i = 1,\, 2,$$

and $\mathbf{O}$ is the 3×3 null matrix and Z is the nine-dimensional vector

$$Z = (\mathbf{e}_1, \mathbf{e}_2, \mathbf{e}_3). \tag{11.1.8}$$

We take into account the potentiality of the fields $\mathbf{e}_i$, which leads to the existence of translators. The translators are quasiaffine quadratic forms of the type (see (8.2.3), (8.2.4))

$$\mathbf{e}_i \cdot T(t_{kl}^{ij})\mathbf{e}_j,$$

where $T_{ij}(t_{kl}^{ij})$ is an arbitrary antisymmetric 3×3 matrix,

$$T_{ij} = \begin{pmatrix} 0 & t_{12}^{ij} & t_{13}^{ij} \\ -t_{12}^{ij} & 0 & t_{23}^{ij} \\ -t_{13}^{ij} & -t_{23}^{ij} & 0 \end{pmatrix},$$

and the t_{kl}^{ij} are arbitrary real parameters. The nine real parameters t_{kl}^{ij} must be optimally chosen to obtain the bound.

The 9×9 translator matrix T associated with the vector Z (11.1.8) has the block form (see (8.2.3), (8.2.4))

$$T = \begin{pmatrix} 0 & T_{12} & T_{13} \\ -T_{12} & 0 & T_{23} \\ -T_{13} & -T_{23} & 0 \end{pmatrix}. \tag{11.1.9}$$

Note that T is symmetric.

Bounds

We use the matrix inequality (10.1.7) to obtain the bound. First, we choose the parameters t_{kl}^{ij} to simplify further calculations. Namely,

$$T_{12} = \begin{pmatrix} 0 & t & 0 \\ -t & 0 & 0 \\ 0 & 0 & 0 \end{pmatrix}, \quad T_{13} = \begin{pmatrix} 0 & 0 & t \\ 0 & 0 & 0 \\ -t & 0 & 0 \end{pmatrix}, \quad T_{23} = \begin{pmatrix} 0 & 0 & 0 \\ 0 & 0 & t \\ 0 & -t & 0 \end{pmatrix}.$$

We also choose the orientation of the coordinate system in such a way that the matrix A_* becomes diagonal. It consists of three equal diagonal blocks, $\mathrm{Diag}(\lambda_1,\ \lambda_2,\lambda_3)$. The 9×9 matrix $U(\boldsymbol{\sigma}_*)$ (see (10.1.7)) can be split into blocks; the first block consists of elements staying in the intersection of the first, fourth, and ninth row and column and has the form

$$U^{1,4,9}(\boldsymbol{\sigma}_*) = \begin{pmatrix} \lambda_1 & t & t \\ t & \lambda_2 & t \\ t & t & \lambda_3 \end{pmatrix}^{-1}$$
$$-m_1 \begin{pmatrix} \sigma_1 & t & t \\ t & \sigma_1 & t \\ t & t & \sigma_1 \end{pmatrix}^{-1} - m_2 \begin{pmatrix} \sigma_2 & t & t \\ t & \sigma_2 & t \\ t & t & \sigma_2 \end{pmatrix}^{-1} \tag{11.1.10}$$

and the other blocks are

$$U^{i,j}(\boldsymbol{\sigma}_*) = \begin{pmatrix} \lambda_1 & -t \\ -t & \lambda_2 \end{pmatrix}^{-1} - m_1 \begin{pmatrix} \sigma_1 & -t \\ -t & \sigma_1 \end{pmatrix}^{-1} - m_2 \begin{pmatrix} \sigma_2 & -t \\ -t & \sigma_2 \end{pmatrix}^{-1}.$$

The strongest inequality is

$$U^{1,4,9}(\boldsymbol{\sigma}_*) \leq 0.$$

The rest of the calculation is similar to the two-dimensional case. We put

$$t = \sigma_1 < \sigma_2$$

to obtain the sharpest bound. The matrix $U^{1,4,9}(\sigma_1)$ has two arbitrarily large negative eigenvalues and one finite eigenvalue that corresponds to the eigenvector

$$\mathbf{f}^T = \frac{1}{\sqrt{3}}(1,\ 1,\ 1). \tag{11.1.11}$$

The bound is obtained by the projection of (11.1.10) onto this direction:

$$\mathbf{f}^T U^{1,4,9}(\boldsymbol{\sigma}_*)\,\mathbf{f} \leq 0.$$

The bound can be written in several equivalent forms. It can be expressed as

$$\frac{I_2^* - 2I_1^*\sigma_1 + 3\sigma_1^3}{I_3^* - I_1^*\sigma_1^2 + 2\sigma_1^3} \leq \frac{m_1}{3\sigma_1} + \frac{m_2}{\sigma_2 + 2\sigma_1}, \tag{11.1.12}$$

where I_i^* are the main invariants of the matrix $\boldsymbol{\sigma}_*$:

$$\begin{aligned}
I_1^* &= \mathrm{Tr}\,\boldsymbol{\sigma}_* = \lambda_1 + \lambda_2 + \lambda_3, \\
I_2^* &= \lambda_1,\lambda_2 + \lambda_2\,\lambda_3 + \lambda_3\,\lambda_1, \\
I_3^* &= \mathrm{Det}\boldsymbol{\sigma}_* = \lambda_1\,\lambda_2\,\lambda_3.
\end{aligned}$$

Notice that the bound is a hyperplane in the space of the main invariants of $\boldsymbol{\sigma}_*$.

The bound can be rewritten in the form

$$\sum_{i=1}^{3} \frac{1}{\lambda_i - \sigma_1} \leq \frac{1}{m_2}\left(\frac{3}{\sigma_2 - \sigma_1} + \frac{m_1}{\sigma_1}\right). \tag{11.1.13}$$

The bound also can be rewritten in the tensor form (11.1.6), which is independent of the number of dimensions.

The isotropic effective conductivity σ_* corresponds to the three-dimensional Hashin-Shtrikman bound

$$\frac{1}{\sigma_* + 2\sigma_1} \leq \frac{m_1}{3\sigma_1} + \frac{m_2}{\sigma_2 + 2\sigma_1}.$$

A Dual Component of the Boundary

The complementary component of the boundary is obtained by estimating the functional equal to the sum of the complementary energies and using the translator (8.2.17) related to three divergencefree fields. The analogue of inequality (11.1.10) is

$$\begin{pmatrix} \frac{1}{\lambda_1} - 2t & t & t \\ t & \frac{1}{\lambda_2} - 2t & t \\ t & t & \frac{1}{\lambda_3} - 2t \end{pmatrix}^{-1} - m_1 U_1^{-1} - m_2 U_2^{-1} \leq 0$$

where

$$U_1 = \begin{pmatrix} z_1(t) & t & t \\ t & z_1(t) & t \\ t & t & z_1(t) \end{pmatrix}, \quad U_2 = \begin{pmatrix} z_2(t) & t & t \\ t & z_2(t) & t \\ t & t & z_2(t) \end{pmatrix},$$

$z_1 = \frac{1}{\sigma_1} - 2t$, and $z_2 = \frac{1}{\sigma_2} - 2t$. The critical value of t is $t = \frac{1}{\sigma_2}$. When t takes this value, U_1 has two improper eigenvalues. The eigenvector of the remaining finite eigenvalue is again equal to $\mathbf{f}_1$ (11.1.11).

After obvious calculations, the bound becomes

$$\sum_{i=1}^{3} \frac{1}{\lambda_i - \sigma_2} \leq \frac{1}{m_1}\left(\frac{3}{\sigma_1 - \sigma_2} + \frac{m_2}{\sigma_2}\right). \qquad (11.1.14)$$

Notice that the bound is similar to (11.1.13).

Additional Bounds

The bounds (11.1.12) and (11.1.14) must be complemented by the Wiener bound (11.1.2). In the three-dimensional case, the Wiener bound provides an independent inequality that determines a component of the boundary of the G_m-closure. If the largest eigenvalue λ_3 reaches its limit $\lambda_3 = m_1\sigma_1 + m_2\sigma_2$ and if (11.1.13) holds, then the eigenvalues λ_1 and λ_2 satisfy equation (11.1.5) for two-dimensional G-closure. This component of the boundary is

$$\frac{1}{\lambda_1 - \sigma_1} + \frac{1}{\lambda_2 - \sigma_1} = \frac{1}{m_2}\left(\frac{2}{\sigma_2 - \sigma_1} + \frac{m_1}{\sigma_1}\right),$$
$$\lambda_3 = m_1\boldsymbol{\sigma}_1 + m_2\boldsymbol{\sigma}_2. \qquad (11.1.15)$$

Notice also that if one eigenvalue of $\boldsymbol{\sigma}_* \in G_m$-closure is equal to the harmonic mean σ_h, then the other two are necessarily equal to σ_a. This point of the boundary of the closure corresponds to simple laminate. Notice that the point $(\sigma_h, \sigma_a, \sigma_a)$ is the vertex where all three boundary components (11.1.13), (11.1.14), and (11.1.15) meet. The G_m-closure is shown in Figure 11.2.

The Structures

It is easy to verify (Lurie and Cherkaev, 1984a) that the matrix laminates of third rank correspond to both components (11.1.13) and (11.1.14) of the G_m-closure. The component (11.1.13) corresponds to the envelope of the first material and the nuclei from the second material, and the component (11.1.14) to the envelope of the second material and the nuclei from the first material.

These structures degenerate into cylindrical matrix laminates of second rank when

$$\lambda_3 = m_1\sigma_1 + m_2\sigma_2.$$

One can check that the eigenvalues λ_1 and λ_2 satisfy the inequalities (11.1.4) for the two-dimensional G_m-closure.

To achieve any point inside the closure, one can consider a composite of the material on its boundary. These composites obviously correspond to all inner points of the closure.

Remark 11.1.5 *The boundary of the G_m-closure corresponds to a plane in the space of the main invariants of $\boldsymbol{\sigma}_*$. To fix the position of that plane,*

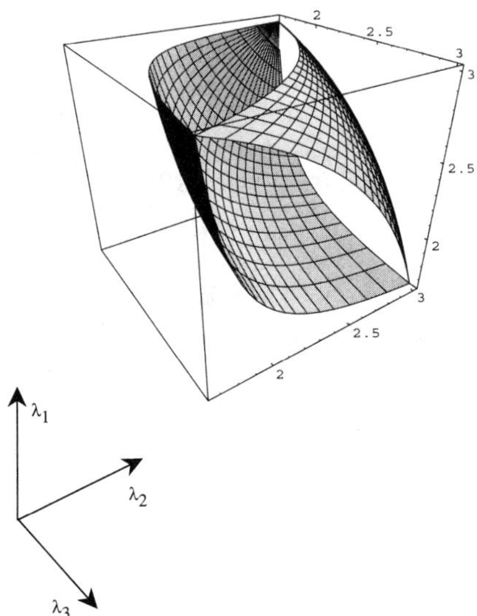

FIGURE 11.2. The G_m-closure in three dimensions. Corner points correspond to effective properties of laminates, the faces correspond to the "cylindrical" matrix laminates, and the smooth surfaces correspond to third-rank matrix laminates.

it is enough to indicate its three characteristic points. One can choose the points that correspond to the laminates, the coated spheres, and the coated cylinders.

Another type of microstructure that realizes the boundary of the G_m-closure (Tartar, 1985) is the structure of "coated ellipsoids." These structures are an obvious generalization of the "coated spheres" described in Chapter 2. Such structures were introduced and investigated in (Milton, 1980).

The Y-transform

Let us find the Y-transform of the bounds of the G_m-closure. For the three-dimensional problem, the bound (11.1.10) becomes

$$\det \begin{pmatrix} Y(\lambda_1) & t & t \\ t & Y(\lambda_2) & t \\ t & t & Y(\lambda_3) \end{pmatrix} = 0,$$

where $|t| \le \sigma_1$. Setting $t = \sigma_1$, we obtain

$$\det Y\left(\frac{\sigma_*}{\sigma_1}\right) - \mathrm{Tr}\, Y\left(\frac{\sigma_*}{\sigma_1}\right) + 2 \ge 0.$$

The dual bound can be obtained similarly. It takes the form

$$\det Y\left(\frac{\boldsymbol{\sigma}_*}{\sigma_2}\right) - \operatorname{Tr} Y\left(\frac{\boldsymbol{\sigma}_*}{\sigma_2}\right) + 2 \le 0, \quad Y(\lambda_i) \ge 0, \ i = 1, 2, 3.$$

The calculation is similar to the two-dimensional case. In both cases the G_m-closures are semi-infinite in the Y-notation. A simple laminate corresponds to an improper point.

11.2 *G*-Closures

11.2.1 Two Isotropic Materials

Two-Dimensional G-Closure

Consider first the simplest two-dimensional G-closure problem, already solved in Chapter 3. Let us obtain the result using the variational method, (10.1.19).

To find the G-closure, we bound the functional J

$$J = \langle \mathbf{j} \cdot \boldsymbol{\sigma}^{-1}\mathbf{j} + \mathbf{e} \cdot \boldsymbol{\sigma}\mathbf{e}\rangle = \frac{1}{\lambda_2}\mathbf{j}^2 + \lambda_1\mathbf{e}^2 \tag{11.2.1}$$

where $\mathbf{e}$ and $\mathbf{j}$ are the mutual orthogonal external field and external current, and λ_1 and λ_2 ($\lambda_1 \le \lambda_2$) are the eigenvalues of the effective tensor. The matrices A_i in formula (10.1.6) for the translation bound have the form

$$A_i = \begin{pmatrix} \frac{1}{\sigma_i} & 0 & 0 & 0 \\ 0 & \frac{1}{\sigma_i} & 0 & 0 \\ 0 & 0 & \sigma_i & 0 \\ 0 & 0 & 0 & \sigma_i \end{pmatrix}, \quad i = 1, 2.$$

No translators are needed for this problem.

To find the G-closure, we project the basic inequality (10.1.6) onto a direction $\mathbf{L}$ such that $\mathbf{L}^T A_1^{-1}\mathbf{L} = \mathbf{L}^T A_2^{-1}\mathbf{L} = K$. A suitable projector $\mathbf{L}$ (see (10.1.18)) is

$$\mathbf{L}^T = \left(1, \quad 0, \quad 0, \quad \sqrt{\sigma_1\sigma_2}\right).$$

Indeed, we have

$$\mathbf{L}^T A_1^{-1}\mathbf{L} = \mathbf{L}^T A_2^{-1}\mathbf{L} = \sigma_1 + \sigma_2 = \mathbf{K}. \tag{11.2.2}$$

The left-hand-side term in bound (11.2.2) is

$$\mathbf{L}^T (A_*)^{-1}\mathbf{L} = \lambda_1 + \frac{1}{\lambda_2}\sigma_1\sigma_2. \tag{11.2.3}$$

Combining (11.2.2) and (11.2.3) we obtain the bound

$$\lambda_1 + \frac{1}{\lambda_2}\sigma_1\sigma_2 \le \sigma_1 + \sigma_2, \tag{11.2.4}$$

which coincides with bound (3.2.4) obtained in Chapter 3.

Three-Dimensional G-Closure

This time we demonstrate the straightforward approach (10.1.14), calculating the union of all G_m-closures.

Observe that the family of the surfaces (11.1.13) or (11.1.14) of the G_m-closure boundary (m_1 is the parameter of the family) does not have an envelope. Therefore, the boundary of the union of all G_m-closures (the G-closure) is drawn by the boundary curves of the surfaces (11.1.13) or (11.1.14). The boundary curve corresponds to the case where one of the eigenvalues (say, λ_3) of $\boldsymbol{\sigma}_*$ reaches its limit:

$$\lambda_3 = \sigma_a = m_1\sigma_1 + (1 - m_1)\sigma_2.$$

The last equality enables us to express m_1 as a function of λ_3 and exclude m_1 from (11.1.13). We obtain the inequality

$$\frac{\sigma_1}{\lambda_1 - \sigma_1} + \frac{\sigma_1}{\lambda_2 - \sigma_1} - \frac{\sigma_1 + \sigma_2}{\lambda_3 - \sigma_1} \geq 1, \quad \sigma_1 \leq \lambda_i \leq \sigma_2. \qquad (11.2.5)$$

The equality in (11.2.5) gives the component of the boundary of the G-closure.

The structures that realize the bound are the cylindrical second-rank laminates. Material σ_1 forms the envelope, and material σ_2 forms the cylindrical inclusions.

The other two components of the G-closure boundary are obtained by cyclic permutation of the eigenvalues λ_i. The intersection of each two components corresponds to laminates (11.2.4). All three boundaries meet at the points $\lambda_1 = \lambda_2 = \lambda_3 = \sigma_1$ and $\lambda_1 = \lambda_2 = \lambda_3 = \sigma_2$ (see Figure 11.3).

To complete the investigation, one checks that the analogous boundary derived from inequality (11.1.14) instead of (11.1.13) is weaker. We leave this to the reader.

Remark 11.2.1 *The projection technique demonstrated earlier on the two-dimensional G-closure problem is also applicable for the three-dimensional problem. This technique shows the difference between the bounds (11.1.13) and (11.1.14). If the estimated functional (an analogue of J in (11.2.1)) depends on two terms of type $\mathbf{e}_i \cdot \boldsymbol{\sigma}_* \mathbf{e}_i$, $i = 1, 2$ and one term of type $\mathbf{j} \cdot \boldsymbol{\sigma}_*^{-1}\mathbf{j}$, then the translator (11.1.9) is involved. The bound coincides with (11.2.5). But if the functional depends on one term of the type $\mathbf{e} \cdot \boldsymbol{\sigma}_* \mathbf{e}$ and two terms of type $\mathbf{j} \cdot \boldsymbol{\sigma}_*^{-1}\mathbf{j}$, no translators exist, because two divergencefree vectors in three dimensions do not correspond to a translator (Chapter 8). The corresponding bound coincides with the arithmetic-harmonic bound (11.2.4). The bound corresponds to the laminates.*

We recommend that the reader conducts the corresponding calculations as an exercise.

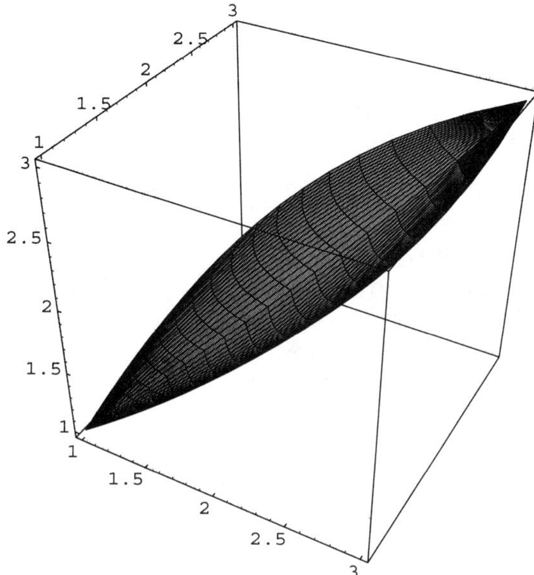

FIGURE 11.3. *G*-closure of the set of two isotropic conductors.

11.2.2 Polycrystals

Consider the homogenization problem for a polycrystal: the conglomerate made of single crystals by disorientation of its fragments. Suppose that the initial crystallite is characterized by the conductivity tensor $\boldsymbol{\sigma}_0$ and that σ_i are the ordered eigenvalues:

$$\boldsymbol{\sigma}_0 = \mathrm{Diag}(\sigma_1, \sigma_2, \sigma_3).$$

We are dealing with a mixture of infinitely many materials that share the same rotationally invariant characteristics. Note that the constraints on the volume fractions of differently oriented fragments are often unnatural. We do not prescribe the volume fractions.

The homogenized material is characterized by an effective tensor $\boldsymbol{\sigma}_*$ with eigenvalues λ_i, $i = 1, 2, 3$. We assume that $\lambda_1 \le \lambda_2 \le \lambda_3$.

The *G*-closure problem describes of the set $\Lambda = \{\lambda_1, \lambda_2, \lambda_3\}$ as a function of parameters σ_1, σ_2, and σ_3.

Using the Wiener inequalities, one easily establishes the bounds for *G*-closure

$$\sigma_1 \le \lambda_1 \quad \text{and} \quad \sigma_3 \ge \lambda_3$$

that ensure that all triplets Λ of eigenvalues of any polycrystal lie in the box determined by two extreme eigenvalues of a crystal.

Similarly, we obtain the two inequalities

$$\mathrm{Tr}\,\boldsymbol{\sigma}_0 \le \mathrm{Tr}\,\boldsymbol{\sigma}_* \quad \text{and} \quad \mathrm{Tr}(\boldsymbol{\sigma}_0)^{-1} \ge \mathrm{Tr}(\boldsymbol{\sigma}_0)^{-1} \tag{11.2.6}$$

by taking the trace of the matrices in the Wiener bounds.

11.2.3 Two-Dimensional Polycrystal

To obtain the G-closure in the two-dimensional case, we use a special method (Lurie and Cherkaev, 1981a; Lurie and Cherkaev, 1981b; Jikov et al., 1994). The isotropic components of the G-closure were described earlier in (Keller, 1964; Dykhne, 1971).

Consider two fields $\mathbf{e}_1$ and $\mathbf{j}_2$ and two corresponding currents $\mathbf{j}_1$ and $\mathbf{j}_2$; each pair of a field and a current corresponds to different external conditions but to the same inhomogeneous medium.

We know from (8.2.7) that the bilinear form $\mathbf{j}_1 \cdot R\mathbf{j}_2$ is quasiaffine,

$$\langle \mathbf{j}_1 \cdot R\mathbf{j}_2 \rangle = \langle \mathbf{j}_1 \rangle \cdot R \langle \mathbf{j}_2 \rangle, \tag{11.2.7}$$

where R is the matrix of rotation through a right angle,

$$R = \begin{pmatrix} 0 & 1 \\ -1 & 0 \end{pmatrix}.$$

This form can be rewritten as

$$\mathbf{j}_1 \cdot R\mathbf{j}_2 = \mathbf{e}_1 \cdot (\sigma R \sigma)\mathbf{e}_2 = \det(\sigma)\mathbf{e}_1 \cdot R\mathbf{e}_2.$$

Here we used the easily verifiable equality

$$\sigma R \sigma = \det(\sigma)R$$

for all 2×2 symmetric matrices σ.

Applying this equality to both sides of (11.2.7) and keeping in mind that $\det(\sigma) = $constant$(\mathbf{x})$ in polycrystals, we obtain the equality

$$\det(\sigma)\langle \mathbf{e}_1 \cdot R\mathbf{e}_2 \rangle = \det(\sigma_*)\langle \mathbf{e}_1 \rangle \cdot R \langle \mathbf{e}_2 \rangle. \tag{11.2.8}$$

Now we use the quasiaffinnes of the bilinear form $\mathbf{e}_1 \cdot R\mathbf{e}_2$ (see (6.1.8)):

$$\langle \mathbf{e}_1 \cdot R\mathbf{e}_2 \rangle = \langle \mathbf{e}_1 \rangle \cdot R \langle \mathbf{e}_2 \rangle.$$

Comparing with (11.2.8), we obtain the equation for the G-closure

$$\det(\sigma_*) = \det(\sigma). \tag{11.2.9}$$

The Wiener bound leads to another constraint: the inequality

$$\mathrm{Tr}(\sigma_*) \leq \mathrm{Tr}(\sigma). \tag{11.2.10}$$

The exceptional nature of this problem is clear: The G_m-closure corresponds to a segment of the curve (11.2.9) in the space of eigenvalues of σ_*. The inequality (11.2.10) reflects the irreversibility of homogenization: The difference between eigenvalues monotonically decreases during homogenization.

Remark 11.2.2 *The described G-closure is an example of a G-closure with empty interior; see Chapter 3. Equality (11.2.9) is an example of the exact relations; all composites satisfy this equality, independently of their structure. A discussion on the exact relations can be found in (Grabovsky, Milton, and Sage, 1999). A more general problem of modeling two-dimensional polycrystal was considered in (Clark and Milton, 1994).*

11.2.4   Three-Dimensional Isotropic Polycrystal

Let us derive the bounds for isotropic conductivity of a three-dimensional polycrystal. We use the method described in Chapter 10 (see (10.1.16), (10.1.17)). We follow (Avellaneda et al., 1988).

Polycrystal from a Transversal Isotropic Crystallite: Lower Bound

Assume that the crystallite $\boldsymbol{\sigma}_0$ is transversal isotropic,

$$
\boldsymbol{\sigma}_0 = \begin{pmatrix} \sigma_1 & 0 & 0 \\ 0 & \sigma_1 & 0 \\ 0 & 0 & \sigma_2 \end{pmatrix} \quad \text{and} \quad \sigma_1 \le \sigma_2;
$$

and that the polycrystal $\boldsymbol{\sigma}_*$ is isotropic, $\boldsymbol{\sigma}_* = \sigma_* I$. The isotropy of the polycrystal implies that it is assembled from at least three differently oriented fragments of the crystallite and that the directions of σ_2 in these fragments are not coplanar.

The lower bound is given by (11.2.6). We can improve this bound (11.2.6) by the translation method. Let us start with the inequality (10.1.15)

$$
(A_* - T)^{-1} \le \sum_{i=1} \left(A(\Phi_i^T \boldsymbol{\sigma}_0 \Phi_i) - T \right)^{-1} \quad \forall T \in \mathcal{T},
$$

where T is defined in (11.1.9) and Φ_i is a rotation matrix in the ith fragment. Let us compute the set $\mathcal{T}$. Let us choose the parameters t_i equal to each other, which makes T isotropic. The permitted values of t are found from the inequality (compare with (10.1.15))

$$
\begin{pmatrix} \sigma_1 & t & t \\ t & \sigma_1 & t \\ t & t & \sigma_2 \end{pmatrix} \ge 0.
$$

Solving this inequality for t, we find that $t \in [t_0, t_1]$, where

$$
t_0 = \frac{\sigma_2 - \sqrt{\sigma_2(8\sigma_1 + \sigma_2)}}{4}, \quad t_1 = \sigma_1.
$$

An extremal translation t_0 corresponds to an infinite eigenvalue in each matrix $\left(A(\Phi_i \boldsymbol{\sigma}_0 \Phi_i^T) - T \right)^{-1}$ because T is isotropic. The eigenvectors of these singular eigenvalues span the eigenspace of the matrices

$$\sum_{i=1} A(\Phi_i^T \sigma_0 \Phi_i) - T$$

and (10.1.16) becomes

$$U_1(\boldsymbol{\sigma}_*) = \begin{pmatrix} \sigma_* & t_0 & t_0 \\ t_0 & \sigma_* & t_0 \\ t_0 & t_0 & \sigma_* \end{pmatrix} \geq 0$$

(compare with (11.1.10)). Solving this inequality for σ_*, we obtain

$$\sigma_* \geq -2t_0 = \frac{1}{2}\left(\sqrt{\sigma_2(8\sigma_1 + \sigma_2)} - \sigma_2\right). \tag{11.2.11}$$

This bound is exact. We have demonstrated (see (7.3.7)) that a laminate structure of infinite rank realizes this bound.

Remark 11.2.3 *The structure (7.3.7) is not unique. Another construction of optimal microstructures was suggested in (Schulgasser, 1976). It is the coated spheres geometry. Each sphere is made of an axially symmetric anisotropic material; the single eigenvalue of σ_2 is directed to the center of the sphere. Interestingly, that the field is singular in the center of the sphere, if $\sigma_2 > \sigma_1$ or is zero. The verification is left to the reader.*

Generalization. Anisotropic Composites. We could look for a natural generalization of the bounds obtained and find the bounds for an anisotropic polycrystal. This problem was studied in (Avellaneda et al., 1988; Nesi and Milton, 1991; Astala and Miettinen, 1998). The rather straightforward generalization of the translation estimates leads to a surface that restricts G_m -closure from below and passes trough the attainable isotropic point σ_* (11.2.11) and through the curves $\lambda_1\lambda_2 = \sigma_1\sigma_2$, $\lambda_3 = \sigma_3$ and $\lambda_2\lambda_3 = \sigma_2\sigma_3$, $\lambda_1 = \sigma_1$. However, the question of attainability of other points of this surface is open, as is the question of the best anisotropic polycrystal structures. Some results can be found in (Nesi and Milton, 1991; Astala and Miettinen, 1998).

Upper Bound

We may try to improve the upper bound (11.2.6) using translations, but this time the optimal value of the translation parameter is zero (we leave the check to the reader); hence the upper bound (11.2.6) cannot be improved by adding translators. Instead, let us demonstrate that this bound is exact.

An ingenious construction of the bound was found in (Schulgasser, 1976; Schulgasser, 1977); it is applicable for the most general case of fully anisotropic crystallite and polycrystal. Let us choose an arbitrary effective tensor with the eigenvalues $\Lambda = (\lambda_1, \lambda_2, \lambda_3)$ corresponding to the upper bound:

$$\lambda_1 + \lambda_2 + \lambda_3 = \sigma_1 + \sigma_2 + \sigma_3,$$
$$\sigma_1 \le \lambda_1 \le \lambda_2 \le \lambda_3 \le \sigma_3. \tag{11.2.12}$$

The microstructures that realize the effective tensor are constructed as follows. Consider a laminate of differently rotated crystallite fragments σ_1 and σ_2 and direct the normal along the first axis:

$$\boldsymbol{\sigma}_1 = \begin{pmatrix} \sigma_2 & 0 & 0 \\ 0 & \sigma_1 & 0 \\ 0 & 0 & \sigma_3 \end{pmatrix}, \quad \boldsymbol{\sigma}_2 = \begin{pmatrix} \sigma_2 & 0 & 0 \\ 0 & \sigma_3 & 0 \\ 0 & 0 & \sigma_1 \end{pmatrix}, \quad \mathbf{n} = \begin{pmatrix} 1 \\ 0 \\ 0 \end{pmatrix}.$$

The laminate has the effective properties $\boldsymbol{\sigma}_L$:

$$\boldsymbol{\sigma}_L = \begin{pmatrix} \sigma_2 & 0 & 0 \\ 0 & l_1 & 0 \\ 0 & 0 & l_3 \end{pmatrix}, \quad \begin{array}{l} l_1 = c\sigma_1 + (1-c)\sigma_3, \\ l_3 = c\sigma_3 + (1-c)\sigma_2, \end{array}$$

where $l_1, l_2 = \sigma_2$ and l_3 are eigenvalues of $\boldsymbol{\sigma}_L$. Indeed, the crystallite properties in tangent directions are averaged and the properties in the normal direction are the same. Hence we have

$$\boldsymbol{\sigma}_L = c\boldsymbol{\sigma}_1 + (1-c)\boldsymbol{\sigma}_2, \quad \mathrm{Tr}\,\boldsymbol{\sigma}_L = \sigma_1 + \sigma_2 + \sigma_3.$$

Choose the fraction c to fit the equality

$$l_1 = c\sigma_1 + (1-c)\sigma_3 = \lambda_2.$$

Two other eigenvalues are:

$$l_2 = \sigma_2, \quad l_3 = \sigma_1 + \sigma_3 - \lambda_2.$$

Repeat the procedure. Laminate two differently oriented fragments of the obtained composite $\boldsymbol{\sigma}_L$, and direct the normal along the common eigenvalue λ_2:

$$\boldsymbol{\sigma}_L^1 = \begin{pmatrix} \sigma_2 & 0 & 0 \\ 0 & \lambda_2 & 0 \\ 0 & 0 & l_3 \end{pmatrix}, \quad \boldsymbol{\sigma}_2 = \begin{pmatrix} l_3 & 0 & 0 \\ 0 & \lambda_2 & 0 \\ 0 & 0 & \sigma_2 \end{pmatrix}, \quad \mathbf{n} = \begin{pmatrix} 0 \\ 1 \\ 0 \end{pmatrix}.$$

The composite has eigenvalues l_1', $l_2' = \lambda_2$, and l_3', which again have a fixed sum:

$$l_1' + \lambda_2 + l_3' = \sigma_1 + \sigma_2 + \sigma_3.$$

Finally, we can choose the fractions c' of mixed material in the second-rank laminates to obtain any triplet of eigenvalues that satisfy (11.2.12).

11.3 Coupled Bounds

11.3.1 Statement of the Problem

Here we apply the method to a more complicated problem of G_m-closure of a media with two scalar conductivities. For example, the medium can conduct thermal flux and electricity. The effective properties are coupled, because both conductivities are determined by the same microstructure. Therefore, they cannot be completely independent. The coupling opens the possibility of estimating one of the effective properties of a composite by measuring another effective property.

The problem is rich enough to demonstrate general properties of the translation method. At the same time the calculations are relatively easy.

The isotropic component of the bounds for such problems has been found in (Milton, 1981c; Milton, 1981a) using the "analytic method" (see (Bergman, 1978; Milton, 1981c; Milton, 1981a)). The problem was further investigated in (Clark and Milton, 1995) where the anisotropic problem was solved for the general case using the analytic method. The problem of coupled bounds was considered in a number of papers in different settings. We mention (Milgrom and Shtrikman, 1989; Milton, 1980; Milton, 1981b; Cherkaev and Gibiansky, 1993; Gibiansky and Torquato, 1993; Gibiansky and Torquato, 1995b; Milgrom, 1997), and the references therein.

The exact bounds for anisotropic coupled conductivities were obtained in (Cherkaev and Gibiansky, 1992) using the translation method. We follow this paper.

Equations and Notation

Consider a periodic two-dimensional composite of two materials. Each material is characterized by two conductivities s^α and s^β (Greek indices indicate the property, for example, $^\alpha$ is used for thermal conductivity and $^\beta$ is used for electrical conductivity). We consider a composite of two isotropic materials with the properties s_1^α, s_1^β and s_2^α, s_2^β; the volume fractions are m_1 and m_2. It is assumed that the tensor conductivities $\mathbf{s}_i^\alpha = s_i^\alpha I$ and $\mathbf{s}_i^\beta = s_i^\beta I$ are piecewise constant:

$$
\begin{aligned}
\mathbf{s}^\alpha(\mathbf{x}) &= \left(s_1^\alpha \chi_1(\mathbf{x}) + s_2^\alpha \chi_2(\mathbf{x})\right) I, \\
\mathbf{s}^\beta(\mathbf{x}) &= \left(s_1^\beta \chi_1(\mathbf{x}) + s_2^\beta \chi_2(\mathbf{x})\right) I
\end{aligned}
\tag{11.3.1}
$$

where I is the unit matrix and $\chi_i(\mathbf{x})$, $i = 1, 2$, are the characteristic functions of the subdomains occupied by corresponding materials. We call $\mathbf{s}_*^\alpha$ and $\mathbf{s}_*^\beta$ the effective properties of the composite; $\mathbf{s}_*^\alpha$ and $\mathbf{s}_*^\beta$ are symmetric, positive definite 2×2-tensors.

The conductivity equations for the materials are

$$
\begin{aligned}
\mathbf{j}^\alpha &= \mathbf{s}^\alpha \mathbf{e}^\alpha, & \mathbf{e}^\alpha &= \nabla w^\alpha, & \nabla \cdot \mathbf{j}^\alpha &= 0, \\
\mathbf{j}^\beta &= \mathbf{s}^\beta \mathbf{e}^\beta, & \mathbf{e}^\beta &= \nabla w^\beta, & \nabla \cdot \mathbf{j}^\beta &= 0,
\end{aligned}
$$

where $\mathbf{j}^\alpha$ and $\mathbf{j}^\beta$ are divergencefree currents, $\mathbf{e}^\alpha$ and $\mathbf{e}^\beta$ are the curlfree fields.

G_m-Closure

Separate bounds for the effective conductivities $\mathbf{s}_*^\alpha$ and $\mathbf{s}_*^\beta$ can be obtained by applying the results of Section 11.1. Here we obtain coupled bounds of the form $f(\mathbf{s}_*^\alpha, \mathbf{s}_*^\beta) \geq 0$.

Let us discuss the structure of the G_m-closure. It is the set of possible values of the tensor pairs $\mathbf{s}_*^\alpha$ and $\mathbf{s}_*^\beta$. Any pair of symmetric tensors in two dimensions is characterized by five invariant to a rotation parameters. They are: Two pairs of eigenvalues λ_1^α, λ_2^α and λ_1^β, λ_2^β of the tensors $\mathbf{s}_*^\alpha$ and $\mathbf{s}_*^\beta$, respectively, and the angle ω between the eigenvectors of these tensors. It is also convenient to use the rational invariants

$$I_1 = \mathrm{Tr}\, \mathbf{s}_*^\alpha, \qquad I_2 = \mathrm{Tr}\, \mathbf{s}_*^\beta, \quad I_3 = \det \mathbf{s}_*^\alpha,$$
$$I_4 = \det \mathbf{s}_*^\beta, \qquad I_5 = \mathrm{Tr}(\mathbf{s}_*^\alpha \, \mathbf{s}_*^\beta),$$

which are polynomial functions of the tensors' elements. The set of these parameters corresponds to a five-dimensional space. Each microstructure corresponds to a point in this space, and the G_m-closure corresponds to a body in it. Recall that G_m-closure is closed, simply connected, and bounded. The boundary of the G_m-closure is a four-dimensional manifold.

Minimizing Functionals

We use the translation method to find boundaries of the $G_m U$ set. The bounds are based on the minimization of the energy of the periodic cell. The energy can be expressed as a quadratic form of the fields $\mathbf{e}^\alpha$ and $\mathbf{e}^\beta$ or currents $\mathbf{j}^\alpha$ and $\mathbf{j}^\beta$. This leaves several choices for the minimizing functional. Let us discuss heuristic arguments for choosing an appropriate functional. We know that separate bounds for each property are realized by different matrix laminate structures, where either the first or second material serves as inclusions. Both structures belong to the general class of multicoated matrices (MCM). They are denoted as MCM(2,1) and MCM(1,2) .

The appropriate functional should correspond to a general class of structures that degenerates into both MCM(2,1) and MCM(1,2). This observation guides us to the appropriate functional.

The form of the functional depends on the relation between the material constants in the first and second phases. For definiteness, we assume that

$$s_1^\alpha < s_2^\alpha,$$

and we distinguish the case where the constants in the second material satisfy the relationship

$$s_1^\beta - s_2^\beta \geq 0 \qquad (\text{case } \mathbf{A})$$

and the case where these constants satisfy the relationship

$$s_1^\beta - s_2^\beta \leq 0 \qquad \text{(case } \mathbf{B}\text{)}.$$

Consider both cases sequentially.

Case $\mathbf{A}$. Let us demonstrate that the appropriate functional for the bound is

$$J_1 = \int_S \mathbf{z}_1^T A_1 \mathbf{z}_1, \tag{11.3.2}$$

where eight-dimensional vector $\mathbf{z}_1$ and the 8×8 matrix A_1 have the following block representation:

$$\mathbf{z}_1 = \begin{pmatrix} \mathbf{e}_1^\alpha \\ \mathbf{e}_2^\alpha \\ \mathbf{e}_1^\beta \\ \mathbf{e}_1^\beta \end{pmatrix}, \quad A_1 = \begin{pmatrix} \mathbf{s}^\alpha & 0 & 0 & 0 \\ 0 & \mathbf{s}^\alpha & 0 & 0 \\ 0 & 0 & \mathbf{s}^\beta & 0 \\ 0 & 0 & 0 & \mathbf{s}^\beta \end{pmatrix},$$

where $\mathbf{s}^\alpha$ and $\mathbf{s}^\beta$ are determined by (11.3.1).

The minimization of J_1 produces a component of the boundary of the G_m-closure. If the β-fields vanish, $\mathbf{e}_{10}^\beta = \mathbf{e}_{20}^\beta = 0$, the solution to the variational problem is attained on the structures MCM(2,1), because $s_1^\alpha < s_2^\alpha$. If the α-fields vanish, $\mathbf{e}_{10}^\alpha = \mathbf{e}_{20}^\alpha = 0$, then the solution corresponds to MCM(1,2) (because $s_2^\beta < s_1^\beta$). In the general case, the bound solution connects the manifolds that correspond to the structures MCM(1,2) and MCM(2,1). We demonstrate later that the optimal structures for this problem are multicoated spheres MCM(1,2,1) (see Chapter 7), which obviously include both MCM(1,2) and MCM(2,1).

Case $\mathbf{B}$. Consider now the functional (11.3.2) in case $\mathbf{B}$. This time, both limits $\mathbf{e}_{10}^\alpha = \mathbf{e}_{20}^\alpha = 0$ and $\mathbf{e}_{10}^\beta = \mathbf{e}_{20}^\beta = 0$ correspond to MCM(2,1) (because $s_1^\beta \leq s_2^\beta$. We expect (and we could prove) that in the general case the solution to the minimization problem for this functional corresponds to MCM(2,1), i.e., to the edge of the G_m-closure.

Remark 11.3.1 *This functional (11.3.2) in case $\mathbf{B}$ could be compared to the functional equal to the sum of the energy and the complementary energy in the problem of a conducting composite (Section 11.1). This functional always corresponds to laminates as optimal structures. Laminates form a component of the G_m-closure, but the component of the low dimensionality or a vertex. Similarly, (11.3.2) corresponds to a component of the G_m-closure of low dimensionality.*

The appropriate functional for case $\mathbf{B}$ is

$$J_2 = \int_S \mathbf{z}_2^T A_2 \mathbf{z}_2,$$

where

$$\mathbf{z}_2 = \begin{pmatrix} \mathbf{e}_1^\alpha \\ \mathbf{e}_2^\alpha \\ \mathbf{j}_1^\beta \\ \mathbf{j}_1^\beta \end{pmatrix}, \quad A_2 = \begin{pmatrix} \mathbf{s}^\alpha & 0 & 0 & 0 \\ 0 & \mathbf{s}^\alpha & 0 & 0 \\ 0 & 0 & \frac{1}{\mathbf{s}^\beta} & 0 \\ 0 & 0 & 0 & \frac{1}{\mathbf{s}^\beta} \end{pmatrix}.$$

Indeed, the solution to the minimization problem for this functional provides the appearance of the matrix composites of either type MCM(2,1) or MCM(1,2), in the asymptotics where $\mathbf{e}_1^\alpha = \mathbf{e}_2^\alpha = 0$ or $\mathbf{j}_1^\beta = \mathbf{j}_2^\beta = 0$, respectively. This follows from the previous arguments.

Conjugate Functionals

By minimization of the functional J_1 or J_2, we determine one component of the G_m-closure boundary. To obtain the complementary component we consider the dual variational problems of minimization of the conjugate to the J_1 or J_2 functionals. The functional conjugate to J_1 (case **A**) is

$$J_3 = \int_S \mathbf{z}_3^T A_3 \mathbf{z}_3,$$

where

$$\mathbf{z}_3 = \begin{pmatrix} \mathbf{j}_1^\alpha \\ \mathbf{j}_2^\alpha \\ \mathbf{j}_1^\beta \\ \mathbf{j}_1^\beta \end{pmatrix}, \quad A_3 = \begin{pmatrix} \frac{1}{\mathbf{s}^\alpha} & 0 & 0 & 0 \\ 0 & \frac{1}{\mathbf{s}^\alpha} & 0 & 0 \\ 0 & 0 & \frac{1}{\mathbf{s}^\beta} & 0 \\ 0 & 0 & 0 & \frac{1}{\mathbf{s}^\beta} \end{pmatrix}.$$

Similarly, in case **B** we consider the functional conjugate to J_2:

$$J_4 = \int_S \mathbf{z}_4^T A_4 \mathbf{z}_4,$$

where

$$\mathbf{z}_4 = \begin{pmatrix} \mathbf{j}_1^\alpha \\ \mathbf{j}_2^\alpha \\ \mathbf{e}_1^\beta \\ \mathbf{e}_1^\beta \end{pmatrix}, \quad A_4 = \begin{pmatrix} \frac{1}{\mathbf{s}^\alpha} & 0 & 0 & 0 \\ 0 & \frac{1}{\mathbf{s}^\alpha} & 0 & 0 \\ 0 & 0 & \mathbf{s}^\beta & 0 \\ 0 & 0 & 0 & \mathbf{s}^\beta \end{pmatrix}.$$

The conjugate functionals J_3, J_4 are the Legendre transforms of the primary functionals J_1, J_2, respectively.

11.3.2 Translation Bounds of G_m-Closure

The Structure of the Translators

We apply the translation method to this problem. Again we examine the basic inequality (10.1.6). All four functionals are minimized by the same procedure; see (Cherkaev and Gibiansky, 1992). We demonstrate the procedure to bound J_1.

Let us determine the translations. The functional J_1 deals with four curlfree fields $\mathbf{e}_1^\alpha$, $\mathbf{e}_2^\alpha$, $\mathbf{j}_1^\beta$, and $\mathbf{j}_s^\beta$. Each pair of these fields corresponds to the translator of the type $\mathbf{e}_1^\alpha \cdot R\,\mathbf{e}_2^\alpha$ and $\mathbf{j}_1^\alpha \cdot R\mathbf{j}_2^\alpha$, where R (4.4.8) is the tensor of rotation through a right angle.

A linear combination of these translators corresponds to an 8×8 translation matrix in block form:

$$
T(t_i) = \begin{pmatrix}
0 & 0 & t_1 R & t_3 R \\
0 & 0 & -t_3 R & t_1 R \\
t_2 R^T & -t_3 R^T & 0 & 0 \\
t_3 R^T & t_2 R^T & 0 & 0
\end{pmatrix}.
$$

Parameters of the Optimal Translator

Inequality (10.1.6) depends on the choice of parameters t_i. We choose these parameters to bring $A_i - T(t_i)$ to the boundary of the permitted domain $A_i - T(t_i) \geq 0$. This choice minimizes the sum of the ranks of the matrices $A_i - T$

$$
\min_t \ \{\mathrm{rank}(A_1 - T(t)) + \mathrm{rank}(A_2 - T(t))\}.
$$

One can check that $\det(A_i - T(t)) = H^2$ where

$$
H = \left(t_1^2 - (s_i^\alpha)^2\right)\left(t_2^2 - \left(s_i^\beta\right)^2\right) - 2\left(s_i^\alpha s_i^\beta - t_1 t_2\right) t_3^2 + t_3^4.
$$

To reduce the rank of $A_1 - T(t)$ let us choose the parameters of translation

$$
\begin{aligned}
t_1 &= -\cos(\theta)s_1^\alpha, \\
t_2 &= \cos(\theta)s_1^\beta, \\
t_3 &= -\sin(\theta)\sqrt{s_1^\alpha s_1^\beta},
\end{aligned}
\tag{11.3.3}
$$

where θ is an arbitrary parameter. We can see that $\mathrm{rank}(A_1 - T(t_i)) = 4$ for any θ. The projector N_1 onto the null subspace of the matrix $A_1 - T$ is

$$
N_1 = \begin{pmatrix}
0 & 0 & 0 & 1 & -\nu\csc\theta & 0 & 0 & -\nu\tan\theta \\
0 & 0 & 1 & 0 & 0 & \nu\csc\theta & -\nu\tan\theta & 0 \\
0 & 1 & 0 & 0 & 0 & -\nu\tan\theta & -\nu\csc\theta & 0 \\
1 & 0 & 0 & 0 & -\nu\tan\theta & 0 & 0 & -\nu\csc\theta
\end{pmatrix},
$$

where $\nu = \sqrt{\dfrac{s^\alpha}{s^\beta}}$.

Next we choose θ to decrease the rank of $A_2 - T(t)$. One can check that $\mathrm{rank}(A_2 - T(t_i)) = 6$ when

$$
\cos(\theta) = \left(\frac{s_1^\alpha s_1^\beta - s_2^\alpha s_2^\beta}{s_1^\alpha s_2^\beta - s_2^\alpha s_1^\beta}\right)^2.
$$

Notice that the absolute value of the right-hand side is less than one for the considered case.

The projector N_2 onto the null subspace of the matrix $A_2 - T$ is

$$N_2 = \begin{pmatrix} 0 & 0 & 0 & 1 & -\nu\csc\theta & 0 & 0 & -\nu\tan\theta \\ 0 & 0 & 1 & 0 & 0 & \nu\csc\theta & -\nu\tan\theta & 0 \end{pmatrix}.$$

Notice that the parameters of the optimal translator depend on the properties of both materials.

The Bound

The projections onto N_1 or N_2 make the inequality (10.1.6) trivial because the right-hand side of (10.1.6) becomes infinitely large.

To obtain the translation bound, we project the matrix inequality (10.1.6) onto a subspace orthogonal to N_1 and N_2. The subspace orthogonal to N_1 and N_2 is spanned by two linearly independent vectors $\mathbf{c}_1$ and $\mathbf{c}_2$, which are obtained from the conditions

$$\mathbf{c}_k \cdot N_1 = \mathbf{c}_k \cdot N_2 = 0, \quad k = 1, 2.$$

The bound takes the form

$$\mathbf{C}^T (A_* - T)^{-1}\mathbf{C} \geq m_1 \mathbf{C}^T (A_1 - T)^{-1}\mathbf{C} + m_2 \mathbf{C}^T (A_2 - T)^{-1}\mathbf{C},$$

where $\mathbf{C} = (\mathbf{c}_1 \oplus \mathbf{c}_2)$ is the 2×8 projector. The results of the calculation of bounds are discussed in the next section.

Y-Transform

The calculation of the bound is simpler when the Y-transform is used. This technique was used in (Cherkaev and Gibiansky, 1992) to derive the bounds.

The matrices A_1 and A_2 commute. Therefore, the bound (10.1.9) becomes:

$$Y(A_*) + T \geq 0.$$

The scalar inequality (see (10.1.10))

$$\Delta = \det(Y(A_*) + T) \geq 0$$

is satisfied in the G_m-closure and is satisfied as an equality at its boundary.

Transforming Δ (see the details in (Cherkaev and Gibiansky, 1992)), we bring it to the form

$$\Delta' = \begin{pmatrix} (d_1 - t_1^2) & t_3^2 \\ (2t_1 t_2 - t_3^2 - d_3) & (d_2 - t_2^2) \end{pmatrix} \geq 0 \tag{11.3.4}$$

where

$$d_1 = \det Y(\mathbf{s}_*^\alpha), \quad d_2 = \det Y(\mathbf{s}_*^\beta), \quad d_3 = \det\left(Y(\mathbf{s}_*^\alpha)\,R\,Y(\mathbf{s}_*^\beta)\right), \tag{11.3.5}$$

and the translation parameters t_1, t_2, t_3 are defined in (11.3.3). The decoupled bounds require the nonnegativeness of the diagonal elements of Δ'. The coupled bound require the positiveness of its determinant of Δ'.

The dual bound is derived in the same way. Case **B** is considered similarly.

The Results

The translation bound (11.3.4) implies the following inequalities. The pair of the effective tensors $\mathbf{s}_*^\alpha, \mathbf{s}_*^\beta$ belong to the G_m-closure if and only if the following conditions hold:

1. The effective tensors $\mathbf{s}_*^\alpha$ and $\mathbf{s}_*^\beta$ belong to their G_m-closures

$$\det Y(\mathbf{s}_*^\alpha) \in \left[(s_1^\alpha)^2, \ (s_2^\alpha)^2 \right],$$

$$\det Y(\mathbf{s}_*^\beta) \in \left[\left(s_1^\beta\right)^2, \ \left(s_2^\beta\right)^2 \right].$$

$$(11.3.6)$$

2. The following coupled inequalities are valid ($i = 1, 2$):

$$\frac{\det\left(\frac{Y(\mathbf{s}_*^\alpha)}{s_i^\alpha} - \frac{Y(\mathbf{s}_*^\beta)}{s_i^\beta} \right)}{\left(1 - \det\left(\frac{Y(\mathbf{s}_*^\alpha)}{s_i^\alpha} \right)\right)\left(1 - \det\left(\frac{Y(\mathbf{s}_*^\beta)}{s_i^\beta} \right)\right)} \leq c \qquad (11.3.7)$$

and

$$\mathbf{s}_1^\alpha \mathbf{s}_1^\beta \leq Y(\mathbf{s}_*^\alpha)\, R\, Y(\mathbf{s}_*^\beta)\, R^T \leq \mathbf{s}_2^\alpha \mathbf{s}_2^\beta. \qquad (11.3.8)$$

Here the constant c is

$$c = \frac{\left(s_1^\alpha s_2^\beta - s_2^\alpha s_1^\beta \right)^2}{\left((s_1^\alpha)^2 - (s_2^\alpha)^2 \right)\left(\left(s_1^\beta\right)^2 - \left(s_2^\beta\right)^2 \right)},$$

and it is assumed that the materials are ordered so that $s_1^\alpha s_1^\beta \leq s_2^\alpha s_2^\beta$ (in case **A**) and $\frac{s_1^\alpha}{s_1^\beta} \leq \frac{s_2^\alpha}{s_2^\beta}$ (in case **B**).

Optimal Structures

In both cases **A** and **B**, the G_m-closure boundary corresponds to a set of pairs of the effective properties tensors $(\mathbf{s}_*^\alpha, \ \mathbf{s}_*^\beta)$ of the structures of multi-coated matrices of the type MCM(1,2,1) or MCM(2,1,2) (see Figure 7.5). The internal points of the $G_m U$ set correspond to their composites. Recall that the structures MCM(1,2,1) are the matrix laminates of the first material that forms an envelope and a composite that forms the nucleus.

This composite is also a matrix laminate, but the second material forms the envelope, and the first material the nucleus (Chapter 7).

A straight but long computation demonstrates that every MCM(1,2,1) and MCM(2,1,2) structure belongs to the complementary component of the boundary of the G_m- closure. Together, the set of these structures and their degenerate cases form a closed surface in the five-dimensional space of invariants of the pair of effective tensors, that is, the boundary of the G_m-closure.

We mention the following properties of the MCM(1,2,1):

1. The effective tensor of MCM(1,2,1) is controlled by four invariant parameters. Two parameters α, β determine the degree of anisotropy of the first and second envelopes, respectively. The third parameter γ controls the arrangement of the first material between inclusions and the external envelope, and the fourth parameter $\mathbf{n}_1 \cdot \mathbf{n}_2$ describes the angle of declination of the normals in the first envelope to the normals of the second envelope. These four parameters correspond to the dimensionality of the boundary of the five-dimensional G_m-closure.

2. The structures MCM(1,2,1) degenerate into structures MCM(1,2) and MCM(1,2). They play the role of vertices of the five-dimensional G_m-closure. The other four-dimensional component of the boundary, MCM(2,1,2), meets MCM(1,2,1) at these structures.

3. The fields in the structure are piecewise constant. The first material is presented in the central nuclei and the second-rank exterior envelope, which gives three independent values of the field in it. The second material $A_2 - T$ forms the second-rank envelope, which gives two different values of the field in it. Recall that the translation bound admits the piecewise constancy of the fields $\mathbf{v}$ in the structure (Chapter 8). The number of different values of the field in a material agrees with the dimensionality of the null space of the matrices $A_1 - T$ and $A_2 - T$.

Special Cases

There are two special cases that correspond to the G_m-closure with an empty interior, or to exact relations between the effective properties:

1. If the constants of the initial materials are related as follows

$$\frac{s_1^\alpha}{s_1^\beta} = \frac{s_2^\alpha}{s_2^\beta} = k, \tag{11.3.9}$$

then the effective tensors $\mathbf{s}_*^\alpha$ and $\mathbf{s}_*^\beta$ satisfy the equality

$$\mathbf{s}_*^\alpha = k\mathbf{s}_*^\beta$$

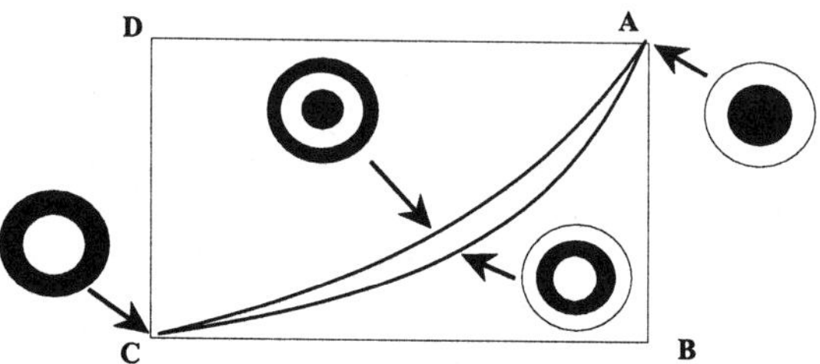

FIGURE 11.4. Coupling: The isotropic component of the G_m-closure and the optimal structures.

and the inequalities (11.3.6).

This equality is physically obvious. Indeed, the relationship (11.3.9) means that both properties of components differ only by a scale coefficient.

2. If the constants of the initial materials are related as follows

$$s_1^\alpha s_1^\beta = s_2^\alpha s_2^\beta = v^2,$$

then the effective tensors $\mathbf{s}_*^\alpha$ and $\mathbf{s}_*^\beta$ satisfy the equality

$$\mathbf{s}_*^\alpha R \mathbf{s}_*^\beta R^T = v^2 I \qquad (11.3.10)$$

and the inequalities (11.3.6).

The equality (11.3.10) is an extension to the anisotropic case of the known relationship obtained in (Keller, 1964; Dykhne, 1970; Dykhne, 1971; Mendelson, 1975). This equality is one more example of the discussed "exact relations" between effective coefficients.

Isotropy. The bounds are simplified when the resulting composite is isotropic. In this case, the effective properties are $\mathbf{s}_*^\alpha = s_0^\alpha I$, $\mathbf{s}_*^\beta = s_0^\beta I$; the invariants (11.3.5) are related as $d_3 = d_1 d_2$; and the bound (11.3.4) becomes

$$(Y(s_*^\alpha) + t_1)(Y(s_*^\beta) + t_2) - t_3^2 \geq 0.$$

The boundaries of the isotropic section of the G_m-closure correspond to the isotropic MCM(1,2,1) or MCM(2,1,2) structures in which $\alpha = \beta = \frac{1}{2}$. Due to isotropy, the effective properties do not depend on the parameter $\mathbf{n}_1 \cdot \mathbf{n}_2$. The structure is controlled by the only parameter γ that varies from 0 to 1. The boundaries of this interval (see Figure 11.4) correspond to the values $\gamma = 0$ and $\gamma = 1$, i.e., to the isotropic matrix composites MCM(1,2) and MCM(2.1).

These isotropic structures are equivalent to the double-coated matrices MCM(1,2,1) and MCM(2,1,2); the isotropic component of the boundary of G_m-closure is also realized by the double-coated spheres introduced in (Milton, 1981c).

Generalization

The obtained results enable us to solve a more general problem of the G_m-closure of an equilibrium described by a coupled elliptic system

$$\nabla \cdot \sum_{j=1}^{n} A_{ij}(\chi)\nabla w_j = f_i, \quad i = 1,\ldots,n. \tag{11.3.11}$$

This time each of the materials may have several properties, and they may be coupled.

This system can be reduced to a decoupled system and investigated by the described method. To reduce the coupled system to a decoupled system, one introduces new potentials $\hat{\mathbf{w}} = G\mathbf{w}$, where G is the appropriate linear transformation of the potentials $\mathbf{w} = [w_1,\ldots w_n]$ (Cherkaev and Gibiansky, 1992). This is always possible because both matrices $\{A_{ij}(0)\}$ and $\{A_{ij}(1)\}$ are positive definite. Therefore, both quadratic forms of the energy of the first and second materials can be simultaneously diagonalized (see, for example, (Strang, 1976)). The transformed decoupled system has the form (11.3.11), where $\hat{A}_{ij} = \hat{A}_i\delta_{ij}$. Hence, the effective coefficients have the form $\hat{A}^*_{ij} = \hat{A}_i^{\;*}\delta_{ij}$; in other term, the transformed effective tensors are uncoupled for all microstructures.

This diagonal form of $\hat{A}^*_{ij}$. implies certain equalities between the effective coefficients $A^*_{ij} = G\hat{A}^*_{ij}G^{-1}$. Again, we observe that the G_m-closure of the set A_{ij} has the empty interior or that there are some exact relations between the coefficients of the effective tensors. This phenomenon was discussed in (Milgrom and Shtrikman, 1989; Cherkaev and Gibiansky, 1992; Milgrom, 1997; Grabovsky and Milton, 1998).

11.3.3 The Use of Coupled Bounds

Indirect Measurements

The inequalities derived suggest the use of indirect measurements of the composite properties. The effective properties are linked through the structure of a composite. Therefore, by measuring one of the properties, we constrain the class of microstructures and therefore constrain the range of the variation of the other property.

For example, one could measure the effective electrical conductivity of a sample and make a conclusion about the range of its porosity.

Assume that the composite is isotropic only with respect to the measured property $\mathbf{s}^\beta_* = s^\beta_0 I$. In this case the general bounds (11.3.7), (11.3.8) for the unknown 2×2-tensor $\mathbf{s}^\alpha_*$ takes the form

$$\frac{A_i}{B_i} \leq c \quad i = 1, 2 \tag{11.3.12}$$

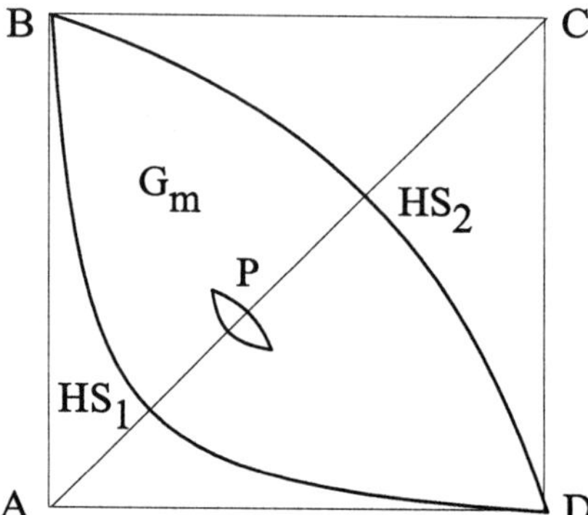

FIGURE 11.5. Coupling: Domain P show the range of the eigenvalues of the conductivity tensor $\mathbf{s}_*^\alpha$ if the other conductivity tensor $\mathbf{s}_*^\beta$ is fixed to be isotropic.

where

$$
A_i = \frac{y^2 \left(s_0^\beta\right)}{\left(s_i^\beta\right)^2} - \frac{y(s_0^\beta)\, \mathrm{Tr}\, Y(\mathbf{s}_*^\alpha)}{s_i^\alpha s_i^\beta} + \frac{\det Y(\mathbf{s}_*^\alpha)}{(s_i^\alpha)^2},
$$

$$
B_i = \left(1 - \left(\frac{y(s_0^\beta)}{s_i^\beta}\right)^2\right)\left(1 - \det\left(\frac{Y(s_0^\alpha)}{s_i^\alpha}\right)\right).
$$

These bounds describe the set of admissible values of the tensor $Y(\mathbf{s}_*^\alpha)$ if the value of s_0^β is fixed (see Figure 11.5). Notice that these bounds are linear-fractional functions of the eigenvalues of $\mathbf{s}_*^\alpha$.

Structural Parameter

Many problems of material science and rock mechanics require an estimation of effective properties of a composite of a special geometrical type. For example, it may be a priori known that one of the phases plays the role of inclusions. The shapes and disposition of those inclusions may randomly and significantly vary, but visually one easily recognizes this type of composite; see Figure 11.6. The problem of how to describe the effective behavior of such composites arises. Usually, the corresponding geometrical constraints are difficult to formulate even in geometrical terms, and even more difficult to formulate in terms of acting fields. Different approaches to this problem are developed in (Torquato and Rubinstein, 1991; Bruno, 1991) using the idea of "secured spheres."

Here we suggest dealing with this problem by introducing an additional structural parameter of the composite. We use the idea of coupled bounds. Consider a composite of two materials σ_1 and σ_2 that are presented in the composite with volume fractions m_1 and m_2. Phase σ_1 forms the envelope

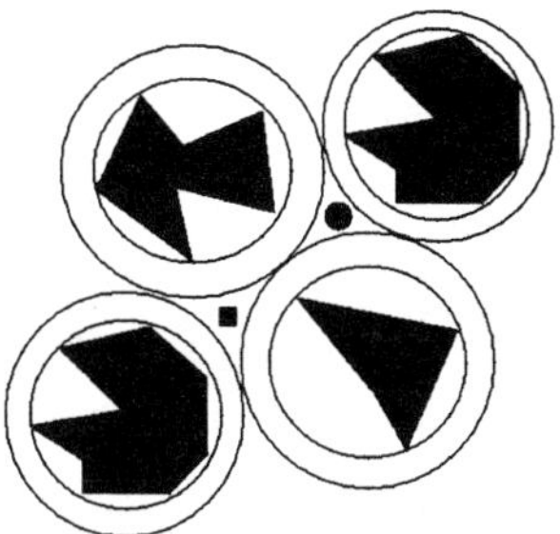

FIGURE 11.6. Geometry of the "secured spheres." The relative radius of a sphere determines the structural parameter κ.

and phase σ_2 forms the inclusions. Suppose that κ is the relative volume of a "secured circle" in which the inclusions are inscribed. This volume must be greater than the volume fraction m_1, $m_1 \leq \kappa \leq 1$.

Let us include the geometrical constraints into the translation method. Consider a composite with the same geometrical configuration filled with a phase with unit conductivity $s_1 = 1$ (envelope) and superconducting inclusions $s_2 = \infty$. The effective conductivity s_* of any composite of this kind is restricted from below by Hashin–Shtrikman bounds, which take the form:

$$\frac{2}{s_* + 1} \geq m_1.$$

This bound corresponds to the geometry of the coated spheres, which at obviously belongs to the class of permitted geometries.

The geometry of the secured circles, together with ideal nature of the material in the inclusion, yields to another nontrivial inequality: The effective conductivity s_* is restricted *from above* by the inequality

$$\frac{2}{s_* + 1} \leq \kappa.$$

Indeed, the maximal conductivity corresponds to the geometry in which the superconducting phase occupies the outside annulus in the secured circles so that the conductivity of the whole secured circle is equal to infinity. This extremal structure is also a coated spheres geometry, but the volume fraction of the material in the inclusion is replaced by the parameter of the secured spheres.

Note that the effective conductivity s_* is still finite, because inclusions do not percolate. The parameter s_* can be treated as a new structural parameter that measures the distance from the percolation point.

To find the bound, we use the formula (11.3.12) for the coupled properties in the following setting:

$$s_1 = 1, \quad s_2 = \infty, \quad s_* = \frac{2}{\kappa} - 1.$$

We leave the derivation of the final formulas to the reader.

11.4   Problems

1. Derive the "dual" bounds (11.1.14), using the same technique as for the primary bounds.

2. Prove that the cross-section of the three-dimensional G_m-closure by the plane $\sigma_3 = \sigma_a$ gives the two-dimensional G_m-closure.

3. Check that the intersection of two boundary components of the G-closure corresponds to the laminates.

4. Prove the statement in Remark 11.2.1.

5. Consider the G_m-closure for two anisotropic materials with fixed codirected eigenvectors. Derive the bounds.

6. Derive the formulas for the conductivity of the secured spheres.

12
Multimaterial Composites

In this chapter, we consider the G_m-closure problems for multiphase conducting composites. These problems are much richer than those for two-phase composites. The topology of optimal microstructures is more diverse and less systematized. In the first section we demonstrate the variety of optimal multicomponent structures, and in the second section we develop a systematic approach to these problems based on necessary conditions.

Optimal multicomponent composites were intensively investigated during the past decade. Mathematically, the problem of optimal multiphase composites can be formulated as the problem of relaxation of a multiwell energy (Firoozye and Kohn, 1993). Several approaches to the problem and some examples of optimal structures can be found in (Hashin and Shtrikman, 1963; Milton, 1981d; Lurie and Cherkaev, 1985; Milton, 1987; Milton and Kohn, 1988; Zhikov, 1991b; Milton and Nesi, 1999; Smyshlyaev and Willis, 1999). The complex conductivity of multiphase composites was studied in (Golden and Papanicolaou, 1985; Golden, 1986; Milton, 1987; Milton and Golden, 1990; Astala and Miettinen, 1998). Recently, Nesi obtained a set of geometrically independent bounds on the conductivity of multiphase two-dimensional composites that improve on the Hashin–Shtrikman results (Nesi, 1995) by the new "weighted" translation method.

The paper (Cherkaev, 1999) develops the necessary conditions approach to the problem. A new approach was suggested (Gibiansky and Sigmund, 1998) that contains a combination of the analytical and numerical methods. However, the whole picture of optimal multicomponent structures is still not completely clear.

The Use of Multicomponent Composites. An understanding of optimal multiphase composites is important for many theoretical and practical applications. Indeed, the majority of natural and artificial composites involve more than two materials, especially those that show unusual properties.

In structural optimization, knowledge of the geometry of two-component optimal composites allows us to solve problems of topology optimization (see (Bendsøe, 1995) for a detailed discussion). Topology optimization corresponds to the special case where one of the mixed "materials" is void. Meaningful consideration of an optimal design of even two-component composites requires knowledge of the geometry of optimal multicomponent composites, because "void" is an always available additional "material" that can be added to the list of given materials.

Optimal Topology. From a physical viewpoint, the problem of the optimal composite of two conducting materials has an intuitively expected solution: To find the best periodic composite assembled from the "good" and "bad" materials with prescribed volume fractions, one must place the "bad" material into compact nuclei surrounded by a continuum of the "good" material that forms the connected construction. This topology increases the influence of the good material and reduces the disadvantages of the bad one. Of course, it is a long way from this intuitive idea to a mathematically rigorous description of optimal microstructures, but the optimal topology is clear from the beginning. The variety of known optimal solutions: the already introduced high-rank laminates and coated spheres or coated ellipsoids (Milton, 1980), as well as the "truly" periodic structures by Vigdergauz (Vigdergauz, 1989) (see below, Chapter 15)–demonstrate different realizations of this intuitive idea.

On the other hand, optimal multicomponent composites can hardly be treated using similar intuitive ideas. The question of where to place a material with intermediate properties is more delicate, and an intuitive answer is not so clear. As we will demonstrate, the optimal topology depends not only on the ordering of the material's properties, but also on the volume fractions of the materials.

Optimality Conditions and Optimal Structures. Sufficient optimality conditions have the form of classical Wiener bounds, Hashin–Shtrikman bounds, or translation bounds; generally, they do not lead to exact bounds for multiphase composites. These conditions a priori parametrize the bounds and are adjusted to a special form of the Lagrangians.

It is hard to speculate on an optimal structure if the sufficient conditions do not lead to exact inequalities. Besides, the problem of an optimal multicomponent composite depends on a large number of prescribed parameters: properties of materials and their fractions. This makes straightforward optimization in a set of arbitrarily chosen laminates ineffective. One needs an a priori hint of optimal topology and a corresponding factorization of the set of "suspicious" structures.

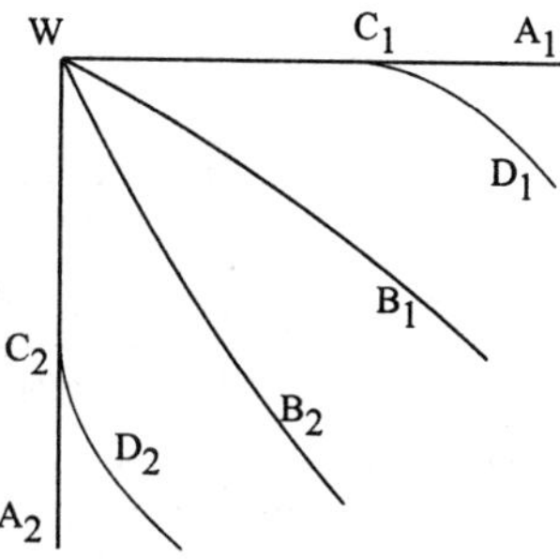

FIGURE 12.1. Attainability of the boundary of the Wiener box $A_1C_1WC_2A_2$. The vertex W corresponds to the simple laminate structures. The curve B_1WB_2 corresponds to the G_m-closure of two materials. The curve $D_1C_1WC_2D_2$ is attainable by the third-rank laminates of three materials.

12.1 Special Features of Multicomponent Composites

Here we demonstrate various optimal multicomponent structures and discuss their common properties. We emphasize the difference between two- and multicomponent structures.

12.1.1 Attainability of the Wiener Bound

Consider again a two-dimensional conducting multicomponent composite made from n isotropic components, where $n \geq 3$. Its eigenvalues λ_A, λ_B[1] satisfy the Wiener bound $\sigma_h \leq \lambda_A$, $\lambda_B \leq \sigma_a$, where σ_h, σ_a are the harmonic and arithmetic means of the conductivity of phases, respectively. The Wiener box corresponds to the smallest rectangle in the λ_A, λ_B plane, which contains the G_m-closure, because two of its corners, (σ_h, σ_a) and (σ_a, σ_h), correspond to laminates.

The two-material G_m-closure has the following feature: The equality $\lambda_A = \sigma_a$ implies that $\lambda_B = \sigma_h$ and vice versa (see Figure 12.1, the curve B_1WB_2). In other words, the G_m-closure set and the Wiener box have only the corner points in common. The picture is different for multicomponent G_m-closures. Following (Cherkaev and Gibiansky, 1996), let us show that a significant part of the boundary of the Wiener box is attainable by some structures, as in Figure 12.1.

More exactly, we show that the equality $\lambda_A = \sigma_a$ implies that $\lambda_B \in [\sigma_h, \lambda_{\max}]$ where $\lambda_{\max}$ is a parameter inside the interval σ_h, σ_a. The G_m-

---

[1]In this chapter, we use the subindices A and B to denote the direction of the orthogonal axes in the $\mathbf{x}$ plane, and we leave the digital subindices to distinguish the materials in the composite.

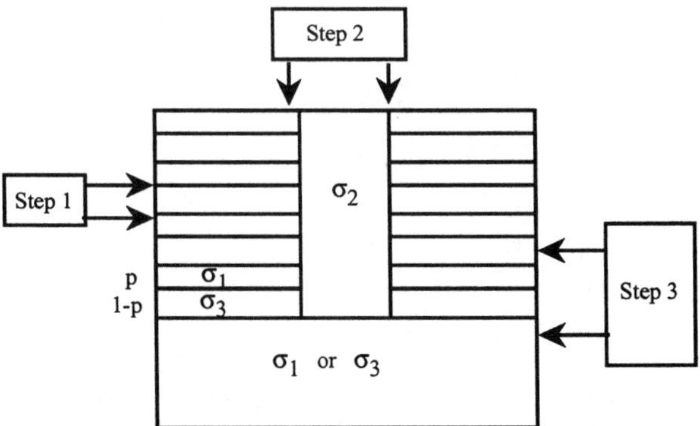

FIGURE 12.2. A structure that belongs to the side of the Wiener box.

closure set has a common component of the boundary with the Wiener box.

Structures

For definiteness, consider three-component structures from materials with conductivities $\sigma_1, \sigma_2, \sigma_3$: $\sigma_1 < \sigma_2 < \sigma_3$. We assume that the volume fractions m_1, m_2, m_3 of the materials are fixed.

Suppose that a structure has the conductivity λ_A equal to the arithmetic mean σ_a of the phase's conductivities in the direction x_A: $\lambda_A = \sigma_a$. The field in such a structure is constant in every point of the structure if the external field is directed along the x_A axis. A laminate structure is an obvious example of such structures: The field along the layers is constant.

If the laminate is submerged into the current field in the orthogonal direction x_B, then the current in the structure is constant everywhere. The effective conductivity λ_B in the x_B -direction is equal to the harmonic mean of the phase's conductivities, $\lambda_B = \sigma_h$.

We are looking for a structure in which the field in the x_A -direction is constant everywhere, because it has the conductivity $\lambda_A = \sigma_a$ in that direction. On the other hand, the current in the direction x_B is not constant in this structure (as in a laminate), because its conductivity in that direction should not be equal to the harmonic mean, $\lambda_B \geq \sigma_h$. Note that there is no such two-component structure, because the boundary of the G_m-closure touches the Wiener box only in the corners.

Let us demonstrate that these structures are special third-rank laminates. The structure is constructed in the following steps:

1. Mix some amounts of the first and third materials in proportions p and $1 - p$ in the laminates parallel to the x_A-axis (see Figure 12.2, step 1). Choose the proportion p such that the conductivity $\lambda_A^{(1)}$ of

this laminate[2] in the direction x_A equals the conductivity of the intermediate material σ_2, i.e.,

$$\lambda_A^{(1)} = p\sigma_1 + (1 - p)\sigma_3 = \sigma_2. \tag{12.1.1}$$

The required proportion p is equal to

$$p = \frac{\sigma_2 - \sigma_3}{\sigma_1 - \sigma_3}.$$

Obviously, $p \in [0, 1]$. The other eigenvalue $\lambda_B^{(1)}$ (in the direction x_B across the laminates) is given by the harmonic mean and is equal to

$$\lambda_B^{(1)} = \left(\frac{p}{\sigma_1} + \frac{1 - p}{\sigma_3}\right)^{-1}.$$

2. At the second step, we use the idea of imitation. We treat the previously obtained composite as a new homogeneous material and mix it with the available amount m_2 of the material σ_2 in the laminates parallel to the x_B-axis (see Figure 12.2, step 2).

Let us calculate the properties of the resulting composite. Both components have the same properties σ_2 and $\lambda_A^{(1)} = \sigma_2$ in the direction x_A. Therefore, the conductivity $\lambda_A^{(2)}$ in the direction x_A (given by the harmonic mean) is equal to σ_2, i.e.,

$$\lambda_A^{(2)} = \left[\frac{\mu}{(\mu + m_2)\lambda_A^{(1)}} + \frac{m_2}{(\mu + m_2)\sigma_2}\right]^{-1} = \sigma_2.$$

Here μ is the amount of the composite of the first and third materials prepared at the first step of the process. We may also say that $\lambda_A^{(2)}$ is equal to the arithmetic mean of the conductivities $\lambda_A^{(1)}$ and σ_2 because the arithmetic and harmonic means of equal quantities trivially coincide:

$$\lambda_A^{(2)} = \frac{\mu}{(\mu + m_2)}\lambda_A^{(1)} + \frac{m_2}{(\mu + m_2)}\sigma_2 = \sigma_2. \tag{12.1.2}$$

The conductivity of the laminate in the other direction x_B is equal to the arithmetic mean of the component's conductivities $\lambda_B^{(1)}$ and σ_2 according to the lamination formula

$$\lambda_B^{(2)}(\mu) = \frac{\mu}{(\mu + m_2)}\lambda_B^{(1)} + \frac{m_2}{(\mu + m_2)}\sigma_2.$$

---

[2]Here and later, the lower index in the notation $\lambda_A^{(k)}, \lambda_B^{(k)}$ denotes the direction; the upper index denotes the rank of the lamination.

This step is the key point of the construction. Indeed, we obtain the composite with effective properties in all directions equal to the arithmetic mean of the properties of the components that enter the process at the second step. We have achieved this by a special choice of the intermediate material prepared at the first step. This material imitates the material σ_2 with respect to the conductivity in the direction x_A. Whereas the arithmetic and the harmonic averages of equal quantities trivially coincide, the resulting effective tensor is the arithmetic mean of the tensors of components.

Remark 12.1.1 *A similar scheme was suggested in (Schulgasser, 1977) for an extremal three-dimensional polycrystal structure. We discussed it in Chapter 11. A similar idea was used in (Milton, 1981d; Lurie and Cherkaev, 1985) to obtain multicomponent isotropic structures. We discuss these examples later in this section.*

3. To finish the construction let us laminate the already obtained amount

$$\nu_2 = \mu + m_2$$

of the described composite with the remaining amounts

$$\nu_1 = m_1 - p\mu, \quad \nu_3 = m_3 - (1-p)\mu \qquad (12.1.3)$$

of the first and third materials, respectively. Now we orient the lamination along the x_A-axis (see Figure 12.2, step 3).

Applying the arithmetic and harmonic mean rules, we find that the conductivity $\lambda_A^{(3)}$ of this composite in the direction x_A is equal to

$$\lambda_A^{(3)} = \nu_1\sigma_1 + \nu_2\lambda_A^{(2)} + \nu_3\sigma_3. \qquad (12.1.4)$$

Using formulas (12.1.1), (12.1.2), and (12.1.4) one can check that the resulting conductivity is given by the arithmetic mean of the initial components:

$$\lambda_A^{(3)} = m_1\sigma_1 + m_2\sigma_2 + m_3\sigma_3. \qquad (12.1.5)$$

This follows from the fact that $\lambda_A^{(3)}$ is a result of three sequential arithmetic averages. Physically, we observe that the external field applied to the described composite along the x_A-axis causes the constant local field. This implies equality (12.1.5).

The other principal conductivity is

$$\lambda_B^{(3)} = \left[\frac{\nu_1}{\sigma_1} + \frac{\nu_2}{\lambda_B^{(2)}} + \frac{\nu_3}{\sigma_3}\right]^{-1} = \lambda_B^{(3)}(\mu),$$

which is equal to the harmonic mean σ_h only if $\mu = 0$; otherwise it lies between σ_h and σ_a. Therefore, we obtain the composite that corresponds to some point of the side of the Wiener box other than its corners.

Effective conductivity of the composite depends on the amount μ of the materials σ_1 and σ_3 involved at the first step of the process. More exactly, the eigenvalue $\lambda_A^{(3)} = \sigma_a$ of the conductivity tensor is independent of μ, but the other eigenvalue $\lambda_B^{(3)}$ depends on it. One can check that $\lambda_B^{(3)}(\mu)$ monotonically increases as μ increases. By changing μ one obtains an interval of attainable points on the side of the Wiener box. The value $\mu = 0$ corresponds to a simple laminate composite, i.e., to the corner points $B = (\sigma_h, \sigma_a)$ or $D = (\sigma_a, \sigma_h)$ of the Wiener box. The maximal value $\mu_{\max}$ corresponds to the other end of the interval (point C_1, Figure 12.1).

The maximum amount μ allowed by this construction is equal to

$$\mu_{\max} = \min\left\{\frac{m_1}{p}, \frac{m_3}{(1-p)}\right\} = \min\left\{m_1\frac{\sigma_1-\sigma_3}{\sigma_2-\sigma_3}, m_3\frac{\sigma_1-\sigma_3}{\sigma_1-\sigma_2}\right\}$$

(see (12.1.3)). Indeed, if

$$\frac{m_1}{p} \le \frac{m_3}{1-p},$$

or, equivalently, if

$$m_1(\sigma_2-\sigma_1) - m_3(\sigma_3-\sigma_2) \le 0, \tag{12.1.6}$$

then μ is restricted by the available amount m_1 of the first phase. In the opposite case,

$$m_1(\sigma_2-\sigma_1) - m_3(\sigma_3-\sigma_2) \ge 0, \tag{12.1.7}$$

$\mu_{\max}$ is restricted by m_3.

The maximum value of the effective conductivity $\lambda_B^{(3)}$ is equal to

$$\lambda_{\max} = \lambda_B^{(3)}(\mu_{\max}) = \begin{cases} \lambda_B^{(3)}(\mu') & \text{if} \quad (12.1.6) \text{ holds,} \\ \lambda_B^{(3)}(\mu'') & \text{if} \quad (12.1.7) \text{ holds.} \end{cases}$$

In summary, we have found the composites corresponding to any point of the intervals

$$\begin{aligned} \lambda_A = \sigma_a, \quad &\lambda_B \in [\sigma_h, \lambda_{\max}]; \\ \lambda_B = \sigma_a, \quad &\lambda_A \in [\sigma_h, \lambda_{\max}], \end{aligned} \tag{12.1.8}$$

on the boundary of the Wiener box (see Figure 12.1).

Dual Extremal Structures

The structures that realize the dual component of the boundary $\lambda_A = \sigma_h$, $\lambda_B \in [\sigma_{\min}, \lambda_a]$ are built in the same way (Cherkaev and Gibiansky, 1996). This time, we choose the fraction p at the first step to make the harmonic mean of the extreme conductivities equal to σ_2. Therefore, the effective conductivity of the composite obtained in the second step is equal to the harmonic mean of the conductivity of the components.

Generalization

The obtained results can also be generalized for the three-dimensional composite assembled of more than two phases. Indeed, the same construction (which exploits the idea of imitation) is directly applicable to the three-dimensional problem. Following the described scheme, one can obtain anisotropic structures that have harmonic mean conductivity in the x_A-direction, arithmetic mean conductivity in an orthogonal x_B-direction, and a conductivity in the third direction that is less than the arithmetic mean. These structures correspond to cylindrical structures with the cross sections identical to the described two-dimensional structures. Similarly, one can find structures that have arithmetic mean conductivity in two orthogonal directions, but with the conductivity in the third direction that belongs to the interval (12.1.8). This enables us to attain all points of the Wiener box in a neighborhood of the corner points $(\lambda_a, \lambda_a, \lambda_h)$, $(\lambda_a, \lambda_h, \lambda_a)$, and $(\lambda_h, \lambda_a, \lambda_a)$.

Remark 12.1.2 *This result says that the class of multicomponent structures, that minimize the sum of energies does not include simple laminates. Indeed, the described structures correspond to a smaller weighted sum of eigenvalues of the effective tensor than the laminate, because one of their eigenvalues is equal to the harmonic mean, and the other is less than the arithmetic mean. Recall that the degenerate problem of minimization of one energy of a multicomponent composite (the example in Chapter 4) has a nonunique solution.*

12.1.2 Attainability of the Translation Bounds

Translation Bound

As we mentioned in Chapter 10, the translation method is applicable to multicomponent composites. The bounds are defined analogously to the two-component bounds. The lower bound has the form:

$$\left[1 + \frac{\sigma_1}{\lambda_A - \sigma_1} + \frac{\sigma_1}{\lambda_B - \sigma_1}\right]^{-1} \geq \sum_{i=2}^{N} m_i \frac{\sigma_i - \sigma_1}{\sigma_i + \sigma_1}, \qquad (12.1.9)$$

where N is the number of materials, and it is assumed that σ_1 is the smallest conductivity. The upper bound has a similar representation, but the phase σ_1 is replaced by σ_N.

The bounds for isotropic composites were obtained by Hashin and Shtrikman, and the anisotropic bounds were obtained in (Milton and Kohn, 1988), where they were called trace bounds. Similar bounds were obtained in (Zhikov, 1986).

We showed in Chapter 10 that the bounds cannot be exact for small values of the volume fraction m_1 of the first phase. However, they are exact for large enough values of m_1.

The Structures

We describe the structures that realize the trace bound following (Milton and Kohn, 1988). This construction for the isotropic components was described earlier in (Milton, 1981d).

Let us divide the amount of the material σ_1 into $n-1$ parts $\mu_1, \ldots, \mu_{n-1}$, where $\mu_1 + \cdots + \mu_{n-1} = m_1$. Consider the matrix laminate from the material σ_1 (envelope) and σ_i, $i = 2, \ldots, n$ (nuclei). The relative fractions of the mixed materials are $\frac{\mu_i}{\mu_i + m_i}$ and $\frac{m_1}{\mu_i + m_i}$, respectively.

Suppose now that the resulting $n-1$ structures have the same effective conductivities λ_A, λ_B (this assumption restricts the value of m_1 from below). Obviously, a mixture of such structures shares the same conductivity no matter what the microstructure of that mixture is.

The eigenvalues λ_A, λ_B of the conductivity of each matrix laminate satisfies the equation

$$N = \left(1 + \frac{\sigma_1}{\lambda_A - \sigma_1} + \frac{\sigma_1}{\lambda_B - \sigma_1}\right) = \left(\frac{m_i}{\mu_i}\right)\left(1 + \frac{2\sigma_1}{\sigma_i - \sigma_1}\right).$$

Multiplying by μ_i and summing over i, we obtain the inequality

$$\sum_{i=2}^{n} \mu_i N = m_1 \left(1 + \frac{\sigma_1}{\lambda_A - \sigma_1} + \frac{\sigma_1}{\lambda_B - \sigma_1}\right) = \sum_{i=2}^{n} m_i \left(1 + \frac{2\sigma_1}{\sigma_i - \sigma_1}\right),$$

which coincides with the bound (12.1.9).

The limiting assumption of the construction is the possibility of assembling composites of the same effective conductivities from the pairs σ_1, σ_i, $i = 2, \ldots, n$. A large amount of σ_1 is needed for this; particularly, the effective conductivities λ_A, λ_B must lie in the interval $[\sigma_1, \sigma_2]$. For a three-component composite, the attainability conditions are (Milton and Kohn, 1988)

$$\sigma_1 \leq \lambda_{\min} \leq \lambda_{\max} \leq \sigma_2, \qquad \frac{\lambda_{\max} - \sigma_1}{\lambda_{\min} - \sigma_1} \leq \frac{\sigma_2}{\lambda_{\min}}, \qquad (12.1.10)$$

where $\lambda_{\max} = \max\{\lambda_A, \lambda_B\}$ and $\lambda_{\min} = \min\{\lambda_A, \lambda_B\}$.

The Variety of Optimal Structures

In contrast with two-material composites, multicomponent optimal structures may correspond to various topologies. Here we describe two optimal isotropic composites that do not look geometrically similar but share the same effective conductivity.

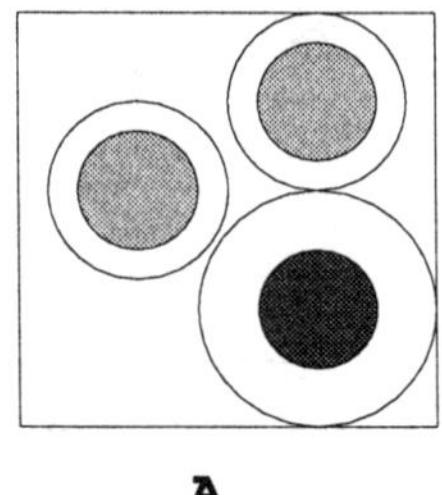 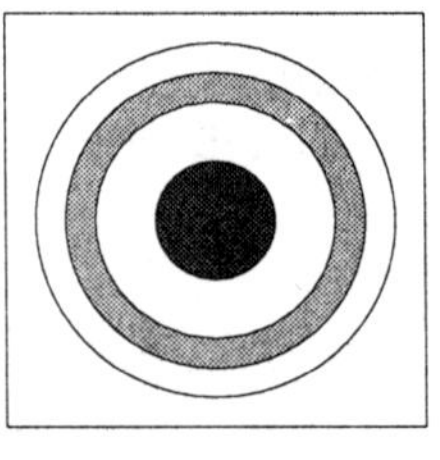

A   **B**

FIGURE 12.3. Isotropic optimal structures: (A) coated spheres of equal conductivity and (B) multicoated spheres.

The scheme suggested in (Milton, 1981d) has already been described. For isotropic composites, one can use the coated spheres construction instead of matrix laminates to mix the pairs of materials, and then mix together the obtained composites with equal isotropic conductivities (see Figure 12.3 A).

The other method was suggested in (Lurie and Cherkaev, 1985), where the following differential scheme was developed to optimize multicomponent composites. Consider the conductivity of an arbitrary multicoated sphere. The conductivity depends on the sequence of materials in the structure or on their layout in annuli around the center of the multicoated sphere. We can treat this layout as a control, as we described in Chapter 7 (the differential scheme). At each infinitesimal step of the construction of a multicoated sphere one decides what material to use to envelop the existing structure. Solving the corresponding optimal control problem (Lurie and Cherkaev, 1985), we find the optimal structure among all possible multicoated spheres. The optimal structure corresponds to the following algorithm (see Figure 12.3 B):

1. Starting with the material σ_n of maximal conductivity, we wrap it into an envelope of σ_1 in the coated sphere geometry. The effective conductivity $\sigma(\mu)$ depends on the added amount μ of σ_1. Here $\sigma(\mu)$ is a continuously decreasing function of μ, and $\sigma(0) = \sigma_n$.

2. At the point μ_{n-1} where this decreasing function reaches the value of σ_{n-1}, $\sigma(\mu_{n-1}) = \sigma_{n-1}$, one adds all the material σ_{n-1} to the composite. Note that the resulting conductivity is still equal to σ_{n-1}, because this material is used in the envelope and σ_{n-1} is the effective conductivity of the nucleus. Again, we observe that the composite in the nucleus imitates the material σ_{n-1} in the envelope.

3. The process is repeated. The material σ_1 is added until the point μ_{n-2} where $\sigma(\mu_{n-2}) = \sigma_{n-2}$; at this point the material σ_{n-2} is added, and so on. One can calculate the effective conductivity of that construction. It is equal to the Hashin–Shtrikman bound.

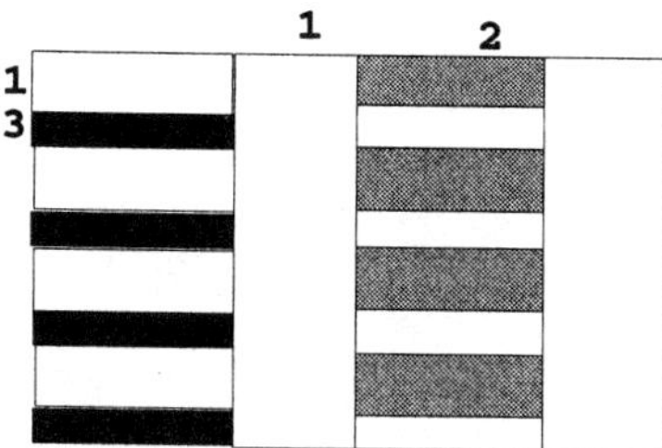

FIGURE 12.4. Optimal three-component structures (large volume fraction of σ_1).

The applicability of this scheme is also restricted: The amount of the material σ_1 must be large enough. This restriction is similar to the one in the previous case. If this amount is smaller, the scheme still gives an answer, but this time effective conductivity can be improved using the nonsymmetric structures (Milton, 1981d); see the next example.

A common feature of these two constructions is a mixture of the microstructures with equal effective properties made from different initial materials. Obviously, this feature does not have an analogue in two-material structures.

Attainability of the "Weighted" Translation Bound

We already mentioned that the translation bounds (12.1.9) cannot be exact for all values of the parameters σ_1, σ_2, σ_3, m_1, m_2, m_3 and that the translation bound of the G_m-closure significantly depends on σ_1, even if the volume fraction of σ_1 tends to zero. When the volume fraction reaches zero, the right-hand side of the translation inequality has a discontinuity, because at this point σ_1 is replaced by σ_2. This discontinuity contradicts the expected continuous dependence of the bounds on the volume fractions; the bound is not exact.

A modification of the translation bounds for two-dimensional conducting composites, suggested in (Nesi, 1995), is free of this defect. To tighten the bounds, an additional inequality is used in the scheme of the translation method (Chapter 8), which is called the *weighted translation method*.

The Bound. Consider a two-dimensional structure that is submerged into two linearly independent external fields, which cause the fields $\mathbf{E} = (\nabla w_1, \nabla w_2)$. Nesi observed and proved that $\psi = \det \mathbf{E}$ never changes sign in the structure. Adding the inequality

$$\det(\nabla w_1, \nabla w_2) \geq 0 \quad \forall \mathbf{x} \in \Omega \tag{12.1.11}$$

to the procedure of the translation method, he ended up with the modified translation bound (Nesi, 1995), see also (Talbot et al., 1995). For the isotropic conductivity σ_*, the upper bound has the form

$$\sigma_* \leq \min_{\lambda \in [\frac{1}{\sigma_1}, \frac{1}{\sigma_N}]} \left(\left(\sum_{i=1}^{N} \frac{m_i}{\frac{1}{\sigma_i} + \max\left\{\frac{1}{\sigma_i}, \lambda\right\}} \right)^{-1} - \lambda \right)^{-1}, \qquad (12.1.12)$$

The lower bound is similar. Notice that the algebraic expressions for the bound depend on the interval of the effective conductivity. For large m_1, when $\sigma_* \in [\sigma_1, \sigma_2]$, it coincides with the translation bound.

Attainability. The modified bound (12.1.12) corresponds to the correct limit when $m_1 \to 0$. However, the bound is generally not attainable if it does not coincide with the translation bound. Indeed, the additional inequality (12.1.11) becomes the active constraints when $\det E = 0$ in the materials σ_i, $i < k$. The vanishing of $\det(\nabla w_1, \nabla w_2)$ implies that the fields ∇w_1 and ∇w_2 are parallel to each other inside a material in the structure, despite the fact that they are caused by linearly independent sources. Consequently, any external field causes the local field to be directed in a fixed direction. Obviously, this condition is too strong, and it is not realizable in a composite of materials with finite positive conductivities.

Optimal Structures. However, there are structures that realize the bound (12.1.12) in the limiting case where

$$\sigma_n = \infty \qquad (12.1.13)$$

and in a limited range of volume fractions of components. These structures were found in (Cherkaev, 1999) by an analysis of the necessary conditions of optimality (see Section 12.2).

Let us describe these structures for the case of three materials. Consider a matrix laminate where the envelope is made of the material σ_2 and the anisotropic inclusion is made of a simple laminate of the ideal material σ_3 (12.1.13) and the best material σ_1; see Figure 12.5.[3] There is no field in the direction along the laminates, because one of the materials is an ideal conductor. Hence, condition (12.1.11) is satisfied for the materials σ_1 and σ_3. The structure in Figure 12.5 has conductivity σ_* of the weighted translation method. The calculations of its effective properties is done in the next section.

Again, these structures are realizable in a limited range of volume fractions. The constraints restrict from below the amount of σ_k. These constraints are: (i) the possibility of obtaining an isotropic composite from second-rank matrix laminates with highly anisotropic nucleus and (ii) constraints similar to (12.1.10); they express attainability of equal effective conductivity of the matrix laminates.

---

[3]If the number of materials is larger than three, the inner layers are made of all materials but σ_2.

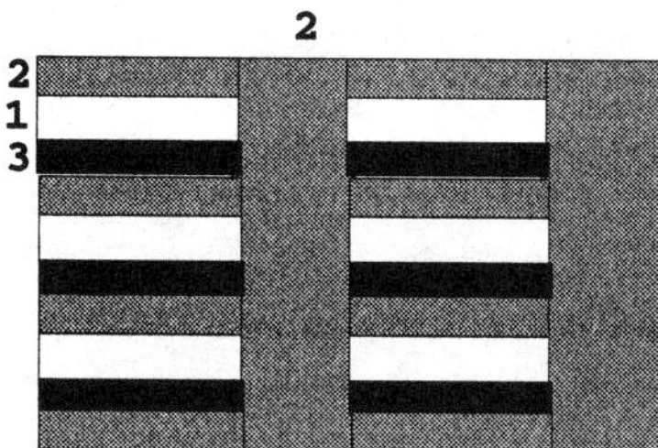

FIGURE 12.5. The optimal structures that satisfy the modified translation bound ($n = 3$). Black color shows the layer of the superconducting material. The current in the parallel layers is aligned across layers.

12.1.3 The Compatibility of Incompatible Phases

The next example highlights more properties of multicomponent composites that have no analogy with two-component structures. The example follows (Bhattacharya et al., 1994).

Consider a composite of four materials that are different in their "eigenstrains." Namely, the materials are characterized by the energies

$$W_i(\nabla \mathbf{w}) = \sigma(\nabla \mathbf{w} - \mathbf{E}_i)^2, \quad i = 1, 2, 3, 4. \tag{12.1.14}$$

For simplicity, let us consider the two-dimensional problem and assume that $\mathbf{w}$ is a two-component vector, $\mathbf{E}_i$ are given constant symmetric matrices called the *eigenstrains*, and σ is a positive constant. Suppose that $\mathbf{E}_i$ has the same eigenvectors, but different eigenvalues

$$\mathbf{E}_i = \begin{pmatrix} s_i^{(1)} & 0 \\ 0 & s_i^{(2)} \end{pmatrix}.$$

This example is related to physical problems of the phase transition. In these problems, $\mathbf{w}$ is the deflection of a point of the body, and W_i is the elastic energy of the ith phase. It is assumed that an initial crystal splits into several forms and that these forms are characterized by different unstressed configurations $\mathbf{E}_i$. For example, the quadratic piece of the initial crystal transforms into one of four parallelograms. $\mathbf{E}_i$ characterizes the transformation: an elongation of the sides and a change in the angle. A detailed discussion of the physical problem and the related formalism can be found in many papers. We mention the collections (Ball, 1988; Kinderlehrer, James, Luskin, and Ericksen, 1993) and (Khachaturyan, 1983; Ball and James, 1987; Lurie and Cherkaev, 1988; Ball, 1989; Fonseca, 1989; Bhattacharya et al., 1994; Truskinovsky and Zanzotto, 1996; Bhattacharya and Kohn, 1997); see also the references therein.

Gibbs' variational principle states that the energy of a natural composites of these four materials minimizes its energy; this leads principle to the variational problem with multiwell Lagrangian

$$L(\nabla \mathbf{w}) = \min_{i=1,\ldots,4} W_i(\nabla \mathbf{w}).$$

Consider the particular question of how to mix the materials (12.1.14) so that the obtained composite has zero energy and a given eigenstrain $\mathbf{E}_0$. More exactly, we want to obtain the composite with the energy of the form (12.1.14) with a given eigenstrain $\mathbf{E}_0$.

The graph of each well W_i is a paraboloid with center shifted to $\mathbf{E}_i$: The energy is zero if $\nabla \mathbf{w} = \mathbf{E}_i$. The convex envelope $\mathcal{C}W = \sum_{i=1}^{4} \alpha_i W_i$ of the four paraboloids has a horizontal component that corresponds to zero energy. This component is characterized by the set $\mathcal{S}_c$ of eigenstrains $\mathbf{E}_0$, which is a convex combination of eigenstrains $\mathbf{E}_i$, $i = 1, \ldots, 4$:

$$\mathcal{S}_c = \left\{ \mathbf{E}_0 : \quad \mathbf{E}_0 = \sum_{i=1}^{4} \alpha_i \mathbf{E}_i, \quad \sum_{i=1}^{4} \alpha_i = 1, \ \alpha_i \geq 0. \right\} \qquad (12.1.15)$$

However, the eigenstrains $\mathbf{E}_i \in \mathcal{S}_c$ are not necessarily realizable by any structure because of the compatibility conditions.

Compatibility. Two materials with eigenstrains $\mathbf{E}_1$ and $\mathbf{E}_2$ are called *compatible* if they can form a structure with the homogenized energy $W_0 = \sigma(\nabla \mathbf{w} - \mathbf{E}_0)^2$ of type (12.1.14). Physically, the compatibility means that the pieces of the materials fit each other and they do not require additional energy to be glued together after transformation. In other words, the material with eigenstrain $\mathbf{E}_1$ can be transformed to the material with eigenstrain $\mathbf{E}_2$ without consuming energy. The continuity of the potential $\mathbf{w}$ says that the compatibility requires the continuity of tangent components of the fields,

$$[(\nabla \mathbf{w}_i - \mathbf{E}_i) - (\nabla \mathbf{w}_j - \mathbf{E}_j)] \cdot \mathbf{t} \qquad (12.1.16)$$

where $\mathbf{t}$ is an arbitrarily oriented unit vector.

The materials are compatible if the eigenstrains $\mathbf{E}_1$ and $\mathbf{E}_2$ have one eigenvalue in common: $\det(\mathbf{E}_1 - \mathbf{E}_2) = 0$. Indeed, the laminate composite of such materials satisfies (12.1.16) if the tangent $\mathbf{t}$ to the laminates is directed along the direction of the common eigenvalue so that $(\mathbf{E}_i - \mathbf{E}_j) \cdot \mathbf{t} = 0$.

The laminate of compatible materials $(\det(\mathbf{E}_1 - \mathbf{E}_2) = 0)$ has the homogenized energy (12.1.14) with the eigenstrain $\mathbf{E}_m$ that is the convex combination of $\mathbf{E}_1$ and $\mathbf{E}_2$

$$\mathbf{E}_0 = c\mathbf{E}_1 + (1 - c)\mathbf{E}_2$$

where $c \in [0, 1]$ is the volume fraction of the mixed phases.

Generally, the eigenstrains $\mathbf{E}_i$ may be not compatible: $\det(\mathbf{E}_k - \mathbf{E}_i) \neq 0$. In this case, one has to deform the pieces of incompatible materials in order to make a solid composition of them. This deformation requires additional energy because it is accompanied by inner strains.

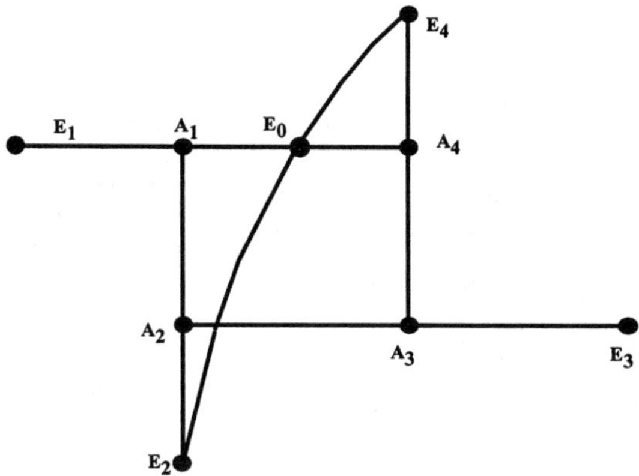

FIGURE 12.6. Fields in a compatible composite from four incompatible materials. The coordinates correspond to the eigenvalues of the phases.

For example, one cannot transform a piece of the material with $\mathbf{E}_1 = 0$ into the material with $\mathbf{E}_2 = I$ without inner strains because the second sample is enlarged in all directions if it is unstressed. Therefore one need to apply an external stress to compress that piece to fit the environment. The energy of compression is called the *residue energy* $W_{\mathrm{res}} > 0$. The required residue energy is proportional to σ and to $|\det(\mathbf{E}_k - \mathbf{E}_i)|$. The energy W of a composite of two incompatible materials has the form

$$W = W_{\mathrm{res}} + \sigma(\nabla\mathbf{w} - \mathbf{E}_0)^2, \quad W_{\mathrm{res}} > 0.$$

Incompatible materials cannot occupy neighboring domains in a structure if this structure has zero total energy. Naturally, one may argue that no point of the convex envelope (12.1.15) of incompatible materials is achievable by a structure, except at its corners. However, there exists a set of structures of zero energy assembled from incompatible materials. These structures have the energy of type (12.1.14) within a set $\mathcal{S}_0$ of eigenstrains (Bhattacharya et al., 1994).

The set $\mathcal{S}_0$ is built in (Bhattacharya et al., 1994) that is smaller than $\mathcal{S}_c$ but it still has nonzero measure. According to private communication by Kohn, the mathematical idea of this construction was put forward earlier by L.Tartar (see (Tartar, 1993)) in a discussion with J.Ball and R.Kohn. From a more general perspective, the discussed method is close to the idea of the self-repeating structures developed in (Nesi and Milton, 1991); see Chapter 7.

Consider the following geometrical construction of the set $\mathcal{S}_0$, using Figure 12.6 and Figure 12.7:

1. Represent the fields as a point with coordinates $\mathbf{E}_i = \left[s_i^{(1)}, \; s_i^{(2)}\right]$ in a plane of the eigenvalues of $\mathbf{E}_0$; Figure 12.6. These points correspond to zero energy of the components.

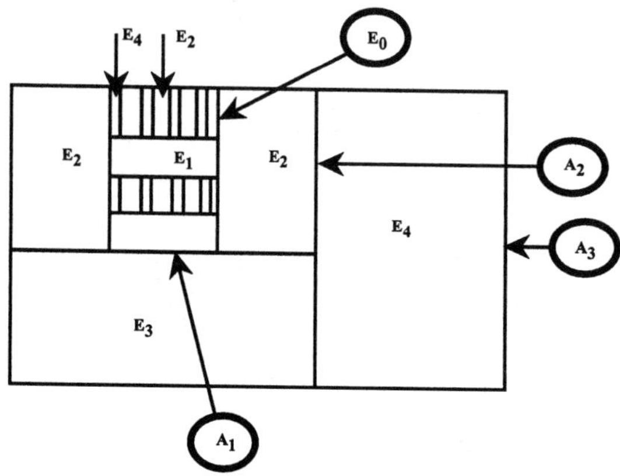

FIGURE 12.7. Compatible composites from four incompatible materials: Geometry. The circled "materials" represent the effective properties of composites.

2. Make a "gambit" move. Mix the incompatible materials $\mathbf{E}_2$ and $\mathbf{E}_4$ to obtain a composite material with eigenstrain $\mathbf{E}_0$ that is compatible with $\mathbf{E}_1$; see Figure 12.6. This composite corresponds to a nonzero residue energy $W_{\mathrm{res}} > 0$.

3. Mix the obtained composite $\mathbf{E}_0$ with material $\mathbf{E}_1$ in the laminate, directing them along the common eigenvalue of eigenstrains $\mathbf{E}_0$ and $\mathbf{E}_1$, Figure 12.7. Choose the volume fraction to bring the eigenstrain of the composite to the point A_1, where it becomes compatible with material $\mathbf{E}_2$; see Figure 12.6.

Notice that this process does not increase the residue energy of the composite $\mathbf{E}_0$ and does not yield to the residue energy in the compatible material $\mathbf{E}_1$. The total residue energy of the composite is equal to $c_0 W_{\mathrm{res}}$, where $c_0 > 0$ is the fraction of the composite $\mathbf{E}_0$ in the composite.

4. Mix the obtained composite $\mathbf{E}(A_1)$ with material $\mathbf{E}_2$ in the laminate, directing layers along the common eigenvalue of eigenstrains; see Figure 12.6. Choose the volume fraction to bring the eigenstrain of the composite to the point A_2, where it becomes compatible with the material $\mathbf{E}_3$ (Figure 12.7). This step does not yield to an additional residue energy, because the materials A_1 and $\mathbf{E}_2$ are compatible. The energy of the composite is equal to $c_1 c_0 W_{\mathrm{res}}$, where $c_1 > 0$ is the fraction of the material A_1 in the composite.

5. Continue the process. Mix the obtained composite $\mathbf{E}(A_2)$ with material $\mathbf{E}_3$ in the laminate, directing layers along the common eigenvalue of eigenstrains, bringing the eigenstrain of the composite to the point

A_3 (Figure 12.6). One of the obtained structures is shown on Figure 12.7.

6. Repeat the process infinitely many times, adding in turn the materials $\mathbf{E}_i$ to make the eigenstrain in the composite move along the sides of the rectangle $A_1 A_2 A_3 A_4$; (Figure 12.6).

   Notice that the volume fraction of the first component $\mathbf{E}_0$ goes to zero. Therefore, the relative amount of the total energy stored in this component goes to zero as well. The rest of the structure does not store additional energy, because it is assembled from compatible materials.

7. All the inner points of $A_1 A_2 A_3 A_4$ can be achieved by mixing the points on the opposite side of it.

This example demonstrates an unexpected and elegant solution to the central problem of optimal composites: how to resolve the contradiction between the algebraic construction of the convex envelope and compatibility. Note that the arguments used are different from the translation method, and are similar to the scheme of the self-repeating structures; see (Nesi and Milton, 1991) and Chapter 7. In both cases, the infinite self-similar procedure resolves the contradiction. Note also that the principle of imitation is applied: At each step, the eigenstrains of the added material and the optimal mixture share a common eigenvalue and eigenvector.

12.2   Necessary Conditions

The Variational Problem

A systematic approach to the optimization is based on Weierstrass-type conditions of optimality, discussed in Chapter 9. Here we apply the technique developed to the problem for three-component composites.

Using the technique described in Chapter 10, we formulate the variational problem to find a component of the lower boundary of the G_m-closure. This component is found from the solution of the variational problem for the multiwell Lagrangian $F(\mathbf{e})$ (see (6.1.5), (7.5.1)):

$$F(\mathbf{e}) = \min_{i=1,2,3} \{\sigma_i \operatorname{Tr} \mathbf{e}^2 + \gamma_i\}$$

where $\mathbf{e}$ is the symmetric matrix, the root of the equation $\mathbf{e}^T \mathbf{e} = (\nabla \mathbf{w})^2$ and $\mathbf{w} = (w_1, w_2)$ is the two-dimentional vector of potentials. The case of the scalar potential ($w_2 = 0$) was described in Chapter 4.

We apply the technique of the Weierstrass variations of the properties to this problem to determine minimal extension of the Lagrangian. We consider the plane problem.

Setting. Consider again a composite of minimal conductivity assembled from three materials with conductivities

$$\sigma_1 < \sigma_2 < \sigma_3.$$

To simplify calculations, assume that

$$\sigma_3 = \infty,$$

i.e., the third material is an ideal conductor. This assumption does not lead to trivial degeneration of the problem, because we are looking for a composite with minimal conductivity. We denote by $\mathcal{O}_1, \mathcal{O}_2$ and $\mathcal{O}_3$ the domains in the periodicity cell occupied with materials σ_1, σ_2 and σ_3 respectively.

We normalize the cost of the materials, and we assign the cost γ_2 of the intermediate material to be between the costs of the extremal materials,

$$\gamma_1 = 1, \quad \gamma_3 = 0, \quad 0 < \gamma_2 < 1. \tag{12.2.1}$$

12.2.1 Single Variations

First we compute the necessary conditions for the fields in each material, using the variations described in Chapter 9. Namely, we place each of the three materials into regions occupied by one of the other two materials and calculate the corresponding inequalities for the admissible fields.

The optimal infinitesimal inclusion of the material σ_2 in $\mathcal{V}_1$ produces the inequalities:

$$F_1(\sigma_2, \sigma_1, \mathbf{e}_1) \geq 0 \text{ in } \mathcal{V}_1,$$

where

$$F_1(\sigma_2, \sigma_1, \mathbf{e}_1) = \begin{cases} \sigma_1(\sigma_2 - \sigma_1)\left(\frac{e_A^2}{\sigma_2} + \frac{e_B^2}{\sigma_1}\right) + \gamma - 1 & \text{if } \frac{e_A}{e_B} \leq \frac{\sigma_1}{\sigma_2}, \\ \frac{\sigma_1(\sigma_2-\sigma_1)}{\sigma_1+\sigma_2}(e_A + e_B)^2 + \gamma - 1 & \text{if } \frac{e_A}{e_B} \geq \frac{\sigma_1}{\sigma_2} \end{cases}$$

(recall that $e_A \leq e_B$ are the eigenvalues of the external field (9.3.3)).

The optimal infinitesimal inclusion of material σ_1 in $\mathcal{V}_2$ produces the inequality

$$F_2(\sigma_1, \sigma_2, \mathbf{e}_2) \geq 0 \text{ in } \mathcal{V}_2, \tag{12.2.2}$$

where

$$F_2(\sigma_1, \sigma_2, \mathbf{e}_2) = \sigma_2(\sigma_1 - \sigma_2)\left(\frac{e_A^2}{\sigma_2} + \frac{e_B^2}{\sigma_1}\right) + 1 - \gamma.$$

These inequalities follow from (9.2.8), (9.2.10) if the conductivities are properly specified.

The optimal infinitesimal inclusion of material σ_3 with infinite conductivity in the regions $\mathcal{V}_1$ and $\mathcal{V}_2$ occupied with materials σ_1 and σ_2, respectively,

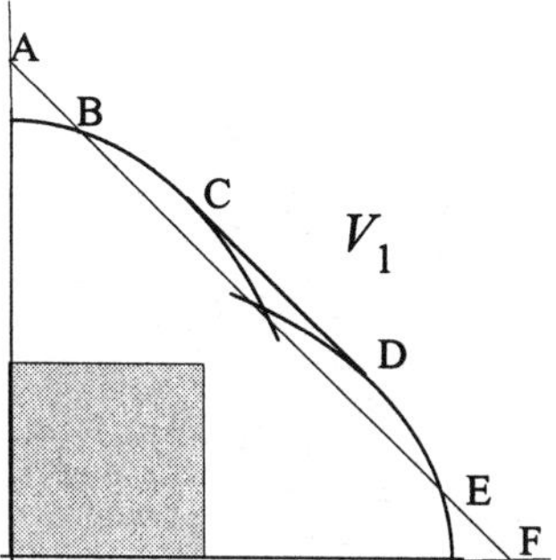
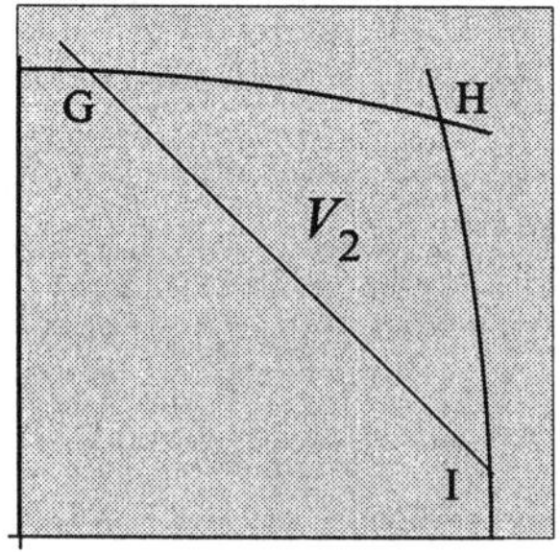

FIGURE 12.8. The permitted regions, based on single variations: (Left) the region V_1 and V_2, (right) magnified picture of the region V_2. The boundary of V_1 is composed of the elliptic segments BC and DE that correspond to the variations with stiplike the inclusions of σ_2, the convex envelope CD of these ellipses that corresponds to the variations with the second-rank inclusions of σ_2, and the straight segments AB and EF that correspond to the variations with the second-rank inclusions of σ_3. The upper boundary of V_2 is composed of the intersection of two ellipses GH and HI that correspond to the variations with the striplike inclusions of σ_1; the lower boundary is straight segment GI that corresponds to the variations with the second-rank inclusions of σ_3. The relative positions of the straight segments AF and GI depend on γ.

corresponds to the inequalities

$$F_1(\infty, \sigma_1, \mathbf{e}_1) \geq 0 \quad \text{or} \quad e_A + e_B \geq \sqrt{\frac{1}{\sigma_1}} \text{ in } V_1, \qquad (12.2.3)$$

$$F_1(\infty, \sigma_2, \mathbf{e}_2) \geq 0 \quad \text{or} \quad e_A + e_B \geq \sqrt{\frac{\gamma}{\sigma_2}} \text{ in } V_2.$$

Finally, the optimality of inclusion of σ_1 in V_3 leads to

$$F_2(\sigma_1, \sigma_3, \mathbf{e}_3) \geq 0 \quad \Rightarrow \quad e_A = e_B = 0 \text{ in } V_3.$$

This condition says that the field in the third (superconducting) phase is always zero, as expected.

The topology of the permitted regions V_i is described as follows. The region V_1 is permitted for fields of great magnitudes, while region V_3 is permitted for zero fields only. The forbidden region V_f lies between these two regions, which makes the picture similar to that of the problem for two materials. Within the forbidden region is located region V_2, where the second (intermediate) material is optimal. The size of this region depends on γ. If this region is not empty, it either divides V_f into two disconnected parts (forms a connected "belt" in V_f), or it leaves V_f connected (forms an "island" in V_f) (see Figure 12.8).

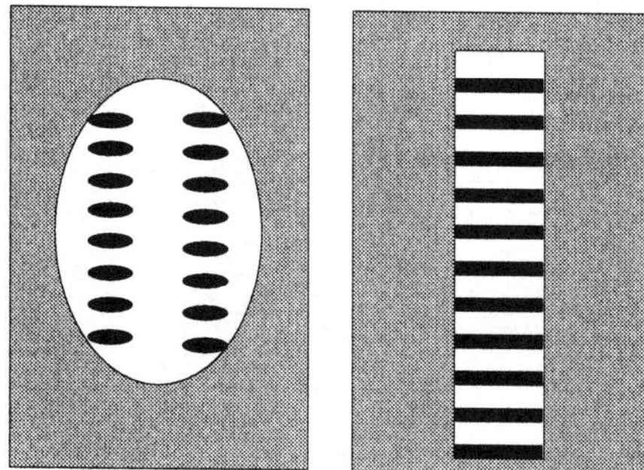

FIGURE 12.9. (Left) The scheme of a composite variation. The composite inclusion of σ_1 (envelope) and σ_3 (nuclei) is inserted into the region of σ_2. (Right) Optimized shape of the inclusion (*tiger's tail*) corresponding to the most sensitive variation.

12.2.2 Composite Variations

Before analyzing the system of necessary conditions, let us discuss whether the variations used are the strongest ones. So far we are dealing with optimality of the boundary between a pair of given materials. However, it is possible that the geometry of optimal multiphase composites includes zones where all three materials are densely mixed together. The dividing curves contain a dense set of points where the domains $\mathcal{O}_1$, $\mathcal{O}_2$, and $\mathcal{O}_3$ meet. Our previous analysis is not applicable to the proximity of these points. The properties of such composites may be significantly different from a two-material composite.

To examine the optimality of such boundaries, one can also look for more complicated types of local variations, which we call the composite variations. The variation is performed as follows: A composite of two available materials is placed in a domain of the third one. A composite inside the inclusion is described by its tensor of effective properties σ_{inc}. We may use knowledge of the bounds of effective properties of any two-component composite (the G_m-closure problem) to solve this problem for three materials.

Improving $\mathcal{V}_2$

The scheme is the following. We form an inclusion of an anisotropic composite of materials σ_1 and σ_3 and place this inclusion in the domain $\mathcal{O}_2$ of the second material (see Figure 12.9). Let c be the volume fraction of material σ_1 (the fraction of σ_3 is obviously $1 - c$).

The change in the materials' cost due to the variation is computed as the difference between the cost of the material σ_2 and the cost of the inserted materials:

$$\Gamma = -\gamma_2 + c\gamma_1 + (1 - c)\gamma_3.$$

In our setting (see (12.2.1)) the cost is

$$\Gamma = -\gamma + c.$$

The increment has the form

$$\delta_2(\alpha, c, \mathbf{e}) = \delta W + \Gamma, \tag{12.2.4}$$

where δW is the increment of energy caused by a composite inclusion with fraction c of σ_1 inserted into the domain of σ_2; α is the parameter of the inclusion.

Let us compute the increment $\delta_2(\alpha, c, \mathbf{e})$. Denote the eigenvalues of the composite in the inclusion by $(l_A(c)$ and $l_B(c)$. Using (9.3.4), we compute the increment of energy:

$$\delta W = e_A^2 \frac{\sigma_2(l_A(c) - \sigma_2)}{\alpha l_A(c) + (1 - \alpha)\sigma_2} + e_B^2 \frac{\sigma_2(l_B(c) - \sigma_2)}{\alpha \sigma_2 + (1 - \alpha)l_B(c)}. \tag{12.2.5}$$

As before, the parameter $\alpha \in [0, 1]$ defines the rate of anisotropy of the second-rank laminate structure or the elongation (eccentricity) of the equivalent elliptical inclusion.

In order to obtain the most sensitive composite variation, the structure of a composite inside inclusion must be optimized. The composite in the inclusion has an extremal anisotropic conductivity and belongs to the boundary of G_m-closure of the two materials mixed in the inclusion. Therefore, the extremal two-component structure is a second-rank laminate. Its conductivity tensor has the eigenvalues $l_A(c, \beta)$ and $l_B(c, \beta)$:

$$l_A(c, \beta) = \sigma_1 + (1 - c)\left(\frac{1}{\sigma_3 - \sigma_1} + \frac{c\beta}{\sigma_1}\right)^{-1},$$

$$l_B(c, \beta) = \sigma_1 + (1 - c)\left(\frac{1}{\sigma_3 - \sigma_1} + \frac{c(1 - \beta)}{\sigma_1}\right)^{-1}.$$

Here $\beta \in [0, 1]$ defines the degree of anisotropy of a composite (the rate of elongation of the inclusions of σ_3 in the matrix material σ_1) and c is the fraction of material σ_1. When β varies in the interval $[0, 1]$, each eigenvalue l_i varies in the interval

$$\left[\left(\frac{c}{\sigma_1} + \frac{1 - c}{\sigma_3}\right)^{-1}, \; (c\sigma_1 + (1 - c)\sigma_3)\right].$$

If $\sigma_3 = \infty$, the formulas become

$$l_A(c, \beta) = \sigma_1 \left(1 + \frac{1 - c}{c\beta}\right), \quad l_B(c, \beta) = \sigma_1 \left(1 + \frac{1 - c}{c(1 - \beta)}\right). \tag{12.2.6}$$

Each eigenvalue l_i varies in the interval $[\frac{\sigma_1}{c}, \; \infty]$.

The increment $\delta_2(\alpha, \beta, c, \mathbf{e})$ is defined by (12.2.4), (12.2.5), and (12.2.6); it depends on three parameters: (i) the volume fraction c of material that forms the composite in the inclusion, (ii) the degree of anisotropy of this composite β, and (iii) the relative elongation (eccentricity) α of the inclusions: Each parameter varies in the interval $[0, 1]$.

To find an extremal variation, we solve the problem

$$\delta_2(\mathbf{e}) = \min_{c \in [0, 1]} \left\{ \min_{\alpha \in [0, 1]} \min_{\beta \in [0, 1]} \delta_2(\alpha, \beta, c, \mathbf{e}) + \Gamma(c) \right\}. \qquad (12.2.7)$$

Laminate Variation

The minimization of (12.2.7) on β and α is independent of the cost of the materials, because these parameters affect only the geometrical structure of an inclusion. Each of the parameters α and β, varies in the interval $[0, 1]$. Therefore, the optimal point corresponds to either an inner point of the square of parameter values $\alpha \in (0, 1)$, $\beta \in (0, 1)$ or its side ($\beta = 0$ or $\beta = 1$, $\alpha \in (0, 1)$) or ($\beta \in (0, 1)$, $\alpha = 0$ or $\alpha = 1$).

First, let us check the case ($\beta = 1$): The inclusion is a laminate composite with eigenvalues

$$l_A(c) = \frac{\sigma_1}{c}, \quad l_B = \infty.$$

The increment (12.2.5) becomes

$$\delta_2(\alpha, 1, c, \mathbf{e}) = \sigma_2 e_A^2 \frac{\sigma_1 - c\sigma_2}{\alpha\sigma_1 + c(1 - \alpha)\sigma_2} + \sigma_2 e_B^2 \frac{1}{(1 - \alpha)} - \gamma + c, \ \alpha \in [0, 1].$$

The stationary points with respect to α make the increment linear with respect to c. Therefore the optimal values of c are either zero or one, which reduces the complex variations to the case of single-phase variations, which was already discussed.

Check the corners of the region of parameters ($\alpha = 0$ or $\alpha = 1$). This case corresponds to a laminate composite placed into a strip-shaped inclusion. The value $\alpha = 1$ (layers are parallel to the elongated inclusion) leads to an infinite increment $\delta_2(1, 1, c, \mathbf{e})$ and is obviously not optimal.

The value $\alpha = 0$ leads to the condition

$$\delta_2(0, 1, c, \mathbf{e}) = \Delta - \gamma + c = (\frac{\sigma_1}{c} - \sigma_2)e_A^2 + \sigma_2 e_B^2 - \gamma + c \geq 0. \qquad (12.2.8)$$

It depends only on the volume fraction $c \in [0, 1]$.

The optimal value c_0 is found from the condition

$$\frac{\partial}{\partial c}(\delta_2(0, 1, c, \mathbf{e})) = 1 - \frac{\sigma_1}{c^2} e_A^2 = 0$$

and is equal to

$$c_0 = \begin{cases} \sqrt{\sigma_1} e_A & \text{if } e_A \leq \frac{1}{\sqrt{\sigma_1}}, \\ 1 & \text{if } e_A \geq \frac{1}{\sqrt{\sigma_1}}. \end{cases}$$

The case $c_0 = 1$ corresponds to the two-phase variation. The case $c_0 \in (0,1)$ leads to the new necessary condition

$$\delta_2(\mathbf{e}) = \delta_2(0,1,c_0,\mathbf{e}) = 2\sqrt{\sigma_1}e_A - \sigma_2 e_A^2 + \sigma_2 e_B^2 - \gamma \geq 0, \qquad (12.2.9)$$

or

$$\left(\sqrt{\sigma_2}e_A - \sqrt{\frac{\sigma_1}{\sigma_2}}\right)^2 - \sigma_2 e_B^2 \leq -\frac{\sigma_1}{\sigma_2} + \gamma. \qquad (12.2.10)$$

This inequality provides an additional restriction on $\mathcal{V}_2$. The condition checks the optimality of a boundary between the phase σ_2 and the laminates of σ_1 and σ_3 orthogonal to the boundary. This boundary is not a dividing line between any two available phases, but all three phases are densely met at it. The jump conditions on this boundary involve all three phases. Notice that the continuity conditions on the boundary of the strip cannot be reduced to conditions on two-material boundaries. Therefore, variations that involve only two materials are not selective enough to permit conclusion about its optimality.

Geometrically, the optimal inclusion is a strip made of laminates made from materials σ_1 and σ_3 and directed across the strip. We call this inclusion the *tiger's tail*. This geometry is the limiting case of the geometry of the T-structure, described in Chapter 7. Similarly to the fields in simple laminates, the fields in the T-structures are constant in each material.

Conditions (12.2.9) and (12.2.10) are supplemented by twin conditions in which the eigenvalues e_A and e_B are interchanged. Finally, we obtain the inequalities (see Figure 12.10)

$$F_3(\sigma_2, \sigma_1, \infty; \mathbf{e}) \geq 0,$$

where

$$F_3 = -\frac{\sigma_1}{\sigma_2} + \gamma$$
$$- \max\left\{ \left(\sqrt{\sigma_2}e_A - \sqrt{\frac{\sigma_1}{\sigma_2}}\right)^2 - \sigma_2 e_B^2, \ \left(\sqrt{\sigma_2}e_B - \sqrt{\frac{\sigma_1}{\sigma_2}}\right)^2 - \sigma_2 e_A^2 \right\}.$$

These inequalities restrict the fields in the set $\mathcal{V}_2$ by the requirement that these fields lie above the intersections of two hyperbolas with asymptotes

$$(\pm e_A \pm e_B) = \frac{\sqrt{\sigma_1}}{\sigma_2}.$$

They supplement the constraints (12.2.3), (12.2.2) obtained by single variations.

Other Variations

It can be shown that no other choice of parameters α, β, and c improves the bound given by the inequality (12.2.9) and the single permutations. The

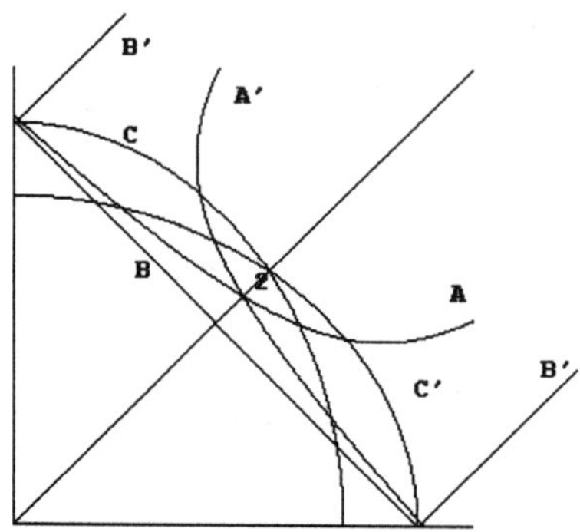

FIGURE 12.10. Permitted region $\mathcal{V}_2$, based on composite variations. A, A': hyperbolic bounds, obtained from composite variations; B, B': the asymptote of the hyperbolas that coincides with the bound obtained from single variation; 2: Region $\mathcal{V}_2$.

formal investigation is routine but long, and Maple is a real help. Instead of presenting here the details of corresponding calculations, let us discuss physical reasons for failure of the other variations to improve the bounds.

Formally, one can check that the stationary points of the increment $\delta_2(\alpha, \beta, c, \mathbf{e})$ inside the square in the variables β, α correspond to saddles and therefore are not optimal. This can be explained by the topology of these variations: If $\beta \in (0, 1)$, then the material σ_3 is placed inside the material σ_1 in the inclusions and the resulting structure is placed into the domain $\mathcal{O}_2$ occupied by σ_2. This construction suggests the existence of the boundary between subdomains occupied by the materials σ_2 and σ_1 and materials σ_1 and σ_3 (inside the inclusions). But the optimality of these boundaries corresponds to the variations of the single permutations. Therefore, the more complicated variations do not produce new, more restricted, inequalities.

The graph of $\mathcal{V}_2$ is represented in Figure 12.10.

Improving $\mathcal{V}_1$

To complete the investigation, we need to consider two other schemes: The composite of materials σ_2 and σ_3 is submerged into the domain $\mathcal{O}_1$ and the composite of materials σ_1 and σ_2 is submerged into the domain $\mathcal{O}_3$. In our setting, the last case is trivial.

The composite variations lead to a decrease of the domain $\mathcal{V}_1$. The scheme of the variations is the same as in the previous case. Note that we can a priori restrict ourselves to the case $\beta = 0$ because we look for inclusions that produce a common boundary between all three materials. The new inequality can be algebraically obtained from (12.2.4) where one replaces σ_1 with σ_2, σ_2 with σ_1 and $\Gamma = -\gamma + c$ with $-1 + c\gamma$. The variation corresponds

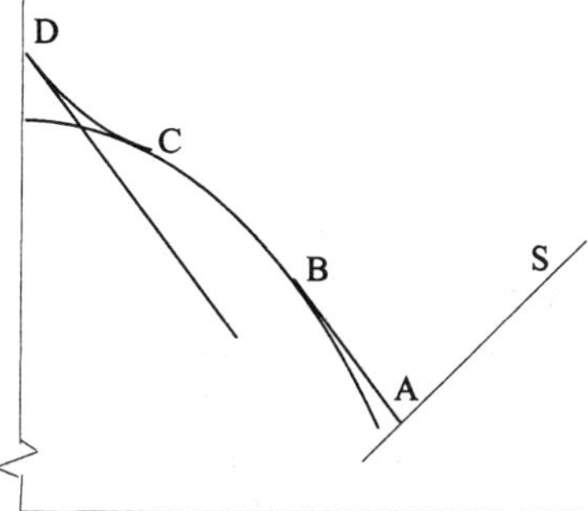

FIGURE 12.11. Upper half of the boundary of $\mathcal{V}_1$ (above the bisector S), based on composite variations. The boundary is composed from the straight segment AB that corresponds to the second rank variations with inclusions from σ_2, the elliptic segment BC that corresponds to the striplike variation from σ_2, and the hyperbolic segment CD that corresponds to the striplike composite variations (tiger's tail) from σ_2 and σ_3. The point D corresponds to the striplike variation from σ_3.

to a striplike inclusion (tiger's tail) assembled with perpendicular layers of σ_2 and σ_3; the fraction c of σ_2 must be optimally chosen. The increment is given by (12.2.8) with the mentioned replacements.

The corresponding graph is shown in Figure 12.11. The new boundary component ABC of the boundary of $\mathcal{V}_1$ corresponds to a hyperbola that joins the corner point A (where $c_0 = 0$ and striplike inclusion of σ_3 is optimal) and the elliptical boundary component C (where $c_0 = 1$ and striplike inclusion of σ_2 is optimal).

Remark 12.2.1 *The optimality conditions depend on the type of variations used. More complicated problems require more sophisticated variations. On the other hand, the easier results correspond to more elementary variations. In this book we describe the simplest variation in a strip (Chapter 4), the more general variation in an ellipse (Chapter 9), and, finally, a composite variation. One can consider more complicated variations; for example, the trial inclusion can be filled with the T-structure. We leave the consideration to the reader. The freedom of choice of the variations agrees with the general concept of Calculus of Variations.*

12.3 Optimal Structures for Three-Component Composites

12.3.1 Range of Values of the Lagrange Multiplier

The Parameters. Let us discuss the link between the cost of materials and their volume fractions The initial optimization problem of a composite structure is characterized by the parameters: the volume fractions m_1 and m_2 of the first and second materials in the mixture and the degree of anisotropy of the resulting composite. The volume fractions are subject to the obvious constraints

$$m_1 \geq 0, \quad m_2 \geq 0, \quad m_1 + m_2 \leq 1.$$

In the process of solution, these parameters are replaced by three other parameters: the magnitude $\|\mathbf{E}\|$ of the external field $\mathbf{E}$, the ratio of the eigenvalues of $\mathbf{E}$, and the relative cost of the second material γ. However, not all values of these three parameters correspond to optimal volume fractions in their intervals $(0,1)$. For some parameters, an optimal solution corresponds to either a two-component composite or a solid material. Let us find the range of parameters $\mathbf{E}$ and γ that leads to an optimal three-component composite.

The mean field $\mathbf{E}$ must belong to the forbidden region

$$\mathbf{E} \in \mathcal{V}_{\mathrm{f}}.$$

Otherwise, the solution is trivial; the field is constant everywhere, and the structure is filled with one of the initially given materials. Two volume fractions out of three are zero.

Let us demonstrate that the range

$$\gamma_2 \leq \gamma \leq \gamma_1, \quad \gamma_1 = \frac{2\sigma_1}{\sigma_1 + \sigma_2}, \quad \gamma_2 = \frac{\sigma_1}{\sigma_2}$$

of γ corresponds to all three-component optimal composites.

The Optimality of Three-Component Composites. The question of the materials' costs that require a three-component composite as the optimal solution is nontrivial. Recall the similar problem of the optimal three-component composite considered in Chapter 4 following (Burns and Cherkaev, 1997). In that problem, we found the composite of three materials that minimizes the energy $\sigma_*(\nabla u)^2$ of a single scalar field ∇u. It was shown that all three-component composites correspond to the unique value of the cost of the intermediate material.

That problem is reduced to the convex envelope of three-well Lagrangian $F = \min_{i=1,2,3}\{\sigma_i(\nabla u)^2 + \gamma_i\}$. All three materials can coexist in an optimal composite if the convex envelope has a straight component supported by all three wells simultaneously. Geometrically, the requirement is that three parabolas have the same tangent. This uniquely determines the parameter γ_2 if the parameters γ_1 and γ_3 are fixed as in (12.2.1).

However, in the problem under consideration, nontrivial three-component composites correspond to a range of γ. The answer to the question what is better to use, σ_2 or a composite of σ_1 and σ_3, depends on the degree of anisotropy of the field $\mathbf{E}$. The closer to the isotropy is $\mathbf{e}$, the more "useful" is σ_2 comparing with the composite of the other two materials. The dependence on the degree of anisotropy yields to the optimality of three-component composites in a range of γ.

Range of γ. Let us review the inequalities that determine the fields in the region $\mathcal{V}_2$. The field $\mathbf{e}_2$ satisfies the inequalities:

$$F_1(\sigma_2, \sigma_1, \mathbf{e}_2) \geq 0, \tag{12.3.1}$$

$$F_2(\sigma_2, \infty; \mathbf{e}_2) \geq 0, \tag{12.3.2}$$

$$F_3(\sigma_2, \sigma_1, \infty; \mathbf{e}_2) \geq 0. \tag{12.3.3}$$

Inequality (12.3.1) demonstrates that the set $\mathcal{V}_2$ belongs to the intersection of two ellipses. This region is obtained by considering inclusions filled with σ_1. Inequality (12.3.2) demonstrates that the set $\mathcal{V}_2$ lies above the straight line. This constraint is obtained from the consideration of inclusions of σ_3. The inequality (12.3.3) says that the set $\mathcal{V}_2$ lies above the intersection of two hyperbolas obtained by the considering inclusions of composite type. The shape of $\mathcal{V}_2$ depends on γ.

• If $\gamma > \gamma_1$ then the set $\mathcal{V}_2$ is empty and the second material is never used in optimal structure. We interpret this as material σ_2 being too "expensive" to use in an optimal composite; it is "cheaper" to use a composite of the first and third materials than the second material. The quasiconvex envelope is supported by the first and third wells and the second well is strictly about it.

• If $\gamma = \gamma_1$ then the set $\mathcal{V}_2$ degenerates to a point,

$$\mathbf{e}_2 = \sqrt{\frac{\sigma_1}{2\sigma_2(\sigma_1 + \sigma_2)}} I;$$

and the optimal structure keeps the field in the second phase constant and isotropic (see Figure 12.12). Here, the quasiconvex envelope is also supported by the first and third wells and the second well touches it at one point.

• If $\gamma \in (\gamma_1, \ \gamma_2)$ then the set $\mathcal{V}_2$ is strongly convex:

$$F_1(\sigma_2, \sigma_1, \mathbf{e}_2) \geq 0,$$
$$F_3(\sigma_2, \sigma_1, \infty; \mathbf{e}) \geq 0;$$

set $\mathcal{V}_2$ is restricted by ellipses and hyperbolas. The domain $\mathcal{V}_2$ forms an "island" in the forbidden region, which leaves open the possibility of optimal

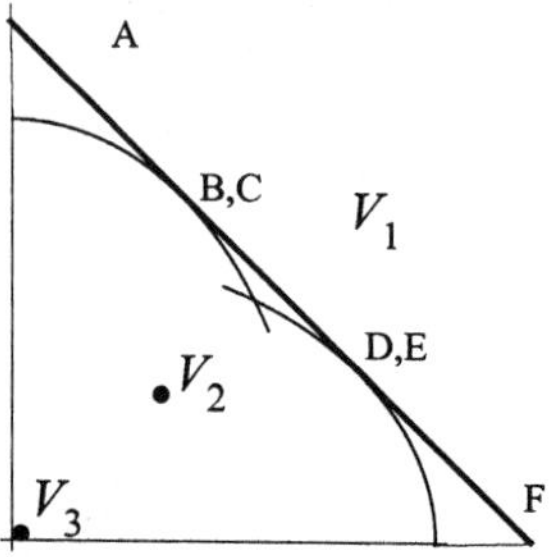

FIGURE 12.12. The permitted regions, $\gamma = \gamma_1$. Region $\mathcal{V}_2$ degenerates into a single point; boundary $\partial\mathcal{V}_1$ becomes a straight line.

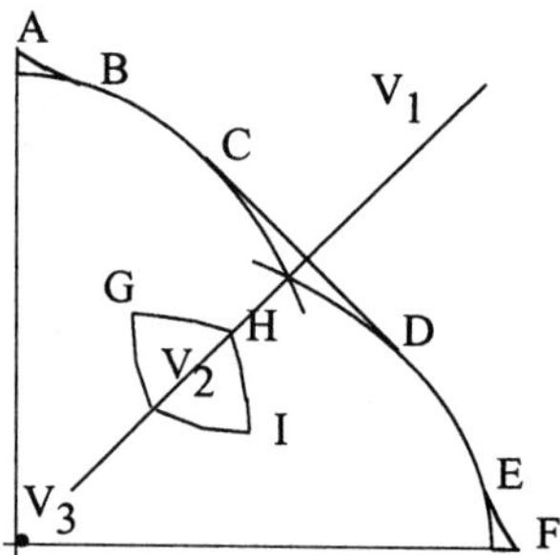

FIGURE 12.13. The permitted regions, $\gamma \in (\gamma_1, \gamma_2)$. Region $\mathcal{V}_2$ is strictly convex.

three-component composites (see Figure 12.13). The quasiconvex envelope is supported by all three wells.

- If $\gamma = \gamma_2$ then the boundary of $\mathcal{V}_2$ has a straight component:

$$F_1(\sigma_2, \infty; \mathbf{e}_2) \geq 0 \Rightarrow e_A + e_B \geq \frac{\sqrt{\sigma_1}}{\sigma_2}.$$

Both hyperbolas F_3 degenerate into their straight asymptotes that coincide with the bound $F_1(\sigma_2, \infty; \mathbf{e}_2) = 0$, which becomes active. The straight component of $\partial\mathcal{V}_2$ (see Figure 12.14) shows that the field in the second phase is not necessarily constant (see the discussion in Chapter 9). The quasiconvex envelope is supported by all three wells.

- If $\gamma < \gamma_2$ then the region $\mathcal{V}_2$ forms a "belt" that divides the forbidden region into the two disconnected parts $\mathcal{V}_f^{12}$ and $\mathcal{V}_f^{23}$. If the mean field belongs to the inner part $\mathcal{V}_f^{23}$ of the forbidden region, then only materials σ_3 and σ_2 can neighbor in an optimal structure. If the mean field belongs to the exterior part $\mathcal{V}_f^{12}$ of the forbidden region, then only materials σ_1 and σ_2 can neighbor.

This range of γ corresponds to two-component composites. The type of optimal composite is determined by the mean field $\mathbf{E}$. When the mean

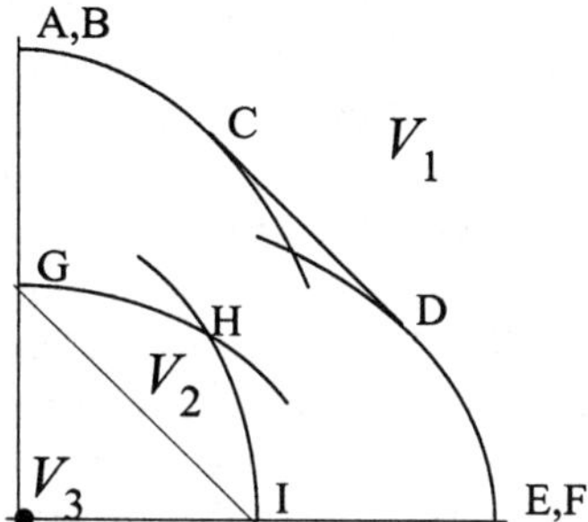

FIGURE 12.14. The permitted regions, $\gamma = \gamma_2$. The lower boundary ∂V_2 becomes straight: Both hyperbolas degenerate and coincide with their asymptotes. The straight component of ∂V_1 disappears.

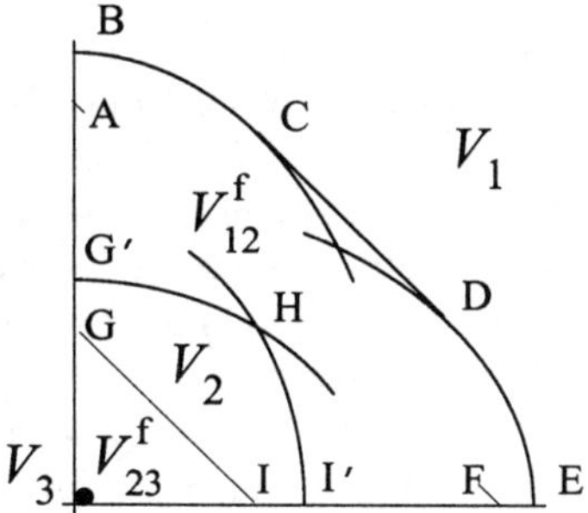

FIGURE 12.15. The permitted regions, $\gamma < \gamma_2$. Three-component composites are not optimal.

field $\mathbf{E}$ belongs to the region $\mathcal{V}_{\mathrm{f}}^{23}$ lying in proximity to the origin, then optimal composites are the matrix laminates of materials σ_3 and σ_2. When the magnitude of the mean field increases, the fraction of σ_3 decreases and the fraction of σ_2 increases. When $\mathbf{E}$ reaches the beltlike region $\mathcal{V}_2$, the optimal composite degenerates into pure material σ_2. A further increase of the magnitude of $\mathbf{E}$ ($\mathbf{E} \in \mathcal{V}_{\mathrm{f}}^{12}$) brings the field into the region $\mathcal{V}_{\mathrm{f}}^{12}$ and leads to optimal matrix laminates of σ_2 and σ_1. The fraction of σ_2 in the composite decreases with an increase of the magnitude of $\mathbf{E}$. When $\mathbf{E}$ reaches the exterior region $\mathcal{V}_1$, it becomes zero.

Note that three-component composites never appear in this process. We interpret this as the second material being too "cheap." It is always better to use this material than a composite of materials σ_1 and σ_3 (see Figure 12.15). Here, the quasiconvex envelope is supported either by the first and second wells or by the second and third wells, but not by three wells simultaneously.

The range of parameters $\mathbf{E} \in \mathcal{V}_{\mathrm{f}}$, $\gamma \in [\gamma_1,\ \gamma_2]$ is sufficient to produce optimal solutions that involve all three materials. A more detailed consideration may couple the ranges of γ and $\mathbf{E}$.

12.3.2   Examples of Optimal Microstructures

Let us now determine some optimal structures that satisfy the necessary conditions as equality everywhere. Our guess is guided by the derived conditions.

Large Values of m_1

Consider the case $\gamma = \gamma_1$ (Figure 12.12). Suppose that the mean field $\mathbf{E}$ belongs to the triangle CDQ. The optimal structures have isotropic fields $\mathbf{e}_2$ and $\mathbf{e}_3$ in the second and third materials, and they may have a varying field $\mathbf{e}_1$ in the first material; the trace of $\mathbf{e}_1$ is constant. Also, one can see from Figure 12.12 that rank-one contacts between σ_2 and σ_1 and between σ_3 and σ_1 are allowed. But a rank-one contact between σ_2 and σ_3 is not allowed. The rank-one contact corresponds to the vertical or horizontal line in Figure 12.12 that join the permitted regions.

The structures in Figure 12.3 and Figure 12.4 satisfy all these requirements. They are described in the previous section. Recall that these structures realize the sufficient (translation) conditions and therefore are surely optimal.

The necessary conditions (Figure 12.12) explain the optimality of the fields in the structures in Figure 12.3 and Figure 12.4. Note that $\mathbf{e}_2$ and $\mathbf{e}_3$ are in rank-one contact with $\mathbf{e}_1$, but they are not in contact with each other. These contacts correspond to laminates of the pair σ_2 and σ_1 and the pair σ_2 and σ_1. The outside layer of $\mathbf{e}_1$ is in rank-one contact with the laminates. The field $\mathbf{e}_1$ takes three values. These values have the common trace if the volume fraction of σ_1 is properly chosen. The fields $\mathbf{e}_2$ and $\mathbf{e}_3$ are isotropic. The fields in the structures in Figure 12.3 and Figure 12.4 belong to the following points of the boundary of the permitted regions (see Figure 12.12): F, D, and a point on the line $[C,\ D]$, their relative fractions are chosen to preserve the trace of the field $\mathbf{e}_1$ in these parts.

The limitations of this construction were discussed in the previous section. In particular, isotropic composites must have the effective conductivity σ_* in the interval $[\sigma_1, \sigma_2]$. The requirement $\sigma_* \in [\sigma_1, \sigma_2]$ leads to the condition

$$m_1 \in \left[2(1 - m_2)\frac{\sigma_1}{\sigma_1 + \sigma_2},\ 1\right]$$

of attainability of the isotropic composites (for a more detailed discussion see (Milton and Kohn, 1988)). In the recent paper (Gibiansky and Sigmund, 1998), new structures were suggested that should increase the rank of m_1.

Small Values of m_1

Next, we demonstrate the optimality of the structure in Figure 12.5. Necessary conditions are satisfied when $\gamma = \gamma_2$ and $\mathbf{E}$ lies between $\mathcal{V}_2$ (the region PQR in Figure 12.14) and the origin. This time, the field $\mathbf{e}_2$ can vary, and $\mathbf{e}_1$ stays constant. The material σ_1 can be in rank-one contact with σ_2 and σ_3 if one eigenvalue of the tensors $\mathbf{e}_1$, $\mathbf{e}_2$, and $\mathbf{e}_3$ is zero. Structures that realize the necessary conditions are easily guessed: They are matrices of σ_2 with inclusions of laminates of σ_1 and σ_3 (see Figure 12.5). These structures are optimal because they correspond to the weighted translation bound (12.1.12).

Let us demonstrate how the necessary conditions are satisfied. Note that the domains $\mathcal{V}_1$ and $\mathcal{V}_3$ are in rank-one contact. The field in $\mathcal{V}_1$ is in contact with $\mathcal{V}_3$ (point F in Figure 12.14); the field in $\mathcal{V}_2$ is piecewise constant (point R and a point on the line $[P, R]$ in Figure 12.14). Furthermore, $\mathcal{V}_2$ is in rank-one contact with either the laminates from σ_1 and σ_3 (point R in Figure 12.14) or the laminate of three materials (the field in an exterior layer, see Figure 12.5).

The effective conductivities are found as follows: The structure in Figure 12.5 can be considered as a second-rank matrix laminate where σ_2 forms the core and laminates from σ_1 and σ_3 the nuclea. The eigenvalues λ_A and λ_B of this matrix laminate satisfy the equation

$$g(\lambda_A, \lambda_B) = \frac{1}{\lambda_A - \sigma_2} + \frac{1}{\lambda_B - \sigma_2} = C, \tag{12.3.4}$$

where C is a constant.

To determine C we observe that the structure in Figure 12.5 degenerate into the *T-structure* (see Figure 7.4 and Example 7.3.5) when all the material σ_2 moves in the exterior layer. Therefore we have $g(\lambda_A, \lambda_B) = g(\lambda_A^T, \lambda_B^T)$, where λ_A^T, λ_B^T are the eigenvalues of the T-structure.

To compute λ_A^T and λ_B^T, we laminate materials σ_1 and $\sigma_3 = \infty$ and obtain a composite with the eigenvalues

$$l_A = \frac{m_1}{m_2 + m_3}\sigma_1 + \frac{m_3}{m_2 + m_3}\sigma_3 = \infty$$

$$l_B = \left(\frac{m_1}{m_2 + m_3}\frac{1}{\sigma_1} + \frac{m_3}{m_2 + m_3}\frac{1}{\sigma_3}\right) = \frac{m_2 + m_3}{m_1}\sigma_1$$

Next, we laminate this composite with σ_2 in orthogonal layers and obtain

$$\lambda_A^T = \frac{\sigma_2}{m_2}, \quad \lambda_B^T = m_2\sigma_2 + \frac{(1 - m_2)^2}{m_1}\sigma_1. \tag{12.3.5}$$

These formulas enable us to calculate the constant C in (12.3.4). Indeed, the curve $g(\lambda_A, \lambda_B)$ passes through the point λ_A^T, λ_B^T; therefore

$$C = \frac{1}{\lambda_A^T - \sigma_2} + \frac{1}{\lambda_B^T - \sigma_2}.$$

The obtained structures realize the bounds given by the weighted translation method. In particular, the isotropic conductivity $\sigma_* = \lambda_A = \lambda_B$ satisfies the relation

$$\frac{1}{\sigma_* + \sigma_2} = \frac{m_1}{2\sigma_1} + \frac{m_2}{2\sigma_2}.$$

Note that these structures have a different topology from the previously discussed ones: (i) the material σ_2 forms a matrix and the laminates of σ_1 and σ_3 form inclusions; (ii) the materials σ_1 and σ_3 are always glued together; (iii) the inclusions are highly anisotropic. The anisotropy of inclusion is compensated by an eccentricity of their shape in the overall isotropic structure.

These structures are optimal over a range of parameters. For isotropic composites, the limiting case corresponds to isotropy of "T-structures": $\lambda_A^T \leq \lambda_B^T$, or (use (12.3.5))

$$m_1 \in \left[0, \frac{m_2(1 - m_2)}{1 + m_2}\left(\frac{\sigma_1}{\sigma_2}\right)\right].$$

Intermediate Case

Similarly, one can check that the structures topologically similar to Figure 12.2 realize an intermediate case for some range of extremal fields.

These structures correspond to an asymptotic case $e_A \to 0$. The rearrangement of σ_1 from the external layer to the inner layer in the structure corresponds to the curve $\lambda_A(c), \lambda_B(c)$, where c is the volume fraction of σ_1 in the external layer. One can check that this curve smoothly touches the Wiener box, so that the tangent is zero, $\frac{\partial \lambda_A(c)}{\partial c}\left(\frac{\partial \lambda_B(c)}{\partial c}\right)^{-1} = 0$ (see Figure 12.1).

Particularly, we conclude that laminates of three or more materials are never optimal for the minimization of the sum of energies, no matter how small the ratio of the applied fields is. Indeed, the maximal or minimal weighted sum of eigenvalues of an effective tensor never corresponds to point W (Figure 12.1) but will stay to the right of point C_1 (maximum of the weighted sum) or below point C_2 (minimum of the weighted sum).

Remark 12.3.1 *These examples do not complete the variety of optimal structures for three materials. However, we have accomplished our goal, which is to demonstrate applications of the necessary conditions to structural optimization and to show diversity of optimal topologies of multicomponent composites.*

12.4 Discussion

Optimal Polycrystal of Isotropic and Anisotropic Materials

The method of necessary conditions is attractive due to its universality. This universality is emphasized if the method is applied to another complex problem of G_m-closure of the composite of an isotropic material with the conductivity σI and an anisotropic material with the conductivity $\Phi^T \mathbf{S} \Phi$; the orientation Φ of the axes in the last material is arbitrary and may depend on the position $\mathbf{x}$. This problem was investigated in (Cherkaev and Miettinen, 1999) using the necessary conditions approach. Previously, the problem was studied in (Nesi and Milton, 1991; Astala and Miettinen, 1998).

The problem is indeed complex. Different methods are applicable to bound the effective conductivity of an optimal composite. As in the discussed problem about three isotropic materials, the exactness of the bound depends on the relation between the parameters: the eigenvalues of the anisotropic phase and the isotropic conductivity. For example, the translation method leads to the exact lower bound if the materials are ordered so that $\sigma \leq \sqrt{\det \mathbf{\Lambda}}$, as it was shown in (Nesi and Milton, 1991). The quasiconformal mapping is applicable in another case; see (Astala and Miettinen, 1998; Milton and Nesi, 1999), where the detailed discussion of the bounds can also be found.

We analyzed the problem using the necessary conditions technique. Remarkably, the optimality of all structures previously obtained by different methods was uniformly confirmed. Moreover, the optimality of new types of structures was demonstrated. The regions $\mathcal{V}$ of optimality of the fields in the materials are obtained by variations in an ellipse. Similarly to the discussed problem, these regions allow us to make an educated guess on the optimal structures. These optimal structures are complicated: They include the laminates of infinite rank made of the anisotropic phase, as well as layers of a finite volume fraction. We do not discuss these structures here; the reader is referred to the original paper (Cherkaev and Miettinen, 1999).

Resume

The hunt for the multicomponent G_m-closures is just beginning. The problem is more complex than that of the two-component G_m-closure, and it requires a larger arsenal of methods. However, the examples we have seen enable us to formulate some common features of extremal composites:

1. The underlying principle is the imitation of properties of an intermediate material by a composite of the extremal materials. This principle was suggested in (Schulgasser, 1976) in the polycrystal problem and exploited in (Milton, 1981d; Milton, 1987) and in (Lurie and Cherkaev, 1985; Cherkaev and Gibiansky, 1996; Gibiansky and Sigmund, 1998) for multiphase composites.

2. Topology of optimal structures is nonunique. The reliable characteristic of the optimality is the range of fields in the materials.

3. In an optimal structure, the fields in the "weak" materials $\sigma_i > \sigma_*$ tend to become isotropic, and the fields in the "strong" materials $\sigma_i < \sigma_*$ are anisotropic. We observe these features in matrix laminates and coated spheres; they are also emphasized in the structures in Figure 12.5.

4. Generally, the components of the convex and quasiconvex envelopes of a multiwell energy are supported by two or more initial wells that represent the energy of pure materials. The case where the quasiconvex envelope is supported by two wells and the third well touches it can be effectively investigated by means of the translation method and the matching microstructures. Recall that the translation bound is built as a convex envelope of the shifted wells.

   In the general case, the quasiconvex envelope is supported by more than two wells. Correspondingly, its algebraic type is different and the translation method fails to describe it.

We do not suggest here special problems to work on. Only a few examples of optimal multiphase structures have been investigated so far. Each new problem could highlight new unexpected features of these structures.

13
Supplement: Variational Principles for Dissipative Media

An interesting variant of the G-closure problem appears in the study of linear processes in dissipative media. Examples of such materials are an electromagnetic conducting medium possessing resistivity, inductance, and capacity and an optical transparent medium (Landau and Lifshitz, 1984), viscoelasticity, etc. Time-periodic fields in such media are described by linear differential equations for complex-valued potentials. The properties of the media are characterized by complex-valued tensors, for example, by complex permeability or complex elasticity tensors.

Methods of Analytic Continuation. In dealing with sets of complex-valued tensors, one could use special properties of complex variable theory to describe the variety of them. An elegant approach based on analytic continuation was suggested in (Bergman, 1978; Milton, 1980) for two-dimensional problems: The effective permeability turns out to be an analytic function of the permeability of the phases. In studying properties of this function one can find a variety of values of the effective tensors. This approach was developed and implemented in (Bergman, 1980; Milton, 1980; Milton, 1981a; Milton and Golden, 1990; Golden and Papanicolaou, 1983; Golden and Papanicolaou, 1985; Golden, 1986; Sawicz and Golden, 1995).

In addition to the mathematical elegance, the approach is able to take into account various restrictions on the set of estimated structures. The exposition of the method and additional references can be found in the mentioned papers.

The Variational Method. To apply the developed variational methods to the problem of complex permeability, one needs to introduce the variational formulations for such problems. This was done in (Cherkaev and Gibiansky, 1994) and (Fannjiang and Papanicolaou, 1994). The method was used in a number of papers (Gibiansky and Milton, 1993; Milton and Berryman, 1997) to derive bounds for complex properties. The variational method is applicable to two- and three-dimensional problems.

In order to apply the variational method, we need to formulate minimal variational principles for complex conductivity or permeability. We derive the variational formulation of the problem, following (Cherkaev and Gibiansky, 1994).

13.1   Equations of Complex Conductivity

13.1.1   The Constitutive Relations

The Process

Consider conductivity in a dissipative medium with inductance and capacity along with resistivity. The current $\mathbf{j}$ and the electric field $\mathbf{e}$ are now functions of time and space coordinates. The current is divergencefree, and the field is curlfree (see Chapter 2):

$$\nabla \cdot \mathbf{j} = 0, \quad \nabla \times \mathbf{e} = 0. \tag{13.1.1}$$

These constraints allow us to introduce a vector potential $\mathbf{a}$ of the current field $\mathbf{j}$ and a scalar potential ϕ of the electrical field $\mathbf{e}$ through the relations

$$\mathbf{j} = \nabla \times \mathbf{a}, \quad \mathbf{e} = -\nabla \phi. \tag{13.1.2}$$

Consider a body Ω occupied by a conducting material and suppose that this body is loaded on the boundary $S = \partial\Omega$. The boundary conditions are similar to those for a conducting material (see Chapter 4)

$$\phi = \phi_0 \text{ on } S_1, \quad \mathbf{n} \cdot \mathbf{j} = j_0 \text{ on } S_2, \quad S_1 \cup S_2 = S, \tag{13.1.3}$$

where $\mathbf{n}$ is the normal.

Assume that the properties of the material are local in space and in time: The current field and its derivatives at a point $\mathbf{x} \in \Omega$ at the moment t depend only on the electrical field and its derivatives at the same point at the same moment of time. Assume that the material is linear in the following sense: A linear combination of the current and its time derivatives linearly depends on a linear combination of the field and its time derivatives:

$$\sum_k a_k \frac{\partial^k \mathbf{j}}{\partial t^k} = \sum_k b_k \frac{\partial^k \mathbf{e}}{\partial t^k}. \tag{13.1.4}$$

Here $a_k = a_k(\mathbf{x})$ and $b_k = b_k(\mathbf{x})$ are some time-independent coefficients, which are scalars (for the isotropic conductors) or symmetric matrices (for the anisotropic ones). The properties of the material (i.e., the scalar or matrix parameters a_k, b_k) do not depend on time.

Monochromatic Excitation

Consider steady-state oscillations in a dissipative medium caused by a monochromatic excitation. The electrical field and current in the material are also monochromatic, i.e.,

$$\mathbf{j}^s(\mathbf{x},t) = (\mathbf{J}(\mathbf{x})e^{i\omega t})' = \mathbf{J}'(\mathbf{x})\cos\omega t + \mathbf{J}''(\mathbf{x})\sin\omega t,$$
$$\mathbf{e}^s(\mathbf{x},t) = (\mathbf{E}(\mathbf{x})e^{i\omega t})' = \mathbf{E}'(\mathbf{x})\cos\omega t + \mathbf{E}''(\mathbf{x})\sin\omega t,$$

where $\Phi_0(s)$, $J_0(s)$, $\mathbf{J}(\mathbf{x})$, and $\mathbf{E}(\mathbf{x})$ are the complex-valued Fourier coefficients of corresponding functions, and s is the coordinate along the boundary. Here, the real and imaginary parts of variables are denoted by the superscripts $'$ and $''$, i.e., $\mathbf{c} = \mathbf{c}' + i\mathbf{c}''$.

The Complex-Valued Conductivity Equations

The linearity of the constitutive relations (13.1.4) leads to a linear relationship between the vectors $\mathbf{J}(\mathbf{x})$ and $\mathbf{E}(\mathbf{x})$:

$$\mathbf{J} = \sigma\mathbf{E}, \qquad (13.1.5)$$

where $\sigma = \sigma(\omega) = \sigma'(\omega) + i\sigma''(\omega)$ is a complex conductivity tensor that depends on the frequency of oscillations (Landau and Lifshitz, 1984). For an isotropic material with state law (13.1.4), the tensor σ is defined by

$$\sigma = \frac{\sum_k(-i\omega)^k a_k}{\sum_k(-i\omega)^k b_k}I,$$

where I is a unit matrix.

The divergencefree nature of the current field and the curlfree nature of the electrical field means that the Fourier coefficients of these fields satisfy relations similar to (13.1.1)

$$\nabla\cdot\mathbf{J} = 0, \quad \nabla\times\mathbf{E} = 0. \qquad (13.1.6)$$

Therefore, they allow the representation (see (13.1.2))

$$\mathbf{J} = \nabla\times\mathbf{A}, \quad \mathbf{E} = -\nabla\Phi, \qquad (13.1.7)$$

where $\mathbf{A}$ and Φ are the Fourier coefficients of the potentials $\mathbf{a}$ and ϕ.

The boundary conditions (13.1.3) lead to the relations

$$\Phi = \Phi_0 \text{ on } S_1, \ \mathbf{n}\cdot\mathbf{J} = J_0 \text{ on } S_2, \ S_1\cup S_2 = S, \qquad (13.1.8)$$

where Φ_0 and J_0 are the Fourier coefficients of the functions ϕ_0 and j_0.

A harmonic oscillation in the conducting media is described by the constitutive relations (13.1.5) and differential equations (13.1.6), (13.1.7) in conjunction with the boundary conditions (13.1.8).

The System of Real First-Order Equations

The complex-valued equations (13.1.5), (13.1.6), (13.1.7), and (13.1.8) describe the conductance of the medium. They look exactly like the equations for the real conductivity; however, they correspond to more complicated processes. Indeed, the complex-valued differential equations (13.1.6) and (13.1.7) form a fourth-order system of differential equations for the real and imaginary parts of the variables $\mathbf{J}'$ and $\mathbf{E}$,

$$\nabla \cdot \mathbf{J}' = 0, \quad \nabla \cdot \mathbf{J}'' = 0, \quad \nabla \times \mathbf{E}' = 0, \quad \nabla \times \mathbf{E}'' = 0. \tag{13.1.9}$$

These equations are identically satisfied if the following potentials are introduced:

$$\mathbf{J}' = \nabla \times \mathbf{A}', \quad \mathbf{J}'' = \nabla \times \mathbf{A}'', \quad \mathbf{E}' = -\nabla\Phi', \quad \mathbf{E}'' = -\nabla\Phi''. \tag{13.1.10}$$

The currents and electrical fields are connected by the constitutive relations (13.1.5)

$$\begin{aligned} -\mathbf{J}' &= -\boldsymbol{\sigma}'\mathbf{E}' + \boldsymbol{\sigma}''\mathbf{E}'', \\ \mathbf{J}'' &= \boldsymbol{\sigma}''\mathbf{E}' + \boldsymbol{\sigma}'\mathbf{E}''. \end{aligned} \tag{13.1.11}$$

The vector form of the last equations is

$$\begin{pmatrix} -\mathbf{J}' \\ \mathbf{J}'' \end{pmatrix} = \mathbf{D}_{EE} \begin{pmatrix} \mathbf{E}' \\ \mathbf{E}'' \end{pmatrix},$$

where

$$\mathbf{D}_{EE} = \begin{pmatrix} -\boldsymbol{\sigma}' & \boldsymbol{\sigma}'' \\ \boldsymbol{\sigma}'' & \boldsymbol{\sigma}' \end{pmatrix} \tag{13.1.12}$$

is the conductivity matrix of the medium.

The boundary conditions (13.1.8) can be rewritten as

$$\Phi' = \Phi'_0 \text{ on } S_1, \tag{13.1.13}$$
$$\Phi'' = \Phi''_0 \text{ on } S_1, \tag{13.1.14}$$
$$\mathbf{n} \cdot \mathbf{J}' = J'_0 \text{ on } S_2, \tag{13.1.15}$$
$$\mathbf{n} \cdot \mathbf{J}'' = J''_0 \text{ on } S_2. \tag{13.1.16}$$

The formulated system of the real-valued differential equations and boundary conditions describes the conductivity of the complex conducting medium. Notice that it has double dimensions compared to the real conductivity problem.

The conductivity is defined by two tensors $\boldsymbol{\sigma}'$ and $\boldsymbol{\sigma}'$. The real part $\boldsymbol{\sigma}'$ is nonnegative,

$$\boldsymbol{\sigma}' \geq 0, \tag{13.1.17}$$

because the dissipation rate is nonnegative. Indeed, the energy dissipation averaged over the period of oscillations is equal to:

$$\frac{\omega}{2\pi} \int_t^{t+\frac{2\pi}{\omega}} \mathbf{j}^s \cdot \mathbf{e}^s \, dt = \frac{1}{2}(\mathbf{J}' \cdot \mathbf{E}' + \mathbf{J}'' \cdot \mathbf{E}'') = \frac{1}{2}(\mathbf{E}' \cdot \boldsymbol{\sigma}' \mathbf{E}' + \mathbf{E}'' \cdot \boldsymbol{\sigma}' \mathbf{E}'') \tag{13.1.18}$$

(see (Landau and Lifshitz, 1984)). The condition (13.1.17) expresses the positiveness of the dissipation rate.

13.1.2 Real Second-Order Equations

The system (13.1.10), (13.1.11) of four first-order differential equations can be rewritten as a system of two second-order equations. We do it in four different ways, and we end up with four equivalent systems. Each of them turns out to be Euler–Lagrange equations for a variational problem.

First, we express the fields though scalar potentials Φ' and Φ'' and take the divergence $(\nabla\cdot)$ of the right- and left-hand sides of (13.1.11). The left-hand-side terms $\nabla\cdot\mathbf{j}', \nabla\cdot\mathbf{j}''$ vanish and we obtain:

$$0 = \nabla\cdot[-\sigma'\nabla\Phi' + \sigma''\nabla\Phi''],$$
$$0 = \nabla\cdot[\sigma''\nabla\Phi' + \sigma'\nabla\Phi''].$$

Thus we obtain two second-order equations for two potentials Φ' and Φ''. The vector form of this system is

$$\begin{pmatrix} 0 \\ 0 \end{pmatrix} = \begin{pmatrix} \nabla\cdot & 0 \\ 0 & \nabla\cdot \end{pmatrix} \mathbf{D}_{EE} \begin{pmatrix} \nabla\Phi' \\ \nabla\Phi'' \end{pmatrix}. \tag{13.1.19}$$

We may also rewrite this system of equations taking any other pair of four scalar and vector potentials (13.1.10) and excluding the other two. For example, let us exclude the fields $\mathbf{E}'$ and $\mathbf{E}''$. First, we solve equations (13.1.11) for $\mathbf{E}'$ and $\mathbf{E}''$:

$$\begin{pmatrix} \mathbf{E}' \\ \mathbf{E}'' \end{pmatrix} = \mathbf{D}_{JJ} \begin{pmatrix} -\mathbf{J}' \\ \mathbf{J}'' \end{pmatrix}, \tag{13.1.20}$$

where

$$\mathbf{D}_{JJ} = \begin{pmatrix} -(\sigma' + \sigma''\sigma'^{-1}\sigma'')^{-1} & (\sigma'' + \sigma'\sigma''^{-1}\sigma')^{-1} \\ (\sigma'' + \sigma'\sigma''^{-1}\sigma')^{-1} & (\sigma' + \sigma''\sigma'^{-1}\sigma'')^{-1} \end{pmatrix}. \tag{13.1.21}$$

(Note that $\mathbf{D}_{EE} = \mathbf{D}_{JJ}^{-1}$.)

Take the curl $(\nabla\times)$ of the right- and left-hand sides of both equations (13.1.20). The left-hand-side terms identically vanish, and we obtain two vector equations:

$$\begin{pmatrix} 0 \\ 0 \end{pmatrix} = \begin{pmatrix} \nabla\times & 0 \\ 0 & \nabla\times \end{pmatrix} \mathbf{D}_{JJ} \begin{pmatrix} -\nabla\times\mathbf{A}' \\ \nabla\times\mathbf{A}'' \end{pmatrix}. \tag{13.1.22}$$

Here we use the representation (13.1.10) of current fields $\mathbf{J}'$ and $\mathbf{J}''$ through the vector potentials $\mathbf{A}'$ and $\mathbf{A}''$.

We may as well solve (13.1.11) for the fields $\mathbf{E}'$ and $\mathbf{J}''$ and obtain

$$\begin{pmatrix} \mathbf{E}' \\ \mathbf{J}' \end{pmatrix} = \mathbf{D}_{JE} \begin{pmatrix} \mathbf{J}'' \\ \mathbf{E}'' \end{pmatrix},$$

where

$$\mathbf{D}_{JE} = \begin{pmatrix} (\sigma')^{-1} & (\sigma')^{-1}\sigma'' \\ \sigma''(\sigma')^{-1} & \sigma' + \sigma''(\sigma')^{-1}\sigma'' \end{pmatrix}. \tag{13.1.23}$$

Recall that $\mathbf{E}'$ is curlfree and $\mathbf{J}''$ is divergencefree. Therefore, by using (13.1.9) and (13.1.10) we arrive at the following system of second-order equations:

$$\begin{pmatrix} 0 \\ 0 \end{pmatrix} = \begin{pmatrix} \nabla\times & 0 \\ 0 & \nabla\cdot \end{pmatrix} \mathbf{D}_{JE} \begin{pmatrix} \nabla\times\mathbf{A}' \\ \nabla\Phi'' \end{pmatrix}. \tag{13.1.24}$$

Similarly, we solve (13.1.11) for $\mathbf{J}'$ and $\mathbf{E}''$ and obtain

$$\begin{pmatrix} \mathbf{J}' \\ \mathbf{E}'' \end{pmatrix} = \mathbf{D}_{EJ} \begin{pmatrix} \mathbf{E}' \\ \mathbf{J}'' \end{pmatrix}, \tag{13.1.25}$$

where

$$\mathbf{D}_{EJ} = \begin{pmatrix} \sigma' + \sigma''(\sigma')^{-1}\sigma'' & -(\sigma')^{-1}\sigma'' \\ -\sigma''(\sigma')^{-1} & (\sigma')^{-1} \end{pmatrix}.$$

(Note that $\mathbf{D}_{JE}^{-1} = \mathbf{D}_{EJ}$.)

Again, the operations $(\nabla\cdot)$ and $(\nabla\times)$ eliminate the corresponding terms on the left-hand side in equations (13.1.25). Applying these operators, we obtain the second-order system

$$\begin{pmatrix} 0 \\ 0 \end{pmatrix} = \begin{pmatrix} \nabla\cdot & 0 \\ 0 & \nabla\times \end{pmatrix} \mathbf{D}_{JE} \begin{pmatrix} \nabla\Phi' \\ \nabla\times\mathbf{A}'' \end{pmatrix}. \tag{13.1.26}$$

We have written four different forms of the same equations. The systems (13.1.19), (13.1.22), (13.1.24), and (13.1.26) are equivalent to each other and to the original system (13.1.9). Each of them in conjunction with the boundary conditions (13.1.13)–(13.1.16) allows us to find the solution that describes the processes in the conducting medium. We now show that each of them represents the Euler equations for a corresponding variational problem.

13.2 Variational Principles

Let us establish variational principles for the problem of complex conductivity. There is no direct complex analogue to the variational principles for the real-valued problem because the inequalities cannot be considered for complex variables. However, the real-valued differential equations just described are the stationary conditions for some real-valued functionals. These functionals lead to variational principles that describe the complex conductivity processes.

First, we formulate two minimax variational principles. They follow naturally from the equations in the form (13.1.19) and (13.1.22). Then we obtain two minimal variational principles based on the equations of the problem in the form (13.1.24) and (13.1.26). Finally, we discuss the relation between these four principles, referring to the procedure of Legendre transform.

13.2.1 Minimax Variational Principles

The Minimax Variational Principle for the Fields

Consider the following variational minimax problem:

$$\min_{\mathbf{E}''} \max_{\mathbf{E}'} U_{EE}, \qquad (13.2.1)$$

where the fields $\mathbf{E}'$, $\mathbf{E}''$ are subject to the constraints

$$\mathbf{E}'' = -\nabla\Phi'', \quad \Phi'' = \Phi_0'' \text{ on } S_1,$$
$$\mathbf{E}' = -\nabla\Phi', \quad \Phi' = \Phi_0' \text{ on } S_1;$$

the functional U_{EE} is

$$U_{EE} = \int_\Omega W_{EE}(\mathbf{E}', \mathbf{E}'') + \int_{S_2} [\Phi'' J_0'' - \Phi' J_0']; \qquad (13.2.2)$$

and

$$W_{EE}(\mathbf{E}', \mathbf{E}'') = \frac{1}{2}\begin{pmatrix}\mathbf{E}' \\ \mathbf{E}''\end{pmatrix}^T \mathbf{D}_{EE} \begin{pmatrix}\mathbf{E}' \\ \mathbf{E}''\end{pmatrix}. \qquad (13.2.3)$$

The matrix $\mathbf{D}_{EE}$ is defined in (13.1.12).

The vanishing of the first variation with respect to $\mathbf{E}'$, $\mathbf{E}''$ of the functional U_{EE} (see (13.2.2)) leads to two Euler–Lagrange equations that coincide with (13.1.11). One can check that they coincide with the original system of equations in the form (13.1.19) and with the boundary conditions (13.1.15), (13.1.16). The boundary conditions (13.1.13), (13.1.14) must be assumed at all admissible fields.

To check the sense of optimality of the stationary solution we examine the sign of the second variation of the functional; see, for example, (Gelfand and Fomin, 1963). The second variation is the main term of the increment of the functional at the perturbed solution of the Euler–Lagrange equation. Whereas the first variation is zero at the solution, the second variation of the cost is proportional to the quadratic form

$$(\delta\mathbf{E}, \delta\mathbf{E}'')^T \mathbf{D}_{EE}(\delta\mathbf{E}, \delta\mathbf{E}'').$$

The functional has a local minimum at the stationary solution if the second variation is positive, and it has a local maximum at the stationary solution if the second variation is negative. The sign of the variation is determined by the matrix $\mathbf{D}_{EE}$.

Here the second variation is neither positive nor negative, because the matrix $\mathbf{D}_{EE}$ is neither positive nor negative definite. The stationary solution corresponds to the saddle point of the functional. The variational problem is of the minimax type.

The Minimax Variational Principle for the Currents. Similarly, one can derive the Euler–Lagrange equations of the variational problem

$$\max_{\mathbf{J}'} \min_{\mathbf{J}''} U_{JJ}, \tag{13.2.4}$$

where the fields $\mathbf{J}', \mathbf{J}''$ are

$$\{\mathbf{J}' : \mathbf{J}' = \nabla \times \mathbf{A}', \quad \mathbf{n} \cdot \mathbf{J}' = J_0' \text{ on } S_2\},$$
$$\{\mathbf{J}'' : \mathbf{J}'' = \nabla \times \mathbf{A}, \quad \mathbf{n} \cdot \mathbf{J}'' = J_0'' \text{ on } S_2\};$$

the functional U_{JJ} is

$$U_{JJ} = \int_\Omega W_{JJ}(\mathbf{J}', \ \mathbf{J}'') + \int_{S_1} [\Phi_0'' \, \mathbf{n} \cdot \mathbf{J}'' - \Phi_0' \, \mathbf{n} \cdot \mathbf{J}']; \tag{13.2.5}$$

and

$$W_{JJ}(\mathbf{J}', \ \mathbf{J}'') = \frac{1}{2} \begin{pmatrix} -\mathbf{J}' \\ \mathbf{J}'' \end{pmatrix}^T \mathbf{D}_{JJ} \begin{pmatrix} -\mathbf{J}' \\ \mathbf{J}'' \end{pmatrix}. \tag{13.2.6}$$

The matrix $\mathbf{D}_{JJ}$ is defined by (13.1.21).

We check that the Euler equations for the functional (13.2.5) coincide with equations (13.1.20) that describe the same problem in different notation.

The matrix $\mathbf{D}_{JJ}$ is neither positive nor negative definite, hence the second variation of the functional U_{JJ} is again neither positive nor negative. We conclude that the variational problem (13.2.5) is of the minimax type.

Remark 13.2.1 *The minimax nature of the variational principles (13.2.1) and (13.2.4) does not allow us to apply the technique developed to the bounds. This technique uses the fact that the energy (i.e., the value of the functional) on any trial field should exceed the actual energy stored in the material. Therefore the energy on any trial field provides an upper bound on the actual energy. For the minimax principles (13.2.1), (13.2.4), however, the situation is different. Consider, for example, the problem (13.2.1) and let us calculate the energy on trial fields of two potentials Φ' and Φ''. The actual energy is increased if the trial field $\nabla\Phi''$ differs from the optimal one and is decreased if the other trial field $\nabla\Phi'$ is not optimal. The value of the functional (13.2.2) on the trial fields can be lower or higher than the actual energy and cannot bound the functional (13.2.2) from either side.*

13.2.2   Minimal Variational Principles

The First Minimal Variational Principle

Consider the following variational problem for the variables $\mathbf{J}'$ and $\mathbf{E}''$:

$$\min_{\mathbf{J}'} \min_{\mathbf{E}''} U_{JE}, \tag{13.2.7}$$

where the fields $\mathbf{J}', \mathbf{E}''$ are

$$\begin{aligned}
\{\mathbf{J}' :\ \mathbf{J}' = \nabla \times \mathbf{A}', &\quad \mathbf{n}\cdot\mathbf{J}' = J_0' \text{ on } S_2\}, \\
\{\mathbf{E}'' :\ \mathbf{E}'' = -\nabla\Phi'', &\quad \Phi'' = \Phi_0'' \text{ on } S_1\};
\end{aligned}$$

the functional U_{JE} is

$$U_{JE} = \int_\Omega W_{JE}(\mathbf{J}', \mathbf{E}'') - \int_{S_1} \mathbf{n}\cdot\mathbf{J}'\Phi_0' + \int_{S_2} \Phi'' J_0''; \tag{13.2.8}$$

and

$$W_{JE}(\mathbf{J}', \mathbf{E}'') = \frac{1}{2}\left(\begin{array}{c}\mathbf{J}'\\ \mathbf{E}''\end{array}\right)^T \mathbf{D}_{JE}\left(\begin{array}{c}\mathbf{J}'\\ \mathbf{E}''\end{array}\right). \tag{13.2.9}$$

The matrix $\mathbf{D}_{JE}$ is defined in (13.2.12). The first variation of (13.2.7) with respect to $\mathbf{J}'$ and $\mathbf{E}''$ coincides with the system of original equations in the form (13.1.24) and the boundary conditions (13.1.14), (13.1.16).

Note that this time the quadratic form (13.2.9) is positive. This follows from the physically clear condition (13.1.17). As we see, this functional is equal to the whole energy dissipated in the body Ω during one period of oscillation (see (13.1.18)).

The second variation $\delta^2 U_{JE}$ of the functional (13.2.8) is positive due to the positivity of the matrix $\mathbf{D}_{JE}$ (for physical reasons we always suppose that $\sigma' \geq 0$ or the dissipation rate is positive). For the quadratic functional (13.2.8) the positivity of the second variation is sufficient to guarantee the global minimum at a stationary point (Gelfand and Fomin, 1963).

The Second Minimal Variational Principle. Similarly, we consider the variational problem

$$\min_{\mathbf{J}''} \min_{\mathbf{E}'} U_{EJ}, \tag{13.2.10}$$

where the fields $\mathbf{J}'', \mathbf{E}'$ are

$$\begin{aligned}
\{\mathbf{J}'' :\ \mathbf{J}'' = \nabla \times \mathbf{A}'', &\quad \mathbf{n}\cdot\mathbf{J}'' = J_0'' \text{ on } S_2\}, \\
\{\mathbf{E}' :\ \mathbf{E}' = -\nabla\Phi', &\quad \Phi' = \Phi_0' \text{ on } S_1\};
\end{aligned}$$

the functional U_{EJ} is:

$$U_{EJ} = \int_\Omega W_{EJ}(\mathbf{E}', \mathbf{J}'') + \int_{S_1} \mathbf{n}\cdot\mathbf{J}''\Phi_0'' - \int_{S_2} \Phi' J_0'; \tag{13.2.11}$$

and

$$W_{EJ}(\mathbf{E}', \mathbf{J}'') = \frac{1}{2}\begin{pmatrix}\mathbf{E}' \\ \mathbf{J}''\end{pmatrix}^T \mathbf{D}_{EJ}\begin{pmatrix}\mathbf{E}' \\ \mathbf{J}''\end{pmatrix}. \qquad (13.2.12)$$

The matrix $\mathbf{D}_{EJ}$ is defined in (13.3.1).

In considering the first variation of the functional (13.2.10), we conclude again that the Euler equations for the functional (13.2.11) coincide with the system of original equations in the form (13.1.26) and the boundary conditions (13.1.13), (13.1.15).

One could also see that the second variation of this functional is positive if $\sigma' \geq 0$.

Remark 13.2.2 *Note that the two variational principles are equivalent:*

$$W_{JE}(\mathbf{J}, \mathbf{E}) = W_{EJ}(\mathbf{E}, \mathbf{J}),$$

This feature is specific for this problem; usually we meet two different variational principles of minimization of the potential energy and the complementary energy (for example, the Dirichlet and Thomson principles).

13.3   Legendre Transform

One can check that the pairs of variational problems, (13.2.1) and (13.2.4), (13.2.7) and (13.2.10), are mutually dual (Gelfand and Fomin, 1963). The matrices associated with the quadratic forms, (13.2.3) and (13.2.6), and (13.2.9) and (13.2.12), are reciprocally inverse, i.e.,

$$\mathbf{D}_{EE} = \mathbf{D}_{JJ}^{-1}, \quad \mathbf{D}_{EJ} = \mathbf{D}_{JE}^{-1}. \qquad (13.3.1)$$

One could pass from the first integrand in each pair to the second one by taking the appropriate Legendre transform (see the discussion in Chapters 1 and 2).

To find the relation between the minimax and minimal variational principles we refer to the duality (Rockafellar, 1967; Rockafellar, 1997). Any saddle function $f(x, y)$ of two variables x and y, corresponds through the Legendre transform x^* over the first variable x to the convex function $f_x^*(x^*, y)$ of the arguments x^*, y.

As we have shown in Chapter 5, see (5.3.1), the saddle function

$$f(x, y) = \frac{a}{2}x^2 - \frac{b}{2}y^2$$

is conjugate in the variable x of the convex function

$$f_x^*(x^*, y) = \max_x[xx^* - f(x, y)] = \frac{1}{2a}(x^*)^2 + \frac{b}{2}y^2.$$

By using a similar idea, we can take the Legendre transform of the functional (13.2.2) over one of its variables (namely, over $\mathbf{E}''$) and obtain the minimal variational principle (13.2.10). Similarly, we can take the Legendre transform of the functional (13.2.5) over the variable $\mathbf{J}''$ and arrive at the minimum variational principle (13.2.7).

The relations between the four described variational problems are illustrated by the following scheme: The minimax variational principle

$$\min_{\mathbf{E}''} \max_{\mathbf{E}'} \{U_{EE}(\mathbf{E}', \mathbf{E}'')\}$$

is transformed by the Legendre transform over the variable $\mathbf{E}''$ into the minimum variational principle

$$\min_{\mathbf{J}''} \min_{\mathbf{E}'} \{U_{EJ}(\mathbf{E}', \mathbf{J}'')\}.$$

The next Legendre transform over the variable $\mathbf{E}'$ leads to the minimax variational problem

$$\min_{\mathbf{J}'} \max_{\mathbf{J}''} \{-U_{JJ}(\mathbf{J}', \mathbf{J}'')\},$$

which is equivalent to (13.2.4). The next transform over the variable $\mathbf{J}''$ leads to the maximization problem

$$\max_{\mathbf{J}'} \max_{\mathbf{E}''} \{-U_{JE}(\mathbf{J}', \mathbf{E}'')\},$$

which is equivalent to (13.2.7). If we take one more Legendre transform over the variable $\mathbf{J}'$, we arrive at a problem that coincides with the one with which we started.

The same method can be used to formulate the minimization problem for other problems described by equations with complex coefficients. For example, the equations of torsion oscillation of a bar made of viscoelastic materials coincide with (13.1.1)–(13.1.4) with some changes in the definitions of the fields and moduli (Christensen, 1971). The other important example of the complex moduli problem is given by viscoelasticity equations. The reader can find a discussion in the papers (Cherkaev and Gibiansky, 1994; Gibiansky and Milton, 1993; Milton and Berryman, 1997).

13.4   Application to G-Closure

The derived minimal variational principle makes the G-closure problem for convex conductivity similar to the problem of coupled bounds for two conductivities. The difference is that the initial equations are coupled. However, the system of translators and the derivation of bounds remain the same, as do the equations for effective properties of laminates. Let us illustrate this by deriving the simplest Wiener-type bounds for complex conductivity.

Consider a composite material assembled of several materials that differ in their complex moduli $\mathcal{D}^k$ and that are taken in prescribed volume fractions m_k. Here k is the index of the material in the composite. The effective properties tensor σ_* is in general an anisotropic second-order tensor that corresponds to an anisotropic conducting media with arbitrary symmetry.

Let us find the bounds on σ_* using the minimal variational principle (13.2.7). The energy of the effective medium

$$(W_*)_{JE}(\mathbf{J}'_0, \mathbf{E}''_0) = \begin{pmatrix} \mathbf{J}'_0 \\ \mathbf{E}''_0 \end{pmatrix}^T (\mathbf{D}_*)_{JE} \begin{pmatrix} \mathbf{J}'_0 \\ \mathbf{E}''_0 \end{pmatrix}$$

stored in a composite cell is equal to the energy stored in an inhomogeneous material:

$$W^0_{JE}(\mathbf{J}'_0, \mathbf{E}''_0) = \min_{\mathbf{J}'} \ \min_{\mathbf{E}''} \ \langle W_{JE}(\mathbf{J}', \mathbf{E}'') \rangle . \tag{13.4.1}$$

Here $\mathbf{E}''$ and $\mathbf{J}'$ are local fields. They satisfy differential constraints and have fixed mean value:

$$\begin{aligned} \mathbf{J}' : & \quad \mathbf{J}' = \nabla \times \mathbf{A}', & \langle \mathbf{J}' \rangle &= \mathbf{J}'_0, \\ \mathbf{E}'' : & \quad \mathbf{E}'' = -\nabla \Phi'', & \langle \mathbf{E}'' \rangle &= \mathbf{E}''_0. \end{aligned}$$

Recall that the Wiener bounds neglect differential constraints. The matrix $\mathbf{D}_{JE}(\mathbf{x})$ is piecewise constant. $(\mathbf{D}_*)_{JE}$ is an effective tensor that depends on the complex effective tensor σ_* as in (13.1.23).

The simplest Wiener-type bounds are obtained by using the constant trial fields in equation (13.4.1). These fields lead to the inequality

$$\mathbf{D}_{*JE} \leq \sum_k m_k \mathbf{D}^k_{JE}, \tag{13.4.2}$$

where $\mathbf{D}^k_{JE}$ are the matrices of the properties of the phases.

The matrix inequality (13.4.2) restricts the diagonal components of the block matrix $\mathbf{D}^0_{JE}$ independently of one another:

$$\begin{aligned} (\sigma'_*)^{-1} &\leq \langle (\sigma')^{-1} \rangle, \\ \sigma_*' + \sigma_*''(\sigma_*')^{-1}\sigma_*'' &\leq \langle \sigma' + \sigma''(\sigma')^{-1}\sigma'' \rangle; \end{aligned} \tag{13.4.3}$$

and produces the coupled bound

$$\det(\langle \mathbf{D}_{JE} \rangle - \mathbf{D}^*_{JE}) \geq 0. \tag{13.4.4}$$

The Dual Bound

The dual minimal variational principle (13.2.10), which is associated by the quadratic form with the matrix $\mathbf{D}_{EJ} = \mathbf{D}^{-1}_{JE}$, leads to the bound

$$(\mathbf{D}^*_{JE})^{-1} \leq \langle \mathbf{D}^{-1}_{JE} \rangle = \sum_k m_k (\mathbf{D}^k_{JE})^{-1}. \tag{13.4.5}$$

Due to the relation $\mathbf{D}_{JE}^{-1} = \mathbf{D}_{EJ}$ the bounds (13.4.5) and (13.4.2) coincide. Note, however, that D is the block matrix of twice larger dimensions than $\boldsymbol{\sigma}'$ and $\boldsymbol{\sigma}''$; therefore, the inequality (13.4.3) or (13.4.4) produces, in fact, lower and upper coupled bounds for the real and imaginary parts of complex-valued effective tensor $\boldsymbol{\sigma}_*$.

Remark 13.4.1 *As with the Wiener bounds, one can apply the translation method to derive more detailed bounds on the effective properties tensor. For all of them it is also sufficient to study only the variational problem (13.4.1); consideration of the dual functional gives exactly the same bounds. Gibiansky and Milton applied the translation method to viscoelasticity (Gibiansky and Milton, 1993), Gibiansky and Torquato solved a number of problems involving coupling of electrical and mechanical properties of composites (Gibiansky and Torquato, 1993; Gibiansky and Torquato, 1995b; Gibiansky and Torquato, 1996a; Gibiansky and Torquato, 1996b).*

Part V

Optimization of Elastic Structures

14

Elasticity of Inhomogeneous Media

14.1   The Plane Problem

In this section, we discuss the properties of the plane problem of linear elasticity for inhomogeneous media. During the discussion we emphasize the properties of the equations of elasticity needed for optimization of elastic structures. For detailed discussion of the theory of linear elasticity we refer to the textbooks (Timoshenko, 1970; Lurie, 1970a; Gurtin, 1972; Sokolnikoff, 1983; Atkin and Fox, 1990; Lurie, 1990a; Parton and Perlin, 1984a; Parton and Perlin, 1984b). More general exposition of continuum mechanics can be found in (Truesdell and Noll, 1965; Truesdell, 1991; Šilhavý, 1997).

14.1.1   Basic Equations

Strain and Compatibility Conditions

Consider an elastic body Ω. Let $\mathbf{x} = (x_1, x_2)$ be the Cartesian coordinates of points of Ω. Suppose that the domain is deformed so that a point with coordinates $\mathbf{x}$ moves to the position $\mathbf{x} + \mathbf{u}$. We call $\mathbf{u} = (u_1, u_2)$ the displacement vector. Here, we deal with linear elasticity. This implies the assumption that $\|\mathbf{u}\| \ll 1$, so that all calculations are rounded up to $o(\mathbf{u})$.

First, we introduce an infinitesimal elastic deformation (strain) $\boldsymbol{\epsilon}$ associated with a displacement $\mathbf{u}$ of the continuum. Notice that a uniform shifting of Ω and a rotation through an infinitesimal angle do not cause any strain:

$$\boldsymbol{\epsilon}(\mathbf{u}) = \boldsymbol{\epsilon}(\mathbf{u} + \mathbf{a}), \quad \boldsymbol{\epsilon}(\mathbf{u}) = \boldsymbol{\epsilon}(\mathbf{u} + \boldsymbol{\omega} \times (\mathbf{x} - \mathbf{x}_0)).$$

Here, $\mathbf{a}$, $\boldsymbol{\omega}$, and $\mathbf{x}_0$ are constant vectors.

The independence of $\boldsymbol{\epsilon}$ on a shift implies that $\boldsymbol{\epsilon}$ depends on the 2×2 matrix ζ,

$$\zeta = \nabla \mathbf{u} = (\nabla u_1, \nabla u_2). \tag{14.1.1}$$

The independence of an infinitesimal rotation implies that $\boldsymbol{\epsilon}$ does not depend on the antisymmetric part of ζ, because a rotation $\boldsymbol{\omega} \times (\mathbf{x} - \mathbf{x}_0)$ through an infinitesimal angle $\boldsymbol{\omega}$ is an antisymmetric part of ζ.

From this consideration, we define the *strain* as the symmetric part of $\zeta = \nabla \mathbf{u}$:

$$\boldsymbol{\epsilon} = \operatorname{Def} \mathbf{u} = \frac{1}{2}(\nabla \mathbf{u} + \nabla \mathbf{u}^T); \tag{14.1.2}$$

we denote this differential dependence by Def. The operator Def maps a vector field $\mathbf{u}$ to a field of symmetric tensors. The definition assumes linearity of the strain as a function of displacement, in accord with the assumptions of linear elasticity.

The Compatibility Conditions. In two dimensions, a symmetric tensor of strains $\boldsymbol{\epsilon}$ (14.1.2) has three different components, but it is determined by a displacement vector that has only two independent components. Therefore its elements are subject to a constraint. One can check that the following differential form

$$^2\mathrm{Ink} : \boldsymbol{\epsilon} = \frac{\partial^2 \epsilon_{11}}{\partial x_2^2} - 2\frac{\partial^2 \epsilon_{12}}{\partial x_1 \partial x_2} + \frac{\partial^2 \epsilon_{22}}{\partial x_1^2}$$

identically vanishes due to (14.1.2). One can also show the inverse inclusion: If $^2\mathrm{Ink} : \boldsymbol{\epsilon} = 0$ then $\boldsymbol{\epsilon}$ can be represented in the form (14.1.2).

The tensor form of the constraint is

$$^2\mathrm{Ink} : \boldsymbol{\epsilon} = 0, \tag{14.1.3}$$

where $(:)$ means the contraction of two indices as follows $\mathbf{a} : \mathbf{b} = \sum_{i,j} a_{ij} b_{ji}$. The second-order differential operator $^2\mathrm{Ink}$ is called the tensor operator of incompatibility (Truesdell and Toupin, 1960; Lurie, 1970a) (Ink is short for the German Inkompatabilität) and the index ^{2} indicates the two-dimensionality of the problem:

$$^2\mathrm{Ink} = \begin{pmatrix} \dfrac{\partial^2}{\partial x_2^2} & -\dfrac{\partial^2}{\partial x_1 \partial x_2} \\ -\dfrac{\partial^2}{\partial x_1 \partial x_2} & \dfrac{\partial^2}{\partial x_1^2} \end{pmatrix}.$$

The $^2\mathrm{Ink}$ operator has the following representation:

$$^2\mathrm{Ink} = R^T \left(\nabla \otimes \nabla \right) R,$$

where R is the tensor of rotation through a right angle (see (4.4.8)). The compatibility condition is similar to the equations of the conservation of mass $(\nabla \cdot \mathbf{j} = 0)$ in conductivity: Both equations express kinematic properties of a continuum.

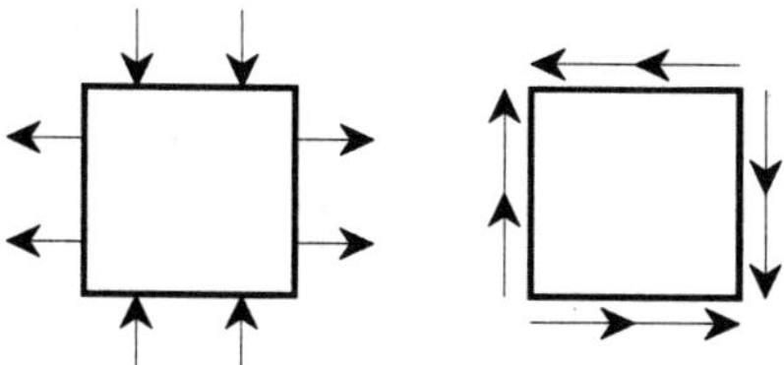

FIGURE 14.1. Stresses, applied to a unit square: (Left) normal stresses (tractions); (Right) tangent stresses (shears).

Stress and Equilibrium Conditions

The second set of elasticity equations determines the equilibrium of stresses in an elastic body. *Stresses* are forces that are applied to the sides of an infinitesimal square inside a body. Suppose that a square is oriented along the reference axes. If we replace the action of the neighboring pieces on the square by applying forces to its sides, we find the following. The applied stresses consist of the normal stresses (tractions) τ_{11} and τ_{22} applied along the normals to its sides and tangential stresses (shears τ_{12} and τ_{21}) applied along its sides. The first index denotes the direction of the applied force; the second index denotes the orientation of the surface to which the force is applied.

Stresses are described by the *stress tensor*[1] $\boldsymbol{\tau}$:

$$\boldsymbol{\tau} = \begin{pmatrix} \tau_{11} & \tau_{12} \\ \tau_{21} & \tau_{22} \end{pmatrix}.$$

The square is in equilibrium, which implies that the total force and the total rotational moment applied to it are zero. The vanishing of the rotational moment requires the symmetry of the stress tensor,

$$\tau_{12} = \tau_{21} \quad \text{or} \quad \boldsymbol{\tau} = \boldsymbol{\tau}^T. \tag{14.1.4}$$

The equilibrium of the forces in two orthogonal directions implies the conditions

$$\frac{\partial \tau_{11}}{\partial x_1} + \frac{\partial \tau_{12}}{\partial x_2} = 0, \quad \frac{\partial \tau_{21}}{\partial x_1} + \frac{\partial \tau_{22}}{\partial x_2} = 0,$$

or

$$\nabla \cdot \boldsymbol{\tau} = 0. \tag{14.1.5}$$

The Airy Potential. Three different components of a symmetric stress tensor $\boldsymbol{\tau}$ satisfy the two differential constraints (14.1.5); this suggests that one

---

[1]The commonly used notation for the stress tensor is $\boldsymbol{\sigma}$, but this symbol is already used in this book for conductivity.

scalar field defines τ. A potential ϕ that guarantees the automatic satisfaction of (14.1.3), (14.1.4), and (14.1.5) is called the *Airy function*. The stress tensor admits the following representation though ϕ:

$$\tau = \begin{pmatrix} \dfrac{\partial^2 \phi}{\partial x_2^2} & -\dfrac{\partial^2 \phi}{\partial x_1 \partial x_2} \\[2ex] -\dfrac{\partial^2 \phi}{\partial x_1 \partial x_2} & \dfrac{\partial^2 \phi}{\partial x_1^2} \end{pmatrix} = {}^2\mathrm{Ink}\,\phi. \tag{14.1.6}$$

Equations (14.1.5) and (14.1.6) are similar to the representations $\nabla \times \mathbf{e} = 0$ and $\mathbf{e} = \nabla w$ for the field that represents forces in the conductivity problem.

If a distributed force $\mathbf{f} = [f_x, f_y]$ such as gravity is acting on a body, then the last equations are modified to the form

$$\nabla \cdot \tau = \mathbf{f}, \quad \tau = \tau^T \quad \text{or } \tau = {}^2\mathrm{Ink}\,\phi + \tau_0.$$

Here τ_0 is a particular solution to the equations $\nabla \cdot \tau_0 = \mathbf{f}$.

Hooke's Law

An elastic material is specified by a relation between τ and ϵ (the constitutive equation); this relation is called *Hooke's law*.

Here we postulate that the material is linear, which means that stresses and strains are linearly related, i.e., each component of the stress tensor is a linear combination of components of the strain tensor:

$$\tau_{ij}(\mathbf{x}) = \sum_{kl} c_{ijkl}(\mathbf{x}) \epsilon_{lk}(\mathbf{x}) \tag{14.1.7}$$

where c_{ijkl} are some coefficients. They form the fourth-rank tensor

$$C = \{c_{ijkl}\}$$

of elastic constants of a material. The tensor C represents material's stiffness. The symmetries of the stress and strain tensor and the existence and positiveness of the energy require the symmetries of coefficients,

$$c_{ijkl} = c_{jikl} = c_{ijlk} = c_{klij},$$

and the inequality,

$$\epsilon : C : \epsilon \geq 0 \quad \forall \epsilon.$$

The tensor form of Hooke's law is

$$\tau = C : \epsilon. \tag{14.1.8}$$

Elasticity Equations

Lamé System. The representations (14.1.2), (14.1.5), and (14.1.8) lead to the elasticity equations, called the *Lamé equations*:

$$\nabla \cdot C : \operatorname{Def} \mathbf{u} = \mathbf{f}. \tag{14.1.9}$$

The weak form of these equations is

$$\int_{\Omega} (\operatorname{Def} \mathbf{v} : C : \operatorname{Def} \mathbf{u} - \mathbf{f} \cdot \mathbf{v}) = 0,$$

$$\forall \mathbf{v} = [v_1, v_2] : v_i \in W_2^1(\Omega), \quad v_i = 0 \text{ on } \partial\Omega \quad i = 1, 2.$$

Remark 14.1.1 *A specific feature of the elasticity equations is the invariance under the rotation. An infinitesimal rotation $\mathbf{u}_R = \boldsymbol{\omega} \times (\mathbf{x} - \mathbf{x}_0)$ corresponds to the antisymmetric matrix $\nabla \mathbf{u}_R$ so that $\operatorname{Def}(\nabla \mathbf{u}_R) \equiv 0$. A displacement $\mathbf{u}$ can be determined from (14.1.9) only up to an infinitesimal rotation and a constant shift, unless the boundary conditions specify these parameters.*

Airy Equation. To rewrite the elasticity equations in a form dual to the Lamé equations, we use the representation (14.1.6) of the stress tensor $\boldsymbol{\tau}$ through the Airy function ϕ, the inverse form of Hooke's law ($\boldsymbol{\epsilon} = C^{-1} : \boldsymbol{\tau}$), and the compatibility condition in the form (14.1.3). Assume for simplicity that $\mathbf{f} = 0$. Accordingly, the dual form of elasticity equations for the plane case becomes a single fourth-order equation for the Airy function,

$$^2\operatorname{Ink} : S : {}^2\operatorname{Ink} \phi = 0, \tag{14.1.10}$$

where the compliance tensor $S = C^{-1}$ is determined by the relationship

$$S = \left\{ s_{ijkl} : \quad \epsilon_{ij} = \sum_{k,l} s_{ijkl} \tau_{kl} \right\}. \tag{14.1.11}$$

Note that the stiffness and compliance tensors are mutual inverses:

$$C : S = S : C = I$$

where I is the fourth-rank identity tensor. As before, we understand the equation (14.1.10) in the weak form

$$\int_{\Omega} \left({}^2\operatorname{Ink} \psi : S : {}^2\operatorname{Ink} \phi \right) = 0 \quad \forall \psi \in W_2^2(\Omega), \ \psi|_{\partial\Omega} = 0.$$

14.1.2   Rotation of Fourth-Rank Tensors

In structural optimization, the tensor C is a control that must be optimally chosen. In particular, it must be oriented optimally. Here we describe the rotation of fourth-rank tensors.

The Basis

It is convenient to represent the second-rank tensors of stresses and strains as vectors in some basis, and the fourth-rank stiffness tensor as a matrix that maps a strain vector to a stress vector.

We introduce (see (Lurie et al., 1982; Lurie and Cherkaev, 1984e; Avellaneda, Cherkaev, Gibiansky, Milton, and Rudelson, 1996)), the following basis in the space of the second-order tensors (we assume that the reference Cartesian coordinates are fixed):

$$\mathbf{a}_1 = \frac{1}{\sqrt{2}} \begin{pmatrix} 1 & 0 \\ 0 & 1 \end{pmatrix}, \quad \mathbf{a}_2 = \frac{1}{\sqrt{2}} \begin{pmatrix} 1 & 0 \\ 0 & -1 \end{pmatrix},$$
$$\mathbf{a}_3 = \frac{1}{\sqrt{2}} \begin{pmatrix} 0 & 1 \\ 1 & 0 \end{pmatrix}, \quad \mathbf{a}_4 = \frac{1}{\sqrt{2}} \begin{pmatrix} 0 & 1 \\ -1 & 0 \end{pmatrix}. \tag{14.1.12}$$

The basis (14.1.12) has many nice properties (see (Avellaneda et al., 1996)). It is orthonormalized with respect to the scalar product (:),

$$\mathbf{a}_i : \mathbf{a}_j = \delta_{ij},$$

where δ_{ij} is the Kroneker symbol. (The coefficients $\frac{1}{\sqrt{2}}$ in (14.1.12) are needed for the normalization.)

The subspace of symmetric second-order tensors is spanned by the orthonormal basis $\mathbf{a}_1$, $\mathbf{a}_2$, $\mathbf{a}_3$. Symmetric strain and stress tensors have the following representation in this basis:

$$\boldsymbol{\epsilon} = \sum_{i=1}^{3} \epsilon_i \mathbf{a}_i, \quad \boldsymbol{\tau} = \sum_{i=1}^{3} \tau_i \mathbf{a}_i,$$

where the coefficients are given by

$$\epsilon_1 = \boldsymbol{\epsilon} : \mathbf{a}_1 = \frac{1}{\sqrt{2}}(\epsilon_{11} + \epsilon_{22}),$$
$$\epsilon_2 = \boldsymbol{\epsilon} : \mathbf{a}_2 = \frac{1}{\sqrt{2}}(\epsilon_{11} - \epsilon_{22}),$$
$$\epsilon_3 = \boldsymbol{\epsilon} : \mathbf{a}_3 = \sqrt{2}\epsilon_{12},$$

and similar formulas for τ_i.

The symmetry of the stress and strain tensors implies that

$$\epsilon_4 = \boldsymbol{\epsilon} : \mathbf{a}_4 = 0, \quad \tau_4 = \boldsymbol{\tau} : \mathbf{a}_4 = 0.$$

The trace and determinant of a symmetric tensor $\mathbf{E}$ in the basis (14.1.12) are

$$\mathrm{Tr}\,\mathbf{E} = \sqrt{2}\mathbf{E} : \mathbf{a}_1,$$
$$\det \mathbf{E} = \frac{1}{2}\left((\mathbf{E} : \mathbf{a}_1)^2 - (\mathbf{E} : \mathbf{a}_2)^2 - (\mathbf{E} : \mathbf{a}_3)^2\right). \tag{14.1.13}$$

In the basis (14.1.12), Hooke's law has the form

$$\tau_j = \sum_{i=1}^{3} c_{ji}\epsilon_i, \quad j = 1, 2, 3,$$

where the coefficients $c_{ij} = \mathbf{a}_i : C : \mathbf{a}_j$ are linear combinations of the coefficients c_{ijkl}. The anisotropic stiffness tensor C is represented by a symmetric positive definite matrix

$$C = \begin{pmatrix} c_{11} & c_{12} & c_{13} \\ c_{12} & c_{22} & c_{23} \\ c_{13} & c_{23} & c_{33} \end{pmatrix}.$$

We observe that the properties of an arbitrary plane elastic material are defined by six constants c_{ij}. One of them fixes the orientation of the reference axes in the plane, and the other five constants represent rotationally invariant properties of an elastic material.

Special types of symmetry, such as orthotropy, cubic symmetry, and isotropy, correspond to special forms of the stiffness matrix.

Relation with the Conventional Representation. Conventionally, the tensor of elastic constants is represented by its coefficients in a Cartesian basis, as in (14.1.11). This representation is convenient for measurements of the coefficients and is associated with natural representation of stresses and strains in the Cartesian basis. The introduced basis (14.1.12) is convenient for use dealing with rotations and other transformations of the fourth-order tensors. We will use this basis for the bounds and for the laminates.

The relations between the conventionally used elastic constants and the introduced ones are easily established by comparing the expressions for the energy in the two bases. We find (Avellaneda et al., 1996) that

$$c_{1111} = \frac{1}{2}\left(c_{11} + c_{22} + 2c_{12}\right), \quad c_{1122} = \frac{1}{2}\left(c_{11} - c_{22}\right),$$

$$c_{2222} = \frac{1}{2}\left(c_{11} + c_{22} - 2c_{12}\right), \quad c_{1212} = \frac{1}{2}c_{33},$$

$$c_{1112} = \frac{1}{2}\left(c_{13} + c_{23}\right), \quad c_{1222} = \frac{1}{2}\left(c_{13} - c_{23}\right)$$

and

$$c_{11} = \frac{1}{2}\left(c_{1111} + 2c_{1122} + c_{2222}\right), \quad c_{12} = \frac{1}{2}\left(c_{1111} - c_{2222}\right),$$

$$c_{22} = \frac{1}{2}\left(c_{1111} - 2c_{1122} + c_{2222}\right), \quad c_{33} = 2c_{1212},$$

$$c_{13} = c_{1112} + c_{1222}, \quad c_{23} = c_{1112} - c_{1222}.$$

Rotations

Let us discuss the transformation of a fourth-rank tensor by a rotation and its symmetries. To start, we find a transformation of the elements of a tensor due to a rotation of the Cartesian axes following (Avellaneda et al., 1996).

Consider two different systems of axes with orthonormal basis vectors $\mathbf{i}, \mathbf{j}$ and $\mathbf{i}^r, \mathbf{j}^r$ and let ϕ denote the angle between $\mathbf{i}$ and $\mathbf{i}^r$, i.e.,

$$\mathbf{i}^r = \mathbf{i}\cos\phi + \mathbf{j}\sin\phi, \quad \mathbf{j}^r = -\mathbf{i}\sin\phi + \mathbf{j}\cos\phi,$$

or, in matrix form,

$$\begin{pmatrix} \mathbf{i}^r \\ \mathbf{j}^r \end{pmatrix} = \begin{pmatrix} \cos\phi & \sin\phi \\ -\sin\phi & \cos\phi \end{pmatrix} \begin{pmatrix} \mathbf{i} \\ \mathbf{j} \end{pmatrix}.$$

Transformation of the Basis. We associate the tensor basis $\mathbf{a}_1^r, \mathbf{a}_2^r, \mathbf{a}_3^r$ with the vectors $\mathbf{i}^r, \mathbf{j}^r$ in the same way that (14.1.12) associates the basis $\mathbf{a}_1, \mathbf{a}_2, \mathbf{a}_3$ with the vectors $\mathbf{i}, \mathbf{j}$. It is easy to verify the relationships

$$\begin{aligned} \mathbf{a}_1^r &= \mathbf{a}_1, \\ \mathbf{a}_2^r &= \cos 2\phi\, \mathbf{a}_2 + \sin 2\phi\, \mathbf{a}_3, \\ \mathbf{a}_3^r &= -\sin 2\phi\, \mathbf{a}_3 + \cos 2\phi\, \mathbf{a}_3, \end{aligned}$$

or, in matrix form, $\mathbf{a}^r = \mathbf{\Phi} : \mathbf{a}$. Here $\mathbf{a}^r = (\mathbf{a}_1^r, \mathbf{a}_2^r, \mathbf{a}_3^r)$, $\mathbf{a} = (\mathbf{a}_1, \mathbf{a}_2, \mathbf{a}_3)$, and $\mathbf{\Phi} = \mathbf{\Phi}(\phi)$ is the rotation matrix

$$\mathbf{\Phi} = \mathbf{\Phi}(\phi) = \begin{pmatrix} 1 & 0 & 0 \\ 0 & \cos 2\phi & \sin 2\phi \\ 0 & -\sin 2\phi & \cos 2\phi \end{pmatrix}.$$

Transformation of the Elasticity Tensor. A rotation of the axes transforms the strain $\boldsymbol{\epsilon} = [\epsilon_1, \epsilon_2, \epsilon_3]$ to the form $\boldsymbol{\epsilon}^r = [\epsilon_1, \epsilon_2^r, \epsilon_3^r]$ where $\epsilon_2^r = \epsilon_2 \cos 2\phi + \epsilon_3 \sin 2\phi$ and $\epsilon_3^r = -\epsilon_2 \sin 2\phi + \epsilon_3 \cos 2\phi$.

Under a rotation of the axes, the matrix C transforms to a new matrix

$$C^r = \mathbf{\Phi}^T : C : \mathbf{\Phi}.$$

This transformation can be rewritten either as

$$\begin{aligned} c_{11}^r &= c_{11}, \\ c_{12}^r &= c_{12}\cos 2\phi + c_{13}\sin 2\phi, \\ c_{13}^r &= -c_{12}\sin 2\phi + c_{13}\cos 2\phi, \\ c_{22}^r &= c_{22}\cos^2 2\phi + c_{33}\cos^2 2\phi + 2c_{23}\sin 2\phi \cos 2\phi, \\ c_{23}^r &= (c_{33} - c_{22})\cos 2\phi \sin 2\phi + c_{23}(\cos^2 2\phi - \sin^2 2\phi), \\ c_{33}^r &= c_{33}\cos^2 2\phi + c_{22}\cos^2 2\phi - 2c_{23}\sin 2\phi \cos 2\phi \end{aligned}$$

or, more concisely, as $c_{11}^r = c_{11}$,

$$\begin{pmatrix} c_{12}^r \\ c_{13}^r \end{pmatrix} = \Psi \begin{pmatrix} c_{12} \\ c_{13} \end{pmatrix}, \qquad \begin{pmatrix} c_{22}^r & c_{23}^r \\ c_{23}^r & c_{33}^r \end{pmatrix} = \Psi \begin{pmatrix} c_{22} & c_{23} \\ c_{23} & c_{33} \end{pmatrix} \Psi^T,$$

where

$$\Psi = \begin{pmatrix} \cos 2\phi & \sin 2\phi \\ -\sin 2\phi & \cos 2\phi \end{pmatrix}$$

is the matrix of rotation through the angle 2ϕ. Hence the action of the rotation on symmetric fourth-rank tensors C can be understood through a rotation by 2ϕ of the 2×2 matrix

$$C_m = \begin{pmatrix} c_{22} & c_{23} \\ c_{23} & c_{33} \end{pmatrix}, \tag{14.1.14}$$

the vector

$$\mathbf{c}_v = (c_{12}, c_{13}), \tag{14.1.15}$$

and the scalar

$$c_s = c_{11}. \tag{14.1.16}$$

The Invariants

General Anisotropy. A fourth-rank elasticity tensor has five invariants to a rotation. Namely, it has two linear invariants that we call the *spherical trace* TrS and the *deviatoric trace* TrD.

$$I_1 = \mathrm{TrS}\, C = c_{11}, \qquad I_2 = \mathrm{TrD}\, C = c_{22} + c_{33}.$$

The first is the scalar (14.1.16) and the second is the trace of the matrix (14.1.14). The tensor C also has two quadratic invariants,

$$I_3 = \|\mathbf{c}_v\|^2 = c_{12}^2 + c_{13}^2,$$
$$I_4 = \det C_m = c_{22}\, c_{33} - c_{23}^2,$$

and one cubic invariant

$$I_5 = \det C.$$

Special classes of the material's symmetries correspond to special relations among these invariants.

Orthotropy. Orthotropic symmetry corresponds to the vector $\mathbf{c}_v$ directed along one of the principal axes of the matrix C_m:

$$C_m \cdot \mathbf{c}_v = \lambda \mathbf{c}_v. \tag{14.1.17}$$

Orthotropy means that the stiffness matrix C has the form

$$C = \begin{pmatrix} c_{11} & c_{12} & 0 \\ c_{12} & c_{22} & 0 \\ 0 & 0 & c_{33} \end{pmatrix}$$

in the specially rotated coordinates that diagonalize the matrix C_m. The reference axes can be chosen in such a way that the tracefree basis tensor $\mathbf{a}_3$ becomes an eigenvector of the matrix C. Shear stress in that direction is caused only by shear strain.

Orthotropy corresponds to a special relation between the elastic moduli. The additional invariant of the orthotropy comes from the representation (14.1.17). It has the form

$$\left(\mathbf{c}_v^{\perp}\right)^T C_m \mathbf{c}_v = 0,$$

where $\mathbf{c}_v^{\perp} \cdot \mathbf{c}_v = 0$. This form can be transferred into an equality for coefficients c_{ij},

$$c_{12}\, c_{13}\, (c_{33} - c_{22}) - c_{23}\, (c_{13}^2 - c_{12}^2) = 0.$$

This equality holds in any coordinate system. It may be used to detect the orthotropy of an elasticity tensor.

"Rectangular" Symmetry. This class corresponds to the case where the matrix C_m (see (14.1.14)) is proportional to the identity:

$$C_m = c_{22}I, \quad \text{or } c_{33} = c_{22}, \quad c_{23} = 0.$$

One can check that this class corresponds to two special relations between the invariants of C:

$$I_2^2 = 4I_4, \quad 4I_5 = I_1 I_2^2 - 2I_2 I_4.$$

Square Symmetry. The class of anisotropic materials with square symmetry (the two-dimensional analogue of the cubic symmetry) corresponds to zero magnitude of the vector $\mathbf{c}_v$ (see (14.1.15)), i.e.,

$$c_{12} = c_{13} = 0.$$

Hence $\mathbf{a}_1$ is an eigenvector of C. Matrix C becomes diagonal in the reference axes that diagonalize the matrix C_m. In these axes, C has the representation

$$C = \begin{pmatrix} c_{11} & 0 & 0 \\ 0 & c_{22} & 0 \\ 0 & 0 & c_{33} \end{pmatrix} = \begin{pmatrix} 2K & 0 & 0 \\ 0 & 2\mu^{(1)} & 0 \\ 0 & 0 & 2\mu^{(2)} \end{pmatrix},$$

where $K = \frac{c_{11}}{2}$ is the two-dimensional bulk modulus and $\mu^{(1)} = \frac{c_{22}}{2}$ and $\mu^{(2)} = \frac{c_{33}}{2}$ are the shear moduli. Square symmetry is characterized by the two equalities

$$I_3 = 0, \quad I_5 = I_1 I_4.$$

Isotropy. Finally, isotropy is the symmetry in which C_m is proportional to the identity and the magnitude of $\mathbf{c}_v$ is zero,

$$c_{22} = c_{33} = 2\mu \quad \text{and} \quad \mathbf{c}_v = 0.$$

The stiffness tensor of an isotropic medium

$$C(K,\mu) = \begin{pmatrix} c_{11} & 0 & 0 \\ 0 & c_{22} & 0 \\ 0 & 0 & c_{22} \end{pmatrix} = \begin{pmatrix} 2K & 0 & 0 \\ 0 & 2\mu & 0 \\ 0 & 0 & 2\mu \end{pmatrix}$$

remains invariant under a rotation. The matrix C is defined by two constants: the bulk modulus $K = \frac{c_{11}}{2}$ and the shear modulus $\mu = \frac{c_{22}}{2} = \frac{c_{33}}{2}$. The bulk modulus K describes the change in the area of an element caused by overall pressure, and the shear modulus μ describes the rate of shear caused by a unit shear force.

The invariants of C are related by

$$I_3 = 0, \quad I_5 = I_1 I_4, \quad I_2^2 = 4I_4.$$

The Hooke's law for an isotropic material has the simplest diagonal form,

$$\tau_1 = 2K\epsilon_1, \quad \tau_2 = 2\mu\epsilon_2, \quad \tau_3 = 2\mu\epsilon_3.$$

The simplicity of this form justifies the use of the basis $\mathbf{a}_i{}^2$.

The compliance tensor for an isotropic material is:

$$S = C^{-1} = \begin{pmatrix} \frac{1}{2K} & 0 & 0 \\ 0 & \frac{1}{2\mu} & 0 \\ 0 & 0 & \frac{1}{2\mu} \end{pmatrix}. \tag{14.1.18}$$

Representation of Differential Operators

The second-order operator $\nabla \otimes \nabla$ is represented in the basis (14.1.12) as an operator vector

$$\nabla \otimes \nabla = \frac{1}{\sqrt{2}} \left(\frac{\partial^2}{\partial^2 x_1} + \frac{\partial^2}{\partial^2 x_2} \right) \mathbf{a}_1$$
$$+ \frac{1}{\sqrt{2}} \left(\frac{\partial^2}{\partial^2 x_1} - \frac{\partial^2}{\partial^2 x_2} \right) \mathbf{a}_2 + \sqrt{2} \left(\frac{\partial^2}{\partial x_1 \partial x_2} \right) \mathbf{a}_3$$

where $\otimes$ denotes the dyadic product: $C = \mathbf{a} \otimes \mathbf{b}$ means that $\{c_{ij} = a_i b_j$.
(Note that $\nabla \otimes \nabla : \mathbf{a}_4 \equiv 0$ due to integrability conditions: $\frac{\partial^2}{\partial x_1 \partial x_2} = \frac{\partial^2}{\partial x_2 \partial x_1}$.)

---

[2] In the next exposition, we use two forms of stresses τ, strains ϵ, and elastic properties C. When doing algebraic manipulations with properties tensors, we view C as 3×3-matrices in a certain basis, and stresses and strains as vectors in that basis. We assume the basis $\mathbf{a}_1, \mathbf{a}_2, \mathbf{a}_3$ unless otherwise stated. However, discussing eigenvectors of stresses or its determinant, we naturally view τ and ϵ as second-rank tensors and C as a fourth-rank tensor. Whereas both representations are equivalent, this should not lead to misunderstanding.

The ^{2}Ink operator is represented as

$$^2\text{Ink} = \mathcal{R} : \nabla \otimes \nabla. \tag{14.1.19}$$

Here $\mathcal{R}$ is the fourth-rank tensor that rotates a symmetric second-rank 2×2 tensor X through a right angle as follows:

$$\mathcal{R} : X = R^T X R$$

where R is the second-order tensor of rotation through a right angle. In the basis (14.1.12), $\mathcal{R}$ is represented by the matrix

$$\mathcal{R} = \begin{pmatrix} 1 & 0 & 0 \\ 0 & -1 & 0 \\ 0 & 0 & -1 \end{pmatrix}. \tag{14.1.20}$$

The rotation through a right angle leaves the trace of a symmetric second-rank tensor X unchanged and alternates the sign of its deviator. Accordingly, tensor $\mathcal{R}$ is formally equal to the stiffness tensor C of an isotropic material with shear and bulk moduli of opposite signs and equal magnitudes.

The representation of the ^{2}Ink-operator allows us to rewrite the Airy equation (14.1.10) in the form

$$P(S, \phi) = \nabla \otimes \nabla : (\mathcal{R} : S : \mathcal{R}) : \nabla \otimes \nabla \phi = 0. \tag{14.1.21}$$

The Eigenbasis. One can always choose an orthonormal basis $\mathbf{c}_i$ in the space of symmetric tensors to make the matrix C orthogonal; (Rykhlevskiĭ, 1988). These tensors are called *eigentensors* and the basis is called the *eigenbasis*. C has the representation

$$C = \sum_{i=1}^{3} \mu_i \mathbf{c}_i \otimes \mathbf{c}_i,$$

where μ_i are nonnegative eigenvalues and $\mathbf{c}_i$ are the eigentensors (Backus, 1970). The three eigenvalues are rotationally invariant. Unlike the eigenvectors, the eigentensors are not completely determined by the orientation of the axes. The system of eigentensors contains two additional invariants to a rotation.

An alternative approach to the representation of the energy based on *unit strain energies* is described in (Taylor, 1998).

Elasticity Equations for Isotropic Materials

For isotropic materials, the basis (14.1.12) becomes the eigenbasis and the Lamé equations take the classical form

$$\frac{\partial}{\partial x_1} \left((K + \mu) \frac{\partial u_1}{\partial x_1} + (K - \mu) \frac{\partial u_2}{\partial x_2} \right) + \frac{\partial}{\partial x_2} \mu \frac{\partial u_1}{\partial x_1} = 0$$

$$\frac{\partial}{\partial x_2} \left((K + \mu) \frac{\partial u_2}{\partial x_2} + (K - \mu) \frac{\partial u_1}{\partial x_1} \right) + \frac{\partial}{\partial x_1} \mu \frac{\partial u_2}{\partial x_2} = 0.$$

Often, these equations are written in a different form through another pair
of elastic constants, the Young modulus E and the Poisson coefficient ν. In
the plane problem, these constants are

$$E = K + \mu, \quad \nu = \frac{K - \mu}{K + \mu}. \tag{14.1.22}$$

The positiveness of the moduli K and μ imply the range of the moduli E
and ν:

$$E > 0, \quad -1 \leq \nu \leq 1.$$

Remark 14.1.2 *Note that the three-dimensional theory of elasticity uses
different definitions for the moduli K and ν, and therefore the range of the
modulus ν is different in the three-dimensional setting.*

Remark 14.1.3 *Media with negative values of Poisson coefficient are ther-
modynamically possible, but the "usual materials" do not show a negative
Poisson coefficient. Examples of structures that have this property have been
demonstrated in (Milton, 1992), see also references therein.*

The isotropy of S implies that $\mathcal{R} : S : \mathcal{R} = S$. Hence the elasticity
equation for an isotropic material has the following representation through
the Airy function,

$$\nabla \otimes \nabla : S : \nabla \otimes \nabla \phi = 0.$$

14.1.3 Classes of Equivalency of Elasticity Tensors

In control problems, we want to know if a layout of the elastic properties
can be uniquely defined from the solution. Here we address this question.

Lamé System as a Degenerate Elliptic System. It is sometimes convenient
to treat the equations of elasticity as an elliptical system

$$\nabla \cdot C' \nabla \mathbf{u} = 0$$

with a 4×4 matrix C' (Milton and Movchan, 1995; Helsing et al., 1997). The
specific feature of the elasticity system is that the matrix C' is degenerate
and its zero eigenvalue corresponds to a special eigenvector $\mathbf{a}_4$. In the basis
(14.1.12), C' has the representation.

$$C' = \begin{pmatrix} C & \mathbf{0^T} \\ \mathbf{0} & 0 \end{pmatrix}, \tag{14.1.23}$$

where $\mathbf{0}$ is the 3×1 zero matrix.

For the matrix C', Hooke's law (14.1.7) becomes $C' : \zeta = [\tau_1, \tau_2, \tau_3, 0]$,
so that the rotation component of the stress is identically zero.

Nonuniqueness of the Constants of Elasticity

The degeneracy of C' leads to the nonuniqueness of the constants of elasticity in the Airy equation. This equation has a remarkable property that says that it is impossible to determine the constants of elasticity from the stresses. This phenomenon was studied in (Lurie and Cherkaev, 1981c; Dundurs, 1989; Cherkaev et al., 1992). For recent development we refer to (Dundurs and Jasiuk, 1997; Norris, 1999) and the references therein. We describe this effect following the paper (Cherkaev et al., 1992).

Observe that the elasticity equation in the form (14.1.21) is satisfied identically if the elasticity matrix S is proportional to $\mathcal{R}$ (see (14.1.20)):

$$P(\mathcal{R}, \phi) = {}^2\text{Ink} : (a\mathcal{R}) : {}^2\text{Ink}\,\phi \equiv 0, \qquad (14.1.24)$$

for all functions ϕ and for all constants a. Indeed, we compute

$$\mathcal{R} : {}^2\text{Ink}\,\phi = \begin{pmatrix} \frac{\partial^2}{\partial x_1^2} & \frac{\partial^2}{\partial x_1 \partial x_2} \\ \frac{\partial^2}{\partial x_1 \partial x_2} & \frac{\partial^2}{\partial x_2^2} \end{pmatrix} \phi$$

and

$$P(\mathcal{R}, \phi) = \frac{\partial^2}{\partial x_2^2}\left(\frac{\partial^2 \phi}{\partial x_1^2}\right) + \frac{\partial^2}{\partial x_1^2}\left(\frac{\partial^2 \phi}{\partial x_2^2}\right) - \frac{\partial^2}{\partial x_1 \partial x_2}\left(\frac{\partial^2 \phi}{\partial x_1 \partial x_2}\right) \equiv 0.$$

Of course, the tensor $\mathcal{R}$ is not positive definite, and it does not represent a real material. Equation (14.1.24) indicates that the quadratic form $\text{Ink}\,\phi : \mathcal{R} : \text{Ink}\,\phi$ is a null-Lagrangian. One can easily check this, because $\text{Ink}\,\phi : \mathcal{R} : \text{Ink}\,\phi = \frac{1}{2}\det(\nabla \mathbf{u})$ where $\mathbf{u} = \nabla\phi$. The last form is a translator; it is used in Chapter 15 to bound the energy of an elastic composite.

The identity (14.1.24) can be used in the following way: Any solution ϕ that satisfies $P(S, \phi) = 0$ (see (14.1.21)) satisfies also the equation

$$P(S + a\mathcal{R}, \phi) = 0 \quad \forall\, a = \text{constant}. \qquad (14.1.25)$$

This demonstrates that if a stress field $\boldsymbol{\tau} = {}^2\text{Ink}\,\phi$ satisfies the equilibrium equations with the properties matrix $S = S(\mathbf{x})$, then the same field satisfies the equilibrium equations with the properties matrix $S + a\mathcal{R}$. Therefore, the compliance tensors $S + a\mathcal{R}$ form an equivalence class. In the basis (14.1.12), they are represented as

$$S + a\mathcal{R} = \begin{pmatrix} S_{11} + a & S_{12} & S_{13} \\ S_{12} & S_{22} - a & S_{23} \\ S_{13} & S_{23} & S_{33} - a \end{pmatrix}.$$

Consider, for example, the isotropic medium that corresponds to the diagonal matrix S. The coefficients $2K = \frac{1}{S_{11}}$ and $2\mu = \frac{1}{S_{22}} = \frac{1}{S_{33}}$ of the isotropic medium S cannot be defined separately from the equilibrium equations. From any measurements of the stress, we determine only the quantities $\frac{1}{2K} + a$ and $\frac{1}{2\mu} - a$, where a is arbitrary. In other words, we could only calculate the quantity $\frac{1}{K} + \frac{1}{\mu}$, but not the constants K and μ separately. Hence, it is impossible to determine the Poisson coefficient (14.1.22) from these measurements.

Particularly, if the medium has the constant shear modulus μ everywhere, one can choose the constant a equal to $\frac{1}{2\mu}$, and thus reduce problem (14.1.25) to the biharmonic equation

$$\left(\frac{\partial^2}{\partial^2 x_1} + \frac{\partial^2}{\partial^2 x_2}\right)\left(\frac{1}{K} + \frac{1}{\mu}\right)\left(\frac{\partial^2}{\partial^2 x_1} + \frac{\partial^2}{\partial^2 x_2}\right)\phi = 0. \qquad (14.1.26)$$

The tensor $S + \frac{1}{2\mu}\mathcal{R}$ describes a medium with infinite shear modulus that can only be deformed by conformal deformations, because any infinitesimal square is transferred to another square. The biharmonic equation is used in the analytic method to obtain solutions to the plane problem of elasticity (Muskhelishvili, 1953).

If the medium has the constant bulk modulus K everywhere, one can choose the constant a equal to $-\frac{1}{2K}$ and thus reduce the problem (14.1.10) to

$$\left[\left(\frac{\partial^2}{\partial^2 x_1} - \frac{\partial^2}{\partial^2 x_2}\right)\left(\frac{1}{K} + \frac{1}{\mu}\right)\left(\frac{\partial^2}{\partial^2 x_1} - \frac{\partial^2}{\partial^2 x_2}\right) \right.$$
$$\left. + 2\frac{\partial^2}{\partial x_1 \partial x_2}\left(\frac{1}{K} + \frac{1}{\mu}\right)\frac{\partial^2}{\partial x_1 \partial x_2}\right]\phi = 0. \qquad (14.1.27)$$

This equation describes the deformation of an incompressible elastic medium.

Obviously, the media (14.1.26) and (14.1.27) have very different strains, but they produce the same stresses if the media are loaded by arbitrary but equal forces. This nonuniqueness will be used to tighten bounds on effective properties.

14.2   Three-Dimensional Elasticity

14.2.1   Equations

Constants of Elasticity

The three-dimensional elasticity problem is described by the same equations (14.1.2), (14.1.5), (14.1.7), where now τ and ϵ are 3×3 symmetric tensors, each determined by six scalar elements. Each 3×3 tensor is represented as a symmetric matrix in a basis $\mathbf{i}_1, \mathbf{i}_2, \mathbf{i}_3$.

The stiffness tensor C for an anisotropic elastic material connects strains and stresses $\tau = C : \epsilon$. The tensor C is represented by a symmetric $6 \times$

6 matrix. It depends on $\frac{6 \times 7}{2} = 21$ constants. Three of these constants determine the orientation of the axes in three-dimensional space, and the 18 remaining constants describe the invariant properties of an anisotropic elastic material. Special symmetries correspond to a smaller number of constants. Particularly, the isotropy of the material reduces the number of elastic constants to two: shear modulus μ and the bulk modulus k.

A bulk modulus is the proportionality factor between the overall stress and the corresponding overall strain, or the proportionality coefficient between the traces of a stress and a strain:

$$\operatorname{Tr} \tau = k \operatorname{Tr} \epsilon.$$

A shear mode of the stress or strain is a tracefree matrix of type $\tau - \frac{1}{3}(\operatorname{Tr} \tau)I$. A shear modulus is the proportionality factor between the shear stress and the corresponding shear strain:

$$\tau - \frac{1}{3}(\operatorname{Tr} \tau)I = \mu\left(\epsilon - \frac{1}{3}(\operatorname{Tr} \epsilon)I\right).$$

The introduced bulk modulus K in planar elasticity is different from a commonly used bulk modulus k in three-dimensional elasticity, because the overall stress in three dimensions and in the plane are differently defined: In the plane problem, the stress in the out-of-plane direction is zero. The relation between these moduli is

$$K = k + \frac{\mu}{3}.$$

It reflects the fact that the true (three-dimensional) hydrostatic stress (or strain)

$$I = \mathbf{i}_1 \otimes \mathbf{i}_1 + \mathbf{i}_2 \otimes \mathbf{i}_2 + \mathbf{i}_3 \otimes \mathbf{i}_3$$

is a weighted sum of a two-dimensional hydrostatic matrix

$$^2I = \mathbf{i}_1 \otimes \mathbf{i}_1 + \mathbf{i}_2 \otimes \mathbf{i}_2$$

and a tracefree (deviatoric) matrix

$$Q = -\mathbf{i}_1 \otimes \mathbf{i}_1 - \mathbf{i}_2 \otimes \mathbf{i}_2 + 2\mathbf{i}_3 \otimes \mathbf{i}_3$$

as follows: $I = \frac{3}{2}\,^2I + \frac{1}{2}Q$.

We use the modulus K to simplify the notations. Notice also that the shear modulus μ is the same in the planar and three-dimensional consideration.

Remark 14.2.1 *The space of 3×3 symmetric tensors can be decomposed into two orthogonal rotationally invariant subspaces. A one-dimensional subspace consists of tensors proportional to the unit matrix I. The orthogonal subspace consists of tracefree tensors (deviators). The deviators, in turn, can be decomposed into five orthogonal components.*

The construction is analogous to the two-dimensional case. However, the subspace of shears in three dimensions lacks an important property of shears in two dimensions: The five basic shear tensors do not transform into each other by a rotation. Indeed, the 3×3 matrix possesses three invariants to a rotation of the axes. The conditions of being tracefree and normalized fix only two of them. The determinant of a normalized shear tensor is not fixed. Therefore, a shear tensor cannot be transferred by a rotation to another arbitrary shear tensor; it may have a different value of the determinant.

Equations of Elasticity

The equations of three-dimensional elasticity take the form (14.1.9) of the Lamé equations. This time the Lamé system consists of three equations for the three-dimensional displacement vector. The strain is the symmetric part of the gradient of the displacement, as in (14.1.2). The stress satisfies the equilibrium conditions identical to (14.1.4). Hooke's law states a linear dependence between stresses and strains, as in (14.1.7).

This system can also be considered an elliptical system for three components of the displacement vector. The matrix of coefficients of this system has three zero eigenvalues corresponding to three rotational degrees of freedom.

Three-Dimensional Ink-*Operator.* The Lamé equations have a dual form, similar to the Airy equation(14.1.10). The dual form uses the representation of the stresses through a tensor of potentials. Such potentials reflect the fact that the six components of the symmetric stress tensor τ are bounded by three constraints $\nabla \cdot \tau = 0$.

The structure of the potentials is more complicated in the three-dimensional case. They are expressed through a symmetric tensor potential Φ of stresses as follows (Truesdell and Toupin, 1960):

$$\tau = \operatorname{Ink} \Phi = \nabla \times (\nabla \times \Phi)^T. \tag{14.2.1}$$

The tensor $\operatorname{Ink} \Phi$ is symmetric. Its diagonal (11) component is

$$(\operatorname{Ink} \Phi)_{11} = \frac{\partial^2 \phi_{33}}{\partial x_2^2} + \frac{\partial^2 \phi_{22}}{\partial x_3^2} - 2\frac{\partial^2 \phi_{23}}{\partial x_2 \partial x_3},$$

and the other two diagonal components are obtained by cyclic permutation. The nondiagonal (12) component is

$$(\operatorname{Ink} \Phi)_{12} = -\frac{\partial^2 \phi_{12}}{\partial x_3^2} + \frac{\partial^2 \phi_{23}}{\partial x_1 \partial x_3} + \frac{\partial^2 \phi_{13}}{\partial x_2 \partial x_3} - \frac{\partial^2 \phi_{33}}{\partial x_1 \partial x_2},$$

and the other two are obtained by cyclic permutation. The representation (14.2.1) defines a symmetric divergencefree tensor τ that satisfies the equilibrium conditions $\nabla \cdot \tau = 0$ and $\tau = \tau^T$.

Variable	Constraints	Potential
Strain ϵ	Ink $\epsilon = 0$	$\epsilon =$ Def $\mathbf{u}$
Stress τ	$\nabla \cdot \tau = 0$	$\tau =$ Ink Φ

TABLE 14.1. Differential constraints and potentials in elasticity.

The symmetric 3×3 tensor Φ is determined up to a tensor Def $\mathbf{u} = \frac{1}{2}(\nabla \mathbf{u} + (\nabla \mathbf{u})^T)$ of an arbitrary vector field $\mathbf{u}$, because Ink Def $\mathbf{u} \equiv 0$:

$$\text{Ink } \Phi = \text{Ink}(\Phi + \text{Def } \mathbf{u}) \quad \forall \mathbf{u}.$$

Therefore, Φ is determined by three independent scalar functions. That number agrees with the number of degrees of freedom of a stress tensor. The six components of the symmetric stress tensor are bounded by three differential restrictions $\nabla \cdot \tau = 0$.

The operator Ink enables us to express the compatibility conditions of the strains ϵ in the form

$$\text{Ink } \epsilon = 0 \quad \forall \epsilon = \frac{1}{2}(\nabla \mathbf{u} + (\nabla \mathbf{u})^T).$$

The inverse form of Hooke's law is $\epsilon = S : \tau$, where $S = C^{-1}$. Combining these equations, we obtain the equations of elasticity in the dual form

$$\text{Ink} : S : \text{Ink } \Phi = 0.$$

One can check that the two-dimensional case corresponds to a special form of the tensor potential Φ:

$$\Phi = \begin{pmatrix} 0 & 0 & 0 \\ 0 & 0 & 0 \\ 0 & 0 & \phi(x_1, x_2) \end{pmatrix}, \tag{14.2.2}$$

which has only one nonzero element $\phi = \Phi_{33}$. This element is the Airy function.

Differential Properties and Potentials

One can check that the operator triplet $(\nabla \cdot, \text{ Ink}, \text{ Def})$ applied to a symmetric tensor $\mathbf{A}$ and the vector $\mathbf{u}$ has properties similar to those of the triplet $(\nabla \cdot, \nabla \times, \nabla)$ applied to a vector $\mathbf{a}$ and a scalar u.

Table 14.1 summarizes differential properties of the strains and stresses and compares them with the properties of fields and currents in conductivity (Table 2.1).

Translators. The Ink operator restricts the linear form of second derivatives of Def $\mathbf{u}$. Still, the compensated compactness method (see Chapter 8) can be used to establish the quasiconvexity or quasiaffineness of the corresponding quadratic forms. Using this method, one can check the following properties that we use in the next chapters.

1. The scalar product ϕ_1 of a symmetric divergencefree tensor and a Inkfree tensor is quasiaffine

$$\phi_1 = A : B, \quad A = \text{Def } \mathbf{u}, \quad B = \text{Ink } \Phi.$$

2. The function $\phi_2 = A : \mathcal{R} : A$, where $A = \text{Def } \mathbf{u}$, is quasiconvex.

These and similar translators are discussed in Chapters 15 and 16, where we deal with the corresponding nonconvex variational problems.

14.2.2 Inhomogeneous Medium. Continuity Conditions

Consider an inhomogeneous elastic medium assembled of several materials with different elasticity tensors C_i:

$$C(\mathbf{x}) = \sum_{(i)} C_i \chi_i(\mathbf{x}),$$

where χ_i is the characteristic function of the subdomain occupied by the ith material. The discontinuity of elastic properties along some lines implies the discontinuity of stresses and the discontinuity of strains along these lines. However, some components of these tensors remain continuous.

Continuity of Stresses. Suppose that the stiffness $C(\mathbf{x})$ takes the values C_1 and C_2 in subdomains separated by a Γ. The equilibrium condition ($\nabla \cdot \boldsymbol{\tau} = 0$) requires that the normal components of the stresses be continuous:

$$[\tau_{nn}] = [\tau_{nt}] = [\tau_{nb}] = 0. \tag{14.2.3}$$

(Compare the continuity condition $[\mathbf{j}_n] = 0$ for the divergencefree vector of current, discussed in Chapter 2.) The other three components of stress can be discontinuous:

$$[\tau_{tt}] \neq 0, \quad [\tau_{bb}] \neq 0, \quad [\tau_{tb}] \neq 0. \tag{14.2.4}$$

The jump of the stress matrix on the surface Φ is

$$[\boldsymbol{\tau}] = \begin{pmatrix} 0 & 0 & 0 \\ 0 & [\tau_{tt}] & [\tau_{tb}] \\ 0 & [\tau_{tb}] & [\tau_{bb}] \end{pmatrix}.$$

Continuity of Strains. The representation (14.1.2) requires continuity of the tangential ($\mathbf{t}$) components of a strain

$$[\epsilon_{tt}] = [\epsilon_{tb}] = [\epsilon_{bb}] = 0. \tag{14.2.5}$$

This follows immediately from the continuity of the tangential components of $\nabla \mathbf{u}$.

The remaining components of strain can be discontinuous:

$$[\epsilon_{nn}] \neq 0, \quad [\epsilon_{nt}] \neq 0, \quad [\epsilon_{nb}] \neq 0.$$

The jump of the strain matrix on the surface Φ is

$$[\epsilon] = \begin{pmatrix} [\epsilon_{nn}] & [\epsilon_{nt}] & [\epsilon_{nb}] \\ [\epsilon_{nt}] & 0 & 0 \\ [\tau_{tb}] & 0 & 0 \end{pmatrix}.$$

To calculate the magnitude of the jumps of discontinuous components, one can use the procedure described in Chapter 2 for conductivity equations. We can use the six relations

$$[\tau] = [C : \epsilon] = C_1 : \epsilon_1 - C_2 : \epsilon_2$$

and six constraints (14.2.3), (14.2.5).

Plane Problem

In the plane problem, the matrices ϵ and τ are the upper-left 2×2 minors of the corresponding 3×3 matrices.

Obviously, we have

$$[\tau] = \begin{pmatrix} 0 & 0 \\ 0 & [\tau_{tt}] \end{pmatrix}, \quad [\epsilon] = \begin{pmatrix} [\epsilon_{nn}] & [\epsilon_{nt}] \\ [\epsilon_{nt}] & 0 \end{pmatrix}.$$

The component τ_{tt} of the stress is discontinuous, and two components $\epsilon_{nn}, \epsilon_{nt}$ of the strain are discontinuous.

14.2.3 Energy, Variational Principles

The elastic energy can be represented as a quadratic form of strain:

$$W_\epsilon = \frac{1}{2}\epsilon : \mathbf{C} : \epsilon = \frac{1}{2}\operatorname{Def}\mathbf{u} : C : \operatorname{Def}\mathbf{u}. \tag{14.2.6}$$

In this case, it is called the *strain energy*. The corresponding variational principle for the strain energy has the form

$$\mathcal{E}_\epsilon = \min_{\mathbf{u} \in \mathcal{U}} \int_\Omega \left(\frac{1}{2}\operatorname{Def}\mathbf{u} : C : \operatorname{Def}\mathbf{u} + \mathbf{f} \cdot \mathbf{u} \right) + \text{ boundary terms}, \tag{14.2.7}$$

where the set of admissible minimizers–the displacement vectors $\mathbf{u}$–consists of the vector fields that satisfy the Dirichlet boundary conditions on a boundary component $\partial\Omega_1$:

$$\mathcal{U} = \{\mathbf{u} : \; u_i \in W_2^1(\Omega), \; i = 1, 2, 3, \quad \mathbf{u}|_{\partial\Omega_1} = \boldsymbol{\rho}_1\}.$$

The Euler–Lagrange equations for this principle coincide with the elasticity equations in the Lamé form(14.1.9).

The elastic energy may also be presented as a quadratic form of stresses (stress energy):

$$W_\tau = \frac{1}{2}\tau : S : \tau, \quad \tau = \operatorname{Ink}\Phi. \tag{14.2.8}$$

The corresponding variational principle for the stress energy) yields to the problem for the potential Φ:

$$\mathcal{E}_\tau = \min_\Phi \frac{1}{2}\int_\Omega \operatorname{Ink}\Phi : S : \operatorname{Ink}\Phi + \text{ boundary terms}, \tag{14.2.9}$$

where the minimizers Φ satisfy the prescribed boundary conditions on the component $\partial\Omega_2$. For simplicity in notation, we assume that the forces are applied only to the boundary. The Euler–Lagrange equations for this principle coincide with the elasticity equations through the Airy function.

14.3   Elastic Structures

14.3.1   Elastic Composites

The theory of composites described in Chapter 2 for conducting media is applicable to elastic materials. One can establish the G-convergence of a sequence of differential operators and the G-limits of sequences of elastic properties layouts.

A composite of elastic materials is characterized by the rapidly oscillating tensor $C(\mathbf{x}) = \sum_{i=1}\chi_i(\mathbf{x})C_i$, where C_i is an elasticity tensor of the ith material, and χ_i is the index function of the subdomain filled with the ith material. Assuming that these materials are mixed in a fine scale, we introduce the average Hooke's law and the effective tensor C_{eff} of the composite.

$$\langle\tau\rangle = \langle C : \epsilon\rangle = C_{\text{eff}}\langle\epsilon\rangle.$$

This tensor connects the stresses and strains averaged over an elementary volume.

We also introduce the effective tensor using the variational principles. The elastic energy of an elementary periodic cell is

$$W_{\text{eff}} = \frac{1}{2}\min_{\langle\epsilon\rangle=\epsilon_0,\operatorname{Ink}\epsilon=0}\langle\epsilon : C : \epsilon\rangle = \frac{1}{2}\epsilon_0 : C_{\text{eff}} : \epsilon_0. \tag{14.3.1}$$

Here C_{eff} is the effective tensor of the elastic composite; ϵ_0 is the averaged strain. Equation (14.3.1) states that the homogeneous material C_{eff} has the same energy as the fine-scale inhomogeneous layout $C\varepsilon$.

Hill's Bounds

The simplest bounds for the effective tensor can be found from (14.3.1).
The constant trial field $\epsilon \equiv \epsilon_0$ leads to inequality

$$W_{\text{eff}} \leq \frac{1}{2}\epsilon_0 : \langle C \rangle : \epsilon_0.$$

The dual variational principle gives the complementary bound

$$W_{\text{eff}} \geq \frac{1}{2}\epsilon_0 : \langle C^{-1} \rangle^{-1} : \epsilon_0.$$

This implies bounds for an arbitrary effective tensor C_{eff}:

$$\langle C^{-1} \rangle^{-1} \leq C_{\text{eff}} \leq \langle C \rangle.$$

These bounds were derived for the elasticity in (Hill, 1963). They are examples of the arithmetic and harmonic mean bounds (see Chapter 3).

Hill's bounds restrict the G-closure of a set of elastic materials. The G-closure is a domain in the space of invariants of elastic tensors. The large number of invariants of elastic tensors (five for the plane problem, eighteen for the three-dimensional problem) and their nonuniform character makes the problem of the description of G-closure technically difficult. Most of the known results deal with a particular setting of the G-closure problem; we discuss them in Chapters 15 and 16.

14.3.2   Effective Properties of Elastic Laminates

Here we compute an effective tensor of elastic laminates. Suppose that laminates are made from materials with stiffness tensors C_1 and C_2. Their compliance tensors are denoted by $S_i = S_i^{-1}$. The normal and the tangent to the laminates are denoted by $\mathbf{n}$ and $\mathbf{t}$, and the fractions of materials in the laminate are denoted by m_1 and m_2.

Consider the plane problem. The theory developed in Chapter 7 is applicable here. To apply the formula (7.2.7), we determine the projector $\mathbf{q}$ to the subspace of discontinuous components of the stresses. We write the tensors of elastic constants as matrices in the basis (14.1.12), and stresses and strains as vectors in this basis. Only one component $\tau_{tt} = \boldsymbol{\tau} : (\mathbf{t} \otimes \mathbf{t})$ of the stress tensor is discontinuous, hence the projector $\mathbf{q}$ to the space of discontinuous components is the 3×1 matrix $\mathbf{T} = \mathbf{t} \otimes \mathbf{t}$. The projector $\mathbf{q}$ has the form

$$\mathbf{q} = \mathbf{T} = (\mathbf{a}_1 + \cos 2\phi\, \mathbf{a}_2 + \sin 2\phi\, \mathbf{a}_3), \qquad (14.3.2)$$

or, in the basis (14.1.12),

$$\mathbf{q} = \mathbf{T} = \frac{1}{\sqrt{2}} \begin{pmatrix} 1 \\ \cos 2\phi \\ \sin 2\phi \end{pmatrix},$$

where ϕ is the angle of declination of the tangential $\mathbf{t}$ to the $\mathbf{i}_1$-axis of the reference system.

In this basis, the compliance is the 3×3 matrix S. Formula (7.2.7) becomes

$$S_{\text{eff}} = m_1 S_1 + m_2 S_2 - m_1 m_2 \, Q, \qquad (14.3.3)$$

where the matrix Q is

$$Q = (S_1 - S_2)\mathbf{T} \left[\mathbf{T}^T (m_1 S_2 + m_2 S_1)\mathbf{T} \right]^{-1} \mathbf{T}^T (S_1 - S_2).$$

Notice that the term

$$\mathbf{T}^T (m_1 S_2 + m_2 S_1)\mathbf{T} = \tilde{s}$$

is a scalar. The formula (14.3.3) becomes

$$S_{\text{eff}} = m_1 S_1 + m_2 S_2 - \frac{m_1 m_2}{2\tilde{s}} \left((S_1 - S_2)\mathbf{T} \right) \otimes \left(\mathbf{T}^T (S_1 - S_2) \right).$$

For isotropic materials, $\tilde{s}$ is equal to

$$\tilde{s} = \frac{1}{2} \left(m_1 \left(\frac{1}{K_2} + \frac{1}{\mu_2} \right) + m_2 \left(\frac{1}{K_1} + \frac{1}{\mu_1} \right) \right)$$

and is independent of the direction $\mathbf{t}$. In the basis (14.1.12), where the axes $\mathbf{i}_1$, $\mathbf{i}_2$ are codirected with $\mathbf{t}$ and $\mathbf{n}$, we have $\mathbf{T} = \frac{1}{\sqrt{2}}(\mathbf{a}_1 + \mathbf{a}_2)$. The term Q in (14.3.3) becomes

$$Q = \frac{1}{\tilde{s}} \begin{pmatrix} c_k^2 & c_k\, c_\mu & 0 \\ c_k\, c_\mu & c_\mu^2 & 0 \\ 0 & 0 & 0 \end{pmatrix},$$

where

$$c_k = \frac{1}{K_1} - \frac{1}{K_2}, \quad c_\mu = \frac{1}{\mu_1} - \frac{1}{\mu_2}.$$

The expression for the effective stiffness C_{eff} can be derived either directly from (14.3.3) as $C_{\text{eff}} = S_{\text{eff}}^{-1}$ or from the basic formula (7.2.7). This time, the projector $\mathbf{q} = \mathbf{q}_\epsilon$ is a 2×3 matrix of the form

$$\mathbf{q}_\epsilon = \left(\mathbf{n} \otimes \mathbf{n}, \frac{1}{\sqrt{2}} (\mathbf{n} \otimes \mathbf{t} + \mathbf{t} \otimes \mathbf{n}) \right)$$

that has the representation

$$\mathbf{q}_\epsilon = \begin{pmatrix} 1 & 0 \\ -\cos 2\phi & -2\sin 2\phi \\ -\sin 2\phi & 2\cos 2\phi \end{pmatrix}$$

in the basis (14.1.12). The reader can check that the two approaches are equivalent.

14.3.3 Matrix Laminates, Plane Problem

Here we derive the effective properties of matrix laminates of the plane elasticity problem. These effective properties were first calculated in (Gibiansky and Cherkaev, 1984; Francfort and Murat, 1986; Gibiansky and Cherkaev, 1987) in connection with the optimization of the energy.

We apply formula (7.3.11) to that problem. The projector $\mathbf{q}_i$ in (7.3.11) is equal to the matrix $\mathbf{T}_i = \mathbf{t}_i \otimes \mathbf{t}_i$, where $\mathbf{t}_i$ is the tangent to the layers of ith rank. Matrix $\mathbf{T}_i$ has the vector form (14.3.2) in the basis (14.1.12).

The term

$$\mathbf{q}^T S_1 \mathbf{q} = H(S_1) = \mathbf{T}^T S_1 \mathbf{T},$$

generally depends on the direction $\mathbf{T}_i$. However, for an isotropic material S_1 with moduli K_1 and μ_1, the value of $H(S_1)$ is independent of $\mathbf{T}_i$:

$$H(S_1) = \frac{1}{2K_1} + \frac{1}{2\mu_1} = \frac{K_1 + \mu_1}{2K_1\mu_1}.$$

The matrix laminates made from two isotropic materials with compliances S_1 and S_2 (S_1 forms the envelope) admit the representation (see (7.3.11))

$$m_2(S_{\text{eff}} - S_1)^{-1} = (S_2 - S_1)^{-1} + m_1 \frac{2K_1\mu_1}{K_1 + \mu_1} G(\alpha_i, \mathbf{T}_i) \qquad (14.3.4)$$

where $\mathbf{T}_i = \mathbf{T}(\phi_i)$,

$$G(\alpha_i, \mathbf{T}_i) = \sum_{i=1}^{p} \alpha_i N_i(\mathbf{T}_i), \quad N_i(\mathbf{T}_i) = \mathbf{T}_i \otimes \mathbf{T}_i = \mathbf{T}_i \mathbf{T}_i^T,$$

$$\sum_{i=1}^{p} \alpha_i = 1, \quad \alpha_i \geq 0 \quad i = 1, \ldots, p. \qquad (14.3.5)$$

The matrix form of N_i (in the basis $\mathbf{a}_i$ (14.1.12)) is

$$N_i = N(\phi_i) = \frac{1}{2} \begin{pmatrix} 1 & \cos 2\phi_i & \sin 2\phi_i \\ \cos 2\phi_i & \cos^2 2\phi_i & \sin 2\phi_i \cos 2\phi_i \\ \sin 2\phi_i & \sin 2\phi_i \cos 2\phi_i & \sin^2 2\phi_i \end{pmatrix}.$$

Example 14.3.1 Consider a matrix laminate of second rank assembled from orthogonal laminates (see Figure 14.2). In this case, $\phi_1 = 0, \phi_2 = \frac{\pi}{2}$, and

$$G = \alpha_1 N(0) + \alpha_2 N\left(\frac{\pi}{2}\right) = \frac{1}{2} \begin{pmatrix} 1 & \alpha_1 - \alpha_2 & 0 \\ \alpha_1 - \alpha_2 & 1 & 0 \\ 0 & 0 & 0 \end{pmatrix}.$$

One can check that the material S_{eff} (14.3.4) is orthotropic if the component materials are isotropic. Its (13)- and (23)-components are equal to

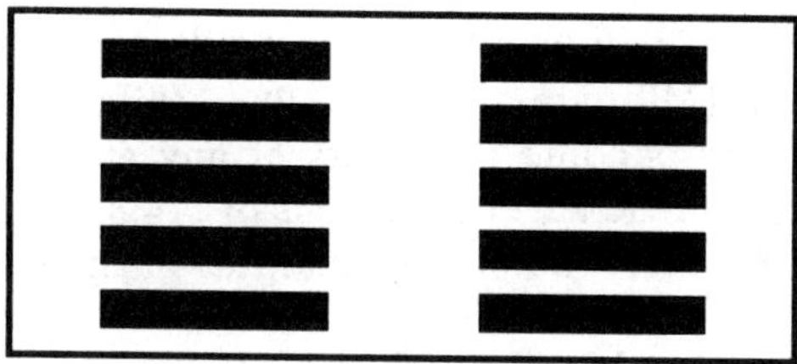

FIGURE 14.2. Orthogonal matrix laminate of the second rank. The "weak" direction is the shear.

zero, and the (33)-component is equal to the arithmetic mean of the shear compliances $\frac{1}{2\mu_1}$ and $\frac{1}{2\mu_2}$,

$$(S_{\text{eff}})_{33} = \frac{m_1}{2\mu_1} + \frac{m_2}{2\mu_2}.$$

If one of the mixed materials is also void, $\mu_2 = 0$, then S_{eff} has zero shear modulus and the structure in Figure 14.2 cannot resist a shear: it collapses under a shear load like a deck of cards.

Invariants of Plane Elastic Matrix Laminates

Let us find what properties of elastic matrix laminates are preserved if their inner parameters vary. Formula (14.3.4) shows that it is enough to find the invariant characteristics of the tensors $N_i(\mathbf{T}_i)$ that are the only terms depending on the inner parameters of structures.

As opposed to the conductivity, here we find two linear invariants of N_i. They are: the (11)-component of N_i

$$\text{TrS}\, N_i = (N_i)_{11} = \frac{1}{2}, \tag{14.3.6}$$

which we call the *spherical trace* of the fourth-rank tensor, and the sum of the (22)- and (33)-components of N_i,

$$\text{TrD}\, N_i = (N_i)_{22} + (N_i)_{33} = \frac{1}{2},$$

which we call the *deviatoric trace*.

Applying now the derived relations to (14.3.4) and using the equality $\sum \alpha_i = 1$, we end up with two scalar invariants of the set of matrix laminates:

$$\text{TrS}\,(S_{\text{eff}} - S_1)^{-1} = \left(\frac{1}{m_2}\right) \text{TrS}\,(S_2 - S_1)^{-1} + \left(\frac{m_1}{m_2}\right)\beta_1 \tag{14.3.7}$$

and

$$\text{TrD}\,(S_{\text{eff}} - S_1)^{-1} = \left(\frac{1}{m_2}\right) \text{TrD}\,(S_2 - S_1)^{-1} + \left(\frac{m_1}{m_2}\right)\beta_1, \tag{14.3.8}$$

where $\beta_1 = \frac{K_1\mu_1}{K_1+\mu_1}$. These expressions coincide with the so-called trace bounds, obtained in (Milton and Kohn, 1988; Zhikov, 1988). It has been proven that the corresponding invariants of any effective tensor S_{eff} are bounded by the right-hand-side expressions of (14.3.7) and (14.3.8).

As in the previous case, these two equalities must be supplemented by the inequalities

$$S_{\text{eff}} \leq m_1 S_1 + m_2 S_2, \tag{14.3.9}$$

which follow from the nonnegativeness of α_i.

One can also see that the matrix laminates of the third and higher rank do not have other restrictions. This inverse statement was proved in (Avellaneda and Milton, 1989a): Any elasticity tensor that satisfies the system of the restrictions (14.3.7), (14.3.8), (14.3.9) is an effective tensor of a third-rank matrix laminate structure.

Isotropic Matrix Laminates. The equations (14.3.7) and (14.3.8) state, in particular, that the isotropic matrix laminates composite is uniquely defined. It has the moduli $K_{\text{HS}}, \mu_{\text{HS}}$:

$$S_{\text{eff}} = \frac{1}{K_{\text{HS}}}\mathbf{a}_1 \otimes \mathbf{a}_1 + \frac{1}{\mu_{\text{HS}}}(\mathbf{a}_2 \otimes \mathbf{a}_2 + \mathbf{a}_3 \otimes \mathbf{a}_3),$$

which are obtained from the representations (14.3.7) and (14.3.8) by substitution (14.3.9) and similar representations for S_1 and S_2 and simplification:

$$\begin{aligned}
K_{\text{HS}} &= K_2 + m_1\left(\frac{1}{K_1-K_2} + \frac{m_2}{K_2+\mu_2}\right)^{-1}, \\
\mu_{\text{HS}} &= \mu_2 + m_1\left(\frac{1}{\mu_1-\mu_2} + \frac{m_2\ (K_2+2\mu_2)}{2\mu_2\ (K_2+\mu_2)}\right)^{-1}.
\end{aligned} \tag{14.3.10}$$

One can check that the third-rank composite with equal angles ($60°$) between the normals to layers and with equal parameters $\alpha_i = \frac{1}{3}$ is isotropic.

Note that these expressions and those obtained by interchanging the materials coincide with the Hashin-Shtrikman bounds (Hashin and Shtrikman, 1962b) that restrict the range of change of bulk and shear moduli of any isotropic mixture of two isotropic compounds. We discuss these bounds in Chapter 16. The effective properties of the structure coincide with the effective properties of the two-dimensional "coated circles" (Hashin and Shtrikman, 1962b).

Remark 14.3.1 *Equalities (14.3.10) show that matrix laminates alone cannot form the surface of the boundary of G_m-closure. Indeed, the moduli of the isotropic matrix laminates are uniquely connected. Generally, G_m-closure must be described by an equality that links these characteristics, but not by two separate equalities (14.3.10). The matrix laminates, however, could play a role similar to the role of the simple laminates in the three-dimensional G-closure: they may form "ribs" of that surface, that is, the subspaces of smaller dimensions, where the analytic components of the boundary meet.*

Moment Formulation

The matrix $G(\alpha_i, \mathbf{T}_i)$ (14.3.5) contains all the information about the structure of matrix laminate composites. It is convenient to represent this matrix through symmetric functions of α_i, ϕ_i. This trigonometric moment representation was described in (Avellaneda and Milton, 1989b). In the basis (14.1.12) the representation has the form

$$
G = \begin{pmatrix} \frac{1}{2} & 0 & 0 \\ 0 & \frac{1}{4} & 0 \\ 0 & 0 & \frac{1}{4} \end{pmatrix} + \frac{1}{2} \begin{pmatrix} 0 & M_1 & M_2 \\ M_1 & \frac{1}{2}M_3 & \frac{1}{2}M_4 \\ M_2 & \frac{1}{2}M_4 & -\frac{1}{2}M_3 \end{pmatrix},
$$

where

$$
M_1 = \sum_{i=1}^{N} \alpha_i \cos 2\phi_i, \quad M_2 = \sum_{i=1}^{N} \alpha_i \sin 2\phi_i
$$

are second-order moments, and

$$
M_3 = \frac{1}{2} \sum_{i=1}^{N} \alpha_i \cos 4\phi_i, \quad M_4 = \frac{1}{2} \sum_{i=1}^{N} \alpha_i \sin 4\phi_i
$$

are fourth-order moments.

It is convenient to use the parameters M_i as control variables in problems of structural optimization. They satisfy the inequalities

$$
M_1^2 + M_2^2 \le 1 \quad M_3^2 + M_4^2 \le 1,
$$

which follow from the definitions of M_i, and the inequalities

$$
5 - M_3^2 - 4(M_1^2 + M_2^2 + M_4^2) \ge 0,
$$
$$
1 - 2(M_1{}^2 + M_2{}^2) + 2\left(M_1{}^2 - M_2{}^2\right) M_3
$$
$$
- M_3{}^2 + 4 M_4(2 M_1 M_2 - M_4) \ge 0,
$$

which express the nonnegativeness of the second main invariant and the determinant of the 3×3 matrix G, respectively. The trace of G is equal to one for all M_i.

The inverse problem: Determination of the parameters of a third-rank composite from the given moments was solved in (Lipton, 1994a). The moments are convenient to use in the control problem, because the energy is easily expressed through them.

14.3.4   *Three-Dimensional Matrix Laminates*

The properties of matrix laminates for the three-dimensional problem are calculated in a similar way. The scheme is the same, but the calculations are more unwieldy. These structures were introduced for three-dimensional elasticity in (Francfort and Murat, 1986; Gibiansky and Cherkaev, 1987) in connection with the optimization of the energy of a three-dimensional elastic composite. Here we follow the last paper.

Basis. The elastic energy is given by the quadratic form (14.2.6), where ϵ is a symmetric 3×3 matrix in a reference system $\mathbf{i}_1$, $\mathbf{i}_2$, $\mathbf{i}_3$. To represent this matrix in vector form, we introduce a basis:

$$
\begin{aligned}
\mathbf{e}_1 &= \mathbf{i}_1 \otimes \mathbf{i}_1, & \mathbf{e}_2 &= \mathbf{i}_2 \otimes \mathbf{i}_2, \\
\mathbf{e}_3 &= \mathbf{i}_3 \otimes \mathbf{i}_3, & \mathbf{e}_4 &= \tfrac{1}{\sqrt{2}} \left(\mathbf{i}_2 \otimes \mathbf{i}_3 + \mathbf{i}_3 \otimes \mathbf{i}_2 \right), \\
\mathbf{e}_5 &= \tfrac{1}{\sqrt{2}} \left(\mathbf{i}_1 \otimes \mathbf{i}_3 + \mathbf{i}_3 \otimes \mathbf{i}_1 \right), & \mathbf{e}_6 &= \tfrac{1}{\sqrt{2}} \left(\mathbf{i}_1 \otimes \mathbf{i}_2 + \mathbf{i}_3 \otimes \mathbf{i}_1 \right).
\end{aligned}
\qquad (14.3.11)
$$

In this basis, the stress is represented by a six-dimensional vector $\tau = [\tau_1, \ldots, \tau_6]$, and the matrix of compliance S of an isotropic material is represented as the following 6×6 matrix:

$$
S = \begin{pmatrix}
s_1 & s_2 & s_2 & 0 & 0 & 0 \\
s_2 & s_1 & s_2 & 0 & 0 & 0 \\
s_2 & s_2 & s_1 & 0 & 0 & 0 \\
0 & 0 & 0 & s_3 & 0 & 0 \\
0 & 0 & 0 & 0 & s_3 & 0 \\
0 & 0 & 0 & 0 & 0 & s_3
\end{pmatrix},
\qquad (14.3.12)
$$

where

$$
s_1 = \frac{1}{3\mu} + \frac{1}{9k}, \qquad s_2 = -\frac{1}{6\mu} + \frac{1}{9k}, \qquad s_3 = \frac{1}{2\mu}.
$$

Only two constants are independent: $3s_1 = 3s_2 + 2s_3$.

Structures

Let us describe the class of third-rank matrix laminates with orthogonal normals to layers, following (Gibiansky and Cherkaev, 1987); see also (Cherkaev and Palais, 1997; Allaire, Bonnetier, Francfort, and Jouve, 1997). The effective properties of these laminates are given by formula (7.3.11). The continuity conditions are given by (14.2.3), (14.2.4).

The effective compliance S_{eff} satisfies equation (14.3.4). The projectors $\mathbf{q}_i$ are 3×6 matrices

$$
\mathbf{q}_1 = \begin{pmatrix}
0 & 0 & 0 \\
1 & 0 & 0 \\
0 & 1 & 0 \\
0 & 0 & 1 \\
0 & 0 & 0 \\
0 & 0 & 0
\end{pmatrix}, \quad
\mathbf{q}_2 = \begin{pmatrix}
1 & 0 & 0 \\
0 & 0 & 0 \\
0 & 1 & 0 \\
0 & 0 & 0 \\
0 & 0 & 1 \\
0 & 0 & 0
\end{pmatrix}, \quad
\mathbf{q}_3 = \begin{pmatrix}
1 & 0 & 0 \\
0 & 1 & 0 \\
0 & 0 & 0 \\
0 & 0 & 0 \\
0 & 0 & 0 \\
0 & 0 & 1
\end{pmatrix},
$$

and $H(S) = \mathbf{q}_i^T S \mathbf{q}_i$ is the 3×3 matrix

$$
H(S) = \begin{pmatrix}
s_1 & s_2 & 0 \\
s_2 & s_1 & 0 \\
0 & 0 & s_3
\end{pmatrix},
$$

which does not depend on the normal to the layer due to the isotropy of C.

The matrix $G = \sum_{i=1}^{3} \alpha_i \mathbf{q}_i H(S_1)^{-1} \mathbf{q}_i^T$ has the form

$$
\begin{pmatrix}
(\alpha_2 + \alpha_3)d_1 & \alpha_3 d_2 & \alpha_2 d_2 & 0 & 0 & 0 \\
\alpha_3 d_2 & (\alpha_1 + \alpha_3)d_1 & \alpha_1 d_2 & 0 & 0 & 0 \\
\alpha_2 d_2 & \alpha_1 d_2 & (\alpha_1 + \alpha_2)d_1 & 0 & 0 & 0 \\
0 & 0 & 0 & \alpha_1 d_3 & 0 & 0 \\
0 & 0 & 0 & 0 & \alpha_2 d_3 & 0 \\
0 & 0 & 0 & 0 & 0 & \alpha_3 d_3
\end{pmatrix},
$$

where

$$
d_1 = 4\frac{\mu(3k + \mu)}{3k + 4\mu}, \quad d_2 = 2\frac{\mu(3k - 2\mu)}{3k + 4\mu}, \quad d_3 = 2\mu.
$$

As before, three parameters α_i determine the relative thickness of differently oriented layers. The properties of the structures are defined by (14.3.4), and one must set the moduli k, μ as $k = k_1$, $\mu = \mu_1$, where k_1, μ_1 are the moduli of the first (enveloping) material.

Arbitrary matrix laminates in three dimensions also have two linear invariants: the spherical trace and the deviatoric trace. To make the structure isotropic, one needs at least six ranks of laminates. It was shown in (Francfort and Murat, 1986) that isotropic matrix laminates correspond to sixth-rank laminates with equal parameters $\alpha_i = \frac{1}{6}$ and the normals that are directed from the center of the icosahedron to its vertices. Isotropic matrix laminates have two moduli k and μ. They coincide with the bounds (Hashin and Shtrikman, 1962b). Therefore the matrix laminates exhibit the extremal properties in both shear and bulk moduli, similarly to the coated spheres obtained in (Hashin and Shtrikman, 1962b).

The anisotropic sixth-rank laminates were studied in (Francfort, Murat, and Tartar, 1995), where it was proved that they minimize the sum of elastic energies stored in an elastic medium.

14.3.5 Ideal Rigid-Soft Structures

The notion of the extremal material was introduced in Chapter 7. In elasticity, the extremal material is a rigid-soft material with zero compliance in certain tensor directions and zero stiffness in an orthogonal direction; see (7.4.4). Such effective properties have laminates that are made from from infinitely soft and infinitely stiff materials. The extremal materials are characterized by a subspace $\mathcal{X}$ (see (7.4.5)) of zero strain and undefined stresses and the complimentary subspace of undefined strains and zero stresses. Here we compute properties of extremal laminates following the method described in Chapter 7, see also (Milton and Cherkaev, 1995). We consider the two-dimensional problem.

Properties of Laminates

First, we repeat the arguments of Chapter 7, applying them to elastic materials. Consider a laminate structure assembled from two extremal materials.

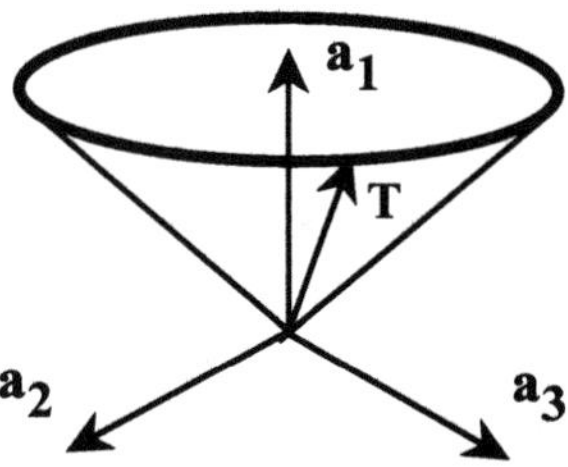

FIGURE 14.3. Representation of tensors $\mathbf{t} \otimes \mathbf{t}$ in the basis (14.1.12).

The strain in a direction is zero in the following cases: (i) The strain in this direction is zero in both mixing materials. (ii) The strain in this direction is zero in one of the mixing materials and the strain in this direction must be continuous within the laminate structure.

To find the subspace of zero strains, one must compute the direct (tensor) sum of the following hyperplanes: the intersection of the subspaces of zero strains of the mixing materials and the intersection of the zero subspace of each of the materials with the subspace of continuous strains in the composite.

The only component of the strain that remains continuous in the laminates is the $\mathbf{T}$ component. In the basis (14.1.12), it corresponds to the vector

$$\mathbf{T}(\gamma) = \frac{1}{2}[1, \ \cos 2\gamma, \ \sin 2\gamma],$$

where γ is the angle of the tangent to the layers $\mathbf{t}$. In the three-dimensional space of symmetric matrices with the basis (14.1.12), any normal corresponds to a vector that rotates along the latitude level of 45° (Figure 14.3).

Simple Laminates from Ideal Rigid and Soft Materials

Let us compute the space $\mathcal{X}_1$ of zero resistance (7.4.8) (the subspace of zeros) of a laminate structure assembled from the ideally stiff (rigid) material $\mathcal{X}_{01} = I$ and the ideally soft material $\mathcal{X}_{00} = 0$

$$\mathcal{X}_1 = (\mathcal{X}_{00} \cap \mathcal{X}_{01}) \oplus (\mathcal{X}_{00} \cap \mathbf{T}) \oplus (\mathcal{X}_{01} \cap \mathbf{T}). \qquad (14.3.13)$$

We have

$$\mathcal{X}_{00} \cap \mathcal{X}_{01} = 0, \quad \mathcal{X}_{01} \cap \mathbf{T} = \mathbf{T}, \quad \mathcal{X}_{00} \cap \mathbf{T} = 0.$$

Therefore, $\mathcal{X}_1 = \mathbf{T}$. This result tells that the only direction of strain that is supported by a laminate is the tangential traction. Any other finite strain cannot exist in the laminate made from the stiff material and the void; under any other impact, the structure simply falls apart and does not offer any resistance.

Second-Rank Laminates from Ideal Materials

Let us compute the subspace of zeros $\mathcal{X}_2$ for a second-rank composite. Mixing the laminate $\mathcal{X}_1 = \mathbf{T}(\gamma_1)$ with the rigid material $\mathcal{X}_{01}$ in a laminate directed along the direction $\mathbf{T}_2 = \mathbf{T}(\gamma_2)$, we obtain by (14.3.13) the matrix laminate characterized by the subspace of zeros:

$$\mathcal{X}_2(\mathbf{T}_1, \mathbf{T}_2) = (I \cap \mathbf{T}_1) \oplus (\mathbf{T}_1 \cap \mathbf{T}_2) \oplus (I \cap \mathbf{T}_2) = \mathbf{T}_1 \oplus \mathbf{T}_2.$$

The second-order laminates support any strain that has the form $c_1 \mathbf{T}_1 + c_2 \mathbf{T}_2$ and does not resist any orthogonal strain; see Figure 14.3. This orthogonal direction $\mathbf{T}_\perp$ is computed as

$$\begin{aligned} \mathbf{T}_\perp = \mathbf{T}_1 \times \mathbf{T}_2 = & \\ & [\sin 2(\gamma_2 - \gamma_1), \ \sin 2\gamma_1 - \sin 2\gamma_2, \ \cos 2\gamma_2 - \cos 2\gamma_1]. \end{aligned} \qquad (14.3.14)$$

In particular, the matrix laminate with orthogonal layers corresponds to the tracefree subspace $\mathbf{T}_\perp$, $\mathrm{Tr}\,\mathbf{T}_\perp = 0$. Indeed, we have from (14.3.14) that if $\gamma_2 - \gamma_1 = \frac{\pi}{2}$, then $\mathbf{T}_\perp$ does not have a projection on $\mathbf{a}_1$; we have $\mathbf{T}_\perp : \mathbf{a}_1 = 0$. This result says again that the structure of orthogonal matrix laminates made of a material and void does not support a shear strain no matter how stiff the material is.

The subspace of zeros is, however, not completely arbitrary even for an arbitrary second-rank matrix laminate. Geometrically, it is clear from Figure 14.3 that a plane supported by two radii pointing to two points on the 45° latitude line is never close to the equator.

Algebraically, this property corresponds to an inequality. We can show that the tensor $\mathbf{T}_\perp$ of zero resistance has a nonpositive determinant. Applying the formula (14.1.13) to $\mathbf{T}_\perp$ (14.3.14), we have

$$2 \det \mathbf{T}_\perp = -(1 + \cos 2(\gamma_2 - \gamma_1))^2 \le 0.$$

This formula shows that there exists a matrix laminate that does not support an arbitrary strain tensor with eigenvalues of different signs. However, all matrix laminates support strains with eigenvalues of the same signs.

Herringbone Structures from Ideal Materials

Finally, let us compute the subspace of zeros of a herringbone structure. These structures are hierarchically built from laminates made of second-rank matrix laminates; see Figure 14.4. These structures correspond to elastic composites of minimal stiffness (Gibiansky and Cherkaev, 1987) and to extremal polycrystals, (Avellaneda et al., 1996). The herringbone structure of ideal elastic materials structures was investigated in (Milton, 1992) to show attainability of composites with negative Poisson ratio.

Let us laminate two structures of second-rank matrix laminates that are characterized by four different angles of laminates. Denote the subspace of zeros of the constituencies as $\mathcal{X}_{00}$ and $\mathcal{X}_{01}$:

$$\mathcal{X}_{00} = \mathbf{T}_1 \oplus \mathbf{T}_2, \quad \mathcal{X}_{01} = \mathbf{T}_3 \oplus \mathbf{T}_4,$$

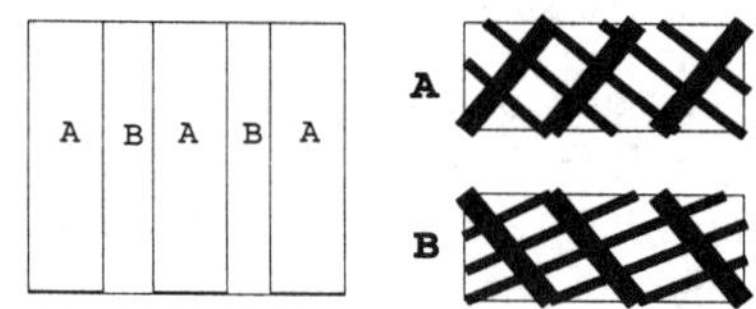

FIGURE 14.4. Herringbone structure.

where $\mathbf{T}_1, \ldots, \mathbf{T}_4$ show the direction of the laminates in the second-rank matrix laminates. Again we use (14.3.13), where we put $\mathbf{T} = \mathbf{T}_5$. The direction $\mathbf{T}_5$ shows the normal to the last layer.

We can easily show that $\mathbf{T}_5$ does not belong to $\mathbf{T}_1 \oplus \mathbf{T}_2$ and to $\mathbf{T}_3 \oplus \mathbf{T}_4$, unless $\mathbf{T}_5$ coincides with one of $\mathbf{T}_1, \ldots, \mathbf{T}_4$. This results in the equalities

$$\mathbf{T}_5 \cap (\mathbf{T}_1 \oplus \mathbf{T}_2) = 0, \quad \mathbf{T}_5 \cap (\mathbf{T}_3 \oplus \mathbf{T}_4) = 0.$$

Further, the formula (14.3.13) is simplified to

$$\mathcal{X} = (\mathbf{T}_1 \oplus \mathbf{T}_2) \cap (\mathbf{T}_3 \oplus \mathbf{T}_4).$$

It shows that the subspace of zeros $\mathcal{X}$ is the direction of the intersection of two planes in the three-dimensional space. To compute $\mathcal{X}$, we compute the cross-product of the normals to the planes $\mathbf{T}_1 \oplus \mathbf{T}_2$ and $\mathbf{T}_3 \oplus \mathbf{T}_4$, that is,

$$\mathcal{X} = \mathbf{T}_1 \times \mathbf{T}_2 \times \mathbf{T}_3 \times \mathbf{T}_4.$$

Using the definition of $\mathbf{T}$, we obtain this direction as a vector $\mathcal{X} = (x_1, x_2, x_3)$ in the basis (14.1.12). The components of $\mathcal{X}$ are

$$\begin{aligned}
x_1 &= -2 \sin \tfrac{\gamma_1 + \gamma_2 - \gamma_3 - \gamma_4}{2}, \\
x_2 &= -\sin \tfrac{\gamma_1 - \gamma_2 - \gamma_3 - \gamma_4}{2} - \sin \tfrac{\gamma_1 + \gamma_2 + \gamma_3 - \gamma_4}{2} \\
&\quad - \sin \tfrac{\gamma_1 + \gamma_2 - \gamma_3 + \gamma_4}{2} + \sin \tfrac{\gamma_1 - \gamma_2 - \gamma_3 + \gamma_4}{2}, \\
x_3 &= -\cos \tfrac{\gamma_1 - \gamma_2 - \gamma_3 - \gamma_4}{2} + \cos \tfrac{\gamma_1 + \gamma_2 + \gamma_3 - \gamma_4}{2} \\
&\quad + \cos \tfrac{\gamma_1 + \gamma_2 - \gamma_3 + \gamma_4}{2} - \cos \tfrac{\gamma_1 - \gamma_2 + \gamma_3 + \gamma_4}{2}.
\end{aligned}$$

The subspace of zeros $\mathcal{X}$ for these structures can be made arbitrary by appropriate choice of the angles γ_i.

Extremal Structures as Mechanisms

The method described allows us to construct simple mechanisms or the structures that transform a given strain to a given stress. A surprising example of such "mechanisms" was demonstrated in (Milton, 1992; Larsen, Sigmund, and Bouwstra, 1997), where the authors found materials with negative Poisson coefficient. The paper (Milton and Cherkaev, 1995) solves the even more general problem assembling an arbitrary tensor of three-dimensional elastic properties by extremal materials. In (Sigmund, 1996; Sigmund and Torquato, 1997) the large variety of hybrids between structures and mechanisms demonstrated solving problems of structural optimization. They are found by numerical solution to structural optimization problems.

14.4   Problems

1. Using representation (14.2.2), derive the equations of plane elasticity. Check the continuity conditions in the basis.

   Check that the operators Ink,Def,$\nabla\cdot$ applied to 3×3 symmetric tensors, and the operators Def, $\nabla\cdot$ applied to three-dimensional vectors, have the orthogonality properties: Ink Def $= 0$, $\nabla \cdot$ Ink $= 0$.

2. Using (14.2.1), show that $\nabla \cdot \boldsymbol{\sigma} = 0$.

3. Verify the representation (14.2.2).

   Consider the elastic energy in the forms (14.2.7) and (14.2.9). Derive the Euler–Lagrange equations for these forms. Show that they coincide with the Lamé system and the Airy equation. Check the duality of the forms: The Legendre transform of the Lagrangian (14.2.7) leads to the Lagrangian in the form (14.2.9).

4. Consider the set of matrix laminates of the second rank in two dimensions. How many invariants do they have? What are they?

5. Consider the set of matrix laminates of the second rank in two dimensions. Show that any matrix laminate of rank greater than three has the same effective properties tensor as a third-rank matrix laminate.

6. Suggest an ideal elastic laminate structure supporting two arbitrary strains.

15
Elastic Composites of Extremal Energy

Here we find the exact bounds of the elastic energy stored in an elastic composite, together with optimal structures that correspond to the "stiffest" and "softest" composites. This problem is similar to the the problem in Chapter 4.

15.1 Composites of Minimal Compliance

15.1.1 The Problem

Notation. Each elastic material with the compliance tensor S is characterized by the energy $W = \tau : S : \tau$, which depends on the stress τ (see (14.2.8)). (We double the energy to avoid repeatedly writing the coefficient $\frac{1}{2}$). We present the energy as a quadratic form in the tensor basis (14.1.12). In that basis, the tensor τ is considered a vector with coordinates $\tau_i = \tau : \mathbf{a}_i$, $i = 1, 2, 3$, where $\mathbf{a}_i$ are the basis tensors (14.1.12).

The isotropic compliance matrices

$$S_1 = S_{\text{isotr}}(\rho_1, \eta_1), \quad S_2 = S_{\text{isotr}}(\rho_2, \eta_2)$$

have the form (14.1.18):

$$S_{\text{isotr}}(\rho_i, \eta_i) = \begin{pmatrix} \rho_i & 0 & 0 \\ 0 & \eta_i & 0 \\ 0 & 0 & \eta_i \end{pmatrix}, \quad i = 1, 2.$$

The eigenvalues by ρ, η are proportional to inverse bulk and shear moduli: $\rho = \frac{1}{2K}$, $\eta = \frac{1}{2\mu}$.

Optimization problem. Consider the problem of the minimization of the compliance of an elastic structure, that occupies the domain $\mathcal{O}$ and is subject to a given external loading. This problem is similar to the problem in Chapter 4 but it is formulated for elastic materials. The stiffest structure minimizes the elastic energy plus the cost of the materials used; it corresponds to the variational problem

$$I = \inf_{\chi} \; \inf_{\tau: \; \tau \cdot \mathbf{n}|_{\partial\mathcal{O}} = \mathbf{f}} \int_{\mathcal{O}} \Phi(\chi, \tau)$$

where

$$\Phi(\chi, \tau) = \chi(W(S_1) + \gamma_1) + (1 - \chi)(W(S_2) + \gamma_2),$$

$W(S_i)$ is the energy of ith material, γ_i is its cost, $\mathbf{f}$ is the external load.

Interchanging the sequence of the extremal operations and calculation the minimum over χ, we reformulate the problem as

$$I = \inf_{\tau: \; \tau \cdot \mathbf{n}|_{\partial\mathcal{O}} = \mathbf{f}} \int_{\mathcal{O}} F(\tau),$$

where

$$F(\tau) = \min\{W(S_1) + \gamma_1, \; W(S_2) + \gamma_2\}. \tag{15.1.1}$$

This nonquasiconvex variational problem requires relaxation.

Optimal composite. Consider a two-dimensional composite made of materials S_1, S_2. Suppose that the materials are mixed in the proportions m_1 and $m_2 = 1 - m_1$. The composite is submerged into a homogeneous stress field τ_0. The compliance of the composite is defined as the stress energy $W(\chi, \tau)$ stored in the unit volume Ω:

$$W(\chi, \tau_0) = \min_{\tau: \; \nabla \cdot \tau = 0, \langle \tau \rangle = \tau_0} \langle \tau : S(\chi) : \tau \rangle, \tag{15.1.2}$$

where

$$S(\chi) = \chi S_{\text{isotr}}(\rho_1, \eta_1) + (1 - \chi) S_{\text{isotr}}(\rho_2, \eta_2).$$

The stiffest composite minimizes $W(\chi, \tau_0)$ among all two-phase composites assembled of the given materials with the given volume fractions. Denote its stiffness tensor by S_{opt}.

The optimal media adjusts itself to an acting field τ, and S_{opt} may depend on the rotationally invariant characteristics of the normalized tensor $\frac{\tau}{\|\tau\|}$. The 2×2 symmetric tensor τ has only one independent parameter. It could be the ratio of eigenvalues τ_A and τ_B of τ, or the ratio of its spherical and deviatoric parts

$$\psi = \frac{\operatorname{Tr} \tau}{\|\operatorname{Dev} \tau\|},$$

where Parameters $\mathrm{Tr}\,\boldsymbol{\tau}$ and $\|\mathrm{Dev}\,\boldsymbol{\tau}\|$ can be expressed either through the elements of the tensor $\boldsymbol{\tau}$ in the natural basis and in the tensor basis $\mathbf{a}_i$ or through its eigenvalues:

$$\mathrm{Tr}\,\boldsymbol{\tau} = \tau_{11} + \tau_{22} = \sqrt{2}\tau_1 = \tau_A + \tau_B,$$

$$\|\mathrm{Dev}\,\boldsymbol{\tau}\| = \tau_2^2 + \tau_3^2 = \sqrt{\frac{(\tau_{11} - \tau_{22})^2}{4} + \tau_{12}^2} = |\tau_A - \tau_B|.$$

Here, τ_{ij} denote the components of the tensor $\boldsymbol{\tau}$ in a reference Cartesian system, and τ_1, τ_2, τ_3 denote the components of $\boldsymbol{\tau}$ in the tensor basis $\mathbf{a}_i$.

The tensor S_{opt} can be found from the following variational problem:

$$W_{\mathrm{opt}}(\boldsymbol{\tau}, m) = \boldsymbol{\tau} : S_{\mathrm{opt}}(m, \psi) : \boldsymbol{\tau} = \min_{\chi : \langle \chi \rangle = m} \langle W(\chi, \boldsymbol{\tau}) \rangle.$$

Note that the stress energy $W_{\mathrm{opt}}(\boldsymbol{\tau}, m)$ of an optimal structure corresponds to the energy of an equivalent nonlinear elastic material. The nonlinearity is due to the dependence of S_{opt} on the parameter ψ of the stress.

We find the best structure, minimizing the Lagrangian over m:

$$F_{\mathrm{opt}}(\boldsymbol{\tau}) = \min_{m} \left(\boldsymbol{\tau} : S_{\mathrm{opt}}(m, \psi) : \boldsymbol{\tau} + \gamma m \right)$$

where γ is the difference in cost of materials S_1 and S_2.

It remains to solve the variational problem

$$\inf_{\boldsymbol{\tau} : \nabla \cdot \boldsymbol{\tau} = 0} \int_{\Omega} F_{\mathrm{opt}}(\boldsymbol{\tau})$$

for $\boldsymbol{\tau}$. The boundary conditions on $\boldsymbol{\tau}$ are assumed. Notice that this problem deals with the unknown arrangement of locally optimal composites, represented by the optimal effective properties tensors.

This problem (in the two-dimensional case) was solved in (Gibiansky and Cherkaev, 1984; Kohn and Lipton, 1988; Lipton, 1988); the three-dimensional case was considered in (Gibiansky and Cherkaev, 1987). The problem was further investigated in numerous papers (Petukhov, 1989; Bendsøe et al., 1993; Allaire and Kohn, 1993a; Allaire and Kohn, 1993b; Lipton, 1993; Lipton, 1994b; Allaire and Kohn, 1994; Bendsøe, Hammer, Lipton, and Pedersen, 1995; Grabovsky, 1996; Allaire et al., 1997), at al.

A numerical approach to the problem was developed and numerical results were obtained in (Gibiansky and Cherkaev, 1984; Bendsøe and Kikuchi, 1988; Suzuki and Kikuchi, 1991; Díaz, Kikuchi, Papalambros, and Taylor, 1993; Haber, Jog, and Bendsøe, 1996; Allaire et al., 1997; Díaz and Lipton, 1997), among others.

15.1.2 Translation Bounds

The Bound. Here we compute the lower bound of the stored energy. We use the translation method to bound the energy (15.1.2) by an expression

$$W(\chi, \boldsymbol{\tau}_0) \geq \max_{T \in \mathcal{T}} \boldsymbol{\tau} : S_p(T) : \boldsymbol{\tau},$$

where S_p is a translated matrix

$$S_p = [m_1(S_{\text{isotr}}(\rho_1, \eta_1) - T)^{-1} + m_2(S_{\text{isotr}}(\rho_2, \eta_2) - T)^{-1}]^{-1} + T, \quad (15.1.3)$$

T is the matrix of the translation, and the set $\mathcal{T}$ is

$$\mathcal{T} = \{T : \ S_1 - T \geq 0, \ S_2 - T \geq 0\}.$$

The Translator. The equilibrium constraints on the stresses $\nabla \cdot \boldsymbol{\tau} = 0$, $\boldsymbol{\tau} = \boldsymbol{\tau}^T$ imply the existence of the translator $t \det \boldsymbol{\tau}$ (see Chapter 8). This translator corresponds to the null-Lagrangian (14.1.24) related to the Airy equation. It has the form:

$$2t \det(\boldsymbol{\tau} : \mathcal{R} : \boldsymbol{\tau}) = \boldsymbol{\tau} : S_{\text{isotr}}(t, -t) : \boldsymbol{\tau}.$$

The translator matrix $T(t) = S_{\text{isotr}}(-t, t)$ is an isotropic matrix of elastic properties with the bulk and shear moduli of equal absolute values and different signs (obviously, this matrix does not correspond to any real material). It is proportional to the fourth-rank rotation tensor $\mathcal{R}$ introduced in Chapter 14 (14.1.20).

Let us compute the shifted matrices $S_{\text{isotr}}(\rho, \eta) - T$ (see (15.1.3)). The matrices T and $S_{\text{isotr}}(\rho, \eta)$ have the same eigentensors(14.1.12). Therefore the shifted matrix has a simple representation

$$S_{\text{isotr}}(\rho, \eta) - tS_{\text{isotr}}(1, -1) = S_{\text{isotr}}(\rho + t, \eta - t). \quad (15.1.4)$$

The shifted matrix is a compliance matrix of an isotropic material with the elastic moduli $\rho + t$ and $\eta - t$.

The translation bound becomes

$$\mathcal{P}W(\boldsymbol{\tau}) = \max_{t \in \mathcal{T}} \left\{ \boldsymbol{\tau} : S_{\text{isotr}}(\rho_p, \eta_p) : \boldsymbol{\tau} \right\} \quad (15.1.5)$$

where $S_{\text{isotr}}(\rho_p, \eta_p)$ is the isotropic matrix with moduli

$$\begin{aligned}
\rho_p &= \left(\frac{m_1}{\rho_1 + t} + \frac{m_2}{\rho_2 + t} \right)^{-1} - t, \\
\eta_p &= \left(\frac{m_1}{\eta_1 - t} + \frac{m_2}{\eta_2 - t} \right)^{-1} + t,
\end{aligned} \quad (15.1.6)$$

and

$$\mathcal{T} = \{t : \ \rho_i + t \geq 0, \ \eta_i - t \geq 0, \ i = 1, 2\}.$$

Both moduli ρ_p, η_p depend on the translation parameter t.

Comparing Media. The stress energy W_p of an arbitrary anisotropic composite is bounded by the stress energy $\mathcal{P}W(\boldsymbol{\tau}_0, t)$ of an isotropic comparing medium with the moduli ρ_p, η_p. The isotropy of the comparing medium is expected because the set of all composites assembled from isotropic materials is isotropic.

The bulk $\rho_p(t)$ and shear $\eta_p(t)$ moduli are connected by a relation that is obtained by excluding the parameter $t \in [-\rho_1, \ \eta_1]$ from (15.1.4). From the variety of the comparing media one should choose the one that leads to the most restrictive bound; the choice depends on the acting field τ. An increase of the bulk modulus K_p of the comparing medium is accompanied by a decrease of the shear modulus η_p.

Optimal Translator. Optimal values of t are found from the conditions

$$\frac{\partial \mathcal{P}W}{\partial t} = 0, \quad \frac{\partial^2 \mathcal{P}W}{\partial t^2} \leq 0, \quad t \in \mathcal{T}. \tag{15.1.7}$$

The equation in (15.1.7) has two roots we call t' and t''. Let us compute one of them; the other is computed similarly. The root t' is a linear-fractional functions of τ_A, τ_B:

$$t' = \frac{A\tau_A + B\tau_B}{C\tau_A + D\tau_B}.$$

Here

$$A = \rho_1\,\eta_1 - \rho_2\,\eta_1 - \rho_1\,m\,\eta_1 + \rho_2\,m\,\eta_1 + \rho_1\,m\,\eta_2 - \rho_2\,m\,\eta_2,$$
$$B = -\rho_1\,\eta_1 + \rho_1\,m\,\eta_1 - \rho_2\,m\,\eta_1 + \ \rho_1\,\eta_2 - \rho_1\,m\,\eta_2 + \rho_2\,m\,\eta_2,$$
$$C = \rho_2 - \rho_1, \quad D = \eta_2 - \eta_1.$$

The expression for t'' is similar.

Suppose for definiteness that the materials' constants are related as

$$\rho_1 \leq \rho_2, \quad \eta_1 \leq \eta_2. \tag{15.1.8}$$

The other case is analyzed similarly; see (Gibiansky and Cherkaev, 1984). In the case (15.1.8), the optimal parameter t_{opt} of translator is equal to t' if $t \in [-\rho_1, \ \eta_1]$, or it takes the boundary values:

$$t_{\mathrm{opt}} = \begin{cases} -\rho_1 & \text{if} & t' \geq -\rho_1, \\ t' & \text{if} & -\rho_1 \leq t' \leq \rho_1, \\ \eta_1 & \text{if} & t' \leq \rho_1. \end{cases}$$

These intervals correspond to the following loadings ψ:

$$t_{\mathrm{opt}} = \begin{cases} -\rho_1 & \text{if} & \psi \leq \Psi_1, \\ t' & \text{if} & \Psi_1 \leq \psi \leq \Psi_2, \\ \eta_1 & \text{if} & \psi \geq \Psi_2, \end{cases} \tag{15.1.9}$$

where the parameters Ψ_1 and Ψ_2 are found by solving (15.1.7) for $t' = -\rho_1$ and for $t' = \eta_1$. They are

$$\Psi_1 = \frac{A + B - \rho_1(C + D)}{A - B - \rho_1(C - D)}, \quad \Psi_2 = \frac{A + B + \eta_1(C + D)}{A - B + \eta_1(C - D)}.$$

In all cases, the optimal value t_{opt} depends only on the ratio ψ. It is independent of the norm of τ_0 and of the orientation of its principle axes.

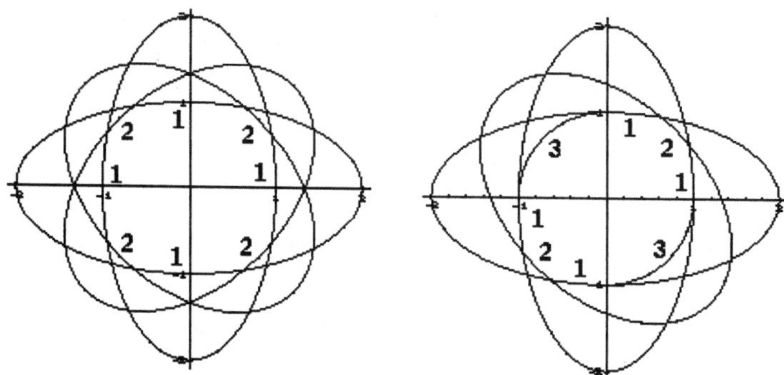

FIGURE 15.1. Bounds on the stored energy. The (τ_A, τ_B)-plane. The levels of the optimal stress energy $QW(\tau_A, \tau_B) = 1$ (left) and the levels of the optimal strain energy $QW'(\epsilon_A.\epsilon_B) = 1$ (right) are the intersection of all ellipses. The digits on the curves show the following cases: (i) The translation parameter t_{opt} belongs to the interior of the admissible interval. Optimal structures are the laminates, codirected with the eigenvector of the tensor of the external loading. (ii) The translator parameter t_{opt} reaches its nonzero limit. Optimal structures are the laminates of the second rank (iii) (Right only) The translator parameter t_{opt} is zero. Optimal structures are the laminates inclined to the eigenvectors of the tensor of the mean field.

The Bounds. The translation bound is given by (15.1.5)–(15.1.6). All the bounds obtained are schematically represented in Figure 15.1 (left) where the coordinates correspond to the eigenvalues τ_A, τ_B of the stress.

Figure 15.1 (left) pictures the level set $QW = 1$ of the minimal stress energy equal to

$$QW(\tau_A, \tau_B) = \rho_p(t_{\mathrm{opt}})(\tau_A + \tau_B)^2 + \eta_p(t_{\mathrm{opt}})(\tau_A - \tau_B)^2 = 1$$

In the (τ_A, τ_B)-plane, the level set is a curved octagon formed as the intersection of four ellipses. Each ellipse corresponds to one of the cases in (15.1.9). The vertically and horizontally oriented ellipses correspond to the limiting values of t_{opt}. The inclined ellipses correspond to the intermediate values of t_{opt}. Two ellipses correspond to two different orders of eigenvalues of τ.

15.1.3 Structures

We can show that the obtained translation bounds for stored energy are exact: There are structures that store the energy $PW(\tau)$ for all values of the tensor τ. These structures are either properly oriented laminates or properly oriented second-rank laminates. The optimal structure varies with the loading τ to minimize the energy of that loading.

Second-Rank Laminates. Let us find the minimal energy that can be stored in the plane elastic laminates and second-rank laminates. The problem can be formulated as an algebraic minimization problem of optimal choice of parameters of the second-rank laminates:

$$\min_{\text{structure}} \ \boldsymbol{\tau} : S_{\text{ml-2}} : \boldsymbol{\tau},$$

where $S_{\text{ml-2}}$ is the compliance of the second-rank matrix laminates.

Consider the second-rank laminates with their normals oriented along the principal axes of the stress tensor $\boldsymbol{\tau}$. The basis tensors $\mathbf{a}_i$ are determined by (14.1.12), and it is assumed that the coordinate axes are codirected to the principal axes of loading: $(\boldsymbol{\tau} : \mathbf{a}_3 = 0)$. We have

$$\tau_1 = \frac{1}{\sqrt{2}}(\tau_A + \tau_B), \quad \tau_2 = \frac{1}{\sqrt{2}}(\tau_A - \tau_B).$$

The effective properties tensor of laminates is given by (14.3.3). The energy is a quadratic form of the vector $[\tau_1, \tau_2, 0]$ or a quadratic form of the two-dimensional vector $[\tau_1, \tau_2]$ with the 2×2-matrix

$$\tilde{S}(\alpha) = \tfrac{1}{2}\begin{pmatrix} \rho_1 & 0 \\ 0 & \eta_1 \end{pmatrix}$$
$$+ \tfrac{m_2}{2}\left(\begin{pmatrix} \frac{1}{\rho_2} - \frac{1}{\rho_1} & 0 \\ 0 & \frac{1}{\eta_2} - \frac{1}{\eta_1} \end{pmatrix} + \frac{m_1}{\rho_1 + \eta_1}\begin{pmatrix} 1 & 2\alpha - 1 \\ 2\alpha - 1 & 1 \end{pmatrix} \right)^{-1},$$

which is the upper-left 2×2 minor of $S_{\text{ml-2}}$ (14.3.4). Here $\alpha \in [0, 1]$ is the inner parameter of the structure that defines its anisotropy. Notice that we set $\alpha_1 = \alpha$, $\alpha_2 = 1 - \alpha$ in (14.3.4). The parameter α must be chosen to minimize the energy.

Optimal Structures. The optimal value of $\alpha \in [0, 1]$ can be found from the equation $\frac{\partial W}{\partial \alpha} = 0$, that has two roots α' and α''.

The first root leads to the energy that coincides with the translation bound in the interval $\psi \leq \Psi_1$, and the second root leads to the energy that matches the translation bound in the interval $\psi \geq \Psi_2$.

The condition α', $\alpha'' \in [0, 1]$ defines the range of applicability of the solution. One can check that $\alpha' = 0$ if $\psi = \Psi_1$ and $\alpha'' = 0$ if $\psi = \Psi_2$. This shows that the translation bound is exact in the intervals $\psi \leq \Psi_1$ and $\psi \geq \Psi_2$ of ψ. The corresponding calculations are obvious but bulky. We refer to the original paper (Gibiansky and Cherkaev, 1984) or leave this calculation as an exercise (Maple is helpful).

In the remaining region, $\Psi_2 \leq \psi \leq \Psi_1$, the optimal value of α is zero. The structure degenerates into simple laminates. One can also check that the energy stored in laminates corresponds to the translation bound if the parameter t of the translator does not reach its limits. This proves that laminates are the optimal structures for this interval.

Discussion. The variety of optimal structures that respond to the stress conditions is presented in Figure 15.2 (A, B, C). The level lines of the energy stored in different optimal structures is shown in Figure 15.1 (left).

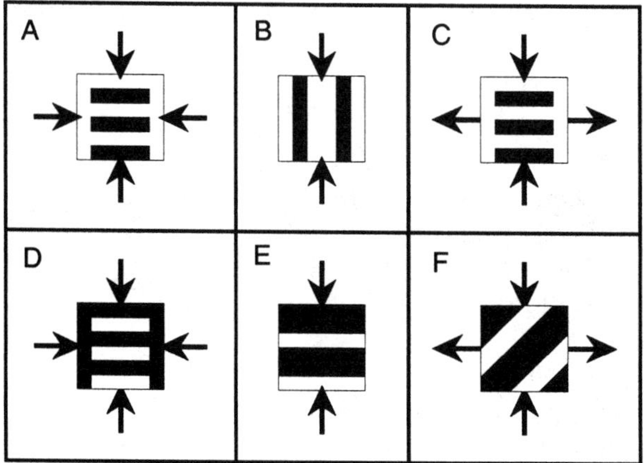

FIGURE 15.2. Structures of minimal compliance (A, B, C) and of minimal stiffness (D, E, F). White indicates a strong material, black indicates a weak material. The arrows show the direction of the applied external stress (A, B, C) or of the applied external strain (D, E, F).

Figure 15.3 compares the upper and lower bounds of the energy. The upper family corresponds to the energy stored in the matrix laminates; the curves correspond to different values of α. The lower family represents the translation bounds; the curves correspond to different values of the translation parameter t. These families are parted for better visibility. One can see the following:

1. The envelopes of the two families coincide.

2. The part of the upper boundary that corresponds to the varying stationary value of the structural parameter α in the second rank laminates matches the lower bound with the fixed value of the parameter t ($t = -\rho_1$ or $t = \eta_1$) lying on the boundary of the admissible interval.

3. The part of the upper boundary that corresponds to the boundary of the admissible interval of the structural parameter α (the structure degenerates into simple laminates) matches the envelope of the lower bounds (which corresponds to the variable stationary value of t).

This completes our consideration of the bounds for the stress energy: The translation bounds are exact; the first- or second-rank laminates realize them.

Asymptotics

Consider the asymptotic case

$$S_2 \to \infty, \quad \text{or} \quad \rho_2 \to \infty, \ \eta_2 \to \infty. \tag{15.1.10}$$

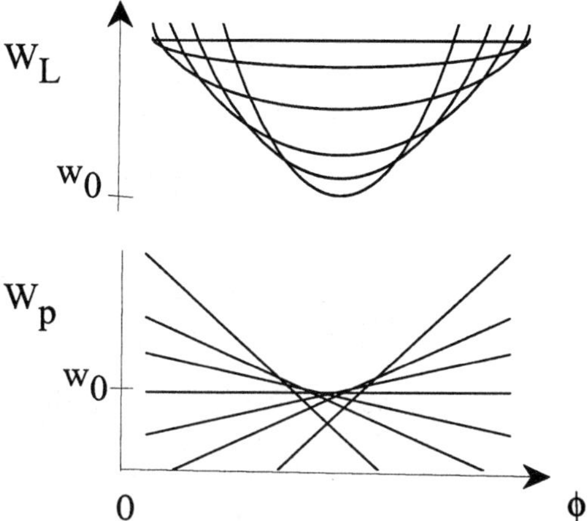

FIGURE 15.3. Families of the upper W_L and lower W_p bounds of the stress energy; dependence on the external field. The curves in the family W_L show the energy of the second-rank laminates with different values of α. The curves in the family W_p show the energy of the translation bound with different values of t.

This case corresponds to zero elastic moduli in the "weak" material; actually, we are dealing with the strongest composite of an elastic material and a material of infinitely compliance (void). This problem is often called the *problem of topology optimization* (Bendsøe, 1995); it asks for the best topology of an elastic skeleton of the construction.

Remark 15.1.1 *The effective properties of a mixture of a material and the void may not exist because the elastic phase could be disconnected by the void phase. Alternatively, some distant points in such composites may be directly connected, but not connected to the points in between; then the force is transferred to distant points passing intermediate points. In these cases, the effective properties are meaningless; the effective elastic moduli of the structure do not exist.*

However, the structures of maximal resistance correspond to finite energy and, therefore, to effective elastic moduli (although some of the moduli could be equal to zero). Indeed, the asymptotic (15.1.10) of the expression for the optimal moduli is regular. Geometrically, the optimal structure does not fall apart because the weak phase is located in the inclusions in the stiff phase.

The moduli ρ_p, η_p of the bound are obtained by passing to the limit (15.1.10) in (15.1.6):

$$\rho_p = \frac{\rho + t}{m_1} - t, \quad \eta_p = \frac{\eta - t}{m_1} + t,$$

where we omit the index $_1$ in the notation for the moduli. The bound $\tau{:}S_{\text{isotr}}(\rho_p, \eta_p) : \tau$ linearly depends on $t \in [-\rho, \eta]$, which implies that the translation bound always corresponds to an extremal value $t = -\rho$ or $t = \eta$. After some calculations, the lower bound becomes

$$PW = \frac{1}{2}\left(\frac{\rho + \eta}{m_1}\left(|\tau_A| + |\tau_B|\right)^2 - 4\beta\,|\tau_A\|\tau_B|\right), \qquad (15.1.11)$$

where

$$\beta = \begin{cases} \eta & \text{if } \tau_A\tau_B \geq 0, \\ \rho & \text{if } \tau_A\tau_B \leq 0. \end{cases} \qquad (15.1.12)$$

Remark 15.1.2 *The quasiconvex and convex envelopes coincide if $\tau_A\,\tau_B = 0$. In this case, the optimal second-rank laminates degenerate into simple laminates. At that point the optimal structure may "fall apart", because a layer of the void appears and the normal stress is not supported. However, this stress is equal to zero.*

15.1.4   The Quasiconvex Envelope

Let us pass to the problem (15.1.1) of the optimal layout of the composites in the elastic structure, i.e., to the construction of the quasiconvex envelope. To make the calculations easier, we consider the asymptotic case (15.1.10) of the mixture of an elastic material with the void.

The Lagrangian of the problem (15.1.1) before relaxation is

$$F(\tau) = \frac{1}{2}\begin{cases} 0 & \text{if } \tau = 0, \\ \gamma + \tau : S : \tau & \text{if } \tau \neq 0, \end{cases} \qquad (15.1.13)$$

where $\frac{1}{2}\gamma$ is the cost of the material and the cost of the void is zero.

To find the quasiconvex envelope, it remains to determine the optimal volume fraction m_{opt} from the equation

$$\frac{\partial}{\partial m}\left(\frac{1}{2}\gamma m + PW(m, \tau)\right) = 0,$$

where $PW(m, \tau)$ is defined in (15.1.11).

It depends on τ as follows:

$$m_{\text{opt}} = \phi(\tau), \quad \phi(\tau) = \min\left\{1, \sqrt{\frac{\rho + \eta}{\gamma}}\left(|\tau_A| + |\tau_B|\right)\right\}.$$

The quasiconvex envelope of the Lagrangian (15.1.13) is

$$QF = \frac{1}{2}\begin{cases} \sqrt{(\rho + \eta)\gamma}\,(|\tau_A| + |\tau_B|) - 4\beta|\tau_A\|\tau_B| & \text{if } \phi(\tau) \leq 1, \\ \gamma + \tau : S : \tau & \text{if } \phi(\tau) \geq 1, \end{cases} \qquad (15.1.14)$$

where β is defined in (15.1.12). The quasiconvex envelope is shown in Figure 15.4.

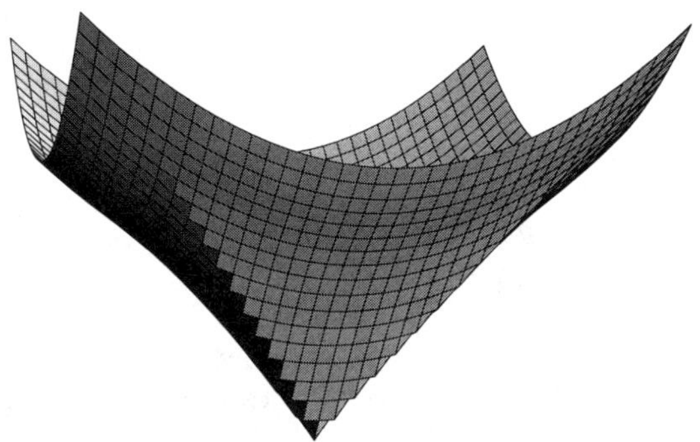

FIGURE 15.4. The quasiconvex envelope $\mathcal{Q}F(\tau_A, \tau_B)$ (15.1.14) of the Lagrangian (15.1.13). The ribs of the quasiconvex envelope correspond to the uniaxial field ($\tau_A \tau_B = 0$) when the quasiconvex envelope coincides with the convex envelope.

The Constitutive Relations in an Optimal Body

Let us analyze the strain–stress relations in an optimal media that has the optimal stress energy $\mathcal{Q}F(\tau)$. The relations can be found from the equation $\epsilon = \frac{\partial}{\partial \tau} \mathcal{Q}F$.

Recall (Section 14.1) that the stress field does not completely determine the strains, because the stress energy is defined up to a null-Lagrangian $t \det \tau = t\tau_A \tau_B$. Therefore, the strain is defined up to the term

$$\epsilon_0 = t\frac{\partial}{\partial \tau} \det \tau = t\mathcal{R} : \tau, \qquad (15.1.15)$$

where t is an arbitrary parameter. One can check that $\nabla \cdot \epsilon_0 \equiv 0$.

Besides the obvious case $\phi(\tau) \geq 1$ when the solid material is optimal, there are two cases to consider:

1. Suppose that $\det \tau \geq 0$, $\phi(\tau) \leq 1$. We rewrite the Lagrangian as

$$\mathcal{Q}F = -2\eta \det \tau + \frac{1}{2}\sqrt{(\rho + \eta)\gamma}\,|\operatorname{Tr} \tau|.$$

The first term on the right-hand side is the null-Lagrangian, and the second term is the L_2 norm of the spherical part of τ. The corresponding constitutive relations are

$$\frac{\partial \mathcal{Q}F}{\partial \tau} = \epsilon = -2\eta\mathcal{R} : \tau + \frac{1}{2}\sqrt{(\rho + \eta)\gamma}\,\operatorname{sign}(\operatorname{Tr} \tau)I.$$

One can use the nonuniqueness of the strain field (see (15.1.15)) and eliminate the first term. The strain-stress dependence becomes

$$\epsilon = \frac{\sqrt{(\rho + \eta)\gamma}}{2}\,\operatorname{sign}(\operatorname{Tr} \tau)I.$$

This dependence implies the following:

1. ϵ is proportional to the unit matrix I; the structure has only overall strain.

2. The constitutive relations are not local; the tensor ϵ stays constant unless the sign of the $\mathrm{Tr}\,\tau$ changes.

3. The strain stays finite, even when the stresses are infinitesimal.

These properties can be explained if we recall that the variation of the load leads to rearrangement of the materials in the optimally designed structure. This rearrangement keeps certain characteristics of the strain and stress constant.

2. Suppose that $\det\tau \leq 0$, $\phi(\tau) \leq 1$. We rewrite the energy as

$$QF = \sqrt{(\rho+\eta)\gamma\,\mathrm{Dev}\,\tau : \mathrm{Dev}\,\tau}$$

(recall that $\mathrm{Dev}\tau$ is the deviatoric (tracefree) part of τ). The last term in the expression for QF is the L_2 norm of $\mathrm{Dev}\,\tau$. The corresponding constitutive relations are

$$\frac{\partial QF}{\partial\tau} = \epsilon = \frac{\sqrt{(\rho+\eta)\gamma}}{2}\frac{\mathrm{Dev}\,\tau}{\|\mathrm{Dev}\,\tau\|}.$$

This equation is similar to the previous case. The term $\frac{\mathrm{Dev}\,\tau}{\|\mathrm{Dev}\,\tau\|}$ is the deviator (the tracefree tensor) with unit magnitude, codirected with τ.

Remark 15.1.3 *It is interesting to compare an optimal elastic medium with an optimal conducting medium, described in Chapter 3. In the optimal conducting medium, the dual vector has a constant magnitude. Here the dual variable ϵ has the constant norm, too. Instead of the constancy of L_2 norm, the preserved norm is equal to the sum of the absolute values of the eigenvalues ϵ_A and ϵ_B of ϵ: $|\epsilon_A| + |\epsilon_B| = constant.$*

The Density of the Energy in the Material. The volume fraction of the material in the optimal composite is proportional to the norm $N(\tau)$ of the stress tensor. The energy density in the material W_{mat} inside the optimal structure is equal to $W_{\mathrm{mat}} = \dfrac{W(m_{\mathrm{opt}},\tau)}{m_{\mathrm{opt}}}$, because no energy is stored in the void phase. We obtain

$$W_{\mathrm{mat}} = \sqrt{(\rho+\eta)\gamma} = \text{constant}. \tag{15.1.16}$$

We note that W_{mat} is independent of the level of the stress applied to the material in the optimal structure. This constancy corresponds to the principle of equally stressed structures, which is often used as an intuitive criterion of optimality; see (Olhoff and Taylor, 1979; Achtziger, Bendsøe, Ben-Tal, and Zowe, 1992). Formula (15.1.16) expands the intuitive principle of "equally stressed material" to an arbitrary anisotropic state of the stress.

15.1.5 Three-Dimensional Problem

These results can be extended to the three-dimensional problem. The method is similar to the two-dimensional case, but calculations are more unwieldy. The problem was solved in (Gibiansky and Cherkaev, 1987); here we review the results following that paper. The problem was further investigated in (Allaire and Kohn, 1994), and the numerical solutions were obtained in (Allaire et al., 1997; Díaz and Lipton, 1997; Olhoff, Scheel, and Rønholt, 1998), and the axisymmetric case was studied in (Cherkaev and Palais, 1997; Cherkaev and Palais, 1998).

We find the optimal structure of a three-dimensional elastic composite, which stores the minimal stress energy in a prescribed tensor field τ. As in the two-dimensional case, we use the translation method to bound the optimal energy. The exposition follows (Gibiansky and Cherkaev, 1987).

Translation Bound

The translation bound takes the form

$$W_{3d} \geq \max_{a_i} \ \tau : S_p(a_i) : \tau,$$

where

$$S_p = \left(m_1 (S_1 - T_\tau(a_i))^{-1} + m_2 (S_2 - T_\tau(a_i))^{-1} \right)^{-1} + T_\tau(a_i),$$

and a_i are the parameters of the translator. The bound is similar to the two-dimensional case. The matrices of compliance S_1 and S_2 of the isotropic materials in the composite have the form (14.3.12) in the basis (14.3.11).

This problem requires use of a translator that nonlinearly depends on the parameters. The constraints $\nabla \cdot \tau = 0$ imply the existence of the translator $\tau : T_\tau : \tau$ (see Chapter 8). This translator (see (8.2.16)) depends on three free parameters a_1, a_2, and a_3. In our basis (14.3.11), the matrix $T_\tau(a_1, a_2, a_3)$ has the form

$$T_\tau = \begin{pmatrix} a_1^2 & -a_1 a_2 & -a_1 a_3 & 0 & 0 & 0 \\ -a_1 a_2 & a_2^2 & -a_2 a_3 & 0 & 0 & 0 \\ -a_1 a_3 & -a_2 a_3 & a_3^2 & 0 & 0 & 0 \\ 0 & 0 & 0 & a_2^2 + a_3^2 & 0 & 0 \\ 0 & 0 & 0 & 0 & a_1^2 + a_3^2 & 0 \\ 0 & 0 & 0 & 0 & 0 & a_1^2 + a_2^2 \end{pmatrix}.$$

The range of the parameters a_i is determined by the inequalities $S_1 - T_\tau(a_i) \geq 0$, $S_2 - T_\tau(a_i) \geq 0$. Note that T_τ becomes nonnegative if one of parameters a_1, a_2, or a_3 vanishes.

It was shown in (Gibiansky and Cherkaev, 1987) that the translation bound is realizable; it corresponds to the structure of the third-rank orthogonal matrix laminates that were described in Chapter 14. The calculations are straight but bulky, and we refer the reader to the original paper. However, we demonstrate the calculation in an especially simple case.

Optimal Structures: Special Case

Consider a mixture of two materials. Suppose that the first material has the constants $k_1 = \frac{1}{3}$, $\mu_1 = \frac{1}{2}$ that corresponds to $S_1 = I$. In other terms, the Poisson coefficient of the material is equal to zero and its Young modulus is equal to one. Suppose also that the second material is void; $(k_2 = \mu_2 = 0)$, which corresponds to $S_2 = \infty$.

The translation bound for the stress energy is

$$W_{3d} \geq \boldsymbol{\tau} : \boldsymbol{\tau} + \frac{m_2}{m_1} \max_{a_i} \ \boldsymbol{\tau} : T_\tau(a_i) : \boldsymbol{\tau}.$$

The range of parameters a_i is determined by the condition $I - T_\tau(a_i) \geq 0$, which leads to inequalities

$$
\begin{array}{ll}
a_1^2 + a_2^2 \leq 1, & a_1^2 + a_3^2 \leq 1, \\
a_2^2 + a_3^2 \leq 1, \quad & a_1^2 + a_2^2 + a_3^2 - 4a_1a_2a_3 \leq 1.
\end{array}
$$

The optimal values of a_i depend on the eigenvalues of the tensor $\boldsymbol{\tau}$. Let us denote them by τ_1, τ_2, τ_3 and let us agree that $|\tau_3| \geq |\tau_1|$ and $|\tau_3| \geq |\tau_2|$. A straightforward calculation demonstrates that the optimal values a_i^{opt} are

$$a_1^{\mathrm{opt}} = \frac{1}{\sqrt{2}}\mathrm{sign}(\tau_1\tau_2), \qquad a_2^{\mathrm{opt}} = \frac{1}{\sqrt{2}}\mathrm{sign}(\tau_2\tau_3),$$

and

$$a_3^{\mathrm{opt}} = \begin{cases} \frac{1}{\sqrt{2}} & \text{if } \ |\tau_1| + |\tau_2| \geq |\tau_3|, \\ \frac{1}{\sqrt{2}}\left(\frac{|\tau_2|}{|\tau_3|} + \frac{|\tau_2|}{|\tau_3|}\right) & \text{if } \ |\tau_1| + |\tau_2| \leq |\tau_3|. \end{cases}$$

The translation bound becomes

$$W_{\mathrm{transl}}(\boldsymbol{\tau}) = \tau^2 + \frac{m_2}{2m_1}C,$$

where $\tau^2 = \boldsymbol{\tau} : \boldsymbol{\tau}$ and

$$C = \begin{cases} 2\left(|\tau_1| + |\tau_2| + |\tau_3|\right)^2 & \text{if} \quad |\tau_1| + |\tau_2| \geq |\tau_3|, \\ \left(|\tau_1| + |\tau_2|\right)^2 + |\tau_3|^2, & \text{if} \quad |\tau_1| + |\tau_2| \leq |\tau_3|. \end{cases}$$

The translation bound for the stress energy is exact. It is attainable on the matrix laminates that are oriented along principal axes of $\boldsymbol{\tau}$. These structures were described in Chapter 14 (see (14.3.4)). The energy stored in such a structure is

$$W_{\mathrm{struct}} = \tau^2 + \frac{m_1}{m_2}\sum_{i=1}^{3}\frac{\tau_i^2}{1 - \alpha_i},$$

where $\alpha_1, \alpha_2, \alpha_3$ are nonnegative parameters of the structure and $\alpha_1 + \alpha_2 + \alpha_3 = 1$.

The optimal values of α_i are

$$
\left.
\begin{aligned}
a_1 &= \frac{-|\tau_1|+|\tau_2|+|\tau_3|}{|\tau_1|+|\tau_2|+|\tau_3|}, \\
a_2 &= \frac{|\tau_1|-|\tau_2|+|\tau_3|}{|\tau_1|+|\tau_2|+|\tau_3|}, \\
a_3 &= \frac{|\tau_1|+|\tau_2|-|\tau_3|}{|\tau_1|+|\tau_2|+|\tau_3|},
\end{aligned}
\right\}
\quad \text{if} \quad |\tau_1| + |\tau_2| \geq |\tau_3| \qquad (15.1.17)
$$

and

$$
\left.
\begin{aligned}
a_1 &= \frac{|\tau_2|}{|\tau_1|+|\tau_2|}, \\
a_2 &= \frac{|\tau_1|}{|\tau_1|+|\tau_2|}, \\
a_3 &= 0
\end{aligned}
\right\}
\quad \text{if} \quad |\tau_1| + |\tau_2| \leq |\tau_3|. \qquad (15.1.18)
$$

The upper and lower bounds again match; this proves the their sharpness.

The case (15.1.17) corresponds to nondegenerate laminates of the third rank. The normals to the layers are codirected with the eigenvectors of the stress τ. The anisotropy of the structure equalizes the response to stresses of different magnitudes acting in the orthogonal directions.

Consider the complementary case $a_3 = 0$, (15.1.18). The optimal structure consists of cylinders of the second-rank laminates. The generator of the cylinders is oriented along the direction of the largest eigenvalue of τ. The normals to the layers are codirected with two smaller eigenvectors of the stress τ. These structures have the maximal possible stiffness in the direction of the generator, and they equalize the response to stresses of different magnitudes applied in the directions across the generator.

We observe that the optimal three-dimensional structures tend to the optimal two-dimensional structures if the absolute value of the largest eigenstress is larger than the sum of the absolute values of the other two eigenstresses.

15.2   Composites of Minimal Stiffness

Similarly in the previous problem, we find the structure of composites of minimal stiffness for the two-dimensional problem. These composites store the minimum of the strain energy: The fixed strain is accompanied by minimal stress. We follow (Gibiansky and Cherkaev, 1984) where this problem was considered.

This time we characterize the materials by the (doubled) strain energy W_ϵ that depends on the strain ϵ (see (14.2.6)) as

$$
W_\epsilon = \frac{K}{2}(\epsilon_{11} + \epsilon_{22})^2 + \frac{\mu}{2}(\epsilon_{11} - \epsilon_{22})^2 + \frac{\mu}{2}\epsilon_{12}^2.
$$

In the tensor basis (14.1.12), the strain energy is presented as a quadratic form

$$
W_\epsilon = K\epsilon_1^2 + \mu(\epsilon_1^2 + \epsilon_3^2) = \epsilon : C(K,\mu) : \epsilon
$$

associated with the stiffness matrix

$$C(K, \mu) = K\mathbf{a}_1 \otimes \mathbf{a}_1 + \mu(\mathbf{a}_2 \otimes \mathbf{a}_2 + \mathbf{a}_3 \otimes \mathbf{a}_3).$$

The eigenvalues K, μ of the matrix C are the bulk and shear moduli of the material, respectively. The vector components ϵ_1, ϵ_2, ϵ_3 are equal to $\epsilon_i = \boldsymbol{\epsilon} : \mathbf{a}_i$.

To find the "soft" composite we minimize the strain energy of a composite over all structures. The composite is built from two isotropic materials with moduli K_1, μ_1 and K_2, μ_2 and it minimizes the energy if is submerged into an external stress field $\boldsymbol{\epsilon}$.

15.2.1   Translation Bounds

The Translator. The general scheme of deriving the bound is similar to the previous case. However, the differential constraints on the strain tensor are different than the discussed ones. The constraints Ink $\boldsymbol{\epsilon} = 0$ are the linear combination of second derivatives of the components ϵ_j of strain $\boldsymbol{\epsilon}$:

$$a_{ijkl}\frac{\partial^2 \epsilon_j}{\partial x_k \partial x_l} = 0, \quad i = 1, \ldots, r. \tag{15.2.1}$$

There are two ways to deal with this situation, and the results are identical. We could consider the strain energy as a function of the nonsymmetric matrix $\boldsymbol{\zeta} = \nabla \mathbf{u}$. The differential constraints on $\boldsymbol{\zeta}$ are familiar: $\nabla \times \boldsymbol{\zeta} = 0$; they depend only on first derivatives. The translator $\phi(\boldsymbol{\zeta})$ is immediately found to be $\phi(\boldsymbol{\zeta}) = \det \boldsymbol{\zeta}$. On the other hand, the properties matrices C_i become degenerate; their eigenvalue that corresponds to the antisymmetric eigentensor is zero. The translation bounds must be adjusted to this case. This method (the modified translation method) will be discussed in Chapter 16, where we will be dealing with bounds for the sum of elastic energies.

Here we demonstrate a straightforward approach (Gibiansky and Cherkaev, 1984). The equilibrium constraint ^{2}Ink $\boldsymbol{\epsilon} = 0$ implies the existence of a translator. The translator $\phi(\boldsymbol{\epsilon})$ for the tensor $\boldsymbol{\epsilon} = \mathrm{Def}\,\mathbf{u}$ is a quasiconvex (not quasiaffine!) function:

$$\phi(\boldsymbol{\epsilon}) = -t \det \boldsymbol{\epsilon} \text{ is quasiconvex} \quad \text{if } t \geq 0.$$

Indeed, this function can be written as a sum of quasiaffine and convex functions of $\nabla \mathbf{u}$:

$$-\det(\mathrm{Def}\,\mathbf{u}) = -\det(\nabla \mathbf{u}) + \frac{1}{4}\left(\frac{\partial u_1}{\partial x_2} - \frac{\partial u_2}{\partial x_1}\right)^2,$$

which implies quasiconvexity. The quasiconvexity of the translator requires the optimal value t_0 of the parameter t to be nonnegative.

As in the previous case of relaxation of the stress energy $W(\tau)$, the translator is associated with the quadratic form $\epsilon : (C_{\text{isotr}}(-t,t)) : \epsilon$, so that the matrix T of the translator is $T = C_{\text{isotr}}(-t,t)$. The only formal difference between this problem and the problem of minimal compliance is the inequality $t \geq 0$.

The Translation Bound. The calculation of the translation bound for this problem is similar the previous case. The shifted matrices are

$$C_{\text{isotr}}(K, \mu) - C_{\text{isotr}}(-t, t) = C_{\text{isotr}}(K + t, \mu - t)$$

and the range of t is

$$0 \leq t \leq \mu_1$$

(recall that in the problem of minimal compliance the range of t includes the negative part).

The translation bound (8.3.3) for the strain energy has the form:

$$\mathcal{P}W(\epsilon) = \max_{t:\, 0 \leq t \leq \mu_1} \{\epsilon : C_{\text{isotr}}(K_p(t),\, \mu_p(t)) : \epsilon\}$$

where

$$K_p(t) = \left(\frac{m_1}{K_1+t} + \frac{m_2}{K_2+t} \right)^{-1} - t,$$
$$\mu_p(t) = \left(\frac{m_1}{\mu_1-t} + \frac{m_2}{\mu_2-t} \right)^{-1} + t.$$

The optimal value of t_{opt} is found from the condition

$$\frac{\partial \mathcal{P}W}{\partial t} = 0, \quad \frac{\partial^2 \mathcal{P}W}{\partial t^2} \leq 0, \quad t \in [0, \mu_1].$$

Again, the parameter t depends only on the ratio ψ_e of the spherical and deviatoric parts of the tensor ϵ_*:

$$\psi_e = \frac{\epsilon_* : \mathbf{a}_1}{|\epsilon_* : \mathbf{a}_2|} = \frac{\epsilon_A + \epsilon_B}{|\epsilon_A - \epsilon_B|},$$

where ϵ_A and ϵ_B are the eigenvalues of ϵ_*.

The results are quite analogous to the previous case. However, here the quasiconvex envelope coincides with the convex envelope in a range of parameters where $t_{\text{opt}} = 0$. For the details, we refer to the original paper (Gibiansky and Cherkaev, 1984), or the reader can carry it out as an exercise.

15.2.2 *The Attainability of the Convex Envelope*

We examine again the laminates and matrix composites. As in the problem of minimal compliance, we can show that the energy in the second-rank laminates (in which the material C_2 forms an envelope and the material

C_1 forms the nuclei) matches the translation bound when $t_{\text{opt}} = \mu_1$. The energy of the simple laminates that are oriented along the axis of maximal strain matches the translation bound when $t_{\text{opt}} \in (0, \mu_1)$.

The new effects occur when $t_{\text{opt}} = 0$, and the quasiconvex and convex envelopes coincide. This happens when ψ_e, the ratio of the deviatoric part of the external field ϵ and its spherical part is larger than a constant c.

The convex envelope has the representation

$$CW(\epsilon) = m\epsilon_1 : C_1 : \epsilon_1 + (1 - m)\epsilon_2 : C_2 : \epsilon_2,$$

where the two supporting strains ϵ_1, ϵ_2 in the first and second materials are subject to the constraints

$$m_1\epsilon_1 + (1 - m)\epsilon_2 = \epsilon, \quad C_1 : \epsilon_1 = C_2 : \epsilon_2.$$

The supporting strains can be expressed through the mean strain ϵ as follows:

$$\epsilon_1 = C_2\tilde{C}^{-1} : \epsilon, \quad \epsilon_2 = C_1\tilde{C}^{-1} : \epsilon, \tag{15.2.2}$$

where $\tilde{C} = m_1 C_2 + m_2 C_1$.

The convex envelope is attainable by a laminate if the supporting strains ϵ_1 and ϵ_2 inside the laminates are compatible. The compatability condition

$$[\epsilon] : \mathbf{t} \otimes \mathbf{t} = [\epsilon]_{tt} = 0 \tag{15.2.3}$$

comes from the differential constrain Ink $\epsilon = 0$. Here $[\epsilon] = \epsilon_1 - \epsilon_2$. Condition (15.2.3) poses additional requirement for the supporting points of the convex envelope.

Condition (15.2.3) can be considered as an equation for the tangent $\mathbf{t}$; it has a solution if the matrix $[\epsilon]$ has eigenvalues λ_1 and λ_2 of different signs. Indeed, the component $[\epsilon_{tt}]$ has the representation

$$[\epsilon]_{tt} = \lambda_1 \cos^2 \theta + \lambda_2 \sin^2 \theta. \tag{15.2.4}$$

Here θ is the angle between the eigenvector of $[\epsilon]$ and $\mathbf{t}$. Condition (15.2.3) is satisfied if the tensor $[\epsilon]$ has the eigenvalues λ_1, λ_2 of different signs, $\lambda_1 \lambda_2 \leq 0$. In this case, there exists a direction $\mathbf{t}$ along which the component $[\epsilon]_{tt}$ is zero.

In turn, the condition $\lambda_1 \lambda_2 \leq 0$ is equivalent to the compatibility condition is $\det[\epsilon] \leq 0$. Using (15.2.2), we transfer the last inequality to the constraint on the applied field ϵ,-

$$\det[(C_2 - C_1) : \tilde{C}^{-1} : \epsilon] \leq 0.$$

If this attainability condition holds, then the quasiconvex envelope is attainable by laminates. In this case, the quasiconvex and convex envelopes coincide.

Compatibility and Differential Constraints of Second Order

Let us comment on the attainability of the convex envelope for a system with the second-order differential constraints.

As we showed in Chapter 7, the attainability of the convex envelope depends on the solvability of the algebraic system determined by the linear differential constraints. In our case, the constraints (15.2.1) are differential forms of second order. However, the consideration of Section 7.1 is still valid.

The algebraic system that determines the attainability of the convex envelope is obtained from the differential constraint $^2\text{Ink}\,\epsilon = 0$ by the formal replacement of ∇ by the vector $\mathbf{n}$ and the vector ϵ by the vector of jumps $[\epsilon]$. This system has the form

$$\sum_{k,l} \mathbf{n}_l \cdot B_{kl}\mathbf{n}_k = 0, \qquad (15.2.5)$$

where the matrix B_{kl} is defined by the tensor $[\boldsymbol{\xi}]$ and by the tensor of differential constraints a_{ijkl} (see (15.2.1)) as

$$B_{kl} = \sum_{i,j} a_{ijkl}[\boldsymbol{\xi}]_{ij}.$$

Here $\mathbf{n}_k$ is the component of the normal to the dividing line where the discontinuity occurs, and $[\ ]$ is the magnitude of the jump over that line.

There is a similarity between the construction of a minimizing sequence for a problem with first-order linear differential constraints (Section 7.1) and the current case, where the constraints are of the second order. In dealing with first-order differential constraints, we replace them with a linear system. The laminate minimizing sequence that realizes the convex envelope can be constructed if the linear system is solvable for some $\mathbf{n}$. It is solvable if rank B is less than d.

Here we face quadratic equations instead. The laminate minimizing sequence that realizes the convex envelope can be constructed if the quadratic system (15.2.5) is solvable for some $\mathbf{n}$. This system is solvable if B is not positive definite. This condition includes, of course, the case (rank $B \leq d$). Generally, the requirements are weaker: B may have nonzero eigenvalues of different signs. The requirement on B to be nonpositive produces constraints on the matrix $[\epsilon]$ because B linearly depends on it. In turn, the value of $[\epsilon]$ depends on the average field ϵ. As a result, the convex envelope becomes attainable in a range of values of mean fields ϵ and not attainable in another range of it.

Structures

The range $t_{\text{opt}} = 0$ when the convex envelope is attainable is realized by laminates turned to the principal axes of the strain ϵ at some angle ϕ. This angle depends on the applied strain ϵ.

The diagonal component $[\epsilon]_{11}$ of the tensor $[\epsilon]$ is expressed through its eigenvalues λ_1, λ_2 as in (15.2.4). The condition

$$[\epsilon]_{11} = 0 \quad \text{or} \quad \tan\theta = \pm\sqrt{-\frac{\lambda_1}{\lambda_2}} \qquad (15.2.6)$$

corresponds to real ϕ if the eigenvalues λ_1, λ_2 are of different signs.

We have shown that there exists a laminate structure with specially oriented laminates that is compatible with the differential constraints, because the tangential component of the difference $[\epsilon]$ is zero. Moreover, such structures are not unique. First, there are two solutions ϕ of (15.2.6) of different signs. They correspond to differently oriented laminates that store the same energy. Consequently, there are composites assembled from these laminates that are also optimal.

The optimal structures are schematically shown in Figure 15.2, and the level lines of the quasiconvex envelope are shown in Figure 15.1.

Three-Dimensional Problem

The three-dimensional problem admits similar consideration. This problem was considered in (Gibiansky and Cherkaev, 1987) for the asymptotic case $D_2 = \infty$. The three-dimensional construction shares the properties of the two-dimensional problem: When the eigenvalues of the tensor ϵ are of the same sign, the problem is similar to the problem of the composite of minimal compliance. If the eigenvalues have different signs, some translators vanish and the optimal structures are cylindrical herringbone structures of laminates of the second rank. The generator $\mathbf{g}$ of cylinders is oriented along the direction where $\epsilon_{gg} = 0$ and the angle between the orthotropic materials in the herringbone structure is optimally chosen (the properties of the herringbone structure are calculated in Section 16.3). The reader is referred to the original paper for details.

15.3   Optimal Structures Different from Laminates

15.3.1   Optimal Structures by Vigdergauz

An interesting type of optimal structures with irregular geometry but with simple fields inside was suggested in the series of papers (Vigdergauz, 1986; Vigdergauz, 1988; Vigdergauz, 1989; Vigdergauz, 1994b; Vigdergauz, 1994a; Grabovsky and Kohn, 1995a; Grabovsky and Kohn, 1995b; Vigdergauz, 1996). The related problem of an optimal shape of a cavity in an elastic plane was considered in many papers. We cite (Cherepanov, 1974; Banichuk, 1977; Banichuk, Bel'skiĭ, and Kobelev, 1984; Vigdergauz and Cherkaev, 1986; Vigdergauz, 1994b; Markenscoff, 1997/98; Cherkaev, Grabovsky, Movchan, and Serkov, 1998).

The Functional. Let us start with the problem of optimization of the shape of a cavity in an infinite plane. This problem is the limiting case of a dilute composite when one phase (the void) has ideal properties and the interaction of the inclusions is neglected.

Consider an infinite elastic plane weakened by a cavity and loaded by a homogeneous loading

$$\tau_0 = a\mathbf{i} \otimes \mathbf{i} + b\mathbf{j} \otimes \mathbf{j} \quad a\,b > 0 \qquad (15.3.1)$$

applied at infinitely distant points.

The minimizing functional is associated with the stress energy stored in that plane. However, the total energy of an infinite plane is infinite. Therefore we consider the functional equal to the difference between the energy stored in a weakened plane and the energy stored in a solid plane. This difference is finite and positive. The problem is to minimize it by choosing the shape of a cavity Ω_{cav} with fixed area $|\Omega_{\mathrm{cav}}|$;

$$I(\tau_0) = \min_{\Omega_{\mathrm{cav}}:|\Omega_{\mathrm{cav}}|=1} \left(\min_{\tau \in \mathcal{S}} H(\tau, \Omega_{\mathrm{cav}}) - \tau_0 : S : \tau_0 \right). \qquad (15.3.2)$$

Here

$$H(\tau, \Omega_{\mathrm{cav}}) = \int_{R^2 - \Omega_{\mathrm{cav}}} W(S, \tau) + \gamma |\Omega_{\mathrm{cav}}|,$$

$$\mathcal{S} = \{ \nabla \cdot \tau = 0, \quad \langle \tau \rangle = \tau_0, \quad \tau_{nn} = \tau_{nt} = 0 \text{ on } \partial\Omega_{\mathrm{cav}} \}, \qquad (15.3.3)$$

γ is the Lagrange multiplier, and $W(S, \tau) = \tau : S : \tau$ is the energy. Obviously, I is proportional to the area of the cavity:

$$I = P(\tau_0)|\Omega_{\mathrm{cav}}|, \qquad (15.3.4)$$

where $P(\tau_0)$ is a dimensionless coefficient that depends on the shape of the cavity. The problem is to minimize P.

Let us derive the necessary conditions of optimality of the shape of the cavity. These conditions are to be satisfied on the boundary of the optimal cavity ∂G. They can be derived using the classical variational scheme (Courant and Hilbert, 1962).

The variational problem compares two nearby boundaries of the cavity Ω_{cav} represented by the normals $\mathbf{n}$ and $\mathbf{n} + \delta\mathbf{n}$, where $|\delta\mathbf{n}| \ll 1$. The stationary condition with respect to $\mathbf{n}$ (Prager, 1968) is

$$W(S, \tau)|_{\partial\Omega_{\mathrm{cav}}} + \gamma = 0. \qquad (15.3.5)$$

This condition states that the stored energy $W(S, \tau)$ is constant along the boundary $\partial\Omega_{\mathrm{cav}}$.

On the other hand, the conditions (15.3.3) on the boundary of the cavity require that two components of τ vanish: $\tau_{nn} = \tau_{nt} = 0$. The energy

$W(S, \tau)$ becomes proportional to the square of the only nonzero component τ_{tt}. The optimality condition (15.3.5) becomes

$$(\tau_{tt})^2|_{\partial\Omega_{\text{cav}}} = \text{constant}; \qquad (15.3.6)$$

two other components of the stress are zero.

Sufficient conditions of optimality were suggested in (Prager, 1968). They consist of the requirement (15.3.6) and the condition that the energy on the unknown boundary be less than the energy in any interior point.

The problem of the optimal cavity can be viewed as the inverse problem for an unknown boundary, or a free boundary problem. Amazingly, that inverse problem has the analytical solution. It was obtained by Vigdergauz using the Kolosov–Muskhelishvili method. This method (Muskhelishvili, 1943) is beyond the scope of this book and the reader is referred to the original papers (Cherepanov, 1974; Vigdergauz, 1988; Vigdergauz, 1989; Vigdergauz, 1994a; Grabovsky and Kohn, 1995a; Grabovsky and Kohn, 1995b) where this inverse problem was investigated in different settings.

The optimality condition allows us to guess the optimal solutions for a dilute optimal composite. The problem is formulated as optimization of the shape of a single cavity in an elastic plane, loaded at infinity by a homogeneous loading. The hydrostatic pressure implies a circular optimal cavity, the pressure of type (15.3.1) requires an elliptical shape as optimal, and the eccentricity of the optimal ellipse is determined by the ratio $\frac{a}{b}$, (15.3.1); see (Cherepanov, 1974; Banichuk et al., 1984; Vigdergauz, 1988).

The optimization problem has a remarkable feature: If one places two or more optimal inclusions near each other, they shapes are different from ellipse, but the functional (the stiffness) stays the same. Any finite number of cavities corresponds to the same value of the functional.

The composite case corresponds to an infinite number of the infinitesimal caverns. Small values of the volume fraction of the void correspond to the array of dilute elliptical inclusions, large values to the truss-type construction; see Figure 15.5. The optimal stiffness is equal to the linearized stiffness of the optimal composite when the volume fraction of the void is neaar zero.

There exists a variety of optimal structures. An optimal structure can be analytically found that corresponds to any ratio of the sides of a periodic cell. Laminate of second rank is the limiting case of extremely elongated cells; see Figure 15.6.

The Translation Bound

Besides the necessary conditions, one can consider the sufficient conditions given by the translation bounds.

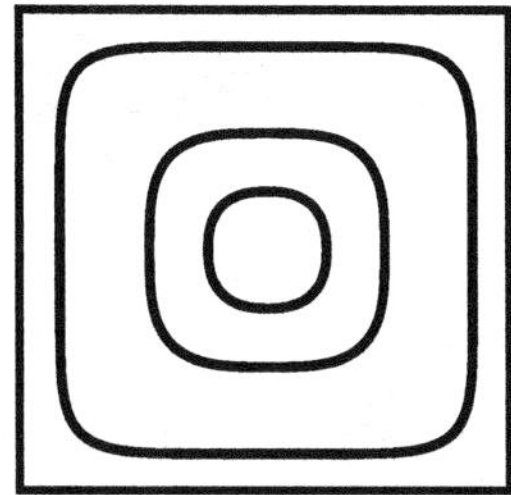

FIGURE 15.5. Optimal periodic structures by Vigdergauz; dependence on the fraction of the void phase. Small values of the fraction of inclusions correspond to ellipses; large values correspond to ovals inscribed in a periodicity cell.

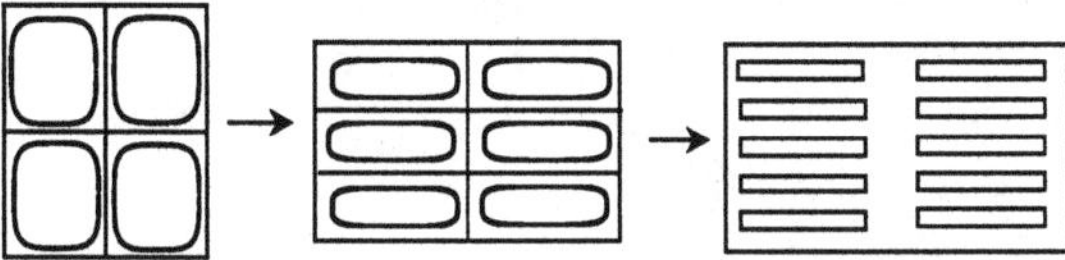

FIGURE 15.6. The optimal structures depend on elongation of the cell of periodicity. The functional stays constant for any rate of elongation. The limiting case of extremely elongated cells corresponds to second-rank laminates.

Consider an optimal plate with several cavities. Obviously, the functional P in (15.3.4) is given by the formula

$$P = \left. \frac{\partial W_{\text{comp}}}{\partial m} \right|_{m=0}. \tag{15.3.7}$$

where W_{comp} is the energy of a composite with inclusions of void and m is the volume fraction of these inclusions.

To obtain P we can use the translation bound W_p of the energy ((15.1.11), where we put $t = -\rho$, because the loading $\boldsymbol{\tau}$ has eigenvalues of the same sign. In our setting, the translation bound of the difference between the energy of a plate with the cavities and a solid plane is

$$\Delta \leq |\Omega_{\text{cav}}| \, (\eta) \, (\text{TrD} \, \boldsymbol{\tau}_0)^2. \tag{15.3.8}$$

These bounds are achievable if a trace of the stress $\boldsymbol{\tau} : \mathbf{a}_1$ is constant everywhere in the material, including the boundary. This condition is satisfied in the above-mentioned structures (see (Vigdergauz, 1988), and it serves to determine the boundary of optimal cavities. The same condition is satisfied in the optimal second-rank laminates.

Remark 15.3.1 *Generally, the constancy of a component of the field decreases the energy of the structure, because the fluctuating component that contains additional energy is eliminated. In the considered case, only the* *trace of the stress is constant in the optimal structures. The deviator of the stress varies, but according to the translation bound, this variation does not influence the functional.*

15.3.2 Optimal Shapes under Shear Loading

Consider the same problem of the optimal shape of the cavities in the case where the applied stress tensor τ_0 (15.3.1) has eigenvalues of different sign: $ab < 0$. Again, the translation bound can be derived similar to (15.3.8):

$$\Delta \leq |\Omega_{\text{cav}}|\,(\rho)\,(\text{Dev}\,\tau_0)^2.$$

This bound corresponds to the optimal structure of the second-rank laminates. This structure is characterized by the constancy of the deviator, $(\tau : \mathbf{a}_2)\mathbf{a}_2 + (\tau : \mathbf{a}_3)\mathbf{a}_3$ of the stress field. The structure corresponds to infinitely many elongated cavities.

However, the optimal solution to the problem of a single cavity is significantly different. Unlike the previous case, there is no solution to the problem of a single cavity that corresponds to the constancy of the deviator of the stress everywhere. Consequently, the optimal value of the functional (15.3.2) depends on the number of cavities.

The Shape of an Optimal Single Cavity

Consider the optimization of the shape of the single cavity. Suppose that a stress

$$\tau_\infty = a\mathbf{i} \otimes \mathbf{i} + b\mathbf{j} \otimes \mathbf{j}, \quad ab < 0$$

is applied at infinity; see Figure 15.7. In this case, the tangent stress $\tau_{tt} = \tau : \mathbf{t} \otimes \mathbf{t}$ necessarily changes its sign on the boundary of the cavity, therefore

$$\text{sign}\,\tau_{tt} \neq \text{constant on } \partial\Omega_{\text{cav}}.$$

Indeed, Figure 15.7 shows that if one cuts the plane through the cavity across the line of $\mathbf{i}_1$ (line AA), the traction on the boundary is positive; if the cut goes through the cavity parallel to the line $\mathbf{i}_2$ (line BB), the traction is negative.

On the other hand, the necessary condition (15.3.6) states that if Ω_{cav} is optimal, then

$$(\tau_{tt})^2 = \text{constant} \quad \text{everywhere on } \partial\Omega_{\text{cav}}.$$

To bring these two conditions together we have to assume that τ_{tt} is only piecewise constant, i.e., it jumps at some points on the contour $\partial\Omega_{\text{cav}}$. Furthermore, the jump points must correspond to irregularities on the boundary. Indeed, a smooth boundary corresponds to a continuous σ_{tt}. Hence, we conclude that an optimal cavity is necessarily bounded by a

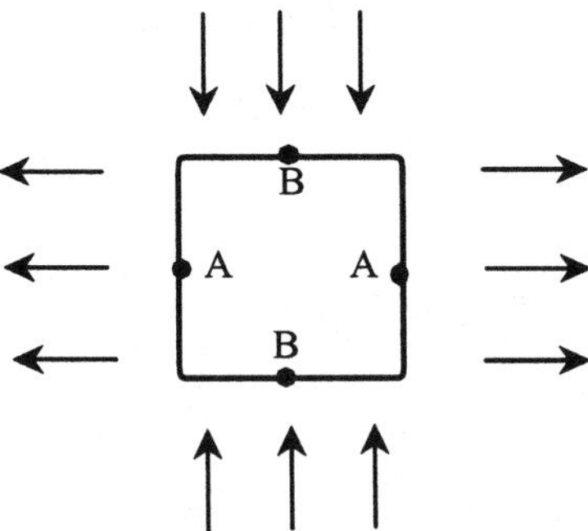

FIGURE 15.7. Why does the optimal cavity have corners? The absolute value of the traction is constant on the boundary of the optimal cavity, but the traction has different signs at points A and B. Therefore, the traction is discontinuous and the boundary has corners.

nonsmooth curve. This conclusion was made in (Vigdergauz and Cherkaev, 1986), where a numerical solution was also given.

This phenomenon raises several questions: Does such a cavity exist? If so, is it possible to satisfy the sufficient Prager conditions everywhere, not only on its boundary? The answer to the first question is affirmative; the answer to the second is negative.

In (Cherkaev, Grabovsky, Movchan, and Serkov, 1998), the solution to the optimal problem was found. It was shown that for the case of pure shear the optimal cavity is a curved quadrilateral, and the angle near the corners is equal to the critical value $\approx 102.6°$. This angle makes the piecewise constant stress on the boundary possible (see the classical paper (Carothers, 1912) and the related papers (Karp and Karal, 1962; Markenscoff and Paukshto, 1998), which discuss the critical behavior of elastic stresses in the neighborhood of the corner).

The minimization algorithm in (Cherkaev, Grabovsky, Movchan, and Serkov, 1998) involves an analytical and numerical component. The complex variable technique is employed and the solution is obtained through Kolosov-Muskhelishvili (Muskhelishvili, 1943) potentials; these are presented in power series. A direct minimization procedure was used to specify the coefficients of the expansion of the conformal mapping function and the optimal shape of the cavity. This direct minimization approach is supplemented with another one: We solve an inverse problem (free boundary problem) for an unknown boundary of the cavity that satisfies necessary conditions of optimality everywhere. The results of the two approaches coincide. Particularly, it is shown that the optimal contour approaches the nonsmooth contour. Using asymptotics near the corners, we found the critical angle that corresponds to finite jump of the τ_{tt}; it equals $102.6°$.

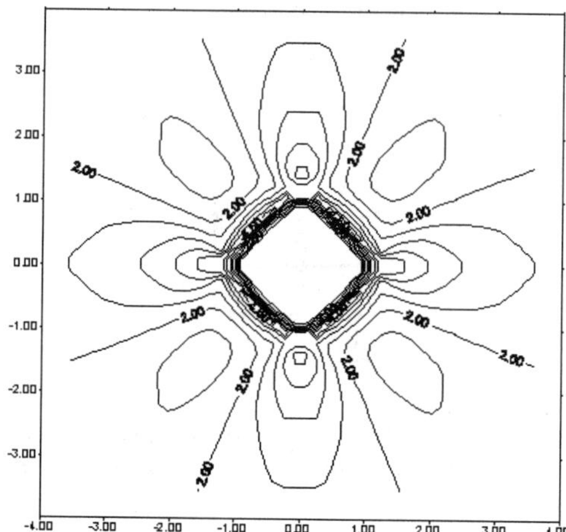

FIGURE 15.8. The picture of the energy outside of the optimal cavity. The minima of the energy are located at a finite distance to the cavity, near the middle of its sides.

However, the obtained stress field outside of an optimal single cavity does not satisfy the sufficient conditions. The functional is larger than the one given by (15.3.7). The results of the calculation are illustrated in Figure 15.8. We observe that the elastic energy has minima outside of the boundary of the cavity in the regions near the middle of its sides.

This observation suggests that the additional indirect constraint–the simple-connectedness of the cavity in each periodic cell–plays an important role in the optimization. Indeed, if a cavity could split into several parts, they would form such a better structure. The region of the minimum of the energy (see Figure 15.8) caused by one cavity would be occupied with another cavity. Therefore, the energy of the system of two cavities would be less than the sum of the energies of two systems with one twice smaller cavity each. This structure should perform better than a single cavity.

It was numerically found in (Cherkaev, Grabovsky, Movchan, and Serkov, 1998) that one simply connected cavity performs 1.83 times worse than the second-rank laminate structure of the same total volume of cavities. The increase of the number of cavities improves the design. Figure 15.9 represents a draft of the sequence of optimal cavities of fixed total area. Notice that the contour of each cavity has four angles of $102.6°$, because the asymptotics of the fields near the corners is that same. The limiting case of extremely elongated cavities corresponds to the second-rank laminates.

The sufficient conditions–the constancy of the shear component of the stress tensor–are satisfied only in the limit when the number of cavities goes to infinity and their shapes become extremely elongated. This feature is in contrast to the system of periodic smooth inclusions that are optimal

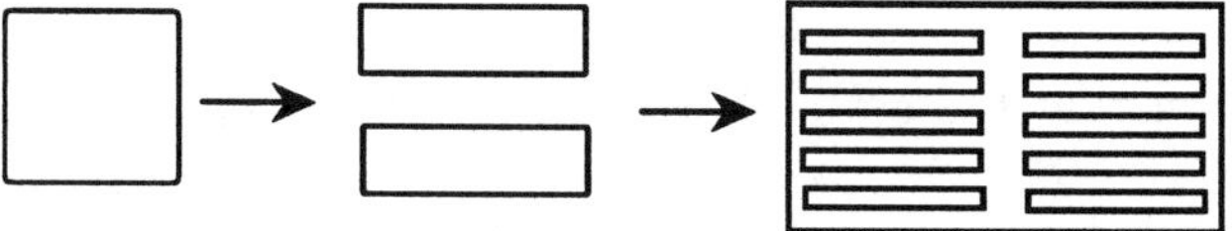

FIGURE 15.9. The draft of a sequence of arrays of optimal cavities; the more cavities the stronger the construction is. The limiting case corresponds to infinitely many cavities, which form the second-rank laminates.

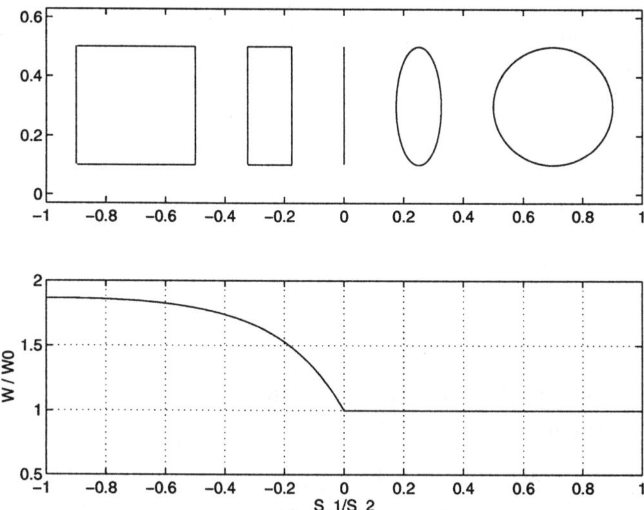

FIGURE 15.10. Drafts of optimal shapes of cavities depending on anisotropy of the loading (the ratio $\frac{a}{b}$, $|a| \leq |b|$), and the ratio of the energy of a single optimal cavity to the energy of the optimal second-rank composite of equal total volume of cavities.

for loading with $ab > 0$. Recall that in the last case the functional is independent of the number of inclusions.

Figure 15.10 shows that the elongated shapes of simple-connected optimal cavities depend on the anisotropy of the loading. This suggests another reason for forming the second-rank composites. Notice that the performance of a single cavity depends on the ratio $\frac{a}{b} < 0$. The closer to zero is this ratio, the closer to the global minimum is the stiffness of the system with one cavity. On the other hand, the elastic field in the inner layer of second-rank laminates is uniaxial due to effect of the inclusions; consequently, optimal cavities are extremely elongated ($\frac{a}{b} \to 0$). This explains why the global minimum is achieved by infinite number of cavities.

Summary

We have seen that the optimal structure of the second-rank laminates is the limit of two different types of minimizing sequences (see Figures 15.6 and 15.9). The first sequence corresponds to the loading by a stress with eigenvalues of equal sign $(a\,b > 0)$. It demonstrates the nonuniqueness of the optimal solution; each term corresponds to the minimum of the functional.

The second case corresponds to loading by a stress with eigenvalues of different signs $(a\,b < 0)$. The minimizing sequence corresponds to increase of the connectedness (number of cavities); the functional reaches the minimum in the limit. The interaction of the cavities brings the local stress field near their contours close to the uniaxial, which increases the effectiveness of each cavity.

15.4   Problems

1. Derive the translation bounds and find optimal structures for the plane structures made of "badly ordered" materials:

$$\rho_1 < \rho_2, \quad \eta_1 > \eta_2.$$

2. Derive and analyze formulas for the quasiconvex envelope for the energy $W(m, \epsilon)$.

3. Derive the formulas for the translation bounds for three-dimensional energy $W(m, \epsilon)$. Does the translation bound degenerate into a harmonic mean bound?

   Hint: Use three two-dimensional quasiconvex translators.

4. Derive the necessary conditions of optimality of the shape of a single cavity in a three-dimensional problem. Analyze the shape of a three-dimensional cavity if the elastic space is loaded at infinity by by the stress with the eigenvalues of different signs. Can the boundary be smooth?

16
Bounds on Effective Properties

Here we survey the results of the bounding of elastic moduli of a two-dimensional two-phase composite. These bounds are physically interesting because they show the limits of improvement in the properties of a composite due to variation of its structure. They are important for the optimal design of materials and structures. The examples herein demonstrate the use and development of the translation method and the laminate technique.

16.1 G_m-Closures of Special Sets of Materials

Hill's Bounds

The first bounds of the effective moduli K_0, μ_0 were found by Hill (Hill, 1964), who in fact applied the Wiener bounds to the elasticity. Hill's bounds have the form

$$\langle K^{-1} \rangle^{-1} \leq K_0 \leq \langle K \rangle, \quad \langle \mu^{-1} \rangle^{-1} \leq \mu_0 \leq \langle \mu \rangle. \tag{16.1.1}$$

For two-material mixtures, the bounds (16.1.1) can be represented as

$$y(K_0) \geq 0, \quad y(\mu_0) \geq 0,$$

where the linear-fractional function y is determined by relation (7.3.18).

Special Sets of Mixed Material

Bounds for elastic properties use the conservation property of G_m-closures (Chapter 3): If the phases have a common eigenvalue and eigenvector, then these elements are preserved in the effective tensor of any structure. For elastic composites, this property has two interesting implementations.

Materials with the Constant Shear Modulus

The solution for this case was obtained in the pioneering paper (Hill, 1964). We follow (Lurie and Cherkaev, 1981c), where the translation method was applied to the problem. The problem was also investigated in (Francfort and Tartar, 1991) by a similar method.

Suppose that isotropic materials in the composite

$$C(K_i, \mu) = K_i \mathbf{a}_1 \otimes \mathbf{a}_1 + \mu(\mathbf{a}_2 \otimes \mathbf{a}_2 + \mathbf{a}_3 \otimes \mathbf{a}_3)$$

have a common value of the shear modulus μ.

Applying the conservation property of G-closures (Chapter 3), we conclude that the shear part of the effective tensor C_* is equal to the shear part of the family of mixed materials. Namely, C_* has the shear modulus μ and the eigenvectors $\mathbf{a}_2$ and $\mathbf{a}_3$ no matter what the microstructure is. The composite is *isotropic* independent of its microstructure:

$$C_* = C(K_*, \mu) = K_* \mathbf{a}_1 \otimes \mathbf{a}_1 + \mu(\mathbf{a}_2 \otimes \mathbf{a}_2 + \mathbf{a}_3 \otimes \mathbf{a}_3) \qquad (16.1.2)$$

where K_* is an effective bulk modulus of the composite. Indeed, the tensor direction $\mathbf{a}_1$ is the only direction orthogonal to $\mathbf{a}_2$ and $\mathbf{a}_3$ simultaneously; therefore, it is the eigenvector of C_*.

Let us demonstrate that K_* depends only on the volume fraction of the materials and is independent of all other structural parameters (Hill, 1964). To find K_*, we exploit the equivalence of the elasticity equations in the Airy form discussed in Chapter 14. Namely, we use the equation (14.1.26), which can be rewritten as the system

$$\Delta\phi = \left(\frac{1}{K(\mathbf{x})} + \frac{1}{\mu} \right)^{-1} N, \quad \Delta N = 0. \qquad (16.1.3)$$

The homogenized form of this equation is

$$\Delta\phi_0 = \left(\frac{1}{K_*} + \frac{1}{\mu} \right)^{-1} N_0 \quad \Delta N_0 = 0. \qquad (16.1.4)$$

Consider the function $N = N(\mathbf{x})$, which depends on the layout χ^s of the materials in composite. Notice that $N(\mathbf{x})$ is a twice differentiable function of $\mathbf{x}$. The continuity of $N(\mathbf{x})$ implies that it tends to a constant everywhere in the cell of periodicity as the size of the cell goes to zero. It can be replaced by its mean value N_0 in the averaging (compare with the case (6.4.2)). We have, by averaging of (16.1.3),

$$\langle \Delta\phi \rangle = \left\langle \left(\frac{1}{K(\mathbf{x})} + \frac{1}{\mu} \right)^{-1} \right\rangle N_0. \qquad (16.1.5)$$

The expression for the effective modulus K_* follows from (16.1.4) and (16.1.5). Modulus K_* of a satisfies the equation

$$\left(\frac{1}{K_*} + \frac{1}{\mu}\right)^{-1} = \sum_i m_i \left(\frac{1}{K_i} + \frac{1}{\mu}\right)^{-1}$$

where K_i are the moduli of the materials in the composite and m_i are their volume fractions. One can see that indeed the effective bulk modulus K_* is completely determined only by the properties of the materials and their volume fractions.

The value of K_* depends on μ as on a parameter; one can check that $K_* = K_*(\mu)$ varies between its arithmetic (K_a) and harmonic (K_h) means, and that $K_* \to K_a$ if $\mu \to \infty$ and $K_* \to K_h$ if $\mu \to 0$.

Similar bounds can be found for three-dimensional elastic composites; see (Hill, 1964). Namely, if the isotropic materials have common shear modulus, then the effective material has this modulus as well. Generally, if the isotropic materials differ by a bulk modulus and have a common shear modulus, then a composite of them is isotropic and has the same shear modulus as the mixing materials. The media can be multiphase or even nonlinear: The bulk modulus could depend on the trace of the stress tensor. This dependence does not change the result because the only feature used is the constancy of the shear modulus.

Materials with Constant Bulk Modulus

Consider now the opposite case: a composite of elastic materials $C(K, \mu_i)$ with the common bulk modulus K:

$$C(K, \mu) = K\mathbf{a}_1 \otimes \mathbf{a}_1 + \mu_i(\mathbf{a}_2 \otimes \mathbf{a}_2 + \mathbf{a}_3 \otimes \mathbf{a}_3).$$

We continue to follow (Lurie and Cherkaev, 1981c).

Again, applying the conservation property of G-closures, we conclude that the bulk part of the effective tensor C_* is equal to the bulk part of the family of the mixed materials. Namely, C_* has the bulk modulus K and the eigenvector $\mathbf{a}_1$ no matter what the microstructure is.

The other two eigentensors are orthogonal to $\mathbf{a}_1$; hence they are tracefree tensors (deviators). Therefore, C_* has the structure

$$C_* = K\mathbf{a}_1 \otimes \mathbf{a}_1 + \mu_{*1}\mathbf{a}_2 \otimes \mathbf{a}_2 + \mu_{*2}\mathbf{a}_3 \otimes \mathbf{a}_3, \tag{16.1.6}$$

where μ_{*1}, μ_{*2} are effective shear moduli. Equation (16.1.6) says that an arbitrary mixture has cubic symmetry.

To find μ_{*1} and μ_{*2}, we use an equivalent form of the elasticity equation in the Airy form (14.1.27). We observe that this equation is similar to the equation of two-dimensional conductivity, where the orthogonal tensors $\mathbf{a}_2$ and $\mathbf{a}_3$ are similar to pairs of orthogonal vectors $\mathbf{i}, \mathbf{j}$.

The Lagrangian associated with equation (14.1.27) is

$$(K + \mu(\mathbf{x}))(\mathbf{e}_2^2 + \mathbf{e}_3^2).$$

The Wiener bound for this quadratic Lagrangian is

$$\left(\sum_i \frac{m_i}{K + \mu_i}\right)^{-1} \leq (K + \mu_*) \leq \sum_i m_i(K + \mu_i). \qquad (16.1.7)$$

To describe the G-closure we can use the elementary technique of Chapter 3. The bounds on the G-closure correspond to one of the shear moduli μ_{*1} of the composite equal to the lower bound, and the other shear modulus μ_{*2} equal to an upper bound:

$$\mu_{*1} = \left(\sum_i m_i(K + \mu_i)^{-1}\right)^{-1} - K, \quad \mu_{*2} = \sum_i m_i\mu_i.$$

Both bounds are attainable by a laminate structure. The shear moduli μ_{*1}, μ_{*2} correspond to these equalities if the eigenvectors of $\mathbf{a}_2$ are aligned with the normal and the tangent to the layers.

Repeating the arguments of Chapter 3, we demonstrate that the G-closure of two materials (16.1.6) corresponds to the relations

$$K_* = K, \quad \mu_{*1} \leq \mu_{*2}, \quad \mu_1 \leq \mu_{*i} \leq \mu_2,$$

$$\mu_{*1} \geq \frac{(\mu_1 + K)(\mu_2 + K)}{\mu_1 + \mu_2 - \mu_{*2} + K} - K,$$

where μ_1 and μ_2 are the minimal and maximal shear moduli of the set of given materials (see (Lurie and Cherkaev, 1981c)).

The derivation of these bounds is similar to the derivation for the two-dimensional G-closure for the conductivity problem. The G-closure is the union of the bounds (16.1.7) that correspond to all volume fractions.

Remark 16.1.1 *In both cases, the bounds use the matrix $\mathcal{R}$ of the translator (see (14.1.20)). They represent the simplest (and the earliest) examples of the translation method: The rate of the translation is fixed, and the translated matrices have one of the eigenvalues equal to zero everywhere.*

Remark 16.1.2 *Both G-closures (16.1.2) and (16.1.6) are examples of the sets with empty interior. One of the moduli is constant for all mixtures. This exact relation significantly simplifies the G-closure problem.*

16.2   Coupled Bounds for Isotropic Moduli

In this section we review an application of the translation method to bound the range of elastic moduli of an isotropic elastic two-phase composite. We mainly follow (Cherkaev and Gibiansky, 1993).

16.2.1   The Hashin–Shtrikman Bounds

In early sixties, Hashin and Shtrikman (Hashin and Shtrikman, 1963) suggested a variational method that was able to predict the range of elastic properties of a composite (the G_m-closure). They originative approach greatly influenced the field of material science; it was suggested about two decades earlier than the translation method.

The method is based on the necessary and sufficient conditions of optimality. Both types of conditions were developed by the authors. The microstructure of coated spheres (see Chapter 2) was suggested to compute the energy of the optimal microstructures that have maximal or minimal values of both bulk and shear moduli. On the other hand, sufficient conditions were suggested that takes into account differential constraints on stress and strain fields, but in a manner different from the translation method.

Instead of the differential constraints, the Hashin and Shtrikman method directly uses the potentials of the variables. Instead of translations, the method operates with the "comparing medium." The fields in an inhomogeneous medium is divided into constant fields and deviations, and the tensor of polarization is introduced, that correspond to the link between deviation of the strains and stresses. This approach was influenced by electromagnetics where the polarization is used to compare magnetic field in vacuum and a material. Instead of permeability of vacuum, Hashin and Shtrikman introduced the comparing medium. This allowed them to choose this medium, which plays a role similar to the role of matrices of translators. The exposition of the method can be found in (Hashin, 1970b; Hashin, 1970a; Christensen, 1979; Nemat-Nasser and Hori, 1993) at al. The method was significantly generalized in the series of papers (Kohn and Strang, 1986b; Avellaneda, 1987b; Kohn and Lipton, 1988; Lipton, 1988; Milton, 1990b).

The bounds have the algebraic form

$$\left(C_* - C_{\text{compar}}\right)^{-1} = m_1 \left(C_1 - C_{\text{compar}}\right)^{-1} + m_2 \left(C_2 - C_{\text{compar}}\right)^{-1},$$

where C_{compar} is a specially chosen "comparing medium." These bounds are applicable to the two- and three-dimensional problem.

For symmetric structures, the method of Hashin and Shtrikman require fewer calculations than the translation method because it can easily incorporate information about the symmetries, particularly isotropy, of the composite.

However, as we demonstrate here, the translation method yields to tighter bounds that include the coupled relations between effective bulk and shear moduli. The comparing medium used in the Hashin–Shtrikman method must have a physical sense; it represents the energy of an elastic material. Hence, some algebraic structure of the tensor C_{compar} is assumed. On the contrary, the translators are purely algebraic constructions and do not require a physical interpretation. The translation method is applicable to a larger variety of problems, such as those discussed in the Chapter 12 problem of coupled properties.

The connection between the Hashin–Shtrikman and the translation methods was investigated in the paper (Milton, 1990b). The detailed exposition of the method is beyond the scope of this book. The reader is referred to the above mentioned papers.

Hashin and Shtrikman considered the case of "well-ordered" materials: both bulk and shear moduli of the first material are bigger than those of the second one:

$$\mu_1 \geq \mu_2, \quad K_1 \geq K_2. \tag{16.2.1}$$

The Hashin–Shtrikman bounds restrict the range of both the bulk and the shear moduli of an isotropic composite. These ranges are independent. The pair K_*, μ_* belongs to a rectangle that is determined by two opposite vertices.

In the two-dimensional case, the bounds are (see (Hashin, 1965))

$$K_{\mathrm{HS}}^{\mathrm{l}} \leq K_* \leq K_{\mathrm{HS}}^{\mathrm{u}}, \quad \mu_{\mathrm{HS}}^{\mathrm{l}} \leq \mu_* \leq \mu_{\mathrm{HS}}^{\mathrm{u}},$$

where

$$K_{\mathrm{HS}}^{\mathrm{l}} = K_2 + m_1 \left(\frac{1}{K_1 - K_2} + \frac{m_2}{K_2 + \mu_2} \right)^{-1},$$

$$K_{\mathrm{HS}}^{\mathrm{u}} = K_1 + m_2 \left(\frac{1}{K_2 - K_1} + \frac{m_1}{K_1 + \mu_1} \right)^{-1},$$

$$\mu_{\mathrm{HS}}^{\mathrm{l}} = \mu_2 + m_1 \left(\frac{1}{\mu_1 - \mu_2} + \frac{m_2 \left(K_2 + 2\mu_2 \right)}{2\mu_2 \left(K_2 + \mu_2 \right)} \right)^{-1},$$

$$\mu_{\mathrm{HS}}^{\mathrm{u}} = \mu_1 + m_2 \left(\frac{1}{\mu_2 - \mu_1} + \frac{m_1 \left(K_1 + 2\mu_1 \right)}{2\mu_1 \left(K_1 + \mu_1 \right)} \right)^{-1}.$$

The Y-Transform of the Bounds

It is convenient to rewrite the bounds in terms of the Y-transform (see (7.3.19)). The coordinates of the opposite corners of the Hashin–Shtrikman rectangle become $y(K_{\mathrm{HS}}^{\mathrm{l}})$, $y(\mu_{\mathrm{HS}}^{\mathrm{l}})$ and $y(K_{\mathrm{HS}}^{\mathrm{u}})$, $y(\mu_{\mathrm{HS}}^{\mathrm{u}})$; they are computed as:

$$y(K_{\mathrm{HS}}^{\mathrm{l}}) = \mu_2, \qquad y(K_{\mathrm{HS}}^{\mathrm{u}}) = \mu_1,$$
$$y(\mu_{\mathrm{HS}}^{\mathrm{l}}) = \frac{K_2 \mu_2}{K_2 + 2\mu_2}, \quad y(\mu_{\mathrm{HS}}^{\mathrm{u}}) = \frac{K_1 \mu_1}{K_1 + 2\mu_1}.$$

Recall that the Y-transform is linear-fractional, and therefore it is invertible if $\det(C_1 - C_2) \neq 0$. Notice, the Y-transform of the bound is independent of the volume fractions.

The Hashin–Shtrikman bounds describe the smallest rectangle in the plane K_*, μ_* that contains the isotropic component of the G-closure. Indeed, two opposite corner points A and C,

$$A = \left(Y \left(K_{\mathrm{HS}}^{\mathrm{l}} \right), Y \left(\mu_{\mathrm{HS}}^{\mathrm{l}} \right) \right), \quad C = \left(Y \left(K_{\mathrm{HS}}^{\mathrm{u}} \right), Y \left(\mu_{\mathrm{HS}}^{\mathrm{u}} \right) \right),$$

of this rectangle are attainable by the matrix laminates (14.3.9) or "coated circles."

The Walpole Bounds

The opposite case of "badly ordered" initial materials when

$$\mu_1 \geq \mu_2, \quad K_1 \leq K_2 \tag{16.2.2}$$

was considered in (Walpole, 1966). Walpole obtained bounds for the effective moduli of an isotropic composite by using an approach similar to the Hashin–Shtrikman variational method. The two-dimensional Walpole's bounds are

$$K_W^l \leq K_* \leq K_W^u, \quad \mu_W^l \leq \mu_* \leq \mu_W^u,$$

where K_W^l, K_W^u and μ_W^l, μ_W^u satisfy the equations

$$
\begin{aligned}
y(K_W^l) &= \mu_2, & y(K_W^u) &= \mu_1, \\
y(\mu_W^l) &= \tfrac{K_1 \mu_2}{K_1 + 2\mu_2}, & y(\mu_W^u) &= \tfrac{K_2 \mu_1}{K_2 + 2\mu_1}.
\end{aligned}
$$

The Walpole bounds also describe a rectangle in the plane K_*, μ_* that contains the isotropic component of the G-closure. This rectangle is determined by its two opposite corners B and D:

$$B = \left(Y\left(K_W^l \right), Y\left(\mu_W^u \right) \right), \quad D = \left(Y\left(K_W^u \right), Y\left(\mu_W^l \right) \right).$$

Let us compare the Hashin–Shtrikman and Walpole bounds. The allowed interval of the bulk moduli stays the same for both cases. The allowed interval of the shear moduli is different. If the materials are well ordered (Figure 16.1), then the Walpole rectangle BD lies inside the Hashin–Shtrikman rectangle AC. If the materials are badly ordered (Figure 16.1), then the Hashin–Shtrikman rectangle AC lies inside the Walpole rectangle BD. If the shear moduli of the materials are equal, then both bounds coincide and the composite has the same shear modulus, as in (16.1.2). In this case, both rectangles degenerate into intervals. In both nontrivial cases, the greater rectangle provides the bounds, and the smaller rectangle does not mean anything special.

Remark 16.2.1 *The cases (16.2.1) and (16.2.2) cover all possible relations between the moduli of two isotropic materials, because one always can number the materials so that $\mu_1 \geq \mu_2$.*

16.2.2   The Translation Bounds

The translation method has been applied to this problem in (Cherkaev and Gibiansky, 1993). The method has led to new *coupled* bounds on the

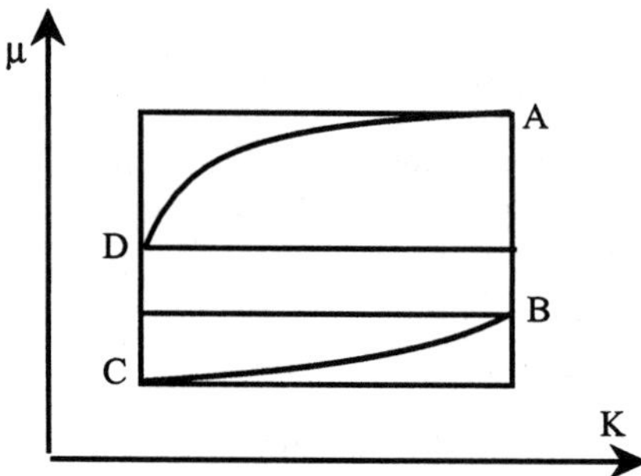

FIGURE 16.1. Well-ordered materials: The rectangle AC corresponds to the Hashin–Shtrikman bounds, the rectangle BD corresponds to the Walpole points; and the curved quadrangle $ABCD$ corresponds to the translation bounds.

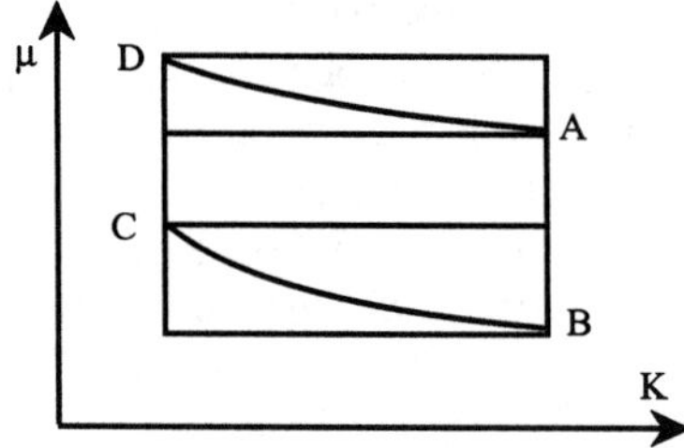

FIGURE 16.2. Badly ordered materials: The rectangle BD corresponds to the Walpole bounds, the rectangle AC corresponds to the Hashin–Shtrikman points; and the curved quadrangle $ABCD$ corresponds to the translation bounds.

moduli. The permitted region is defined for both well-ordered and badly ordered materials. It lies inside the Hashin–Shtrikman or the Walpole rectangles. Moreover, the new bounds demonstrate the connection between the Hashin–Shtrikman and Walpole bounds, because the corners of both rectangles are used in each case of ordering of materials. The earlier and less restrictive coupled bounds by Milton are discussed in the mentioned paper.

The results are the following: The G_m-closure is bounded by a curved quadrangle. Two opposite sides are linear-fractional curves that join the corners corresponding to the Hashin–Shtrikman and the Walpole bounds. Two other sides are straight lines.

The results are shown in Figure 16.1 and Figure 16.2. The resulting inequalities are represented in terms of the Y-transform of the original moduli as follows: All the pairs $(y(K_*), y(\mu_*))$ of values of the Y-transform of the bulk K_* and shear μ_* moduli for an elastic isotropic composite are restricted by the inequalities (Hashin–Shtrikman–Walpole bounds of bulk modulus)

$$\mu_2 \leq y(K_*) \leq \mu_1.$$

The new bounds are given by the inequalities

$$F_1(y(K_*)) \leq y(\mu_*) \leq F_2(y(K_*))$$

in the well-ordered case (16.2.1) or by the inequalities

$$F_3(y(K_*)) \leq y(\mu_*) \leq F_4(y(K_*)) \tag{16.2.3}$$

in the badly ordered case (16.2.2).

All the curves are represented as linear-fractional curves $v = F_i(u)$ in the coordinates $v = y(\mu_*)$, $u = y(K_*)$. Each of these curves is uniquely determined by the coordinates of any three points it passes through as follows:

$$\left(\frac{y(\mu_3) - y(\mu_2)}{y(\mu_3) - y(\mu_1)}\right) \frac{v - y(\mu_2)}{v - y(\mu_1)} = \left(\frac{y(K_3) - y(K_2)}{y(K_3) - y(K_1)}\right) \frac{u - y(K_2)}{u - y(K_1)}.$$

Namely, the curve $v = F_1(u)$ (line EAD, Figure 16.1) passes through the corner point A of the Hashin–Shtrikman bound, the corner point D of the Walpole bound, and an outside point E:

$$A = (y(K_{\mathrm{HS}}^{\mathrm{l}}), y(\mu_{\mathrm{HS}}^{\mathrm{l}})), \quad D = (y(K_{\mathrm{W}}^{\mathrm{u}}), y(\mu_{\mathrm{W}}^{\mathrm{l}})), \quad E = (-K_1, -\mu_1).$$

The curve $v = F_2(u)$ (line FBC, Figure 16.1) passes through the corner point C of the Hashin–Shtrikman bound, the corner point B of the Walpole bound, and an outside point F:

$$B = (y(K_{\mathrm{W}}^{\mathrm{l}}), y(\mu_{\mathrm{W}}^{\mathrm{u}})), \quad C = (y(K_{\mathrm{HS}}^{\mathrm{u}}), y(\mu_{HS}^{\mathrm{u}})), \quad F = (0, -\mu_1).$$

Similarly, the bounds for the badly ordered case are described as follows. The curve $F_3(y(K_*))$ (line ADG, Figure 16.2) passes through the corner

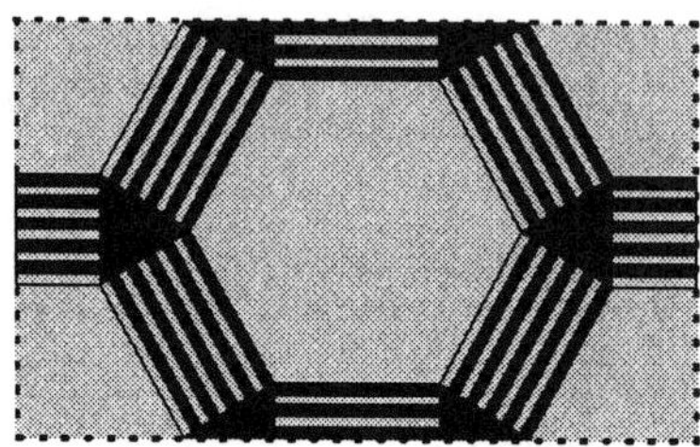

FIGURE 16.3. Hexagonal structures; (Sigmund, 1988): The numerically obtained effective properties of these structures practically coincide with the Walpole bounds.

point A of the Hashin–Shtrikman bound, the corner point D of the Walpole bound, and an outside point G:

$$A = (y(K_{\mathrm{HS}}^{\mathrm{l}}), y(\mu_{\mathrm{HS}}^{\mathrm{l}})), \quad D = (y(K_{\mathrm{W}}^{\mathrm{u}}), y(\mu_{\mathrm{W}}^{\mathrm{l}})), \quad G = (0, -\mu_2).$$

The curve $F_4(y(K_*))$ (line BCH, Figure 16.2) passes through the corner point C of the Hashin–Shtrikman bound, the corner point B of the Walpole bounds, and an outside point H:

$$B = (y(K_{\mathrm{W}}^{\mathrm{l}}), y(\mu_{\mathrm{W}}^{\mathrm{u}})), \quad C = (y(K_{\mathrm{HS}}^{\mathrm{u}}), y(\mu_{HS}^{\mathrm{u}})), \quad H = (-K_2, -\mu_2).$$

The bounds are expressed in terms of the Y-transform, and in such representation the bounds turn out to be independent of the volume fractions of the initial materials.

The Y-transform is itself the linear-fractional function of the effective moduli. Therefore, the corresponding bounds in the plane of the effective moduli (K_*, μ_*) are also given by fraction-linear functions. We refer to (Cherkaev and Gibiansky, 1993) for details.

Optimal Structures

It is not yet known whether the translation bounds are exact. Some points on the boundary of the permitted set are known, however. We already mentioned that the corners of Hashin–Shtrikman bounds correspond to the structures of coated spheres or to the third-rank laminates.

Recently, Sigmund developed a method to numerically find optimal structures. He came out with a new type of structure (Sigmund, 1998) that are extremely close to the Walpole corners. These structures are shown in Figure 16.3 (courtesy of O. Sigmund). The structure includes laminate elements, together with triangles and hexagons.

Remark 16.2.2 *The Hashin-Shtrikman and Walpole bounds are also valid in three-dimensional case. Two corners of the Hashin-Shtrikman bounds correspond to the coated spheres geometry. The three-dimensional Walpole* *bounds can be improved (Milton, 1981b) and therefore are not optimal. The coupled bounds for isotropic three-dimensional structures were obtained in (Milton and Berryman, 1997).*

16.2.3 Functionals

We use the translation method to prove both the Hashin–Shtrikman–Walpole bounds and the coupled bounds. Although we consider the bounds for isotropic composites, the inequalities obtained are also valid for the anisotropic effective tensors.

We find a lower bound of the functional I,

$$I = \sum_{i=1}^{N} W_i,$$

equal to the sum of the values of the elastic energy W_i. The energy is stored in the element of periodicity of a composite that is submerged into N linearly independent external stress or strain fields with fixed mean values applied sequentially.

It is convenient to express the energy as a function of the gradient $\zeta = \nabla \mathbf{u}$ (14.1.1) of the displacement vector $\mathbf{u}$ instead of the strain ϵ. The energy is associated with the 4×4 matrix $\tilde{C}$ (14.1.23), which has zero eigenvalue corresponding to the direction $\mathbf{a}_4$ in the basis (14.1.12).

To obtain a lower bound for the bulk modulus one submerges the composite into external hydrostatic strain field $\epsilon_h = \epsilon_1 \mathbf{a}_1$ because the strain energy of an isotropic composite under the action of this field is proportional to the effective bulk modulus K_*. The corresponding functional has the form

$$I^\zeta = \langle \zeta^T \tilde{C} \zeta \rangle = 2K_* \epsilon_1^2 \quad \text{if} \quad \langle \zeta \rangle = \epsilon_1 \mathbf{a}_1, \qquad (16.2.4)$$

where ϵ_1 is a given constant. A lower bound of the functional 16.2.4) gives a lower bound of the effective bulk modulus K_*, because the magnitude ϵ_1 of the hydrostatic strain field is assumed to be fixed.

To obtain an upper bound of this modulus we submerge the composite into the hydrostatic stress field $\tau_h = \tau_1 \mathbf{a}_1$; this makes the stress energy proportional to $\frac{1}{K_*}$. We minimize the functional

$$I^\tau = \langle \tau^T S \tau \rangle = (2K_0)^{-1} \tau_1^2, \quad \text{if} \quad \langle \tau \rangle = \tau_1 \mathbf{a}_1, \qquad (16.2.5)$$

where τ_1 is a given constant. A lower bound of the corresponding functional gives us an upper bound of K_*. We demonstrate that the exact bounds of these functionals provide the Hashin–Shtrikman bounds for the bulk modulus.

Similarly, to obtain a lower bound for the shear modulus of a composite one can examine the energy stored in the composite submerged into a shear-type trial strain field. This way we obtain a bound on one of the two shear moduli of the composite, which is anisotropic in general. However, the other shear modulus can have arbitrary value, and the energy functionals of the types (16.2.4), (16.2.5) are not sensitive to this value. To provide the isotropy of the composite we should also care about the reaction of the composite on the orthogonal shear field. Hence, to bound the shear modulus of an isotropic composite we minimize the functional equal to the sum of two values of the strain energy stored by the medium under the action of two orthogonal trial orthogonal shear fields $\epsilon^{(1)} = \epsilon a_2$ and $\epsilon^{(2)} = \epsilon a_3$ of equal magnitude ϵ. We express the strain energy in terms of the corresponding gradients $\zeta^{(1)} = \nabla u^{(1)}$ and $\zeta^{(2)} = \nabla u^{(2)}$ of the displacements $u^{(1)}$ and $u^{(2)}$:

$$I^{\zeta\zeta} = \left\langle \left(\zeta^{(1)}\right)^T \tilde{C} \zeta^{(1)} + \left(\zeta^{(2)}\right)^T \tilde{C} \zeta^{(2)} \right\rangle = 4\mu_* \epsilon^2$$
$$\text{if } \langle \zeta^{(1)} \rangle = \epsilon a_2, \quad \langle \zeta^{(2)} \rangle = \epsilon a_3.$$

To find an upper bound of the shear modulus, we consider the sum of two energies caused by two orthogonal shear stress fields $\tau^{(1)} = \tau a_2$ and $\tau^{(2)} = \tau a_3$ of equal magnitude τ:

$$I^{\tau\tau} = \left\langle \left(\tau^{(1)}\right)^T S \tau^{(1)} + \left(\tau^{(2)}\right)^T S \tau^{(2)} \right\rangle = \frac{1}{2\mu_*}\tau^2$$

We will demonstrate that the translation bounds for the last two functionals lead to both the Hashin–Shtrikman and Walpole bounds for the shear modulus.

In order to obtain coupled shear-bulk bounds, we submerge the composite into three different fields: the hydrostatics and two orthogonal shears. We have a choice between stress and strain trial fields (two shear fields are supposed to be of the same nature to provide isotropy of the composite). Therefore, the following functionals are considered. The functional

$$I^{\zeta\zeta\zeta} = I^{\zeta} + I^{\zeta\zeta}$$

minimizes the energy of three external strain fields. The functional

$$I^{\tau\tau\tau} = I^{\tau} + I^{\tau\tau}$$

minimizes the energy of three external stress fields. The functionals

$$I^{\tau\zeta\zeta} = I^{\tau} + I^{\zeta\zeta},$$
$$I^{\zeta\tau\tau} = I^{\zeta} + I^{\tau\tau}$$

minimize the sum of the stress and strain energies. A lower bound of each of these functionals gives some component of the boundary.

We should distinguish the cases of well- and badly ordered materials. The lower bounds of the functionals $I^{\zeta\zeta\zeta}$ and $I^{\tau\tau\tau}$ provide lower and upper bounds of the convex combination of the effective bulk and shear moduli, $I^{\tau\tau\tau} = \alpha_1^2 K_* + \alpha_2^2 \mu_*$. Here α_1, α_2 are real coefficients. The minimum of these functionals is achieved on the structures with maximal and minimal values of both moduli. In the well-ordered case (16.2.1), the points of maximal and minimal values of the bulk and shear moduli (the Hashin–Shtrikman points A and C in Figure 16.2) are attainable by the third-rank laminates. Clearly, the bounds of these functionals cannot improve the Hashin–Shtrikman inequalities.

On the other hand, the minimization of $I^{\tau\varsigma\varsigma}$ or $I^{\varsigma\tau\tau}$ demands the minimization of the combination $I^{\tau\varsigma\varsigma} = \alpha_1^2 K_*^{-1} + \alpha_2^2 \mu_*$ and $I^{\varsigma\tau\tau} = \alpha_1^2 K_* + \alpha_2^2 \mu_*^{-1}$ or the minimization of one of the moduli and maximization of the other. We will demonstrate that for well-ordered materials it leads to coupled bounds of the moduli that are more restrictive than the Hashin–Shtrikman bounds.

In the badly ordered case (16.2.2) we face the opposite situation. The bounds of the functionals $I^{\varsigma\varsigma\varsigma}$ and $I^{\tau\tau\tau}$ improve the Walpole bounds, and the bounds of the functionals $I^{\tau\varsigma\varsigma}$ and $I^{\varsigma\tau\tau}$ leave them unchanged.

16.2.4 Translators

Here we find a set of bilinear quasiaffine functions of stresses τ and gradients of displacement ς. We determine translators depending on two matrices of the type of ς, on two stresses τ, and a bilinear translator depending on τ and ς.

Strain-Strain Quasiaffine Functions

We are looking for the quasiaffine functions of gradients of displacements. We use the quasiaffineness of the Jacobian $J = \det(\nabla w^{(1)}, \nabla w^{(2)})$ of two potentials $w^{(1)}$ and $w^{(1)}$.

Consider the gradients $\varsigma^{(1)} = \nabla \mathbf{u}^{(1)}$, $\varsigma^{(2)} = \nabla \mathbf{u}^{(2)}$ for any two displacement vectors

$$\mathbf{u}^{(1)} = (u_1^{(1)}\mathbf{i} + u_2^{(1)}\mathbf{j}), \quad \mathbf{u}^{(2)} = (u_1^{(2)}\mathbf{i} + u_2^{(2)}\mathbf{j}).$$

Consider the bilinear form

$$\phi(\varsigma^{(1)}, \varsigma^{(2)}) = \begin{pmatrix} \nabla u_1^{(1)} \\ \nabla u_2^{(1)} \end{pmatrix}^T \begin{pmatrix} \hat{t}_1 R & \hat{t}_2 R \\ \hat{t}_3 R & \hat{t}_4 R \end{pmatrix} \begin{pmatrix} \nabla u_1^{(2)} \\ \nabla u_2^{(2)} \end{pmatrix} \qquad (16.2.6)$$

where R is the tensor or rotation through a right angle.

The coefficient by each parameter $\hat{t}_i$ is a Jacobian; the form (16.2.6) is a sum of Jacobians of pairs of components $u_1^{(1)}, u_2^{(1)}, u_1^{(2)}, u_2^{(2)}$. Therefore (see Chapter 8), $\phi(\varsigma^{(1)}, \varsigma^{(2)})$ is a quasiaffine function for any values of the parameters $\hat{t}_i$, $i = 1, 2, 3, 4$, of this linear combination.

It will be convenient to rewrite the quadratic form (16.2.6) the basis $\mathbf{a}_1,\ldots,\mathbf{a}_4$ (see (14.1.12)) as:

$$\phi(\boldsymbol{\zeta}^{(1)},\boldsymbol{\zeta}^{(2)}) = \left(\boldsymbol{\zeta}^{(1)}\right)^T \Phi^{\zeta\zeta}(t_1,t_2,t_3,t_4)\,\boldsymbol{\zeta}^{(2)},$$

where

$$\boldsymbol{\zeta}^{(1)} = \sum_{i=1}^{4} \zeta_i^{(1)}\mathbf{a}_i, \quad \boldsymbol{\zeta}^{(2)} = \sum_{i=1}^{4} \zeta_i^{(2)}\mathbf{a}_i;$$

the matrix $\Phi^{\zeta\zeta}$ is

$$\Phi^{\zeta\zeta}(t_1,\,t_2,\,t_3,\,t_4) = \begin{pmatrix} t_1 & t_2 & t_3 & t_4 \\ -t_2 & -t_1 & t_4 & t_3 \\ -t_3 & -t_4 & -t_1 & -t_2 \\ -t_4 & -t_3 & t_2 & t_1 \end{pmatrix};$$

and the parameters t_i, $i = 1,2,3,4$, are linearly independent linear combinations of the parameters $\hat{t}_i$, $i = 1,2,3,4$.

Remark 16.2.3 *Particularly, for the bounds of the energy of a composite (the functional I^ζ) it is sufficient to use the quadratic form $\phi(\boldsymbol{\zeta},\boldsymbol{\zeta})$ of gradients $\boldsymbol{\zeta}$ instead of the bilinear form (16.2.6). This quadratic form is associated with the symmetric part of the matrix $\Phi^{\zeta\zeta}$. It has the form*

$$\mathbf{T}^\zeta(t_1) = \Phi^{\zeta\zeta}(t_1,\,0,\,0,\,0). \tag{16.2.7}$$

This translator was used in Chapter 15 for the bound of strain energy.

Stress-Stress Quasiaffine Functions

Recall that any stress tensor has the following representation (14.1.6), (14.1.19)

$$\mathcal{R}\boldsymbol{\tau} = \nabla\mathbf{v}, \quad \mathbf{v} = \nabla\phi$$

where the vector $\mathbf{v}$ is the gradient of the Airy function ϕ, $\mathbf{v} = \nabla\phi$; and $\mathcal{R}$ is the fourth-rank tensor of rotation of a matrix through a right angle (see (14.1.20)).

The bilinear form

$$\left(\nabla\mathbf{v}^{(1)}\right)^T \Phi^{\zeta\zeta}\left(\nabla\mathbf{v}^{(2)}\right)$$

is quasiaffine for any vector fields $\mathbf{v}^{(1)}$, $\mathbf{v}^{(2)}$, as we mentioned in the previous paragraph. Therefore, the bilinear form of the stresses $\boldsymbol{\tau}^{(1)}$, $\boldsymbol{\tau}^{(2)}$

$$\left(\boldsymbol{\tau}^{(1)}\right)^T \left(\mathcal{R}^T\Phi^{\zeta\zeta}\mathcal{R}\right)\boldsymbol{\tau}^{(2)} \tag{16.2.8}$$

is quasiaffine.

Recall that τ is a symmetric matrix due to the equilibrium conditions; therefore $\tau : \mathbf{a}_4 \equiv 0$. Hence, the quasiaffine function (16.2.8) can be represented as a bilinear function

$$\left(\tau^{(1)}\right)^T \Phi^{\tau\tau}(t_1, t_2, t_3, t_4)\, \tau^{(2)}$$

of the first three components of $\tau_i^{(1)}$ and $\tau_i^{(2)}$.

The matrix $\Phi^{\tau\tau}$,

$$\Phi^{\tau\tau}(t_1, t_2, t_3, t_4) = \begin{pmatrix} t_1 & t_2 & t_3 \\ -t_2 & -t_1 & t_4 \\ -t_3 & -t_4 & -t_1 \end{pmatrix}, \qquad (16.2.9)$$

is an upper-left 3×3 minor of the matrix $\mathcal{R}^T \Phi^{\zeta\zeta} \mathcal{R}$ in the basis $\mathbf{a}_i$.

Strain-Stress Quasiaffine Functions

Let us consider again the bilinear quasiaffine function

$$\zeta^T \Phi^{\zeta\zeta}(\nabla \mathbf{v})$$

of the gradients $\zeta = \nabla \mathbf{u}$ and $\nabla \mathbf{v}$. We represent this form as a function of the symmetric stress tensor $\tau = \mathcal{R}\nabla \mathbf{v}$ and the asymmetric tensor ζ. After obvious calculations we obtain the quasiaffine bilinear form:

$$\phi^{\tau\zeta} = \zeta^T \Phi^{\tau\zeta} \tau,$$

where the 3×4 matrix $\Phi^{\tau\zeta}$ has the form

$$\Phi^{\tau\zeta} = \Phi^{\tau\zeta}(t_1, t_2, t_3, t_4) = \begin{pmatrix} t_1 & t_2 & t_3 & t_4 \\ t_2 & -t_1 & -t_4 & -t_3 \\ t_3 & t_4 & -t_1 & t_2 \end{pmatrix}.$$

Here the three-dimensional vector τ and the four-dimensional vector ζ are the coefficients of the tensors τ and ζ in the basis $(\mathbf{a}_1, \mathbf{a}_2, \mathbf{a}_3)$ and $(\mathbf{a}_1, \mathbf{a}_2, \mathbf{a}_3, \mathbf{a}_4)$, respectively.

16.2.5 Modification of the Translation Method

Now we are ready to apply the translation method to derive the bounds. However, the method needs a modification because the Y-transform is undefined if $\det(C_1 - C_2) = 0$.

In the case under consideration, both matrices $\tilde{C}_1$ and $\tilde{C}_2$ have a common eigenvalue equal to zero. It corresponds to the same eigenvector $\mathbf{a}_4$. Recall that the stain energy is expressed as a quadratic form of the four-dimensional vector $\zeta = \nabla \mathbf{w}$ instead of the three-dimensional vector $\epsilon = \mathrm{Def}\,\mathbf{w}$. The fourth coordinate ζ_4 corresponds to a rotation: the antisymmetric part of ζ. The use of the variable ζ instead of ϵ simplifies the search for translators, but simultaneously it poses a new difficulty that we address now.

Modified Translation Bound

Consider, for example, the translation bound for the functional $I^{\zeta\zeta}$. The consideration of the functionals I^{ζ}, $I^{\zeta\zeta}$, $I^{\zeta\tau}$, $I^{\zeta\zeta\tau}$, and $I^{\zeta\tau\tau}$ is analogous.

The translation bound has the form (10.1.6)

$$\zeta^T A(\tilde{C}_*)\zeta \geq \zeta^T Q(\tilde{C}_i, T)\zeta \quad \forall T \in \mathcal{T}, \tag{16.2.10}$$

where ζ is a 12-dimensional vector of gradients of displacements $\mathbf{u}_1$, $\mathbf{u}_2$, $\mathbf{u}_3$ in the basis $\mathbf{a}_i^{(k)}$, where $i = 1, 2, 3, 4$, $k = 1, 2, 3$. In this basis, $A(\tilde{C}_*)$ is the 12×12 block-diagonal matrix

$$A(\tilde{C}_*) = \mathrm{diag}(\tilde{C}_*, \tilde{C}_*, \tilde{C}_*) \tag{16.2.11}$$

and

$$Q(\tilde{C}_i, T) = \langle (A(\tilde{C}_i) - T)^{-1} \rangle^{-1} + T. \tag{16.2.12}$$

The set $\mathcal{T}$ of admissible translators T is defined by the condition

$$A(\tilde{C}_i) - T \geq 0, \quad i = 1, 2.$$

The left-hand side of (16.2.10) is independent of fourth components

$$\zeta_4 = \left[\zeta^{(1)} : \mathbf{a}_4, \zeta^{(2)} : \mathbf{a}_4, \zeta^{(3)} : \mathbf{a}_4 \right]$$

of ζ^k that are equal to the antisymmetric parts (rotations) of these non-symmetric matrices. Indeed, the effective tensor $\tilde{C}_*$ has the form (14.1.23)

$$\tilde{C}_* = \begin{pmatrix} C_* & \mathbf{0} \\ \mathbf{0}^T & 0 \end{pmatrix}$$

(the elasticity tensor offers no resistance to the rotation), therefore

$$\zeta^T A(\tilde{C}_*)\zeta = \epsilon^T (P^T A(\tilde{C}_*)P)\epsilon = \mathrm{constant}(\zeta_4)$$

where P is the projector on the subspace of strains. However, the right-hand side of (16.2.10) depends on the components ζ_4^k, because the bound is computed by an algebraic procedure unrelated to the properties of the elasticity operator.

We rewrite the quadratic form in the right-hand side of (16.2.10) by using a 12×9 projector P and the 12×3 projector $P_\perp$. The projector P projects the 12-dimensional vector ζ on the subspace of strains and the projector $P_\perp$ project ζ on the subspace of rotations:

$$P\zeta = \epsilon, \quad P_\perp^T \zeta = \zeta_4,$$

where ϵ is a 9-dimensional vector composed of the elements of strains ϵ^k and ζ_4^k is a three-dimensional vector of the rotational components of ζ.

The right-hand side of (16.2.10) is represented as a quadratic function of ζ_4,

$$\zeta^T Q \zeta = \epsilon^T \Gamma_1(Q)\epsilon + 2\left(\zeta_4^k\right)^T \Gamma_3(Q)\epsilon + \left(\zeta_4^k\right)^T \Gamma_2(Q)\zeta_4^k \qquad (16.2.13)$$

where Γ_1, Γ_2, and Γ_3 are the corresponding minors of Q (see (16.2.12)):

$$\Gamma_1(Q) = PQP, \quad \Gamma_2 = P_\perp Q P_\perp, \quad \Gamma_3(Q) = P_\perp QP.$$

As we mentioned, the left-hand side of (16.2.10) depends only on the strains. Therefore, the vector ζ_4 on the right-hand side of (16.2.10) is not fixed, and the translation inequality (16.2.10) must hold for all ζ_4. To deal with the dependence on ζ_4, we find the optimal value of ζ_4 that delivers the minimum of the right-hand side of (16.2.13). Note that this value is finite because (16.2.13) is the nonnegative quadratic form.

We have

$$\zeta_4^{\text{opt}} = -\Gamma_2^{-1}(Q)\,\Gamma_3(Q)\,\epsilon.$$

The bound becomes

$$\epsilon\left(P^T A(\tilde{C}_*)P\right)\epsilon \le \min_{T\in\mathcal{T}}\left\{\epsilon^T\left(\Gamma_1 - \Gamma_3^T\,\Gamma_2^{-1}\,\Gamma_3\right)\epsilon\right\}, \qquad (16.2.14)$$

where Γ_i depend on the translator T; $\Gamma_i = \Gamma_i(Q(T))$. Equation (16.2.14) implies the bound for the effective tensor C_*:

$$P^T A(\tilde{C}_*)P \in \bigcap_{T\in\mathcal{T}}\left\{\Gamma_1(Q) - \Gamma_3^T(Q)\,\Gamma_2^{-1}(Q)\,\Gamma_3(Q)\right\}. \qquad (16.2.15)$$

We call this bound the *modified translation bound*.

Projection of the Y-Tensor

Let us reformulate the approach in terms of the Y-transform. Recall that the Y-transform of the effective properties matrix $A(D_*)$ is defined by the equality

$$A(D_*) = m_1 A(D_1) + m_2 A(D_2)$$
$$-m_1 m_2(\Delta A)(m_2 A(D_1) + m_1 A(D_2) + Y)^{-1}(\Delta A), \qquad (16.2.16)$$

where $A(D_1)$ and $A(D_2)$ are properties of the phases and

$$\Delta A = A(D_1) - A(D_2).$$

If the matrix ΔA is not degenerate, then the translation bound has the simple form (10.1.9),

$$Y(A(D_*)) + T \ge 0 \quad \forall T \in \mathcal{T}. \qquad (16.2.17)$$

However, here we face the situation where some of the eigenvectors and eigenvalues of the matrices $A(D_1)$ and $A(D_2)$ are equal. In the case (16.2.11), both $A(D_1)$ and $A(D_1)$ have zero eigenvalues with eigenvectors in the subspace ζ_4, which corresponds to rotations.

Again, we are dealing with a case of degenerate G-closure. Any effective tensors have zero eigenvalues with the eigenvectors in ζ_4, independently of the microstructure of the composite. Recall the conservation property of G-closures: If the properties tensors of components have the same eigenvalues and eigenvectors, then the effective tensor of a composite also has the same eigenvalues and eigenvectors (Chapter 3).

The tensor Y-transform is undefined in the whole space because it includes the ratio of two zeros; see (16.2.16)). However, this transform can be defined in the subspace of the nondegenerate difference $P^T(A(D_1)P - P^T A(D_2)P = \Gamma_1 \Delta A$ of ΔA. The bound (16.2.17) becomes (Cherkaev and Gibiansky, 1993)

$$R = Y(\Gamma_1(A(D_*))) + \Gamma_1(T) - \Gamma_3(T)(\Gamma_2(T)^{-1}\Gamma_3(T)^T \geq 0. \qquad (16.2.18)$$

We use the definition of Y-transform to derive the preceding inequality. A long, but straightforward, calculation (see (Cherkaev and Gibiansky, 1993) for details) yields to the result. Note that the structure of the bound (16.2.18) in terms of the Y-transform corresponds to the structure of the bound (16.2.15).

Remark 16.2.4 *This technique has allowed for a variety of problems to be solved. We mention the series of papers (Gibiansky and Torquato, 1993; Gibiansky and Torquato, 1995a; Gibiansky and Torquato, 1995b; Gibiansky and Torquato, 1996a), where the technique was developed in many aspects, simplified, and applied to a variety of physical problems that deal with the coupled elastic and conducting properties of a material. The properties are coupled through the structure of the composite.*

16.2.6 Appendix: Calculation of the Bounds

Here we derive the translation bounds for the sum of elastic energies. All the bounds are obtained by the same procedure. The exposition is rather technical. However, we want to give the reader an example of the calculations for the derivation of bounds. We present the details of the calculations following (Cherkaev and Gibiansky, 1993).

All the bounds are calculated in the four steps:

1. Choose the functional that specifies the matrices of properties.

2. Choose the appropriate translator T.

3. Compute the admissible set $\mathcal{T}$ from (16.2.17).

4. Compute the translator bound or the modified translator bound.

First, let us obtain the Hashin–Shtrikman and Walpole bounds.

A Lower Bound for the Bulk Modulus

1. We find a lower bound for the functional I^ς. The matrices D_i, $i = 1, 2, *$, are equal to $D_i = D_i^\varsigma = C_i$, $i = 1, 2, *$.

2. The translator is the symmetric part of the strain-strain translation matrix $T^\varsigma(t, 0, 0, 0)$ (see (16.2.7)). Indeed, the asymmetric part of D does not affect the quadratic form associated with it.

3. The set $\mathcal{T}$ of admissible t corresponds to the matrix inequality

$$D_i^\varsigma + T = \begin{pmatrix} 2K_i - t & 0 & 0 & 0 \\ 0 & 2\mu_i + t & 0 & 0 \\ 0 & 0 & 2\mu_i + t & 0 \\ 0 & 0 & 0 & -t \end{pmatrix} \geq 0,$$

where $i = 1, 2$. This inequality is equivalent to the two scalar inequalities

$$t \leq 0, \quad t \geq -\min\{\mu_1, \mu_2\} = -\mu_2. \tag{16.2.19}$$

4. The effective moduli K_*, μ_* satisfy the modified translation inequality (16.2.18). Here $\Gamma_3 = 0$, because both T and D_i are diagonal. The bound takes the form

$$\begin{pmatrix} y(2K_*) & 0 & 0 & 0 \\ 0 & y(2\mu_*) & 0 & 0 \\ 0 & 0 & y(2\mu_*) & 0 \\ 0 & 0 & 0 & y(0) \end{pmatrix} + \begin{pmatrix} -t & 0 & 0 & 0 \\ 0 & t & 0 & 0 \\ 0 & 0 & t & 0 \\ 0 & 0 & 0 & -t \end{pmatrix} \geq 0$$

for all $t \in \mathcal{T}$. The bound for the bulk modulus K_* is

$$y(2K_*) \geq -t, \quad \forall\, t \text{ as in (16.2.19)}.$$

Choosing $t = -\mu_2$, we obtain the Hashin–Shtrikman and Walpole lower bound for the bulk modulus,

$$y(2K_*) \geq \mu_2.$$

An Upper Bound for the Bulk Modulus

1. An upper bound for the bulk modulus can be obtained analogously using the functional I^τ instead of I^ς. The matrices D_i, $i = 1, 2, *$, are equal to

$$D_i = D_i^\tau = S_i, \quad i = 1, 2, *.$$

2. The translator matrix T is the symmetric part of the stress-stress translation matrix $\Phi^{\tau\tau}$: $\Phi^{\tau\tau}(t_1, 0, 0, 0)$.

3. The set $\mathcal{T}$ of admissible values of t is

$$-\frac{1}{2\mu_1} = -\min\left\{\frac{1}{2\mu_1}, \frac{1}{2\mu_2}\right\} \leq t \leq \min\left\{\frac{1}{2K_1}, \frac{1}{2K_2}\right\}.$$

4. A bound for the bulk modulus K_* is found from the condition

$$y\left(\frac{1}{2K_*}\right) + t \geq 0, \quad \forall\, t \in \mathcal{T}$$

as

$$y\left(\frac{1}{2K_*}\right) = \frac{1}{2\mu_1}.$$

Using properties (7.3.25), (7.3.24) of the Y-transform, we transform the last equation to the form

$$y(K_*) \leq \mu_1.$$

This bound coincides with the Hashin–Shtrikman and Walpole upper bounds for the bulk modulus.

A Lower Bound for the Shear Modulus

1. We estimate the functional $I^{\zeta\zeta}$. Here $\mathbf{e}$ is an eight-dimensional vector

$$\mathbf{e} = [\zeta_1^{(1)}, \zeta_2^{(1)}, \zeta_3^{(1)}, \zeta_4^{(1)}, \zeta_1^{(2)}, \zeta_2^{(2)}, \zeta_3^{(2)}, \zeta_4^{(2)}],$$

where $\zeta_i^{(1)}$, $\zeta_i^{(2)}$, $i = 1, 2, 3, 4$, are the components of gradients $\boldsymbol{\zeta}^{(1)}$, $\boldsymbol{\zeta}^{(2)}$ in the basis (14.1.12). The matrices D_i, $i = 1, 2, *$, have the block-diagonal form

$$D_i = D_i^{\zeta\zeta} = \begin{pmatrix} \tilde{C}_i & 0 \\ 0 & \tilde{C}_i \end{pmatrix}, \quad i = 1, 2, *.$$

The projector P on the nonzero subspace of the matrix $(D_1 - D_2)$ and the complementary projector $P_\perp$ have the block forms

$$P = P^{\zeta\zeta} = \begin{pmatrix} P^\zeta & 0 \\ 0 & P^\zeta \end{pmatrix}, \quad P_\perp = P_\perp^{\zeta\zeta} = \begin{pmatrix} P_\perp^\zeta & 0 \\ 0 & P_\perp^\zeta \end{pmatrix},$$

where P^ζ is the projector on the first three components of $\boldsymbol{\zeta}$, and $P_\perp^\zeta$ is the projector on the fourth component of $\boldsymbol{\zeta}$.

2. Let us find an appropriate translator. Consider the quasiaffine quadratic form of $\mathbf{e}$ associated with the block matrix

$$T^{\zeta\zeta} = \begin{pmatrix} \Phi^{\zeta\zeta}(t_1^1, t_2^1, t_3^1, t_4^1) & \Phi^{\zeta\zeta}(t_1^2, t_2^2, t_3^2, t_4^2) \\ \Phi^{\zeta\zeta}(t_1^3, t_2^3, t_3^3, t_4^3) & \Phi^{\zeta\zeta}(t_1^4, t_2^4, t_3^4, t_4^4) \end{pmatrix}.$$

This function is quasiaffine as a sum of quasiaffine functions. The matrix $T^{\zeta\zeta}$ depends on 16 parameters $t_i^j, i,j = 1, 2, 3, 4$. However, the matrix of a quadratic form has to be symmetric, which reduces the number of parameters to 6. We will also assume that $t_1^1 = t_1^4$ because of the isotropy of the composite. It turns out that it is sufficient for the shear modulus bounds to set three of the five remaining parameters t_i^j to zero and to deal with the translation matrix

$$T^{\zeta\zeta} = \begin{pmatrix} \Phi^{\zeta\zeta}(t_1, 0, 0, 0) & \Phi^{\zeta\zeta}(0, t_2, 0, 0) \\ \Phi^{\zeta\zeta}(0, t_2, 0, 0) & \Phi^{\zeta\zeta}(t_1, 0, 0, 0) \end{pmatrix},$$

which depends only on the two parameters t_1, t_2.

3. The set T is found from the matrix inequality

$$D_i^{\zeta\zeta} - T^{\zeta\zeta}(t_1, t_2) \geq 0, \quad i = 1, 2,$$

as follows:

$$(2K_{\min} + t_1)t_1 - t_2^2 \geq 0, \quad (2\mu_{\min} - t_1)^2 - t_2^2 \geq 0.$$

Here

$$K_{\min} = \min\{K_1, K_2\}, \quad \mu_{\min} = \min\{\mu_1, \mu_2\} = \mu_2.$$

The limiting values t_1^*, t_2^* of the parameters t_1, t_2 correspond to the equalities in the preceding relations; they are equal to

$$t_1^* = \frac{2\mu_2^2}{K_{\min} + 2\mu_2}, \quad t_2^* = \frac{2\mu_2(K_{\min} + \mu_2)}{K_{\min} + 2\mu_2}.$$

4. Consider the modified translation bound (16.2.18). Substituting the values of t_1^*, t_2^* we obtain the scalar inequality

$$2y(\mu_*) + t_1^* - t_2^* = 2y(\mu_*) - \frac{2K_{\min}\mu_2}{K_{\min} + 2\mu_2} \geq 0.$$

It coincides with the Hashin–Shtrikman bound on the shear modulus in the well-ordered case and with the Walpole bound in the badly ordered case.

An Upper Bound for the Shear Modulus

1. We bound the functional $I^{\tau\tau}$. In this case

$$\mathbf{e} = [\tau_1^{(1)}, \tau_2^{(1)}, \tau_3^{(1)}, \tau_1^{(2)}, \tau_2^{(2)}, \tau_3^{(2)}]$$

is a six-dimensional vector consisting of the components of the two stress tensors $\boldsymbol{\tau}^{(1)}$ and $\boldsymbol{\tau}^{(2)}$; the block-diagonal 6×6 matrices D_i are

$$D_i = D_i^{\tau\tau} = \begin{pmatrix} S_i & 0 \\ 0 & S_i \end{pmatrix}, \quad i = 0, 1, 2.$$

2. We construct the matrix $T^{\tau\tau}$ of a quasiaffine quadratic function of the vector $\mathbf{e}$ in the same way as $T^{\zeta\zeta}$ using the bilinear quasiaffine form $\Phi^{\tau\tau}$ instead of the form $\Phi^{\zeta\zeta}$:

$$T^{\tau\tau} = \begin{pmatrix} \Phi^{\tau\tau}(t_1,0,0,0) & \Phi^{\tau\tau}(0,t_2,0,0) \\ \Phi^{\tau\tau}(0,t_2,0,0) & \Phi^{\tau\tau}(t_1,0,0,0) \end{pmatrix}.$$

3. The set of admissible values of the parameters t_i is found from the matrix inequality

$$D_i^{\tau\tau} - T^{\tau\tau}(t_1, t_2) \geq 0, \quad i = 1, 2.$$

It leads to the scalar inequalities

$$t_1 \geq -\frac{1}{2K_{\max}}, \quad \left(\frac{1}{2\mu_{\max}} - t_1\right)^2 - t_2^2 \geq 0,$$

where

$$K_{\max} = \max\{K_1, K_2\}, \quad \mu_{\max} = \max\{\mu_1, \mu_2\} = \mu_1.$$

The critical values t_1^*, t_2^* of the parameters t_1, t_2 are equal:

$$t_1^* = -\frac{1}{2K_{\max}}, \quad t_2^* = \frac{1}{2\mu_1} + \frac{1}{2K_{\max}}.$$

4. The bound becomes

$$Y(D_*^{\tau\tau}) + T^{\tau\tau}(t_1^*, t_2^*) \geq 0.$$

The last matrix inequality leads to the scalar inequality

$$y\left(\frac{1}{2\mu_*}\right) + t_1^* - t_2^* = y\left(\frac{1}{2\mu_*}\right) - \frac{2}{K_{\max}} - \frac{1}{\mu_1} \geq 0.$$

Simplifying the last inequality, we obtain the bound

$$y(\mu_*) \leq \frac{K_{\max}\,\mu_1}{K_{\max} + 2\mu_1}.$$

This bound coincides with the Hashin–Shtrikman bound in the well-ordered case and with the Walpole bound in the badly ordered case.

A Lower Coupled Bound, the Well-Ordered Case

Continuing the calculation, we now prove the translation bounds. Consider the well-ordered materials first:

1. The functional $I^{\tau\zeta\zeta}$ allows us to obtain the coupled bound. Here $\mathbf{e}$ is the following 11-dimensional vector:

$$\mathbf{e} = [\tau_1,\, \tau_2,\, \tau_3,\, \zeta_1^{(1)},\, \zeta_2^{(1)},\, \zeta_3^{(1)},\, \zeta_4^{(1)},\, \zeta_1^{(2)},\, \zeta_2^{(2)},\, \zeta_3^{(2)},\, \zeta_4^{(2)}].$$

The matrices D_i, $i = 1, 2, *$, have the block-diagonal form

$$D_i = D_i^{\tau\zeta\zeta} = \begin{pmatrix} S_i & 0 & 0 \\ 0 & \tilde{C}_i & 0 \\ 0 & 0 & \tilde{C}_i \end{pmatrix},$$

and the projectors $P = P^{\tau\zeta\zeta}$ and $P_\perp = P_\perp^{\tau\zeta\zeta}$ are the block matrices

$$P = \begin{pmatrix} I & 0 & 0 \\ 0 & P^\zeta & 0 \\ 0 & 0 & P^\zeta \end{pmatrix}, \quad P_\perp = \begin{pmatrix} 0 & P_\perp^\zeta & 0 \\ 0 & 0 & P_\perp^\zeta \end{pmatrix},$$

where I is the unit 3×3 matrix and P^ζ, $P_\perp^\zeta$ are as defined earlier.

2. We use the translation matrix $T^{\tau\zeta\zeta}(t_1,\, t_2,\, t_3,\, t_4)$ of the block form

$$T^{\tau\zeta\zeta}(t_1,\, t_2,\, t_3,\, t_4) =$$
$$\begin{pmatrix} \Phi^{\tau\tau}(-t_1,0,0,0) & \Phi^{\tau\zeta}(0,-t_3,0,0) & \Phi^{\zeta\tau}(0,0,-t_3,0) \\ \Phi^{\zeta\tau}(0,-t_3,0,0) & \Phi^{\zeta\zeta}(-t_2,0,0,0) & \Phi^{\zeta\zeta}(0,0,0,-t_4) \\ \Phi^{\zeta\tau}(0,0,-t_3,0) & \Phi^{\zeta\zeta}(0,0,0,-t_4) & \Phi^{\zeta\zeta}(-t_2,0,0,0) \end{pmatrix},$$

where $\Phi^{\tau\zeta} = (\Phi^{\zeta\tau})^T$.

3. The set of admissible values of t_i is found from

$$D_i^{\tau\zeta\zeta} - T^{\tau\zeta\zeta}(t_1,\, t_2,\, t_3,\, t_4) \geq 0, \quad i = 1, 2.$$

A rather long but straightforward calculation allows us to find critical values of t_i. We mention first that each of the matrices $[D_i^{\tau\zeta\zeta} - T^{\tau\zeta\zeta}]$, $i = 1, 2$ is divided into the direct sum of four blocks

$$D_i^{\tau\zeta\zeta} - T^{\tau\zeta\zeta} = A_i^{1,5,10} \oplus A_i^{2,4,11} \oplus A_i^{3,7,8} \oplus A_i^{6,9}.$$

The matrix $A_i^{a,b,c}$ here denotes a diagonal minor of the matrix $[D_i^{\tau\zeta\zeta} - T^{\tau\zeta\zeta}]$; it consists of the elements of intersection of rows and columns with indices a, b, c.

The matrices on the right-hand side of (3) have the following form:

$$A_i^{1,5,10} = \begin{pmatrix} \frac{1}{2K_i} + t_1 & t_3 & t_3 \\ t_3 & 2\mu_i - t_2 & t_4 \\ t_3 & t_4 & 2\mu_i - t_2 \end{pmatrix},$$

$$A_i^{2,4,11} = \begin{pmatrix} \frac{1}{2\mu_i} - t_1 & t_3 & -t_3 \\ t_3 & 2K_i + t_2 & t_4 \\ -t_3 & t_4 & t_2 \end{pmatrix},$$

$$A_i^{3,7,8} = \begin{pmatrix} \frac{1}{2\mu_i} - t_1 & t_3 & t_3 \\ t_3 & t_2 & -t_4 \\ t_3 & -t_4 & 2K_i + t_2 \end{pmatrix},$$

$$A_i^{6,9} = \begin{pmatrix} 2\mu_i - t_2 & -t_4 \\ -t_4 & 2\mu_i - t_2 \end{pmatrix}.$$

The critical values t_i^* of the parameters t_i are determined from the following system

$$\det A_1^{1,5,10} = 0, \quad \det A_2^{1,5,10} = 0,$$
$$\det A_2^{2,4,11} = 0, \quad \det A_2^{6,9} = 0.$$

These values are

$$t_1^* = -\frac{1}{2K_1} + \frac{1}{H_1}\mu_2^2(K_1 - K_2)(K_1 + \mu_1)(K_1 + \mu_2),$$

$$t_2^* = \mu_1 + \mu_2 - \frac{H_2\,H_3}{H_4},$$

$$t_3^* = \sqrt{\left(\frac{1}{2\mu_1} - t_1^*\right)\left(t_2^* - \frac{\mu_2^2}{K_1 + 2\mu_2}\right)},$$

$$t_4^* = \mu_2 - t_2^*,$$

where

$$\begin{aligned} H_1 &= K_1 H_5(K_1\mu_1 + K_1\mu_2 + 2\mu_1\mu_2) \\ &\quad - K_1\mu_2 H_3(K_1 + 2\mu_2)(K_1 + \mu_1), \\ H_4 &= H_5(K_1\mu_1 + K_1\mu_2 + 2\mu_1\mu_2) \\ &\quad - \mu_2 H_3(K_1 + 2\mu_2)(K_1 + \mu_1), \\ H_2 &= K_1(\mu_1 - \mu_2)(K_1\mu_1 + K_1\mu_2 + 2\mu_1\mu_2) \\ H_3 &= (K_2\mu_1 + K_2\mu_2 + 2\mu_1\mu_2), \\ H_5 &= \mu_1(K_2 + 2\mu_2)(K_1 + \mu_2). \end{aligned}$$

One can check that such a choice of the parameters $t_i, i = 1, \ldots, 4$, provides the minimal sum of the ranks of the nonnegative matrices

$$D_1^{\tau\zeta\zeta} - T^{\tau\zeta\zeta}(t_1^*, t_2^*, t_3^*, t_4^*), \quad D_2^{\tau\zeta\zeta} - T^{\tau\zeta\zeta}(t_1^*, t_2^*, t_3^*, t_4^*)$$

and the nonnegativeness of these matrices.

4. A bound for the isotropic tensors D_* is also expressed as a direct sum of four blocks of the type

$$G = A_*^{1,5,9} \oplus A_*^{2,4} \oplus A_*^{3,7} \oplus A_*^{6,8},$$

and this bound is found from the nonpositiveness of each block:

$$A_*^{1,5,9} = \begin{pmatrix} y(\frac{1}{2K_*}) - t_1^* & -t_3^* & -t_3^* \\ -t_3^* & y(2\mu_*) + t_2^* & -t_4^* \\ -t_3^* & -t_4^* & y(2\mu_*) + t_2^* \end{pmatrix} \leq 0,$$

$$A_*^{3,7} = \begin{pmatrix} y(2\mu_*) + \frac{t_1^* t_2^* + (t_3^*)^2}{t_2^*} & -\frac{t_3^*(t_2^* + t_4^*}{t_2^*)} \\ -\frac{t_3^*(t_2^* + t_4^*}{t_2^*)} & y(2K_*) + \frac{(t_4^*)^2 - (t_2^*)^2}{t_2^*} \end{pmatrix} \leq 0,$$

$$A_*^{2,4} = A_*^{3,7}, \quad A_*^{6,8} = \begin{pmatrix} y(2\mu_*) + t_2^* & t_4^* \\ t_4^* & y(2\mu_*) + t_2^* \end{pmatrix} \leq 0.$$

One can check that the most restrictive bounds on the set of pairs $y(K_*)$, $y(\mu_*)$ come from the inequality

$$\det(A_*^{1,5,9}) \geq 0,$$

which yields to the coupled lower bound. The boundary of the set of possible pairs $y(K_*), y(\mu_*)$ that satisfy this bound is defined by the linear-fractional curve $y(\mu_*) = F_1(y(K_*))$.

An Upper Coupled Bound, the Well-Ordered Case

1. We bound the functional $I^{\zeta\tau\tau}$. The vector $\mathbf{e}$ is

$$\mathbf{e} = [\zeta_1, \zeta_2, \zeta_3, \zeta_4, \tau_1^{(1)}, \tau_2^{(1)}, \tau_3^{(1)}, \tau_1^{(2)}, \tau_2^{(2)}, \tau_3^{(2)}].$$

The matrices D_i, $i = 1, 2$, have the block-diagonal form

$$D_i = D_i^{\zeta\tau\tau} = \begin{pmatrix} \tilde{C}_i & 0 & 0 \\ 0 & S_i & 0 \\ 0 & 0 & S_i \end{pmatrix}, \quad i = 1, 2,$$

The projectors P and $P_\perp$ have the block forms

$$P = P^{\zeta\tau\tau} = \begin{pmatrix} P^\zeta & 0 & 0 \\ 0 & I & 0 \\ 0 & 0 & I \end{pmatrix}, \quad P_\perp = P_\perp^{\zeta\tau\tau} = \begin{pmatrix} P_\perp^\zeta & 0 & 0 \end{pmatrix}.$$

2. The matrix $T = T^{\zeta\tau\tau}(t_1, t_2, t_3, t_4)$ is chosen in the following block form:

$$T = \begin{pmatrix} \Phi^{\zeta\zeta}(-t_1, 0, 0, 0) & \Phi^{\zeta\tau}(0, -t_3, 0, 0) & \Phi^{\zeta\tau}(0, 0, -t_3, 0) \\ \Phi^{\tau\zeta}(0, -t_3, 0, 0) & \Phi^{\tau\tau}(-t_2, 0, 0, 0) & \Phi^{\tau\tau}(0, 0, 0, t_4) \\ \Phi^{\tau\zeta}(0, 0, -t_3, 0) & \Phi^{\tau\tau}(0, 0, 0, t_4) & \Phi^{\tau\tau}(-t_2, 0, 0, 0) \end{pmatrix}.$$

Each of the 10×10 matrices $D_i^{\zeta\tau\tau} - T^{\zeta\tau\tau}$ can be represented as a direct sum of four diagonal minors,

$$D_i^{\tau\zeta\zeta} - T^{\tau\zeta\zeta}(t_1, t_2, t_3, t_4) = A_i^{1,6,10} \oplus A_i^{2,5} \oplus A_i^{3,8} \oplus A_i^{4,7,9},$$

where

$$A_i^{1,6,10} = \begin{pmatrix} 2K_i + t_1 & t_3 & t_3 \\ t_3 & \frac{1}{2\mu_i} - t_2 & t_4 \\ t_3 & t_4 & \frac{1}{2\mu_i} - t_2 \end{pmatrix},$$

$$A_i^{2,5} = A_i^{3,8} = \begin{pmatrix} 2\mu_i - t_1 & t_3 \\ t_3 & \frac{1}{2K_i} + t_2 \end{pmatrix},$$

$$A_i^{4,7,9} = \begin{pmatrix} t_1 & t_3 & -t_3 \\ t_3 & \frac{1}{2\mu_i} - t_2 & -t_4 \\ -t_3 & -t_4 & \frac{1}{2\mu_i} - t_2 \end{pmatrix}.$$

3. The critical values t_i^* of the four parameters t_i of the translation matrix are determined from the system of two equations

$$\det A_1^{2,5} = 0, \quad \det A_2^{2,5} = 0,$$

and two scalar equations following from the condition

$$\mathrm{rank}A_1^{4,7,9} = 1.$$

They are given by

$$t_1^* = \frac{2\mu_1^2\mu_2(K_1 - K_2)}{K_1\mu_1(K_2 + \mu_1) - K_2\mu_2(K_1 + \mu_1)},$$

$$t_2^* = \frac{\mu_2(K_1 + \mu_1) - \mu_1(K_2 + \mu_1)}{2K_1\mu_1(K_2 + \mu_1) - 2K_2\mu_2(K_1 + \mu_1)},$$

$$t_3^* = \sqrt{t_1^*\left(\frac{1}{2\mu_1} - t_2^*\right)},$$

$$t_4^* = \frac{1}{2\mu_1} - t_2^*.$$

4. We obtain an upper bound from the inequality (16.2.18). The matrix R (see (16.2.18)) is split into four uncoupled blocks for isotropic tensors D_*; the most restrictive estimate comes from the condition $\det R^{1,6,9} \geq 0$. This inequality gives the bound. The boundary linear-fractional curve in the coordinates K_*, μ_* corresponds to the equality $\det R^{1,6,9} = 0$.

A Lower Coupled Bound, the Badly Ordered Case

The procedure remains the same in this case; the only difference is in the expressions for the parameters in the translation method.

1. We bound the functional $I^{\zeta\zeta\zeta}$. The matrices D_i, $i = 1, 2$, have the form

$$D_i = D_i^{\zeta\zeta\zeta} = \begin{pmatrix} \tilde{C}_i & 0 & 0 \\ 0 & \tilde{C}_i & 0 \\ 0 & 0 & \tilde{C}_i \end{pmatrix}, \quad i = 1, 2.$$

2. The translations $T^{\zeta\zeta\zeta}$ depend on the four parameters

$$T^{\zeta\zeta\zeta}(t_1, t_2, t_3, t_4) =$$
$$\begin{pmatrix} \Phi^{\zeta\zeta}(-t_1, 0, 0, 0) & \Phi^{\zeta\zeta}(0, -t_3, 0, 0) & \Phi^{\zeta\zeta}(0, 0, -t_3, 0) \\ \Phi^{\zeta\zeta}(0, -t_3 0, 0) & \Phi^{\zeta\zeta}(-t_2, 0, 0, 0) & \Phi^{\zeta\zeta}(0, 0, 0, -t_4) \\ \Phi^{\zeta\zeta}(0, 0, -t_3, 0) & \Phi^{\zeta\zeta}(0, 0, 0, -t_4) & \Phi^{\zeta\zeta}(-t_2, 0, 0, 0) \end{pmatrix}.$$

Each of the 12×12 matrices $D_i^{\zeta\zeta\zeta} - T^{\zeta\zeta\zeta}$, $i = 1, 2$, is a direct sum of the diagonal minors:

$$D_i^{\zeta\zeta\zeta} - T^{\zeta\zeta\zeta} = A_i^{1,6,11} \oplus A_i^{2,5,12} \oplus A_i^{3,8,9} \oplus A_i^{4,7,10},$$

where

$$A_i^{1,6,11} = \begin{pmatrix} 2K_i + t_1 & t_3 & t_3 \\ t_3 & 2\mu_i - t_2 & t_4 \\ t_3 & t_4 & 2\mu_i - t_2 \end{pmatrix},$$

$$A_i^{2,5,12} = \begin{pmatrix} 2\mu_i - t_1 & -t_3 & t_3 \\ -t_3 & 2K_i + t_2 & t_4 \\ t_3 & t_4 & t_2 \end{pmatrix},$$

$$A_i^{3,8,9} = \begin{pmatrix} 2\mu_i - t_1 & -t_3 & -t_3 \\ -t_3 & t_2 & -t_4 \\ -t_3 & -t_4 & 2K_i + t_2 \end{pmatrix},$$

$$A_i^{1,6,11} = \begin{pmatrix} t_1 & t_3 & -t_3 \\ t_3 & 2\mu_i - t_2 & -t_4 \\ -t_3 & -t_4 & 2\mu_i - t_2 \end{pmatrix}.$$

3. The critical values t_i^* of the parameters t_i of the translation matrix $T^{\zeta\zeta\zeta}$ are determined by the conditions

$$\det A_1^{2,5,12} = 0, \quad \det A_2^{2,5,12} = 0,$$
$$\det A_2^{4,7,10} = 0, \quad I_2(A_2^{4,7,10}) = 0,$$

where I_2 is the second main invariant of the 3×3 matrix $A_2^{4,7,10}$ (the conditions show that the rank of this matrix is equal to one). The critical parameters are

$$t_1^* = \frac{2\mu_1\mu_2^2(K_2 - K_1)}{\mu_1(K_2 + \mu_2)(K_1 + 2\mu_2) - \mu_2(K_1 + \mu_2)(K_2 + 2\mu_2)},$$

$$t_2^* = \frac{2\mu_2^2(\mu_1(K_2 + \mu_2) - \mu_2(K_1 + \mu_2))}{\mu_1(K_2 + \mu_2)(K_1 + 2\mu_2) - \mu_2(K_1 + \mu_2)(K_2 + 2\mu_2)},$$

$$t_3^* = \sqrt{t_1^*(2\mu_2 - t_2^*)}, \quad t_4^* = 2\mu_2 - t_2^*.$$

The parameters t_i^* provide the minimal sum of the ranks of the matrices

$$D_1^{\zeta\zeta\zeta} - T^{\zeta\zeta\zeta}(t_1^*, t_2^*, t_3^*, t_4^*), \quad D_2^{\zeta\zeta\zeta} - T^{\zeta\zeta\zeta}(t_1^*, t_2^*, t_3^*, t_4^*)$$

and the nonnegativeness of these matrices.

4. The bound is given by (16.2.18). A calculation demonstrates that R has a block-diagonal structure for the isotropic tensors D_*. One could check that the most restrictive bounds correspond to vanishing of the determinant of its block $A_*^{1,4,9}$,

$$\det A_*^{1,4,9} = 0.$$

This condition leads to the inequality (16.2.3).

An Upper Coupled Bound, the Badly Ordered Case

1. For this bound we consider the functional $I^{\tau\tau\tau}$. The matrices D_i, $i = 1, 2$, are defined by the expression

$$D_i = D_i^{\tau\tau\tau} = \begin{pmatrix} S_i & 0 & 0 \\ 0 & S_i & 0 \\ 0 & 0 & S_i \end{pmatrix}, \quad i = 1, 2.$$

2. The matrix T is given by

$$T^{\tau\tau\tau}(t_1, t_2, t_3, t_4) =$$
$$\begin{pmatrix} \Phi^{\tau\tau}(-t_1, 0, 0, 0) & \Phi^{\tau\tau}(0, -t_3, 0, 0) & \Phi^{\tau\tau}(0, 0, 0, -t_3) \\ \Phi^{\tau\tau}(0, -t_3, 0, 0) & \Phi^{\tau\tau}(-t_2, 0, 0, 0) & \Phi^{\tau\tau}(0, 0, -t_4, 0) \\ \Phi^{\tau\tau}(0, 0, 0, -t_3) & \Phi^{\tau\tau}(0, 0, -t_4, 0) & \Phi^{\tau\tau}(-t_2, 0, 0, 0) \end{pmatrix}.$$

The difference $D_1 - D_2$ is nondegenerate. Therefore, we can use the unmodified translation bound. Each of the 9×9 matrices $[D_i^{\tau\tau\tau} - T^{\tau\tau\tau}(t_1, t_2, t_3, t_4)]$, $i = 1, 2$, is represented as a direct sum of four blocks,

$$D_i^{\zeta\zeta\zeta} - T^{\zeta\zeta\zeta} = A_i^{1,5,9} \oplus A_i^{2,4} \oplus A_i^{3,7} \oplus A_i^{6,8},$$

where

$$A_i^{1,5,9} = \begin{pmatrix} \frac{1}{2K_i} + t_1 & t_3 & t_3 \\ t_3 & \frac{1}{2\mu_i} - t_2 & t_4 \\ t_3 & t_4 & \frac{1}{2\mu_i} - t_2 \end{pmatrix},$$

$$A_i^{2,4} = \begin{pmatrix} \frac{1}{2\mu_i} - t_1 & -t_3 \\ -t_3 & \frac{1}{2K_i} + t_2 \end{pmatrix},$$

$$A_i^{3,7} = \begin{pmatrix} \frac{1}{2\mu_i} - t_1 & -t_3 \\ -t_3 & \frac{1}{2K_i} + t_2 \end{pmatrix},$$

$$A_i^{6,8} = \begin{pmatrix} \frac{1}{2\mu_i} - t_2 & -t_4 \\ -t_4 & \frac{1}{2\mu_i} - t_2 \end{pmatrix}.$$

3. The critical values t_i^* of the parameters t_i of the translation matrix T^{TTT} are obtained from the equations

$$\det A_2^{1,5,9} = 0, \quad \det A_1^{2,4} = 0,$$
$$\det A_2^{2,4} = 0, \quad \det A_1^{6,8} = 0.$$

The exact expressions for t_i^* can be found in (Cherkaev and Gibiansky, 1993).

4. In this case, the unmodified translation bound is applicable. The translation bound comes from the inequality

$$\det \left(Y(D_*^{TTT}) + T^{TTT}(t_1^*, t_2^*, t_3^*, t_4^*) \right) \geq 0.$$

The most restrictive ones correspond to vanishing of the determinant $\det A_*^{1,5,9}$ of the block $A_*^{1,5,9}$ of that matrix. This condition leads to the other inequality in (16.2.3)

16.3   Isotropic Planar Polycrystals

As a last example of the application of the translation method, consider the problem of the range of parameters of an elastic polycrystal assembled from the fragments of an orthotropic crystallite. We follow (Avellaneda et al., 1996) and demonstrate optimal structures that correspond to isotropic components of the closure. A special case was investigated in the earlier paper (Lurie and Cherkaev, 1981c) by means of the translation method.

This problem derives from the more general problem of an optimal orientation of axes of an anisotropic elastic tensor. The paper (Fedorov and Cherkaev, 1983) investigates the problem of minimization of the stress and strain energy of an anisotropic elastic material and addresses an inner contradiction in the necessary conditions, similar to those discussed in Chapter 4. These contradictions lead to the necessary appearance of composites in the optimal structures. The optimal material tends to become more isotropic when an external field is close to the isotropic uniform stress. It was shown in (Fedorov and Cherkaev, 1983) that the necessary conditions of optimality require the appearance of laminates in the problem of energy minimization.

The problem of optimization of the orientation of the axes of anisotropy has been considered many times in different settings. The series of papers (Pedersen, 1989; Pedersen, 1990; Pedersen, 1991; Bendsøe, Hammer, Lipton, and Pedersen, 1995) treats various mathematical and computational aspects of this problem in the two- and three-dimensional settings. The paper (Bendsøe and Lipton, 1997) discusses methods of relaxation. Additional references can be found in the book (Bendsøe and Mota Soares, 1992) and in the survey (Rozvany, Bendsøe, and Kirsch, 1995). Bounds of the energy of elastic crystals were considered in (Fonseca, 1987).

16.3.1   Bounds

Setting

We use the following representation for the compliance tensor S of the orthotropic crystal:

$$S = \begin{pmatrix} s_{11} & s_{12} & 0 \\ s_{12} & s_{22} & 0 \\ 0 & 0 & s_{33} \end{pmatrix}. \tag{16.3.1}$$

The isotropic polycrystal is characterized by the isotropic tensor S_0 of the form

$$S_0 = \begin{pmatrix} \rho & 0 & 0 \\ 0 & \eta & 0 \\ 0 & 0 & \eta \end{pmatrix}, \tag{16.3.2}$$

where ρ and η are isotropic bulk and shear compliances. Recall that they are expressed through the shear μ and bulk K moduli as $\rho = \frac{1}{2K}$ and $\eta = \frac{1}{2\mu}$. The range of possible isotropic polycrystals is described by a set in the ρ, η plane.

Results

The range of moduli ρ and η of an effective tensor corresponds to a rectangle domain $ABCD$ in the ρ, η plane. The corner points

$$A = (\rho^{\mathrm{u}}, \eta^+), \ B = (\rho^{\mathrm{l}}, \eta^+), \ C = (\rho^{\mathrm{l}}, \eta^-), \ D = (\rho^{\mathrm{u}}, \eta^-)$$

are determined by the coefficients c_{ij} of the crystallite as follows:

$$\rho^{\mathrm{u}} = s_{11}, \tag{16.3.3}$$

$$\rho^{\mathrm{l}} = s_{11} - s_{22} + \sqrt{(s_{11} + s_{22})^2 - 4s_{12}^2}, \tag{16.3.4}$$

$$\eta^+ = -s_{11} - s_{12} + \sqrt{(s_{11} + s_{33})(s_{11} + s_{22} + 2s_{12})}, \qquad (16.3.5)$$

$$\eta^- = -s_{11} + s_{12} + \sqrt{(s_{11} + s_{33})(s_{11} + s_{22} - 2s_{12})}, \qquad (16.3.6)$$

and it is assumed that $c_{12} \geq 0$ (the sign of c_{12} depends on the ordering of axes).

Bounds for Materials with Square Symmetry. The bounds are especially simple if the original material has square symmetry ($c_{12} = 0$). In this case, the moduli of an isotropic polycrystal are uniquely defined (see (Lurie and Cherkaev, 1981c)):

$$\rho^{\mathrm{u}} = \rho^{\mathrm{l}} = s_{11}$$

and

$$\eta^+ = \eta^- = -s_{11} + \sqrt{(s_{11} + s_{22})(s_{11} + s_{33})}. \qquad (16.3.7)$$

This property corresponds to the *G*-closure with empty interior. The formula (16.3.7) is similar to the *G*-closure of a two-dimensional conducting polycrystal (Chapter 11). These two problems have much in common, as follows from the discussion in Section 15.1.

Derivation of the Bounds

The translation bounds for the polycrystal have the form (10.1.16). To obtain the bounds, we use the isotropic translators T. The isotropy of T ensures that the translated matrix (10.1.16) of a disoriented crystallite

$$\Phi^T(\phi(\mathbf{x}))^T D \Phi(\phi(\mathbf{x})) - T$$

is positive semidefinite for all $(\phi(\mathbf{x}))$ simultaneously. Here D is the matrix of properties of the anisotropic crystallite, and $\phi(\mathbf{x})$ is the angle of rotation of the crystallite and Φ is the matrix of rotation through an angle ϕ.

Such translators produce the bound

$$D_0 - T \geq 0 \qquad (16.3.8)$$

where D_0 is the matrix of properties of the isotropic polycrystal. We use the translation matrices that were constructed in the previous section using bilinear quasiaffine functions.

Bounds on the Bulk Modulus. Following (Avellaneda and Milton, 1989b), an upper bound for the bulk modulus can be obtained by using the functional J^σ and the quasiaffine function associated with the translation matrix $\Phi^{\tau\tau}(t_1, 0, 0, 0)$; see (16.2.9). The bound (16.3.8) for an isotropic effective tensor S_0 is presented by the following inequality:

$$\begin{pmatrix} \rho - t_1 & 0 & 0 \\ 0 & \eta + t_1 & 0 \\ 0 & 0 & \eta + t_1 \end{pmatrix} \geq 0, \qquad (16.3.9)$$

where, according to (10.1.16), the parameter t_1 must satisfy the constraint

$$\begin{pmatrix} s_{11} - t_1 & s_{12} & 0 \\ s_{12} & s_{22} + t_1 & 0 \\ 0 & 0 & s_{33} + t_1 \end{pmatrix} \geq 0. \qquad (16.3.10)$$

The inequality (16.3.9) implies the following inequality for the bulk modulus ρ:

$$\rho - t_1 \geq 0. \qquad (16.3.11)$$

This inequality is most restrictive when the parameter t_1 is chosen as large as possible, i.e., when it is chosen as the positive root

$$t_1 = \frac{1}{2}\left[s_{11} - s_{22} + \sqrt{(s_{11} + s_{22})^2 - 4s_{12}^2} \right] \qquad (16.3.12)$$

of the quadratic equation

$$(s_{11} - t_1)(s_{22} + t_1) - s_{12}^2 = 0.$$

This equation follows from (16.3.10) treated as an equality. By combining (16.3.11) and (16.3.12) we arrive at an upper bound (16.3.3), (16.3.4) on the bulk modulus of the polycrystal.

A lower bound for ρ is derived similarly, using the functional I^ς and the translation $\Phi^\varsigma(t_1, 0, 0, 0)$.

Bounds on the Shear Moduli. The bounds (16.3.5)–(16.3.6) are obtained similarly to the translation bounds derived for a composite of two isotropic materials. We use the functionals $I^{\varsigma\varsigma}$ and $I^{\tau\tau}$ and the corresponding translators that depend on the two parameters t_1 and t_2. The calculation is similar to the one in Section 15.2, but the details are sometimes tricky. Particularly, we have to solve a fourth-order system for the optimal parameters t_1 and t_2. Fortunately, it is possible to factorize this system and split it into two second-order equations with rational coefficients. We refer to the original paper (Avellaneda et al., 1996) for details.

The Optimal Shear and Bulk Bounds are Uncoupled. In order to obtain coupled shear-bulk bounds one needs to submerge the composite into three different fields: one hydrostatic and two orthogonal shear ones, each chosen to be a stress or strain as appropriate, as was done in the previous section. However, the surprising feature of the problem is that the optimal bounds on the bulk and shear moduli are *uncoupled*, i.e., they form a rectangle in the (ρ_0, η_0) plane. Therefore, this consideration is not needed.

16.3.2 Extremal Structures: Differential Scheme

Extreme Points

The bounds (16.3.3)–(16.3.6) form a rectangle $ABCD$ in the plane of effective constants (ρ, η). We want to show that each point in this rectangle corresponds to effective properties of some polycrystal.

Let us demonstrate that it is enough to find microstructures with isotropic effective tensors that correspond to the corner points $A = (\rho^{\mathrm{u}}, \eta^{+})$, $B = (\rho^{\mathrm{l}}, \eta^{+})$, $C = (\rho^{\mathrm{l}}, \eta^{-})$, and $D = (\rho^{\mathrm{u}}, \eta^{-})$. The points inside this rectangle can be easily obtained as composites of the corner points. Again we use the conservation property of G-closure (see Chapter 3 and Section 15.1).

Consider an isotropic composite assembled from two given isotropic materials sharing a common bulk modulus $\rho_1 = \rho_2$ but having different shear moduli η_1 and η_2. All isotropic composites of that kind have the same bulk modulus, i.e., $\rho_0 = \rho_1 = \rho_2$, and they have a shear modulus η_0 that can lie anywhere in the interval $[\eta_1, \eta_2]$. Similarly, the composites assembled from isotropic materials with a common shear modulus $\eta_1 = \eta_2$ but having different bulk moduli are such that $\eta_0 = \eta_1 = \eta_2$, and they have a bulk modulus ρ_0 that varies in the interval $[\rho_1, \rho_2]$.

Let us start with isotropic polycrystals A with moduli $(\rho^{\mathrm{u}}, \eta^{+})$, and B with moduli $(\rho^{\mathrm{l}}, \eta^{+})$, that share a common shear modulus η^{+}. One can assemble isotropic materials with the same shear modulus and the bulk modulus that corresponds to any point of the interval AB.

In the same way we obtain isotropic composites that fill the interval CD by mixing two isotropic polycrystals C, with moduli $(\rho^{\mathrm{l}}, \eta^{-})$, and D, with moduli (ρ_l, η^{-}). Again we choose a pair of materials, one along the interval AB and the other along the interval CD, so that they share the same bulk modulus $(\rho_{AB} = \rho_{CD})$. We mix them and generate all the materials in the interval

$$\rho_0 = \rho_{AB}, \quad \eta_0 \in [\eta^{-}, \eta^{+}] = [\eta_l, \eta_u],$$

because the composites have a common value of the bulk modulus. This construction enables us to obtain all points in the rectangle as a composite of the materials A, B, C, and D.

It remains to demonstrate that the corner points are the effective tensors of some microstructures. The construction of these points uses two elegant schemes.

Microstructures of the Extremal Points B and C

The corner points B and C can be realized using a laminate composite material of infinite rank obtained by a differential scheme (see Chapter 7). First, we compute the increment dS_0 in the effective tensor of elasticity S_0 caused by adding an infinitesimal amount dm of the material S_b in a layer with tangential $\mathbf{T}$. Using (14.3.3), we have

$$dS_0 = \frac{dm}{m}\left((S_b - S_0) - \frac{(S_b - S_0) : \mathbf{T} \otimes \mathbf{T} : (S_b - S_0)}{\mathbf{T} : S_b : \mathbf{T}}\right),$$

where $\mathbf{T} = \mathbf{t} \otimes \mathbf{t}$ and $\mathbf{t}$ is the tangent to the layer.

Consider the process of constructing a polycrystal shown in Figure 16.4. Start with an isotropic material S_0 with the moduli ρ, μ, which we call the seed. We add an infinitesimal proportion dm of anisotropic crystallite

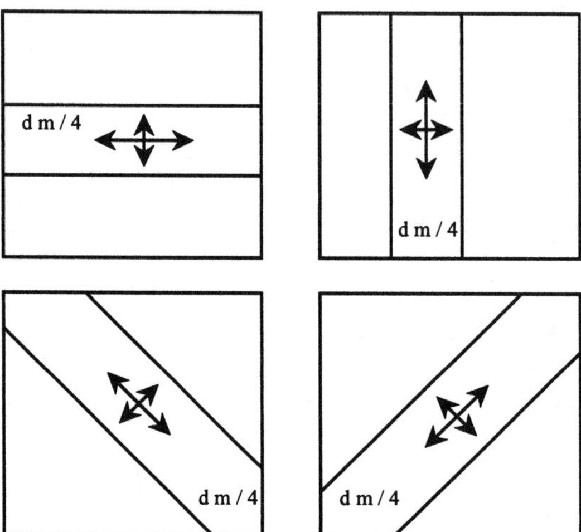

FIGURE 16.4. Scheme of constructing an optimal polycrystal. Note that the orientation of the added anisotropic material matches the orientation of the layer.

to the seed material, layering it in equal proportions $\frac{dm}{4}$ in four different directions $\mathbf{t}_1$, $\mathbf{t}_2$, $\mathbf{t}_3$, and $\mathbf{t}_4$ each $45°$ apart, with the crystal in each layer oriented so that the layering direction is a principal axis of the crystal in that layer.

First we add the tensor $S_1 = S$ (i.e., the original crystal, (16.3.1)),

$$
S_1 = \begin{pmatrix} s_{11} & s_{12} & 0 \\ s_{12} & s_{22} & 0 \\ 0 & 0 & s_{33} \end{pmatrix},
$$

in layers with tangential vector $\mathbf{t}_1 = \mathbf{i}$. Here we use the representation $\mathbf{T}_1 = \mathbf{t}_1 \otimes \mathbf{t}_1$ or $\mathbf{T}_1 = \sqrt{2}(\mathbf{a}_1 - \mathbf{a}_2)$. The corresponding increment is

$$
d_1 S_0 = \left(S_1 - S_0 - \frac{1}{s_{11} + s_{22} - 2s_{12}} H_1 \right) d\frac{m}{4}, \qquad (16.3.13)
$$

where

$$
H_1 = \begin{pmatrix} h_1^2 & h_1 h_2 & 0 \\ h_1 h_2 & h_2^2 & 0 \\ 0 & 0 & 0 \end{pmatrix}, \qquad \begin{aligned} h_1 &= (s_{11} - s_{12} - \rho), \\ h_2 &= (s_{22} - s_{12} - \eta). \end{aligned}
$$

Further, we add the tensor $S_2 = \Phi\left(\frac{\pi}{4}\right) : S : \Phi^T\left(\frac{\pi}{4}\right)$ (i.e., the crystal rotated through the angle $\frac{\pi}{4}$; here $\Phi(\phi)$ is the tensor of rotation through ϕ). We place this material in layers that are rotated by $\frac{\pi}{4}$. The tangential vector of these laminates is $\mathbf{t}_2 = \frac{1}{\sqrt{2}}(\mathbf{i} + \mathbf{j})$.

Then we add the tensor $S_3 = \Phi\left(\frac{\pi}{2}\right) : S : \Phi^T\left(\frac{\pi}{2}\right)$, which is the original crystal rotated through the angle $\frac{\pi}{2}$) in layers with tangential vector $\mathbf{t}_3 = \mathbf{j}$.

Finally, we add the crystal $S_4 = \Phi\left(\frac{3\pi}{4}\right) : S : \Phi^T\left(\frac{3\pi}{4}\right)$ (i.e., original crystal rotated through the angle $\frac{3\pi}{4}$) in layers with the tangential vector $\mathbf{t}_4 = \frac{1}{\sqrt{2}}(-\mathbf{i} + \mathbf{j})$.

The differential step describing this process has the form

$$dS_0 = \frac{dm}{4m} \sum_{i=1}^{4} \left[S_i - S_0 - \frac{(S_i - S_0) : \mathbf{T}_i \otimes \mathbf{T}_i : (S_i - S_0)}{\mathbf{T}_i : S_i : \mathbf{T}_i} \right]. \tag{16.3.14}$$

The isotropy of the original seed tensor S_0 and the symmetry of the differential process ensure that the matrix dS_0 remains isotropic.

We compute two linear invariants $\mathrm{Tr}S$ and $\mathrm{Tr}D$ of the fourth-rank tensor on the right-hand side of (16.3.14) noticing that all four terms on the right-hand side of (16.3.14) differ only by a rotation. Using the representation (16.3.13), we obtain

$$\mathrm{Tr}S\left(m\frac{d}{dm}S_0\right) = s_{11} - \rho_0 - \frac{(s_{11} - s_{12} - \rho_0)^2}{s_{11} + s_{22} - 2s_{12}},$$

$$\mathrm{Tr}D\left(m\frac{d}{dm}S_0\right) = s_{22} + s_{33} - \eta_0 - \frac{(s_{12} - s_{22} + \eta_0)^2}{s_{11} + s_{22} - 2s_{12}}.$$

We iterate this process, adding more and more crystallite in the envelope, until the seed material occupies an infinitesimal portion of the composite. The fixed point of the process corresponds to the condition $dS_0 = 0$.

The condition states that the process is stable to the addition of new portions of crystallite; it defines the parameters of the isotropic polycrystals.

We have

$$\mathrm{Tr}S\left(m\frac{d}{dm}S_0\right) = 0, \quad \mathrm{Tr}D\left(m\frac{d}{dm}S_0\right) = 0,$$

which gives the equations for ρ_0 and η_0:

$$s_{11} - \rho_0 - \frac{(s_{11} - s_{12} - \rho_0)^2}{s_{11} + s_{22} - 2s_{12}} = 0$$

and

$$s_{22} + s_{33} - \eta_0 - \frac{(s_{12} - s_{22} + \eta_0)^2}{s_{11} + s_{22} - 2s_{12}} = 0.$$

These equations have a unique positive solution $\rho_0 \geq 0$ and $\eta_0 \geq 0$. The calculation demonstrates that

$$\rho_0 = \rho^1, \quad \eta_0 = \eta^-,$$

i.e., the pair ρ_0, η_0 corresponds to the corner point C; see (16.3.3)–(16.3.6).

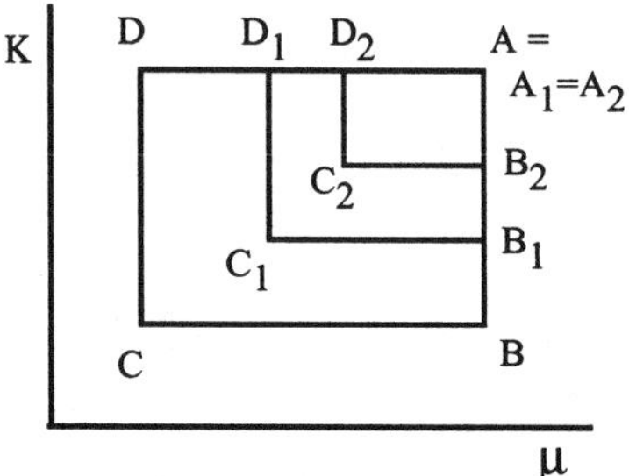

FIGURE 16.5. A scheme for obtaining an optimal polycrystal. The sequence of the bounds $A_n B_n C_n D_n$ converges to the point A.

To find the structure for the matching point B we consider a similar construction but with the crystals in the laminates rotated by 90°. Instead of the tensor S we now consider the tensor $S_\perp$:

$$S_\perp = \begin{pmatrix} s_{11} & -s_{12} & 0 \\ -s_{12} & s_{22} & 0 \\ 0 & 0 & s_{33} \end{pmatrix}.$$

Notice that the crystal in each layer is still oriented so that the layering direction is a principal axis of the crystal in that layer. By comparing the tensors $S_\perp$ and S it is clear that all the previous formulas apply but with s_{12} replaced by $-s_{12}$.

In particular, the effective tensor $S_{0\perp}$ of the structure has a bulk modulus $\rho_\perp$ and a shear modulus $\eta_\perp$ that correspond to the point B:

$$\rho_\perp = \rho^1, \quad \eta_\perp = \eta^+;$$

see (16.3.3)–(16.3.6). Thus we have described the microstructures that correspond to both corner points B and C of the rectangular box defined by the bounds.

16.3.3   Extremal Structures: Fixed-Point Scheme

Now we describe microstructures corresponding to the points A and D of the bounding rectangle. The construction involves composites of infinite rank and demonstrates an interesting use of the stable point principle in optimal structures.

We demonstrate a sequence of orthotropic composites that tend to the isotropic material with bulk and shear moduli corresponding to the extremal point A. A similar construction is used to approximate the point D.

The Procedure

The steps of the procedure are as follows:

1. Assemble an orthotropic polycrystal S_1 of the original material S_0 and compute the bounds (16.3.3)–(16.3.6) for this material (see Figure 16.5). The bounds correspond to a rectangle $A_1B_1C_1D_1$ in the plane of the moduli. The new rectangle $A_1B_1C_1D_1$ lies inside the original rectangle $ABCD$:

$$A_1B_1C_1D_1 \subset ABCD.$$

This inclusion reflects the irreversible character of the homogenization; there is less freedom in varying the parameters of a composite of the composite than of the original material.

2. Continue this process, and observe that the degree of anisotropy decreases at each step of the procedure. The corresponding rectangles of bounds form a nested sequence:

$$A_nB_nC_nD_n \subset A_{n-1}B_{n-1}C_{n-1}D_{n-1}.$$

3. Finally, after an infinite number of steps, we may end up with an isotropic polycrystal. The bounds for the "polycrystal" of the isotropic material trivially coincide with the moduli of this material. In this case, the sequence of rectangles $A_nB_nC_nD_n$ tends to a point that is common for all rectangles $A_nB_nC_nD_n$.

4. Returning to the first step, we construct a polycrystal with a special property. We require that the bound for that polycrystal shares the corner point A_1 with the original rectangle: $A_1 = A$. We keep this requirement, and construct the next polycrystal that corresponds to the bound $A_2B_2C_2D_2$, and so on. Thus we end up with a sequence of bounds $A_nB_nC_nD_n$ for which $A_n = A_{n-1} = \cdots = A$. Clearly, this nested sequence shrinks to the point A when the composite tends to isotropy. The resulting isotropic composite has the properties described by the point A, because the bound for the moduli of the isotropic material coincides with these moduli.

This procedure realizes the fixed-point principle of a contraction operator. At each step, we find an orthotropic structure that belongs to the boundary of the G-closure. The sequence of these structures tends to an isotropic structure.

Calculations

To construct $S^{(1)}$ we use structures similar to the "herringbone" (Chapters 7 and 14). Namely, we laminate a pair of materials $S(\phi)$ and $S(-\phi)$ obtained by rotating S (16.3.2) through the angles ϕ and $-\phi$, respectively. We choose the tangent to the laminates as $\mathbf{i}$ and equal volume fractions of the components $m_1 = m_2 = \frac{1}{2}$; see Figure 16.6.

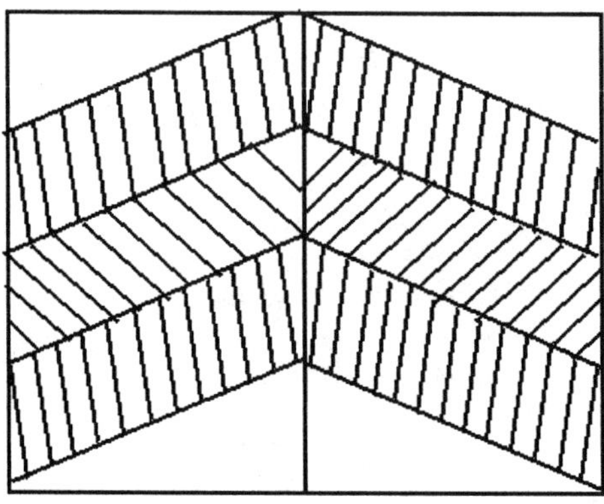

FIGURE 16.6. Fractal structure that tends to the point A.

Remark 16.3.1 *Similar structures are optimal in the problem of the three-dimensional composite of the minimal stiffness (Gibiansky and Cherkaev, 1987) discussed in Chapter 15.*

Let us compute the effective tensor $S^{(1)}$ of this laminate, using (14.3.3). In our setting, the formula becomes

$$S_{\text{lam}} = \frac{1}{2}(S(\phi) + S(-\phi) - \frac{1}{2s}[S(\phi) - S(-\phi)] \cdot \mathbf{T} \otimes \mathbf{T} \cdot [S(\phi) - S(-\phi)],$$

where

$$s = \mathbf{T} \cdot [S(\phi) + S(-\phi)]\mathbf{T}, \quad \mathbf{T} = \mathbf{t} \otimes \mathbf{t} = \frac{1}{\sqrt{2}}(\mathbf{a}_1 + \mathbf{a}_2),$$

and

$$S(\phi) = \begin{pmatrix} s_{11} & s_{12}\cos 2\phi & s_{12}\sin 2\phi \\ s_{12}\cos 2\phi & s_{22}\cos^2\phi + s_{33}\sin^2\phi & (s_{22} - s_{33})\sin 2\phi \cos 2\phi \\ s_{12}\sin 2\phi & (s_{22} - s_{33})\sin 2\phi \cos 2\phi & s_{33}\cos^2\phi + s_{22}\sin^2\phi \end{pmatrix}.$$

A calculation demonstrates that S_{lam} is an orthotropic tensor:

$$S_{\text{lam}}(\phi) = \begin{pmatrix} s_{11} & s_{12}\cos 2\phi & 0 \\ s_{12}\cos 2\phi & s'_{22} & 0 \\ 0 & 0 & s'_{33} \end{pmatrix}, \tag{16.3.15}$$

where the coefficients s'_{22} and s'_{33} are

$$s'_{22} = s_{22}\cos^2 2\phi + s_{33}\sin^2 2\phi,$$

$$s'_{33} = s_{33}\cos^2 2\phi + s_{22}\sin^2 2\phi$$

$$- \frac{[s_{12} + (s_{22} - s_{33})\cos 2\phi]^2 \sin^2 2\phi}{s_{11} + 2s_{12}\cos 2\phi + s_{22}\cos^2 2\phi + s_{33}\sin^2 2\phi}.$$

Bound for the Bulk Modulus. Calculate the bounds on the effective bulk and shear moduli of an isotropic polycrystal assembled from this new orthotropic material $S_{\mathrm{lam}}(\phi)$. Because $s_{11}^{\mathrm{lam}} = s_{11}$, an upper bound on the bulk modulus $\rho^{\mathrm{u}} = s_{11}'$ coincides with the bound $\rho^{\mathrm{u}} = s_{11}$ (see (16.3.4)). This brings the rectangle of bounds for the "herringbone" to the side AD of the original rectangle $ABCD$, no matter what the angle ϕ is,

$$\rho^+(S) = \rho^+(S_{\mathrm{lam}}(\phi)) \quad \forall \phi \in (0, \pi/2).$$

The Point A. Choose the angle ϕ to preserve one bound for the shear modulus bounds, namely, the bound η^+; see (16.3.5). The angle ϕ is found from the requirement:

$$\eta^+(S) = \eta^+(S_{\mathrm{lam}}(\phi)). \qquad (16.3.16)$$

If this equation has a solution, it guarantees that the rectangle $A_1 B_1 C_1 D_1$ associated with $S^{(1)}$ has the same corner $A_1 = A$ as the rectangle $ABCD$ associated with S. A calculation demonstrates that equation (16.3.16) indeed has the real solution $\cos 2\phi_* = c_*$, where

$$c_* = \frac{-s_{11} - s_{33} + \sqrt{(s_{11} + s_{33})(s_{11} + s_{22} + 2s_{12})}}{s_{22} - s_{33} + 2s_{12}}.$$

Because the value c_* should be the cosine of some angle, we want to check if $|c_*| \le 1$. Indeed, one can see that

$$c_* = \frac{\sqrt{pq} - p}{q - p} = \frac{\sqrt{p}}{\sqrt{q} + \sqrt{p}},$$

where the parameters

$$p = s_{11} + s_{33} > 0, \quad q = s_{11} + s_{22} + 2s_{12} > 0$$

are both strictly positive for any positive definite matrix S; we conclude that

$$0 < c_* < 1.$$

Therefore, we can construct an orthotropic material from the original crystal to preserve the corner point A of the bounds.

The Point D. Alternatively, we could choose the angle ϕ such that

$$\eta^-(S) = \eta^-(S_{\mathrm{lam}})(\phi).$$

Calculations demonstrate that this equation has the solution $\cos 2\phi_{**} = c_{**}$, where

$$c_{**} = \frac{-s_{11} - s_{33} + \sqrt{(s_{11} + s_{33})(s_{11} + s_{22} - 2s_{12})}}{s_{22} - s_{33} - 2s_{12}},$$

which in turn provides a solution for ϕ_{**}. The positive definiteness of S implies

$$0 < c_{**} < 1.$$

Therefore, we can construct an orthotropic material from the original crystal that preserves the other corner point $D = (\rho^{\mathrm{u}}, \eta^-)$.

Isotropy of the Limiting Composite. By repeating the process of lamination, we end up with an isotropic material. Indeed, the absolute values of both c_* and c_{**} are strictly less than one. The relations (16.3.15) show that the absolute value of s_{12} must decrease, unless it is already zero. After an infinite number of steps we end up with a material that has a compliance tensor S_0 with at least square symmetry, i.e., with $s_{12} = 0$. This limiting material has square symmetry, hence the bounds for the shear moduli (see (16.3.7)) collapse to a single point. This point obviously coincides with either A or D.

Remark 16.3.2 *These structures are hardly describable in purely geometrical terms: They correspond to a strange fractal geometry. The first iterations of the geometrical construction are presented in Figure 16.6.*

17
Some Problems of Structural Optimization

17.1 Properties of Optimal Layouts

Here we discuss global features of optimally assembled bodies and various formulations of optimization problems. So far, we have mainly discussed the best microstructures of composites. Now we comment on the optimal layout of these composites in an optimally designed body. According to our main concept, the optimal composites have a quasiperiodic structure that varies smoothly with the stress field (see, for example, (Armand et al., 1984)). In dealing with composites, we assume that the scale of the periodicity cells is so small that the external field can be treated as homogeneous, hence the composite is represented by its effective tensor. We keep this concept, but now we consider variation in the effective tensor.

Let us consider the optimal composite structure in the large scale and find how the effective properties of structure vary in the design domain. This part of the investigation requires a numerical solution. There are many numerical methods for structural optimization. We do not discuss the numerical algorithm in this book, referring the reader to the books (Prager, 1974; Banichuk, 1977; Haug, Choi, and Komkov, 1986; Papalambros and Wilde, 1988; Rozvany, 1989; Haftka and Gürdal, 1992; Rozvany, 1989; Haug et al., 1986; Bendsøe, 1995), the papers (Díaz et al., 1993; Allaire and Kohn, 1993c; Tortorelli and Wang, 1993; Jog, Haber, and Bendsøe, 1994; Bendsøe, Guedes, Haber, Pedersen, and Taylor, 1994; Rozvany et al., 1995; van Keulen, Polynkine, and Toropov, 1997; Cherkaev and Palais, 1998; Rodrigues, Soto, and Taylor, 1999), and the references therein.

In these sources, one can also find a variety of approaches to structural optimization.

17.1.1 Necessary Conditions

Let us analyze qualitatively some properties of optimal layouts. Consider the minimization of the total energy of a body:

$$I = \inf_{\chi} \min_{\mathbf{v}} \int_{\Omega} \left(W(D(\chi), \mathbf{v}) + \gamma\chi \right).$$

Recall that the optimal layouts are composites. The energy of the optimal composites $QW(D_{\mathrm{opt}}(m, \mathbf{v})\mathbf{v})$ corresponds to the quasiconvex envelope of the original Lagrangian. After relaxation, the energy becomes

$$I = \inf_{\mathbf{v}} \int_{\Omega} \min_{m} \left\{ QW(D_{\mathrm{opt}}(m, \mathbf{v})\mathbf{v}) + \gamma m \right\}.$$

An optimal layout of structures expresses optimal parameters of the effective tensor of the optimal composite D_{opt} through the acting field $\mathbf{v}$ as: $D_{\mathrm{opt}} = D_{\mathrm{opt}}(\mathbf{v})$. We use necessary conditions of optimality,

$$\frac{\partial}{\partial m} QW(D_{\mathrm{opt}}(m, \mathbf{v})\mathbf{v}) + \gamma = 0,$$

to find the optimal volume fraction $m_{\mathrm{opt}}(\mathbf{v})$ and exclude it. This way, we find the quasiconvex envelope $QF(\mathbf{v})$ of the original Lagrangian, and we end up with a variational problem for the state of the optimal body:

$$\min_{\mathbf{v}} \int_{\Omega} QF(\mathbf{v}).$$

The Euler–Lagrange equations of the quasiconvex envelope have some specific features; we illustrate them on optimal design problems that we have already discussed.

The Conducting Medium of Minimal Conductivity

Consider the problem of a conducting body with minimal conductivity (Chapter 4). In this case, the quasiconvexity coincides with convexity. The optimal composites are the properly oriented laminates.

Recall that the convex envelope $CF(\mathbf{e})$ has the form

$$CF(\mathbf{e}) = a|\mathbf{e}| + b,$$

where a and b are some constants.

The current $\mathbf{j}$ in an optimal body is defined as

$$\mathbf{j} = \frac{\partial}{\partial \mathbf{e}} CF(\mathbf{e}) = a\frac{\partial |\mathbf{e}|}{\partial \mathbf{e}}.$$

This says that $|\mathbf{j}| = C = \text{constant}(\mathbf{x})$ everywhere in the composite zone. In this zone the following first-order equation is valid:

$$\nabla \cdot \mathbf{j} = f, \quad |\mathbf{j}| = C. \tag{17.1.1}$$

Note that the constitutive equations of an optimally designed body are on the boundary of ellipticity. We note several consequences of this (Cherkaev, 1993):

1. The state of the optimal body is described by nonlocal equations. Indeed, $|\mathbf{j}| = C$ in the composite zone. Hence, any local perturbation of the optimal field leads to a global variation of the optimal layout. It changes everywhere in order to preserve the absolute value of the current. One could observe this effect by noticing that a perturbation of the solution to the first-order equation (17.1.1) propagates along its characteristics.

2. The magnitude of the field $|\mathbf{e}|$ inside each material in the structure is constant everywhere: $|\mathbf{e}_i| = \frac{1}{\sigma_i}|\mathbf{j}|$, where i is the number of the material. The magnitude of the mean field in the composite is varied only due to variation of the volume fraction of the components.

3. The Lagrangian of the optimally designed body is on the boundary of the region of stability, and any infinitely small perturbation of it could lead to the nonconvexity of the perturbed energy and nonexistence of a solution to the Euler–Lagrange equation. This poses a serious problem for a suitable approximation of the energy of the optimal media.

These properties are typical for the optimal composites that represent the convexification of the energy. Indeed, the constancy of $|\mathbf{j}| = \frac{\partial \mathcal{C}W}{\partial |\mathbf{e}|}$ follows from the geometrical construction of the convex envelope. This constancy represents a nonlocal property of composite optimal design. We expect that certain function of the variables remain constant in the domain occupied by the composites. The constancy of the field $|\mathbf{e}_i|$ within each material reflects the Carathéodory theorem.

Optimization of the Sum of Conducting Energies

This problem is similar to the previous one. Its relaxation is discussed in Part III (Chapters 6–9). The optimal composites are properly oriented matrix laminates or simple laminates. Notice the differences from and similarities to the previous problem:

1. The energy of the optimally designed body cannot be reduced to an energy of a conducting nonlinear material. The dependence of the effective properties on all the external fields couples them.

2. The state of the optimal body is described by nonlocal equations. Any local perturbation of the optimal field $\mathbf{E}$ leads to variation of the composite parameters everywhere. Particularly, the variation of the volume fraction is nonlocal.

3. A norm $\|\mathbf{E}\|$ of the field $\mathbf{E}$ inside each material in the optimal structure is constant everywhere (see Chapter 9, Figure 9.2). The field in each material keeps the corresponding norm constant everywhere in the composite zone.

Optimal Structures in Plane Elasticity

Consideration of the plane problem of elasticity leads to similar results. It was shown in Chapter 15 that the optimal elastic body consists of second-rank laminates, single laminates, or solid material. Consider the asymptotic case of the "topology optimization" when the weak material becomes void. In this case, the optimal structures are either second-rank laminates or solid material. The normals to the layers in the optimal structure are oriented along the directions of external principal stresses.

We list several properties of the optimal design:

1. The effective compliance tensor S_{opt} has minimum compliance $\boldsymbol{\tau}$: $S_{\mathrm{opt}} : \boldsymbol{\tau}$ in the direction of the acting stress field τ. At the same time the compliance $\boldsymbol{\tau}_{\mathrm{shear}} : S_{\mathrm{opt}} : \boldsymbol{\tau}_{\mathrm{shear}}$ in the direction of the orthogonal shear, $\boldsymbol{\tau}_{\mathrm{shear}} = \mathbf{n}_1 \otimes \mathbf{n}_2 + \mathbf{n}_2 \otimes \mathbf{n}_1$, is infinite.

2. The optimal volume fraction m_0 of the material in the structure is proportional to the norm of the stress tensor

$$m_0 = \sqrt{\frac{\rho_1 + \eta_1}{\gamma}}(|\tau_1| + |\tau_2|).$$

3. The norm of the stress in the material inside the optimal structure is constant everywhere, regardless of the applied stress. This shows the two-dimensional realization of the one-dimensional concept of equally stressed construction. The material in an optimal project is never "understressed" (of course, the material can be "overstressed" in the zones where the volume fraction reaches its limit $m = 1$, because of the lack of further control).

4. The structure has two functions. First, it maintains constancy of the norm of the stress in the material in an optimal structure. The volume fraction of the material in the structure controls the rate of magnification of the mean stress. Due to the optimal choice of the volume fraction, the norm of the stress inside the material is kept constant; it is larger than the norm of the mean stress applied to the whole structure. Second, the structure transforms an applied anisotropic

stress tensor to the tensor of an optimal anisotropy degree inside the material layers.

The requirement of the constancy of the norm yields to appearing of thin materials elements (layers) in an optimal design: The constancy is achieved in the limit when the thickness of an element does to zero.

5. Unless the boundary of Ω is specially prescribed, the optimal project is unbounded. Even if the loading/support is concentrated in a finite region, the composite fills all space with an infinitesimal dense "web." The material is equally stressed at each point of that web no matter what the loading is. The volume fractions at distant points are arbitrarily small but still positive.

17.1.2 Remarks on Instabilities

A relaxation replaces the initial ill-posed variational problem by a new one that definitely has a solution. However, the relaxed solution remains unstable against variations of external conditions: loading, shape of the domain, etc. We mention the following properties of optimal layouts:

1. The optimal structures concentrate their capacity of resistance along the direction of the expected field; because of this they show little or no resistance to a stress field directed orthogonal to the expected one. Therefore, the optimal construction is extremely sensitive to variation in loading and in the orientation of the microstructures at each point of the body: Small perturbations of these parameters could lead to essential reduction of its stiffness.

2. A numerical solution to an optimization problem involves a finite-dimensional approximation. In computations, the parameters of microstructures, such as orientation and volume fractions, are not absolutely precisely calculated. Any error of this kind could make the homogenized energy not quasiconvex, because the exact solution corresponds to the boundary of the quasiconvexity.

   Formally, an approximate relaxation of the problem achieves nothing: The homogenized energy remains not quasiconvex, as was the case with the energy of the initial problem before its relaxation. The new problem may again need relaxation. A minimizing sequence may not converge. However, from a practical point of view, the relaxation works even if the numerical methods do not converge. The rate of divergence could be very small and a the suboptimal solution could be arbitrarily close to the true minimum. This theoretical difficulty, however, may manifest itself in the appearance of a "checkerboard structure" and other irregularities.

Remark 17.1.1 *The appearance of checkerboard structures is, definitely, a more complicated phenomenon. It was systematically studied starting from (Jog, Haber, and Bendsøe, 1993). An explanation was suggested based on the properties of the finite elements model, (Jog and Haber, 1996; Díaz and Sigmund, 1995; Sigmund and Petersson, 1998). It was emphasized that the finite elements model, applied to rapidly oscillating properties tensors, adds additional instabilities, because it fails to correctly approximate a checkerboard-type structures; see (Díaz and Sigmund, 1995; Sigmund and Petersson, 1998). Particularly, the computed effective stiffness may even be outside of the Hashin and Shtrikman bounds.*

3. The necessary conditions lead to a nonlinear Euler equation for the state of the optimal body. This equation belongs to the boundary of ellipticity if only the structure does not degenerates into a pure material (in that case, the (quasi)convex envelope of the Lagrangian coincides with the Lagrangian itself). Therefore, it becomes impossible to investigate the convergence of a numerical scheme based on the standard arguments of the convergence of numerical algorithms. The rate of convergence of algorithms for elliptic differential equations usually depends on the constant of ellipticity, which here is zero. The instabilities have been frequently discussed in the literature. We cite here several papers (Jog et al., 1993; Díaz and Sigmund, 1995; Ding, 1986; Rozvany, Olhoff, Bendsøe, et al., 1987; Rozvany et al., 1995; Arora, 1998; Sigmund and Petersson, 1998), where further references can be found.

4. One may use a lower bound of the energy stored in a composite to compute the relaxed energy. Even if the bounds are not exact, the homogenized energy is quasiconvex. However, the solution may not correspond to any structure if the bounds are not exact.

Because of these arguments, the results of a computational procedure strongly depend on the numerical scheme.

Many modifications of the relaxation procedure have been suggested. They mainly deal with additional penalties imposed on the composite structure that push the construction toward a "black-and-white" design, that is, to the design without composite zones. We mention here several papers (Bendsøe et al., 1993; Šverák, 1993; Haber, Jog, and Bendsøe, 1994; Rozvany et al., 1995; Haber et al., 1996; Chiheb and Panasenko, 1998; Taylor, 1998) where further references can be found.

An alternative approach to relaxation is developed in the series of papers (Bendsøe et al., 1994; Guedes and Taylor, 1996; Rodrigues et al., 1999), where it was suggested to choose optimal stiffness tensor from the set of all stiffness matrices with a constrained trace. This approach emphasizes the

FIGURE 17.1. A third-rank laminate. Notice that the structure supports every stress, unlike second-rank laminates.

optimal layout of materials with the best degree of anisotropy and tracks the dependence of the properties on the stress field. The postulated type of integral constraint reflects the penalization for the use of stiffer materials.

17.2   Optimization of the Sum of Elastic Energies

Stable Structures

Consider the following problem: Find the structures of composites that minimize the total energy of a body under some given loading, with the restriction that the energy stored in the body under some other loadings should be bounded. This restriction may express, for example, the degree of our knowledge of the loading conditions. On the other hand, it may correspond to the situation where the loading changes over time. Here we follow (Cherkaev, Krog, and Kucuk, 1998).

We use here the results (Avellaneda and Milton, 1989a) that the matrix composites of third-rank provide the minimum value of the sum of energies stored in the elastic media under several linearly independent loadings. The effective properties of this set of composites for plane elasticity are described by formula (14.3.4).

Note that if none of the parameters α_1, α_2, α_3, is zero then the compliance tensor of the structure is finite, even if the compliance of the enveloped material D_2 is infinite. These structures are more stable against a variation in the loading. Physically speaking, the layers of strong material form triangles instead of rectangles, which provides a uniform stiffness of microstructures (see Figure 17.1). For the same reason, the three-dimensional problem requires sixth-rank laminates (see (Francfort et al., 1995). The layers in that structure form tetrahedrons instead of parallelepipeds.

17.2.1 Minimization of the Sum of Elastic Energies

Let us find the optimal properties of an elastic structure that is submerged into several elastic fields. We minimize the mean stiffness of the structure, that is, the sum of the stored energies.

The Problem

Recall that the stress energy can be represented as

$$W = \boldsymbol{\tau} : S : \boldsymbol{\tau} = \mathrm{Tr}[S : (\boldsymbol{\tau} \otimes \boldsymbol{\tau})],$$

where S is a fourth-rank tensor of the effective elastic properties. This representation allows us to rewrite the Lagrangian of the considered problem as

$$W = \mathrm{Tr}(S\mathbf{U}), \tag{17.2.1}$$

where $\mathbf{U}$ is the tensor

$$\mathbf{U} = \sum_k \boldsymbol{\tau}^{(k)} \otimes \boldsymbol{\tau}^{(k)} \quad \text{or} \quad \mathbf{U}_{ijkl} = \sum_k \sigma_{ij}^{(k)} \sigma_{kl}^{(k)}$$

that corresponds to all loadings applied to the structure. We see that $\mathbf{U}$ is a symmetric positive definite fourth-rank tensor that contains all the information about the loading. We assume in (17.2.1) that the tensors Δ and $\mathbf{U}$ are represented as 3×3 symmertic matrices.

Optimal structures are the third-rank laminates (Avellaneda and Milton, 1989a). Their effective compliance tensor S has the representation (Chapter 14)

$$S = S_1 + \Delta^{-1}.$$

Here

$$\Delta = \frac{1}{m_2}(S_2 - S_1)^{-1} + \frac{m_1}{m_2} \sum_{i=1}^{K} \alpha_i N(\mathbf{T}_i), \tag{17.2.2}$$

and

$$N(\mathbf{n}_i) = \frac{2K_1\mu_1}{K_1 + \mu_1} \mathbf{T}_i \otimes \mathbf{T}_i, \quad \mathbf{T} = \mathbf{t} \otimes \mathbf{t}.$$

The parameters α_i represent the relative thickness of the laminate of ith rank; they satisfy the relations $\sum_{i=1}^{K} \alpha_i = 1$, $\alpha_i \geq 0$.

Recall (Chapter 14) that there are two linear invariants of Δ,

$$\begin{aligned}
\mathrm{TrS}\,\Delta &= \tfrac{1}{m_2}\tfrac{2K_1 K_2}{K_1 - K_2} + c, \\
\mathrm{TrD}\,\Delta &= \tfrac{1}{m_2}\tfrac{2\mu_1\mu_2}{\mu_1 - \mu_2} + c,
\end{aligned} \tag{17.2.3}$$

where

$$c = \frac{m_1}{m_2}\frac{K_1\mu_1}{K_1 + \mu_1}.$$

They completely describe the set of third-rank composites: If (17.2.3) holds, then Δ has the representation (17.2.2) (Avellaneda and Milton, 1989a; Avellaneda and Milton, 1989b).

Topology Optimization

Consider the asymptotic case where the second material becomes void $k_2 \to 0$, $\mu_2 \to 0$. In this case, the optimal structures of third-rank laminates do not degenerate into simpler structures of the second- and first-rank laminates. Indeed, such degeneracy would imply that at least one of the eigenvalues of Δ is zero. Consequently, the matrix $S = S_1 + \Delta^{-1}$ would have one infinite eigenvalue, and the Lagrangian W would become infinitely large for all nondegenerate loadings $\mathbf{U}$; but the optimal structure minimizes W, and therefore it cannot degenerate.

The formal consequence of this is that all coefficients α_i are positive in an optimal structure, $\alpha_i > 0$, and the equalities (17.2.3) are the only constraints on the set of admissible effective properties.

Optimal Energy, Fixed Volume Fractions

To find the structure that minimizes W for a given loading $\mathbf{U}$, we add the constraints (17.2.3) with positive Lagrange multipliers λ_1^2 and λ_2^2. The augmented Lagrangian is

$$W = \mathrm{Tr}(S_1\mathbf{U}) + \mathrm{Tr}(\Delta^{-1}\mathbf{U}) + \lambda_1^2\,\mathrm{TrS}\ \Delta + \lambda_2^2\,\mathrm{TrD}\ \Delta.$$

The optimal Δ satisfies the equation $\frac{\partial W}{\partial \Delta} = 0$.

To compute the derivative, we use the verifiable formulas for symmetric matrices in the basis $\mathbf{a}_1$, $\mathbf{a}_2$, $\mathbf{a}_3$:

$$\frac{\partial}{\partial \Delta}\mathrm{Tr}(\Delta^{-1}\mathbf{U}) = -\Delta^{-1}\mathbf{U}\Delta^{-1},$$

and

$$\frac{\partial}{\partial \Delta}\mathrm{TrS}\ \Delta = \begin{pmatrix} 1 & 0 & 0 \\ 0 & 0 & 0 \\ 0 & 0 & 0 \end{pmatrix}, \qquad \frac{\partial}{\partial \Delta}\mathrm{TrD}\ \Delta = \begin{pmatrix} 0 & 0 & 0 \\ 0 & 1 & 0 \\ 0 & 0 & 1 \end{pmatrix}.$$

The equation $\frac{\partial W}{\partial \Delta} = 0$ leads to

$$\frac{\partial W}{\partial \Delta} = -\Delta^{-1}\mathbf{U}\Delta^{-1} + AA^T = 0 \qquad (17.2.4)$$

where A is the diagonal matrix

$$A = \begin{pmatrix} \lambda_1 & 0 & 0 \\ 0 & \lambda_2 & 0 \\ 0 & 0 & \lambda_2 \end{pmatrix} \qquad (17.2.5)$$

in the basis $\mathbf{a}_1$, $\mathbf{a}_2$, $\mathbf{a}_3$.

We introduce the tensor V by the equality

$$\mathbf{V}\mathbf{V}^T = \mathbf{U}.$$

Note that $\mathbf{V}$ is defined up to an arbitrary rotation Φ, because of the polar representation of a matrix, $\mathbf{V}\Phi\Phi^T\mathbf{V}^T = \mathbf{V}\mathbf{V}^T$.

Equation (17.2.4) allows us to compute the optimal compliance tensor as a function of $\mathbf{V}$. We rewrite (17.2.4) in the form

$$(\Delta^{-1}\mathbf{V})(\Delta^{-1}\mathbf{V})^T = AA^T,$$

which implies that

$$\Delta = \mathbf{V}A^{-1}. \tag{17.2.6}$$

To eliminate the uncertainty of the polar representation and determine Φ in the definition of V, we require the symmetry of the matrices Δ and A.

Computing linear invariants of (17.2.6) and using (17.2.5), we obtain

$$\mathrm{TrS}\,\Delta \;=\; \mathrm{TrS}(\mathbf{V}A^{-1}), \quad \text{or} \quad \lambda_1 = \frac{\mathrm{TrS}(\mathbf{V})}{c},$$

$$\mathrm{TrD}\,\Delta \;=\; \mathrm{TrS}(\mathbf{V}A^{-1}), \quad \text{or} \quad \lambda_2 = \frac{\mathrm{TrD}(\mathbf{V})}{c}.$$

where linear operators TrS and TrD are defined by equations (14.3.6) and eq14.2.28, respectively.

Substituting the values of λ_i and using the definition (17.2.3) for c, we obtain the optimal energy in the form

$$W_0 = \frac{1-m}{m}B + \mathrm{Tr}(S_1\mathbf{U}),$$

where

$$B = \frac{K_1\mu_1}{K_1 + \mu_1}\left(\mathrm{TrS}(\mathbf{V})\right)^2 + \left(\mathrm{TrD}(\mathbf{V})\right)^2.$$

The optimal stress energy W_0 is not quadratic, but a homogeneous second-degree function of the fourth-rank tensor $\mathbf{V}$ that represents the stresses. The energy W_0 does not represent the energy of any nonlinear elastic material. On the contrary, it is a nonlinear function of all stresses acting in the medium.

The strains

$$\epsilon^{(k)} = \frac{\partial W_0}{\partial \tau^{(k)}} = \epsilon^{(k)}(\tau_1,\ldots,\tau_N)$$

depend on all stress fields, not only on the field $\tau^{(k)}$. Indeed, a variation in any of the acting fields leads to changes in the structure; this way it affects all strains.

Quasiconvex Envelope of the Lagrangian

Minimize over m the energy W_0 plus the cost of the material used. Assume that this cost is equal to one. The variational problem for the best material becomes

$$J = \inf_{\tau_i} \min_m \int_S B^2 \frac{1-m}{m} + \mathrm{Tr}(S_1\mathbf{U}) + m, \quad \mathbf{U} = \mathbf{V}\mathbf{V}^T$$

where

$$B = \sqrt{\frac{K_1\mu_1}{K_1 + \mu_1}\,(\mathrm{Tr}S(\mathbf{V}))^2 + (\mathrm{Tr}D(\mathbf{V}))^2}$$

is a rotationally invariant norm of the symmetric fourth-rank tensor $\mathbf{V}$.

Further, we find that the optimal volume fraction m_{opt} is equal to that norm

$$m_{\mathrm{opt}} = B.$$

We substitute m_{opt} into the expression for the Lagrangian and obtain the formula for the quasiconvex envelope QW:

$$QW = \begin{cases} 1 - (1 - B)^2 + \mathrm{Tr}(S_1(\mathbf{V}\mathbf{V}^T)), & \text{if } m_{\mathrm{opt}} < 1, \\ \mathrm{Tr}(S_1(\mathbf{V}\mathbf{V}^T)) + 1 & \text{if } m_{\mathrm{opt}} = 1. \end{cases}$$

Remark 17.2.1 *The quasiconvexified Lagrangian QW is expressed as a rational function of $\mathbf{V}$. In turn,*

$$\mathbf{V} = \left(\sum_i \boldsymbol{\tau}_i \otimes \boldsymbol{\tau}_i\right)^{\frac{1}{2}}$$

is an irrational function of all stress fields $\boldsymbol{\tau}_i$. The expression of the quasiconvex envelope through the original fields would be awkward; this shows that $\mathbf{V}$ is an adequate tensor variable.

Let us compute the density of the energy W inside the material within the microstructure. Clearly, the void phase does not contain any energy. Therefore the specific energy W_m in the material is $\frac{W_m}{m_{\mathrm{opt}}} = \frac{W_m}{B}$ if $m_{\mathrm{opt}} < 1$.

17.2.2 Optimal Design of Periodic Structures

Another problem naturally arising is the optimal design of composites with a *periodic* structure of maximum stiffness. Suppose we need to find a constant tensor of effective compliance $S = \mathrm{constant}(\mathbf{x})$ of a composite and the corresponding microstructure so that $S = \mathrm{constant}(\mathbf{x})$ corresponds to the maximum stiffness of the body.

This problem could also be reduced to the problem of finding a microstructure with minimum value of the sum of energies. Indeed, an effective tensor resists different stresses in different points of the body.

The total stress energy W of a periodic composite with constant anisotropic tensor $C = c_{ijkl}$ of stiffness can be written as

$$W = \int_\Omega \mathrm{Tr}(S : \mathbf{U}) = \sum_{ijkl} S_{ijkl} U_{ijkl},$$

where

$$\mathbf{U} = \int_\Omega \boldsymbol{\tau} \otimes \boldsymbol{\tau} \quad \text{or} \quad U_{ijkl} = \int_\Omega \tau_{ij}\tau_{kl}.$$

This problem also has a solution within the three-rank laminates. The optimization problem is finite-dimensional.

A Stable Iterative Scheme for Minimization of Compliance

Let us discuss an approximation procedure for solving the optimal design problem; the procedure is stable against perturbations of the external parameters, tends to the optimal solution, and corresponds to the known structures at each step. The procedure, suggested in (Cherkaev, 1993), was realized in (Cherkaev, Krog, and Kucuk, 1998), where numerical examples are also given.

Consider a topology optimization problem for the minimization of the total stiffness of the body. Let us part the domain Ω into N subdomains Ω_k with areas v_k. Suppose that the compliance tensor $S^k = s^k_{ijkl}$ remains constant within each subdomain Ω_k. It depends on the periodic composite structure and the volume fraction m_k in the domain Ω_k: $S^k = S^k(\text{structure})$. We assume that the total volume M of the material S_1 is fixed:

$$\sum_{k=1}^N v_k m_k = M.$$

We come to the problem of the optimal layout of the microstructures:

$$W_N = \min_{m_k}\left\{\sum_{k=1}^N \min_{\text{structure}} \text{Tr}\left(S(\text{structure}, m_k)\int_{\Omega_k}\mathbf{U}^k\right)\right\}. \qquad (17.2.7)$$

Obviously, the solution to this problem is given by the third-rank laminates described in the previous section, because the problem coincides with the previous one in each subdomain Ω_k. Note that the second-rank laminates do not occur as optimal elements if the stress field is not constant in Ω.

The optimal layout of the materials is determined as a solution to the finite-dimensional optimization problem (17.2.7). The sequence of the partitions of the domain Ω corresponds to a sequence of optimization problems. This sequence approximates the initial problem when the maximal diameter of subdomains Ω_k tends to zero. The advantage of the proposed procedure is that any finite partition Ω_k leads to a stable solution that corresponds to known composite structures.

A numerical solution to the optimization problem requires a discretization of the initial continuum model. The procedure corresponds to two scales of discretization. The fine scale provides the discretization of the elasticity problem with a locally constant layout C^k. The rough scale determines $C(\mathbf{x})$. This numerical scheme determines a constant (in the rough

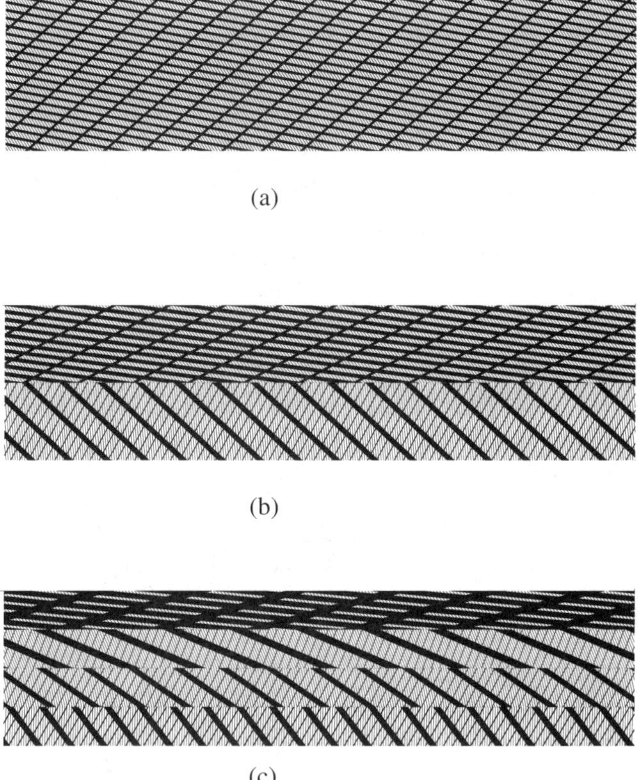

(a)

(b)

(c)

FIGURE 17.2. Optimal console with pice-wise constant properties. The project consists of third-rank laminates. The console is supported at the left and is loaded at the right by a vertical concentrated force. The three projects differ by the partitions Ω_k. Computation is performed by Lars Krog.

scale) microstructure that minimizes the energy of the varied fields, calculated on a fine scale. If these two scales do not coincide, the optimal layout is given by the matrix laminate of third rank. If the fine scale degenerates into the rough scale, then an optimal layout degenerates into second-rank laminates.

The numerical results of optimal structure of a console (Cherkaev, Krog, and Kucuk, 1998) obtained by Lars Krog are shown on Figure 17.2

17.3 Arbitrary Goal Functionals

17.3.1 Statement

The technique developed in Chapter 4 is applicable to problems of elastic constructions. Following (Cherkaev, 1994; Cherkaev, 1998), consider the problem of minimization of an arbitrary weakly lower semicontinuous functional

$$I(\mathbf{w}) = \int_\Omega f(\mathbf{w}, \epsilon)$$

of deflection $\mathbf{w}$ and strain ϵ of an elastic body (see the discussion in Chapter 5 and (Dacorogna, 1982)). For example, one may minimize a weighted norm of the displacement vector if the loading is known.

The equilibrium of the body Ω is described by the Lamé system (14.1.9). Some boundary conditions are fixed on the boundary $\partial\Omega$ of the body Ω. The augmented functional I_A is constructed by adding to the minimizing functional I the differential equation of the state (Lamé system) (14.1.9) with a vector Lagrange multiplier $\boldsymbol{\lambda} = \boldsymbol{\lambda}(\mathbf{x})$:

$$I_A = \min_C \min_{\mathbf{w}} \max_{\boldsymbol{\lambda}} \int_\Omega (f(\mathbf{w}, \epsilon) + \boldsymbol{\lambda}(L(C_*)\mathbf{w} - \mathbf{g})). \tag{17.3.1}$$

Applying the usual arguments we conclude again that the cost of the augmented problem is equal to the value of the functional I.

The equation for the Lagrange multipliers is found in the standard way by variation of the augmented functional (17.3.1) with respect to $\mathbf{w}$, $\delta I_A = R\delta\mathbf{w} = 0$, where

$$R = L^*(C)\boldsymbol{\lambda} - \frac{\delta I_A}{\delta\mathbf{w}} + \nabla \cdot \frac{\partial I_A}{\partial\nabla\mathbf{w}} = 0, \quad \boldsymbol{\lambda}|_{\partial\Omega}.$$

Since the elasticity operator is self-adjoint ($L = L^*$), the Lagrange multiplier satisfies an equation $R = 0$ of elasticity:

$$L(C)\boldsymbol{\lambda} = \frac{\delta I}{\delta\mathbf{w}} - \nabla \cdot \frac{\partial I}{\partial\nabla\mathbf{w}}.$$

The boundary conditions and right-hand-side terms are different from in the original problem (14.1.9). Physically, λ may be considered as a displacement field; it corresponds to the same inhomogeneous medium $C(\mathbf{x})$ and the loading that depends on the functional.

To find the best structure C_*, we proceed as in the conductivity problem considered in Chapter 5. Integrating by parts the term $\int_\Omega \lambda(L(C_*)\mathbf{w}$ of (17.3.1), we formulate the local problem

$$J = \min_{C \in G-\text{closure}} \int_\omega \mathbf{u} : C : \mathbf{v} \qquad (17.3.2)$$

where $\mathbf{u}$, $\mathbf{v}$ are the 2×2 tensors of strains:

$$\mathbf{u} = \text{Def } \mathbf{w}, \quad \mathbf{v} = \text{Def } \lambda.$$

The local problem asks for an optimal periodic layout of material in a small neighborhood ω of a point $\mathbf{x}$ in Ω if the strain fields $\mathbf{u}$, $\mathbf{v}$ also are periodic.

17.3.2 Local Problem

We transform the problem (17.3.2), following the technique of Chapter 5. Normalize the strains $\mathbf{u}$ and $\mathbf{v}$ as follows $\mathbf{u} : \mathbf{u} = \mathbf{v} : \mathbf{v} = 1$ and transform (17.3.2) by introducing the new variables

$$\epsilon = \mathbf{u} + \mathbf{v}, \quad \epsilon' = \mathbf{u} - \mathbf{v} \quad (\epsilon' : \epsilon' = 0).$$

The new form of (17.3.2) is

$$\min_{C} \min_{\epsilon} \max_{\epsilon'} \int_\omega \left(\epsilon : C : \epsilon - \epsilon' : C : \epsilon' \right). \qquad (17.3.3)$$

The problem is to find such a structure of composite that minimizes the difference of the weighted energy density caused by the two orthogonal strain fields ϵ and ϵ'.

The whole problem of the corresponding optimal structures is still open. However, let us demonstrate examples of the optimal structures.

The laminates and the second-rank matrix laminates that are optimal for minimization of the energy are optimal in some cases of this more general problem, as follows.

Example 17.3.1 Consider the following problem of structural optimization. Determine the optimal structure of a thin circular cylindrical shell pushed by a uniform strain at its edges that is directed along the generator (see Figure 17.3). It is required to maximize the deflection of the cylinder in the circumferential direction or to maximize the increase of the radius of the shell.

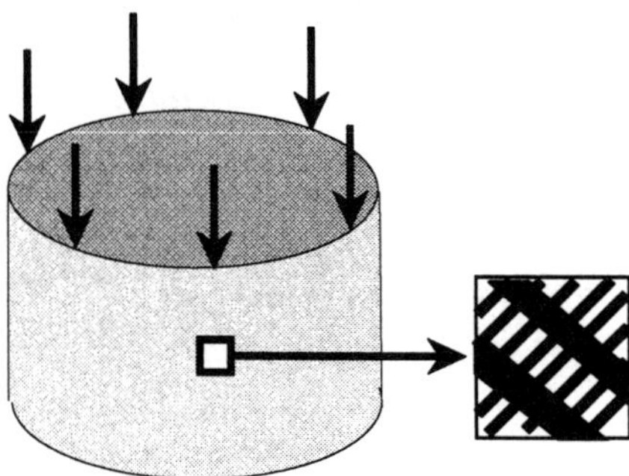

FIGURE 17.3. A cylindrical shell that maximally expands in the radial direction under a force applied in the direction of the generator.

Due to symmetry of the problem we look for a uniform structure that does not depend on the coordinates of the surface. In the scale of the microstructure we also neglect the curvature of the surface. The problem becomes to find the composite that maximizes its displacement in the (circumferential) direction orthogonal to the one-axis loading.

Find the structure that maximizes the strain $\mathbf{v}$ in the $\mathbf{i}_2$ direction when a strain v in the $\mathbf{i}_1$ direction is applied. We put

$$\mathbf{u} = \mathbf{i}_1 \otimes \mathbf{i}_1, \quad \mathbf{v} = \mathbf{i}_2 \otimes \mathbf{i}_2.$$

In this setting, we have

$$\boldsymbol{\epsilon} = \frac{1}{\sqrt{2}}(\mathbf{i}_1 \otimes \mathbf{i}_1 + \mathbf{i}_2 \otimes \mathbf{i}_2), \quad \boldsymbol{\epsilon}' = \frac{1}{\sqrt{2}}(\mathbf{i}_1 \otimes \mathbf{i}_1 - \mathbf{i}_2 \otimes \mathbf{i}_2).$$

The optimal structure has the minimal bulk compliance and the maximal shear compliance. This structure is the second-rank matrix laminate with the inner parameters $\alpha_1 = \alpha_2 = \frac{1}{2}$.

Indeed, the structure minimizes compliance of the bulk load $\boldsymbol{\epsilon}$ independently of its orientation. The optimal laminate is rotated 45^0 to the axes $\mathbf{i}_1$, $\mathbf{i}_2$, which aligns the shear $\boldsymbol{\epsilon}'$ to the tensor $\mathbf{a}_3$ (see (14.1.12)). Recall, that the second-rank matrix laminate with orthogonal layers has the maximal compliance in the direction $\mathbf{a}_3$. This compliance is equal to the arithmetic means of the compliances of components (see Chapter 15).

In summary, the optimal structure of the cylindrical shell is the second-rank laminate turned through an angle of 45° to the generator of the cylinder. The reinforcement can be imitated by two families of mutual orthogonal spirals winded on the cylinder.

The Minimal Variational Problem

The transformation of the min-min-max variational problem (17.3.3) to the minimal problem can be done, as in Chapter 5, by Legendre transform.

The transform leads to the following problem: Find the microstructure of a composite, assembled from two given materials, that minimizes the sum of stored energy generated by some uniform strain ϵ and the complementary energy generated by an orthogonal uniform stress ($\tau = C\epsilon'$):

$$J^{\epsilon\tau} = \min_{\epsilon} \min_{\tau} \min_{C \in \mathcal{C}} \int_{\Omega} W^{\epsilon\tau}(\epsilon, \tau, C),$$
$$W^{\epsilon\tau}(\epsilon, \tau, C) = \epsilon : C : \epsilon + \tau : C^{-1} : \tau. \tag{17.3.4}$$

An optimal microstructure minimizes the sum of its stiffness under one loading and its compliance under another orthogonal loading.

The corresponding translation bounds are applicable, analogous to the bounds discussed in Section 15.2. However, we do not discuss the bounds here. Instead, let us give an example of the optimal structure.

Example 17.3.2 Example 17.3.1 can be easily generalized. Any matrix laminate with orthogonal layers stores the minimal energy in a tracefree strain field ϵ (the compliance in that direction is equal to the arithmetic mean of the compliances of the phases). At the same time, they store minimal elastic energy in a stress field τ orthogonal to ϵ. Therefore, they provide a solution to the general problem (17.3.4) if the field ϵ is tracefree but the field τ is arbitrary.

17.3.3   Asymptotics

Let us pass to an asymptotic. Suppose that one of the materials is infinitely soft, and the other is infinitely stiff: $\|C_1\| \to 0$, $\|C_2^{-1}\| \to 0$. The bound (17.3.4) becomes trivial:

$$J = \int_{\Omega} \left(\tau : C_*^{-1} : \tau + \epsilon : C_* : \epsilon \right) \geq 0.$$

The optimal structures have the effective tensor $C_* = C_{\text{opt}}$ with zero stiffness against any stress τ and zero compliance (or infinite stiffness) against an orthogonal strain ϵ.

The microstructures with the required properties are described in Chapter 14. They are the herringbone structures (see Figure 14.4) that have arbitrary stiff and an arbitrary soft direction (see also (Milton, 1992; Cherkaev, 1994; Milton and Cherkaev, 1995)).

Remark 17.3.1 *In the asymptotic case, the volume fractions of materials in a composite is of no importance; the problem becomes a purely geometrical one. Physically, this can be explained as follows: An infinitely thin element of a structure made from an absolutely rigid or absolutely soft material may make the structure's effective compliance infinitely large or infinitely small independently of the volume fraction of this element.*

Topology Optimization Problem

Interestingly, the same construction works in a less degenerate case. It is enough to assume that one material is void; the second material may have finite properties. The simplest example of an optimal structure is a system of parallel cracks; the cracks have infinitesimal total volume fraction. The stiffness in the direction along the cracks is not affected by their presence, but the stiffness across the cracks is obviously zero. This observation can be generalized to two arbitrary directions of stiffness.

Consider the topology optimization problem $C_1 = 0$, but $C_2 = C$ is finite. The effective tensor C_* of such structures satisfies the obvious inequalities $0 \leq C_* \leq C$. This time we must count the dependence on volume fractions, which has been of no importance in the previous asymptotics. Let us demonstrate that the herringbone structures again are optimal.

Notice that the laminate structures from the void and a material with a finite stiffness have zero effective stiffness in a direction $\mathbf{z}$ independently of the volume fraction $m < 1$ of the stiff phase,

$$\mathbf{z} : C_*^{-1}(m) : \mathbf{z} = 0 \quad \forall m \in [0, 1).$$

For example, laminates made of a void and a nonzero material do not support a shear no matter how small the fraction of the void is. The same is true for the orthogonal second-rank laminates; they do not support a shear. Similarly, the herringbone structures has zero resistance in an arbitrary tensor direction $\mathbf{z}$.

On the other hand, consider the stiffness $\mathbf{z}' : C_* : \mathbf{z}'$ of these structures in directions $\mathbf{z}'$ orthogonal to $\mathbf{z}$. The stiffness continuously depends on the volume fraction $m < 1$ of the solid phase, and therefore it tends to the stiffness of the solid when this fraction tends to one:

$$\mathbf{z}' : C_*(m) : \mathbf{z}' \rightarrow \mathbf{z}' : C : \mathbf{z}' \quad \text{if} \quad m \rightarrow 1.$$

Hence the functional (17.3.4) tends to its maximal value equal to

$$\lim_{m \to 1} \boldsymbol{\tau} : C_*(m)^{-1} : \boldsymbol{\tau} + \boldsymbol{\epsilon} : C_*(m) : \boldsymbol{\epsilon} = \boldsymbol{\epsilon} : C : \boldsymbol{\epsilon}.$$

These structures are asymptotically optimal because they have zero stiffness in an arbitrary tensor direction, and simultaneously they have maximal stiffness in the orthogonal directions. Notice that void is presented in infinitesimal amount, but it still makes the stiffness in certain directions equal to zero due to presence of a system of infinitesimal cracks. Due to their extreme anisotropy, the optimal structures are able to uncouple stresses and strains.

Optimal Composites and Optimal Constructions

The optimal requirements thus formulated determine the structure of the optimal body. Optimal structures have a maximal decree of anisotropy and they support the stress in certain tensor directions only. The mechanical role of these structures is the following: Due to their anisotropy they disband stresses and strains, allowing them to be as "independent" as possible. Informally speaking, the stresses in the structures are restricted by the given loading, while the deflections stay independent of stresses to minimize the cost functional. The cylindrical shell in Example 17.3.1 illustrate this.

An optimally designed continuum body consists of microstructures that behave as simple mechanisms at each point of it. Numerous examples of such structures were numerically created in (Sigmund, 1994; Sigmund, 1996; Sigmund and Torquato, 1997; Sigmund, 1997).

17.4   Optimization under Uncertain Loading

In a rather extensive literature on optimal design, the most attention has been paid to optimization of constructions that are submerged into a fixed loading. However, the typical situation for the practical use of an optimal design is different: Acting forces are either varying in time, varying from one sample to another, or unpredictable. This motivates a reformulation of the problem to account for possible variations and uncertainties in loading. One can foresee a significant change in the reformulated design if the loading is not completely known. The optimality requirement forces the structure to concentrate its resistivity against applied loading; hence its ability to resist other loadings is limited. This section reviews the approach suggested in (Cherkaeva and Cherkaev, 1998; Cherkaeva and Cherkaev, 1999). Other approaches can be found in (Díaz and Bendsøe, 1992; Krog and Olhoff, 1995; Bendsøe, Díaz, Lipton, and Taylor, 1995; Bendsøe, Guedes, Plaxton, and Taylor, 1996).

We formulate a stable optimal design problem with a solution that is stable against small variations in the loadings. We demonstrate that this stable reformulation of the problem leads to optimization of Steklov eigenvalues (Bandle, 1980; Kuttler, 1982). Such formulation makes the problem similar to the well-investigated problem of optimization of the principal eigenfrequency of the vibration of a construction; see (Olhoff, 1974; Bratus' and Seyranian, 1983; Bratus', 1986; Cox and Overton, 1992; Seyranian, Lund, and Olhoff, 1994).

The used approach is similar to one developed for the problem of optimization of boundary excitations in nondestructive testing and electrical tomography (Cherkaeva and Cherkaev, 1995) and to structural optimization (Fuchs and Hakim, 1996). Alternative approaches are developed in (Bendsøe, Hammer, Lipton, and Pedersen, 1995; Díaz and Bendsøe, 1992), among others.

17.4.1 The Formulation

The Problem. Consider again the minimization of the overall compliance of an elastic construction. Recall that compliance is characterized by the mechanical work produced by an applied loading. This work is equal to the total energy stored in the loaded construction. It is found from the variational problem

$$H(\mathbf{p},\mathbf{f}) = \min_{\tau \in \Sigma} \left\{ \int_\Omega W(\mathbf{p},\tau) \right\}, \qquad (17.4.1)$$

where W is the elastic energy

$$W(\mathbf{p},\tau) = \tau : S(\mathbf{p},\mathbf{x}) : \tau; \qquad (17.4.2)$$

τ is the stress tensor; the set Σ is (see (17.4.1), (17.4.2))

$$\Sigma = \left\{ \tau : \quad \begin{array}{ll} \nabla \cdot \tau = 0 \ \text{in} \ \Omega, & \tau = S^{-1}\,\text{Def}\,\mathbf{u}, \\ \tau = \tau^T, & \mathbf{n} \cdot \tau = \mathbf{f} \ \text{on} \ \partial\Omega \end{array} \right\};$$

$\mathbf{f}$ is the vector of applied boundary forces; and $S(\mathbf{p},\mathbf{x})$ is the tensor of elastic compliance. The structural parameter $\mathbf{p}$ defines the material's properties.

The stress τ linearly depends on the loading $\mathbf{f}$. This implies that the stored stress energy $H(\mathbf{p},\mathbf{f})$ is a second-degree homogeneous functional of the loading $\mathbf{f}$ that depends on the layout of the material's properties $\mathbf{p}$.

The problem of minimization of the compliance by a structure becomes as follows: Minimize H with respect to the layout $\mathbf{p}$,

$$\min_{\mathbf{p} \in \mathcal{P}} H(\mathbf{p},\mathbf{f}),$$

where $\mathcal{P}$ is the admissible set of design variables. The set $\mathcal{P}$ can be the set of effective moduli of the composite, or it could describe the shape of the body, the thickness of a thin construction, and so on.

Instabilities. The optimal layout is determined by loading $\mathbf{f}$ and is very sensitive to its variation. The following example demonstrates the instability of the optimal structure and suggests ways of reformulating the problem in order to stabilize the design.

Example 17.4.1 Consider a square domain $a\,b\,c\,d$ (Figure 17.4) filled with a composite material and loaded by a uniaxial loading. Suppose, for simplicity, that the composite is assembled from the material with unit compliance tensor $S_1 = I$ (the Poisson coefficient is equal to zero and the Young modulus is equal to one) and from the void with infinite compliance: $S_2 = \infty$. Suppose also that the fractions m_1 of the material and m_2 of the void are equal to one-half each:

$$m_1 = m_2 = \frac{1}{2}.$$

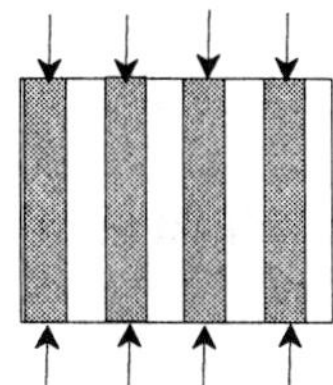

FIGURE 17.4. The optimal composite under homogeneous axial loading. Notice the instability of the design.

Let the domain be loaded by a prescribed loading

$$
\mathbf{f}_0 = \begin{cases} \mathbf{i}_1 & \text{on } a\,b, \\ 0 & \text{on } b\,c, \\ -\mathbf{i}_1 & \text{on } c\,d, \\ 0 & \text{on } a\,d. \end{cases} \tag{17.4.3}
$$

The optimal design is homogeneous. The loading $\mathbf{f}_0$ creates a stress field τ_0,

$$
\tau_0 = \begin{pmatrix} 1 & 0 \\ 0 & 0 \end{pmatrix}, \quad \text{or} \quad \tau_0 = \mathbf{i}_1 \otimes \mathbf{i}_1, \tag{17.4.4}
$$

inside the domain.

Find the composite that minimizes the stress energy of the structure under the loading $\mathbf{f}_0$. Obviously, the best structure is a simple laminate, with layers oriented along the loading (see Figure 17.4).

Let us compute the effective compliance of laminates, using the conventional basis (14.1.11). The effective compliance s_{1111} in the direction $\mathbf{i}_1$ of the loading is equal to the harmonic mean of the (unit) material's compliance $s_m = 1$ and the (infinite) compliance of the void $s_v = \infty$:

$$
s_{1111} = \left(\frac{m_1}{s_m} + \frac{m_2}{s_v} \right)^{-1} = 2.
$$

The minimal energy and the problem cost are

$$
W(\tau_0) = \tau_0 : S_* : \tau_0 = s_{1111}\tau_{11}^2 = 2, \quad H(\tau_0) = 2.
$$

This solution, however, is not satisfactory from a commonsense viewpoint. Indeed, the laminate structure is extremely unstable, and its compliance tensor is singular. The structure does not resist any loading but the prescribed one; its compliance is infinitely large for all other loadings. Simply speaking, the structure falls apart under any infinitesimally small applied stress that has either a shear component or a component along the axis $\mathbf{i}_2$. This instability is typical for projects that are designed to optimally resist a prescribed loading, at the expense of resistivity in other directions.

Optimization Against the Worse Loading

Let us consider a problem of energy optimization of an elastic body Ω loaded by unknown forces $\mathbf{f}$ applied on the boundary $\partial\Omega$. Define the compliance Λ of a construction as the maximum of compliances over all admissible loadings, $\mathbf{f} \in \mathcal{F}$,

$$\Lambda(\mathbf{p}) = \max_{\mathbf{f}\in\mathcal{F}} H(\mathbf{p},\mathbf{f}). \qquad (17.4.5)$$

This way we come to the problem of the optimal design against the "worst" loading:

$$\min_{\mathbf{p}\in\mathcal{P}} \Lambda = \min_{\mathbf{p}\in\mathcal{P}} \left\{ \max_{\mathbf{f}\in\mathcal{F}} H(\mathbf{p},\mathbf{f}) \right\}.$$

Set $\mathcal{F}$ of loading must be constrained. Otherwise, the problem becomes trivial: The worst loading has an infinitely large magnitude. We impose constraints on acting forces and formulate a problem for a design that offers minimal compliance in a set of loadings.

Integral Constraints on the Loading. Let the set $\mathcal{F}$ be characterized by an integral constraint. It is convenient to consider the constraints as a quadratic form of the loading. This form leads to rather simple equations and has a needed generality and flexibility. Suppose that an unknown loading by normal forces $\mathbf{f} \in \mathcal{F}$ is constrained as follows:

$$\mathcal{F} = \left\{ \mathbf{f}: \quad \oint_{\partial\Omega} \mathbf{f}\cdot\Psi(S)\mathbf{f} = 1 \right\},$$

where $\Psi(S)$ is a positive definite weight matrix,

$$\Psi(S) > 0 \quad \forall\, S \in \partial\Omega.$$

Here Ψ expresses a priori assumptions about the unknown loading. For example, the case where all loadings are equally possible corresponds to $\Psi = \text{constant}(S)$.

The compliance of the design, introduced in (17.4.5), corresponds to maximization of the energy stored in the design with respect to the applied loadings $\mathbf{f} \in \mathcal{F}$:

$$\Lambda = \max_{\mathbf{f}} H(\mathbf{p},\mathbf{f}) = \max_{\mathbf{f}} \min_{\tau} \frac{\int_\Omega W(\mathbf{p},\tau)}{\oint_{\partial\Omega} \mathbf{f}\cdot\Psi(S)\mathbf{f}}. \qquad (17.4.6)$$

Problem (17.4.6) is an eigenvalue problem. The value $\Lambda(\mathbf{p})$ corresponds to the first eigenfunction or to the set of eigenfunctions that generate the most "dangerous" loading(s) from the considered set. Hence we formulate the optimal design problem as a problem of eigenvalue optimization:

$$J = \min_{\mathbf{p}\in\mathcal{P}} \Lambda(\mathbf{p}) = \min_{\mathbf{p}\in\mathcal{P}} \max_{\mathbf{f}\in\mathcal{F}} \min_{\tau\in\Sigma} \frac{\int_\Omega W(\mathbf{p},\tau)}{\oint_{\partial\Omega} \mathbf{f}\cdot\Psi\mathbf{f}}. \qquad (17.4.7)$$

17.4.2 Eigenvalue Problem

Saddle Point Case

The optimal eigenvalue J could correspond to either a single eigenvector or to several orthogonal eigenvectors. Whether the multiple eigenvalue case is taking place depends on the power of the control. If the control $\mathbf{p}$ is "weak," i.e., if the control cannot change the sequence of eigenvalues, then we are dealing with a saddle point situation. In this case, the minimum over the control $\mathbf{p}$ eigenvalue corresponds to a unique eigenfunction $\mathbf{f}(\mathbf{p})$. In this case the functional Λ (17.4.6) is a saddle function and the operations of $\max_{\mathbf{f}}$ and $\min_{\tau}$ can be switched. By varying the functional, we find the Euler–Lagrange equations for the most dangerous loading. The following Example 17.4.2 illustrates this situation.

Let us find the most dangerous loading. Variation of (17.4.6) with respect to $\mathbf{f}$ gives

$$\delta \Lambda_{\mathbf{f}} = - \left(\oint_{\partial\Omega} \mathbf{f} \cdot \Psi(S)\mathbf{f} \right)^{-1} (\mathbf{u} - \Lambda\Psi\mathbf{f})\, \delta\mathbf{f},$$

which implies the following relation between the optimal loading and the boundary deflection:

$$\mathbf{f}(S) = \frac{1}{\Lambda}\Psi^{-1}\mathbf{u}(S) \quad \forall S \in \partial\Omega.$$

1. We observe that the most dangerous loading is proportional to the deflection.

2. It is easy to see that the stationary condition corresponds to the *maximum* not the minimum of the functional using the second variation.

3. The problem of the most dangerous loading $\mathbf{f}_0$ becomes an eigenvalue problem:

$$\frac{1}{\Lambda} = \min_{\tau} \frac{\int_\Omega W(\mathbf{p}, \tau)}{\oint_{\partial\Omega} \mathbf{u} \cdot \Psi(S)^{-1}\mathbf{u}}. \tag{17.4.8}$$

   The cost Λ corresponds to the minimal eigenvalue, given by the Rayleigh ratio(17.4.8), and the most "dangerous" loading corresponds to the first eigenfunction of this problem.

Remark 17.4.1 *One can also consider the problem of the most "favorable" loading, that is,*

$$\Lambda_- = \min_{\mathbf{f}} \min_{\tau} \frac{\int_\Omega W(\mathbf{p}, \tau)}{\oint_{\partial\Omega} \mathbf{f} \cdot \Psi\mathbf{f}}.$$

However, Λ_- is zero. Clearly, the spectrum of the operator is clustered at zero. A minimizing sequence is formed from rapidly oscillatory forces which produce infinitesimal energy.

Euler–Lagrange equations

The Euler–Lagrange equations (with respect to τ) are

$$\nabla \cdot \tau = 0, \quad \tau = S^{-1}(\mathbf{p}) : \mathrm{Def}\, \mathbf{u} \quad \text{in } \Omega,$$
$$\mathbf{u} = \Lambda \Psi \tau \cdot \mathbf{n} \qquad\qquad \text{on } \partial\Omega. \tag{17.4.9}$$

They describe the vibration of a body with inertial elements concentrated on $\partial\Omega$.

The problem admits the following physical interpretation: The optimal loading forces are equal to a layout of inertial elements (concentrated masses) on the boundary component $\partial\Omega$. The specific inertia is described by the tensor Ψ, so it could include resistance to turning as well. The vibration of such a loaded system excites forces that are proportional to the deflection $\mathbf{u}$. The compliance is proportional to the eigenfrequency of vibrations. One can see that the introduced quantity Λ characterizes the domain or the construction itself. It represents the maximum of possible stored energy under any loading from the set $\mathcal{F}$.

These equations form an eigenvalue problem that has infinitely many solutions. We choose the pair $\{\Lambda_1,\ \tau_0\}$ that corresponds to the maximal eigenvalue $\Lambda_1 = \max\{\Lambda_k\}$.

The problem (17.4.8) with unit matrix Ψ is called the *Steklov eigenvalue problem*; it considers the ratio of integrals of different dimensions. The corresponding Euler–Lagrange equation (17.4.9) has an eigenvalue Λ in the boundary condition. Similar optimality conditions were derived in (Cherkaeva and Tripp, 1996; Cherkaeva, 1997) for the optimal boundary sources in the electrical tomography problem.

Example 17.4.2 Problems for beams and bending plates admit a loading distributed in the whole domain of the definition: on the interval in the case of the beam and in the plane domain in the case of the bending plate or shell. In these problems the loaded surface $\partial\Omega$ coincides with the domain Ω itself.

Consider an elastic beam whose energy is

$$W = p(w'')^2 - 2fw,$$

where $p \geq 0$ is the material's stiffness that can be varied from point to point and that is subject to the integral constraint

$$\int_0^l p = V, \tag{17.4.10}$$

which expresses the limits on resources; f is the intensity of the normal loading. The constraint

$$\int_0^l f^2 = 1$$

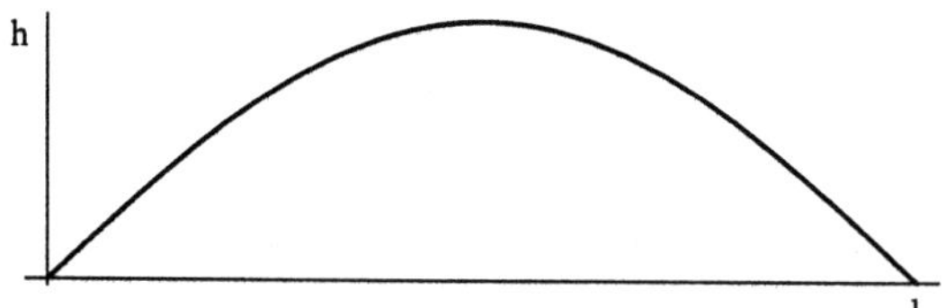

FIGURE 17.5. The stiffness $p(x)$ of the optimal beam under the "worst" loading.

expresses the limits of loading.

Consider an optimization problem of choosing a stiffness $p(x)$ that maximally resists the most dangerous loading f:

$$\min_{p \geq 0,\, p \text{ as in } (17.4.10)} \; \max_f \min_w \; \mu,$$

where

$$\mu = \frac{\int_0^l \left(p(w'')^2 - 2fw\right)}{\int_0^l (f^2)}.$$

The stationary conditions are

$$
\begin{aligned}
\delta w: & \quad (p\,w'')'' - f = 0, \\
\delta f: & \quad f + \tfrac{w}{\mu} = 0, \\
\delta p: & \quad (w'')^2 = \gamma,
\end{aligned}
$$

where γ is the Lagrange multiplier for the constraint (17.4.10). The boundary conditions are $pw''|_{x=0} = pw''|_{x=l} = 0$.

The system admits the solution

$$
\begin{aligned}
w &= -\mu f = -\gamma x(l - x)/2, \\
f &= \frac{\sqrt{\gamma}}{\mu} x(l - x)/2, \\
p &= \frac{1}{\mu}\frac{x}{24}(x - l)\left((x - \tfrac{l}{2})^2 - \frac{5l}{4}\right).
\end{aligned}
$$

Accounting for the constraints, we obtain

$$
\begin{aligned}
\mu &= \frac{l^5}{5!\,V}, \\
p &= \frac{5V}{4l^5} x(l - x)\left(5l^2 - (2x - l)^2\right), \\
w &= -\frac{1}{V}\sqrt{\frac{l^5}{5!}}\frac{x(l - x)}{2}.
\end{aligned}
$$

The optimal stiffness $p(x)$ of the beam is shown in Figure 17.5. Interestingly, the optimal solution is found analytically.

A Min-Max Problem for a Partly Known Loading

The formulation can be generalized to the case of a partly known loading. Suppose that the applied loading $\mathbf{f}$ is composed of the known component $\mathbf{f}_0$ and some unknown component $\mathbf{f}_1$ that describes an uncertainty in the loading. The constraint restricts the norm of $\mathbf{f}$:

$$\oint_{\partial\Omega} (\mathbf{f} - \mathbf{f}_0) \cdot \Psi(S)(\mathbf{f} - \mathbf{f}_0) \leq 1.$$

The problem is no longer homogeneous. The functional is

$$I = \min_{\mathbf{p}\in\mathcal{P}} \Lambda(\mathbf{p})$$

where

$$\Lambda(\mathbf{p}) = \max_{\mathbf{f}_1} \min_{\boldsymbol{\tau}\in\Sigma} \frac{\int_\Omega W(\mathbf{p},\boldsymbol{\tau})}{\oint_{\partial\Omega}(\mathbf{f} - \mathbf{f}_0) \cdot \Psi(S)(\mathbf{f} - \mathbf{f}_0)}.$$

Depending on $\mathcal{P}$ and $\mathcal{F}_1$, we can encounter two different situations corresponding to single or multiple "most dangerous" loadings. In the first case, the problem has a saddle point and the extremal operations $\min_{\mathbf{p}\in\mathcal{P}}$ and $\max_{\mathbf{f}_1\in\mathcal{F}_1}$ are interchangeable. The stationary conditions with respect to $\mathbf{f}$ and $\boldsymbol{\tau}$ lead to the Euler–Lagrange equations

$$\nabla \cdot S^{-1} \operatorname{Def} \mathbf{u} = 0 \quad \text{in } \Omega,$$
$$\mathbf{u} - \Lambda \Psi \cdot S^{-1} \operatorname{Def} \mathbf{u} \cdot \mathbf{n} = \mathbf{f}_0 \quad \text{on } \partial\Omega.$$

It describes an inhomogeneous boundary value problem, which has a unique solution. The optimality condition

$$\frac{\partial W(\mathbf{p},\boldsymbol{\tau})}{\partial \mathbf{p}} \, \delta\,\mathbf{p} \geq 0$$

has different forms depending on the type of control used and possible isoperimetric restrictions.

The other case of multiplicity of the worst loadings $\mathbf{f}_1,\ldots,\mathbf{f}_k$ deals with a nondifferentiable functional $\max\{H(\mathbf{p},\mathbf{f}_1),\ldots,H(\mathbf{p},\mathbf{f}_k)\}$.

17.4.3 Multiple Eigenvalues

Multiplicity of the Worst Loadings

We return to the discussion of the problem of minimizing the stored energy in the most unfortunate situation. The problem has the form (17.4.7). The specific effect of the min-max problem is the possibility of the appearance of multiple eigenvalues. The mechanism of this phenomenon is the following. The minimization of the maximal eigenvalue likely leads to the situation where it meets the second largest eigenvalue of the problem. After this,

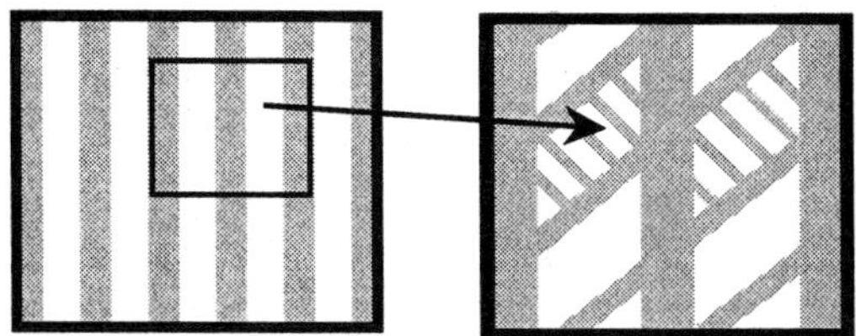

FIGURE 17.6. Changing in the optimal structure due to multiple loading. The structure becomes a third-rank laminate.

both eigenvalues must be minimized together, until their common value reaches the third eigenvalue, and so on. The multiplicity means that two or more loadings correspond to the same stiffness.

We will introduce an example demonstrating this phenomenon: The resistance of an optimal design to five different loadings is the same. The next example illustrates the situation where the eigenvalue could either be multiple or stay single, depending on the set of loadings $\mathcal{F}$ and the set of controls $\mathcal{P}$.

Remark 17.4.2 *A similar min-max problem with multiple eigenvalues has been considered in (Cherkaeva and Cherkaev, 1995) for nondestructive testing that detects the worst possible location of damage inside a body by applying optimal boundary currents.*

There is an extensive literature on eigenvalue optimization. It was previously understood in a different setting: the maximization of the fundamental frequency. We refer to (Olhoff and Rasmussen, 1977; Seyranian, 1987; Cox and Overton, 1992; Seyranian et al., 1994) and references therein.

Consider again the problem of minimal compliance (17.4.5). We find the most resistant structure **p** of the composite. For definiteness, consider the two-dimensional elasticity problem. A domain made of a two-phase composite material of an arbitrary structure is loaded by an uncertain loading f_0.

We do not know a priori how many loadings form the class of equally dangerous loadings. Clearly, it is sufficient to enlarge the set of admissible composites to those that minimize the sum of elastic energies caused by any number of different loadings. These composites were described earlier; they are the matrix laminates of the third-rank (see Figure 17.6).

The effective property tensors of these composites admit an analytical expression of their structural parameters. In this calculation, we use the natural tensor basis (compare with (14.1.11)).

$$\mathbf{e}_1 = \begin{pmatrix} 1 & 0 \\ 0 & 0 \end{pmatrix}, \quad \mathbf{e}_2 = \begin{pmatrix} 0 & 0 \\ 0 & 1 \end{pmatrix}, \quad \mathbf{e}_3 = \frac{1}{\sqrt{2}} \begin{pmatrix} 0 & 1 \\ 1 & 0 \end{pmatrix}. \qquad (17.4.11)$$

Any stress and strain matrices are represented as vectors in their basis, and the effective compliance S_* of matrix laminates is given by the 3×3 matrix (see (14.3.4))

$$S_* = S_1 + m_2 \left((S_2 - S_1)^{-1} + \frac{2m_1}{E_1} G \right)^{-1}, \qquad (17.4.12)$$

where S_1 and S_2 are the compliance matrices of the first and second materials, m_1 and m_2 are the volume fractions, and E_1 is the Young modulus of the first material (which forms the envelope). The matrix G depends on the structural parameters: on the angles θ_i between the tangent to the laminates and the axis $\mathbf{i}_1$, and on the relative thickness α_i (see Figure 17.6). In the basis (17.4.11), G has the representation

$$G = \sum_{i=1}^{3} N_i, \qquad (17.4.13)$$

where

$$N_i = \alpha_i \begin{pmatrix} \cos^4 \theta_i & \cos^2 \theta_i \sin^2 \theta_i & \cos^3 \theta_i \sin \theta_i \\ \cos^2 \theta_i \sin^2 \theta_i & \sin^4 \theta_i & \cos \theta_i \sin^3 \theta_i \\ \cos^3 \theta_i \sin \theta_i & \cos \theta_i \sin^3 \theta_i & \cos^2 \theta_i \sin^2 \theta_i \end{pmatrix}.$$

The structural parameters $\alpha_i \geq 0$ ($\alpha_1 + \alpha_2 + \alpha_3 \equiv 1$) and $\theta_i \in [0, \pi]$ form the control vector $\mathbf{p}$.

Stable Designs

We revisit Example (17.4.1). In discussing the instabilities of the optimal project in Example (17.4.1) we considered the optimization problem

$$\min_{\mathbf{p} \in \mathcal{P}} H(\mathbf{p}, \mathbf{f}_0), \qquad (17.4.14)$$

where $\mathbf{f}_0$ is given by (17.4.3), and the set $\mathcal{P}$ constrains the parameters of the composite α_i and θ_i. The optimal solution is a laminate, which is easily found from (17.4.12), (17.4.13). It corresponds to the parameters $\alpha_1 = 1$, $\theta_1 = 0$, $\alpha_2 = \alpha_3 = 0$. This structure is shown in Figure 17.4.

This solution may be unstable against the variation of the loading. The compliance tensor S_* of a third-rank composite becomes ($m_1 = m_2 = \frac{1}{2}$)

$$S_* = S_1 + \frac{1}{2} G^{-1}. \qquad (17.4.15)$$

For the optimal choice of the parameters α_i and θ_i, the matrix G (see (17.4.13)) has two zero eigenvalues.

The next example shows how to reformulate the design problem (17.4.14) to obtain a stable project.

Example 17.4.3 Suppose the loading is not exactly known. Namely, the loading field τ can take one of the six values $\tau_0 + \tau_i$, $i = 1, \ldots, 6$, where τ_0 is given by (17.4.4) and the additional terms are

$$\tau_{1,2} = \pm r e_1, \quad \tau_{3,4} = \pm r e_2, \quad \tau_{5,6} = \pm r e_3.$$

Here $r > 0$ is a real parameter, and e_i are the tensors of the basis (17.4.11).

The six loadings are considered perturbations of the "main" loading τ_0. They correspond to all linearly independent directions of the symmetric tensor τ. Even if r is small, the perturbation of the functional (17.4.14) is infinitely large if S_* is chosen to optimally resist τ_0.

Reformulate the optimization problem. Now we find a structure that minimizes the maximum of compliances $H(\mathbf{p}, \tau_0 + \tau_i)$ over all considered loadings:

$$\min_{\mathbf{p} = \{\alpha_i, \, \theta_i\}} \left\{ \max_{i=1,\ldots,6} H(\mathbf{p}, \tau_0 + \tau_i) \right\}. \tag{17.4.16}$$

This min-max problem asks for the minimal compliance in the case of the most dangerous loading. To construct the solution to the optimization problem, we introduce a variable z that is greater than any of $H(\mathbf{p}, \tau_0 + \tau_i)$,

$$z \geq H(\mathbf{p}, \tau_0 + \tau_i), \; i = 1, \ldots, 6. \tag{17.4.17}$$

The problem (17.4.16) is formulated as follows (see (Dem'yanov and Malozëmov, 1990)):

$$\min_{\mathbf{p}} \left\{ z + \sum_{i=1}^{6} \lambda_i (z - H(\mathbf{p}, \tau_0 + \tau_i)) \right\}, \tag{17.4.18}$$

where $\lambda_i \geq 0$ are the nonnegative Lagrange multipliers by the constraints (17.4.17). The Lagrange multiplier is equal to zero if the constraint is satisfied as a strong inequality and is nonzero if it is satisfied as an equality (Dem'yanov and Malozëmov, 1990):

$$\begin{aligned} \lambda_i = 0 \quad &\text{if} \quad z > H(\mathbf{p}, \tau_0 + \tau_i), \\ \lambda_i > 0 \quad &\text{if} \quad z = H(\mathbf{p}, \tau_0 + \tau_i). \end{aligned}$$

The problem requires the minimization of the weighted sum of energies of the "dangerous" loadings (τ_i), $i \in \mathcal{I}$. Here $\mathcal{I}$ is the set of such "dangerous" loadings. Other loadings lead to the smaller energies $H(\mathbf{p}, \tau_0 + \tau_j)$: $H(\mathbf{p}, \tau_0 + \tau_j) < H(\mathbf{p}, \tau_0 + \tau_i)$, if $i \in \mathcal{I}$, $j \notin \mathcal{I}$, and therefore to $\lambda_j = 0$. This leads to the equalities

$$\begin{aligned} z = H(\mathbf{p}, \tau_0 + \tau_i) \quad &\text{if} \quad i \in \mathcal{I}, \\ z > H(\mathbf{p}, \tau_0 + \tau_i) \quad &\text{if} \quad i \notin \mathcal{I}. \end{aligned}$$

Applying (17.4.18) to the problem, we argue that the set of dangerous loadings in this example consists of five elements, $\mathcal{I} = \{1, 3, 4, 5, 6\}$:

$$H(\mathbf{p}, \tau_0 + \tau_1) > H(\mathbf{p}, \tau_0 + \tau_2), \tag{17.4.19}$$

$$H(\mathbf{p}, \tau_0 + \tau_3) = H(\mathbf{p}, \tau_0 + \tau_4),$$
$$H(\mathbf{p}, \tau_0 + \tau_5) = H(\mathbf{p}, \tau_0 + \tau_6). \tag{17.4.20}$$

The inequality (17.4.19) is explained by the observation that an additional loading, if aligned with the main load, will either increase or decrease its magnitude independently of the composite structure. Clearly, the energy that corresponds to the more intensive loading is greater. The symmetry of the loadings 3 and 4 and loadings 5 and 6 together with the symmetry of the set of admissible structural tensors $\mathcal{P}$, suggests that these "twin" loadings lead to the same cost of the problem. In other words, the same project $\mathbf{p}$ minimizes both $H(\mathbf{p}, \tau_0 + \tau_3)$ and $H(\mathbf{p}, \tau_0 + \tau_4)$, keeping them equal to each other; the same holds for the other pair of loadings. To achieve the equalities (17.4.20), we require symmetry of the would-be optimal structure S_* (see (17.4.12), (17.4.13)):

$$\alpha_2 = \alpha_3 = a/2, \quad \theta_1 = 0, \quad \theta_2 = -\theta_3 = \theta, \tag{17.4.21}$$

where a and θ are two parameters (note that the parameter α_1 is: $\alpha_1 = 1 - a$). Physically, we require the orthotropy of S_*.

Under the conditions (17.4.21), the matrix G (see (17.4.13)) takes the form

$$G = \begin{pmatrix} 1-a & 0 & 0 \\ 0 & 0 & 0 \\ 0 & 0 & 0 \end{pmatrix} + a \begin{pmatrix} \cos^4\theta & \cos^2\theta\sin^2\theta & 0 \\ \cos^2\theta\sin^2\theta & \sin^4\theta & 0 \\ 0 & 0 & \cos^2\theta\sin^2\theta \end{pmatrix}$$

and, from (17.4.15)

$$S_*(\alpha, \theta) = I + \frac{1}{2a(1-a)} \begin{pmatrix} a & -a\cot^2\theta & 0 \\ -a\cot^2\theta & T_1 & 0 \\ 0 & 0 & T_2 \end{pmatrix}, \tag{17.4.22}$$

where

$$T_1 = \tfrac{1}{8}(8 - 5a + 4a\cos 2\theta + a\cos 4\theta)\csc^4\theta,$$
$$T_2 = (1-a)\csc^2\theta\sec^2\theta.$$

Note that the matrix becomes singular when $a \to 0$, which corresponds to unstable design.

The described set of symmetric composites is defined by two parameters θ and a. The symmetry of the project eliminates the necessity to compare the loadings except those with numbers 1, 3, and 5. It turns out that these loadings are equally "dangerous":

$$H(a, \theta; \tau_0 + \tau_1) = H(a, \theta; \tau_0 + \tau_3) = H(a, \theta; \tau_0 + \tau_5). \tag{17.4.23}$$

The two equalities (17.4.23) allow us to compute the optimal values of θ and a. One can easily see that the problem is always solvable. The optimal

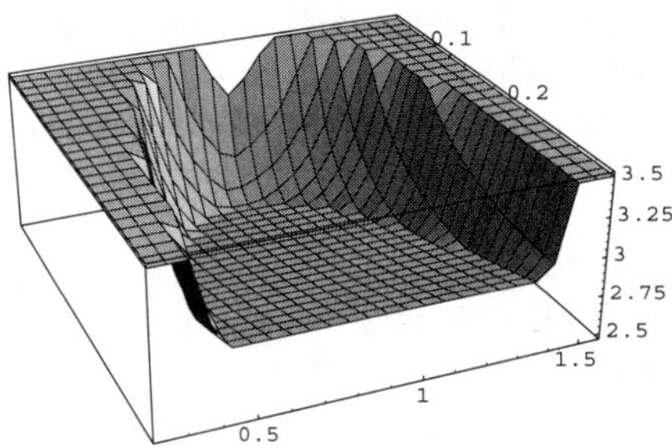

FIGURE 17.7. Graph of the cost function $H_{\max}$. Dependence on parameters of the structure. Observe the nonsmooth minimum.

values of the parameters θ and a correspond to the solution to the min-max problem

$$J = \min_{a,\theta} H_{\max}(a, \theta),$$

where

$$H_{\max} = \max_{i=1,3,5} \left\{ H(a, \theta; \boldsymbol{\tau}_0 + \boldsymbol{\tau}_i) \right\}.$$

Illustration

Set $r = 0.1$. The graph of the function $H_{\max}$ is presented in Figure 17.7, where the viewpoint is changed for better visibility. The optimal values of the parameters are $\theta = 0.889$, $a = 0.0496$, and $J = 2.483$. We see that the compliance is bigger than the compliance of the construction optimal for a single load. On the other hand, the construction is stable against all loadings, unlike the original design.

An optimal structure is shown in Figure 17.6. Note that a part of the material is removed from the laminates that resist the main load. This material is placed in "reinforcements" that reduce the compliance in all directions.

Remark 17.4.3 *Project (17.4.22) is not optimal for any single loading, but it is optimal for the set of them. The solution provides an example of a mixed strategy in the game "loadings versus design."*

Sensitivity

The nonsmoothness of the min-max formulation leads to strongly nonlinear sensitivity to perturbations. Particularly, the optimal design may be totally insensitive to variation in the loading, which is stay the same independently of finite perturbations of the loading, as in the following example.

Example 17.4.4 Consider the previous optimization problem, but suppose that both materials in the composite have finite stiffness. The optimal structure for this problem is again a third-rank laminate. The scheme of the solution is similar to the case discussed. However, there is an important difference. The maximal compliance in any direction is no longer an infinity, but it is estimated by the arithmetic mean of the compliances of the materials.

Assume for definiteness that the compliance matrices are equal to $s_1 I$ and $s_2 I$, and the volume fractions are equal to one-half. The compliance of the laminate material is

$$S_{\text{lam}} = \begin{pmatrix} \frac{2s_1 s_2}{s_1 + s_2} & 0 & 0 \\ 0 & \frac{s_1 + s_2}{2} & 0 \\ 0 & 0 & \frac{s_1 + s_2}{2} \end{pmatrix}.$$

Let us demonstrate that the maximum of compliance can correspond to one loading only. Other loadings do not influence the design, provided that the perturbation parameter r is small enough (but still finite).

Indeed, the energy of the simple laminate $\mathbf{p}_1$ corresponding to the loading $\boldsymbol{\tau}_0 + \boldsymbol{\tau}_1$ is equal to

$$W_1(\mathbf{p}_1) = (1 + r)^2 \frac{2s_1 s_2}{s_1 + s_2}.$$

The loading $\boldsymbol{\tau}_0 + \boldsymbol{\tau}_2$ is obviously not optimal.

For each of the loadings $\boldsymbol{\tau}_0 + \boldsymbol{\tau}_i$, $i = 3, \ldots, 6$, there is a structure $\mathbf{p}_i$ that minimizes its energy. The energy $W_i(\mathbf{p}_1)$ of the simple laminate corresponding to the loadings $\boldsymbol{\tau}_i$, $i = 3, \ldots, 6$, is clearly greater than the energy $W_i(\mathbf{p}_i)$ corresponding to the optimal structure $\mathbf{p}_i$, $W_i(\mathbf{p}_i) \leq W_i(\mathbf{p}_1)$. Therefore, $W_i(\mathbf{p}_i)$ is bounded as

$$W_i(\mathbf{p}_i) \leq W_i(\mathbf{p}_1) = \frac{2s_1 s_2}{s_1 + s_2} + r^2 \frac{s_1 + s_2}{2}, \quad i = 3, \ldots, 6.$$

If the upper bound of $W_i(\mathbf{p}_i)$ is still smaller than $W_1(\mathbf{p}_1)$:

$$W_i(\mathbf{p}_1) < W_1(\mathbf{p}_1), \quad i = 3, \ldots, 6$$

then the laminate is optimal, because W_1 reaches its minimum on it. The corresponding condition for the magnitude r is

$$0 \leq r < \frac{8s_1 s_2}{(s_1 - s_2)^2}. \tag{17.4.24}$$

If r satisfies this inequality, then the laminate structure remains optimal: The design is insensitive to sufficiently small but finite perturbations of the loading. If (17.4.24) holds, then the set of dangerous loadings consists of the first loading only.

Symmetry

The next examples of multiplicity of the most dangerous loadings demonstrate the appearance of symmetric projects in a min-max optimal design problem.

Example 17.4.5 Consider the design of a periodic structure in a completely unknown field. Suppose we do not have any information about the loading stress field τ, and we are looking for a structure of a two-material composite that optimally resists the worst loading;

$$\min_{\text{structure}} \max_{\tau} \frac{W(\tau)}{\|\tau\|^2},$$

where $\|.\|$ is a rotationally invariant matrix norm.

The optimal composite is obviously isotropic. Additionally, it has the maximal effective bulk and shear moduli. Both the bulk and the shear effective moduli are bounded by the Hashin–Shtrikman bounds. These bounds are realized by one structure. This optimal structure is an isotropic laminate composite of third rank with normals n_i uniformly oriented in the plane: $n_i \cdot n_j = -\frac{1}{2}$ and with $\alpha_i = \frac{1}{3}$. Another optimal structure consists of the coated circles.

The corresponding three-dimensional structure is the matrix laminate of sixth rank, with the normals to the layers oriented toward the vertices of a icosahedron, as was shown in (Francfort and Murat, 1986). Another optimal structure is the coated spheres geometry; see (Hashin and Shtrikman, 1963).

In the next example, the symmetry allows us to find an optimal design.

Example 17.4.6 Consider the problem of a design of an optimal wheel. A circular domain is loaded by a nonaxisymmetric loading that consists of a pair of radial forces applied to the rim and hub. These forces can move circumferentially, which corresponds to the revolution of the wheel. If a loading $\mathbf{f}(S)$ is admissible, then any shifted loading $\mathbf{f}(S + \theta)$ is admissible, too. Here S is the circumferential coordinate and θ is an arbitrary real number.

Suppose that it is required to minimize the maximal compliance of the wheel loaded by the pair of forces applied toward each other to the hub and rim. When the wheel turns through an angle θ, the position of the applied pair varies. These pairs form the set of applied forces.

The layout that minimizes the maximal compliance is obviously axisymmetric, although a particular loading is not. The symmetry comes from the min-max requirement of equal resistance to all forces $\mathbf{f}(S+\theta)$; the structure is independent of the angle θ. The solution locally is again a third-rank laminate, symmetric with respect to the angular coordinate θ. The properties of the structure vary with the radius. In the large, it can be represented as a periodic system of radii and two symmetric spirals (see Figure 17.8). The volume fraction of the material varies with the radius.

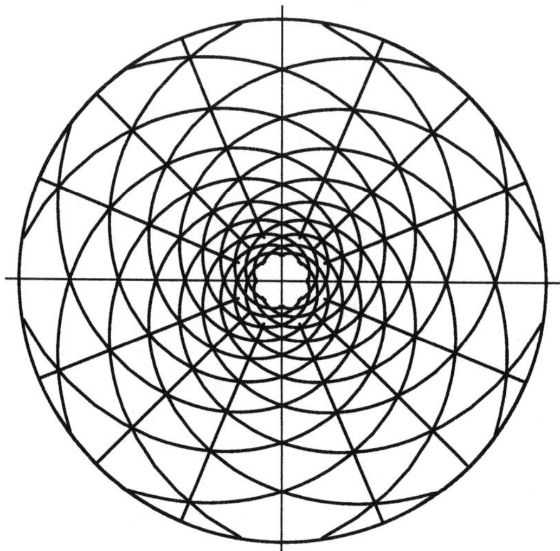

FIGURE 17.8. Optimal structure of a wheel.

17.5   Conclusion

Let us summarize the features of structural optimization.

Mathematical Tools

Structural optimization problems can be considered as a special type of variational problem. Typically, they require the minimization of an unstable (not weakly lower semicontinuous) functional. Therefore, relaxation methods are needed to describe highly oscillating solutions to these problems. These methods include:
- sufficient conditions methods, particularly the translation method;
- minimization sequences, particularly for laminates of a high rank; and
- necessary conditions and minimal extensions.

Stability of Optimal Layouts

Optimal structures are generally unstable if they are designed against a fixed loading. Realistic formulations of structural optimization should find a compromise effectiveness and stability. These formulations either restrict the class of structures (for example, by the requirement of the constancy of the effective properties) or enlarge the class of admissible loadings.

Necessary Conditions

Necessary conditions highlight properties of the fields in an optimal structure and its phases. They demonstrate qualitative properties of optimal design. They are also used to find instabilities in the project. Computa-

Functional	Loading	Local minimizing feature	Structures	Global features
Energy	Single, fixed	Compliance in a chosen direction	Second-rank ML or Vigdergauz structures in 2D, Third-rank ML in 3D	Energy equals the energy of a non-linear material
Sum of energies	Several, fixed	Sum of compliances in chosen directions	third-rank ML in 2D, sixth-rank ML in 3D	Optimal distribution of directions of resistance
Energy of a periodic composite	Single or several, fixed	Mean compliance	As above	Structure adjusts itself to a mean field
Energy	Unknown	Overall compliance	As above	Multiple eigenvalues
Eigenfrequency	None	Overall compliance	As above	Multiple eigenvalues
Arbitrary	Single, fixed	Difference in energies of two loadings	Herringbone, Hexagonal structures by Sigmund	Structure transforms the loading as a mechanism

TABLE 17.1. Types of optimal design problems and the corresponding optimal structures.

tional schemes could use these conditions to find an optimal or suboptimal design.

Practical Usefulness

A mathematically correct solution of a structural optimization problem is not necessarily of immediate practical utility. In practice, several features such as fabrication limitations, failure criteria, and dynamical behavior must be considered. The solution to the mathematical problem provides guidance for a designer rather than serving as a blueprint for manufacturing. A needed step between structural optimization and production is a practical suboptimal project. A theoretically optimal project demonstrates how much room is left for further improvement.

The Types of Optimization Problems

We summarize the properties of different types of variational problems in Table 17.5.

The problems are classified by the cost functionals and the load conditions (the first and second rows). The third row refers to the local problem or the meaning of the relaxed Lagrangian. The fourth row lists the type of optimal minimizing sequences (microstructures). The fifth row corresponds to the meaning of necessary conditions of optimality or the global features in the entire optimal structure.

References

Achtziger, W., Bendsøe, M. P., Ben-Tal, A., and Zowe, J. (1992). Equivalent displacement based formulations for maximum strength truss topology design, *Impact of Computing in Science and Engineering* **4**(4): 315–345. {*402*}

Allaire, G., Bonnetier, E., Francfort, G. A., and Jouve, F. (1997). Shape optimization by the homogenization method, *Numerische Mathematik* **76**(1): 27–68. {*383, 393, 403*}

Allaire, G., and Kohn, R. V. (1993a). Explicit optimal bounds on the elastic energy of a two-phase composite in two space dimensions, *Quarterly of Applied Mathematics* **LI**(4): 675–699. {*393*}

Allaire, G., and Kohn, R. V. (1993b). Optimal bounds on the effective behavior of a mixture of two well-ordered elastic materials, *Quarterly of Applied Mathematics* **LI**(4): 643–674. {*393*}

Allaire, G., and Kohn, R. V. (1993c). Optimal design for minimum weight and compliance in plane stress using extremal microstructures, *European Journal of Mechanics, A, Solids.* **12**(6): 839–878. {*459*}

Allaire, G., and Kohn, R. V. (1994). Optimal lower bounds on the elastic energy of a composite made from two non well-ordered isotropic materials, *Quarterly of Applied Mathematics* **LII**(2): 311–333. {*393, 403*}

Armand, J., Lurie, K. A., and Cherkaev, A. V. (1984). Optimal control theory and structural design, in E. Atrek, R. H. Gallagher, K. M. Ragsdell, and O. C. Zienkeiwicz (eds), *New Directions in optimal structural design*, John Wiley and Sons, New York; London; Sydney, pp. 211–229. {*67, 459*}

In this bibliography, each entry is followed by a braced list of page numbers in small *italic* type, showing where in the book the entry is cited.

Arora, J. S. (1998). Optimization of structures subject to dynamical loads, in C. T. Leondes (ed.), *Structural Dynamical Systems: Computational Technologies and Optimization*, Gordon and Breach Science Publ., pp. 1–73. {*464*}

Astala, K., and Miettinen, M. (1998). On quasiconformal mappings and 2-dimensional *G*-closure problems, *Archive for Rational Mechanics and Analysis* **143**(3): 207–240. {*292, 307, 338*}

Atkin, R. J., and Fox, N. (1990). *An Introduction to the Theory of Elasticity*, Longman, London. {*357*}

Avellaneda, M. (1987a). Iterated homogenization, differential effective medium theory, and applications, *Communications on Pure and Applied Mathematics (New York)* **40**: 803–847. {*183*}

Avellaneda, M. (1987b). Optimal bounds and microgeometries for elastic two-phase composites, *SIAM Journal on Applied Mathematics* **47**(6): 1216–1228. {*423*}

Avellaneda, M., Cherkaev, A. V., Gibiansky, L. V., Milton, G. W., and Rudelson, M. (1996). A complete characterization of the possible bulk and shear moduli of planar polycrystals, *Journal of the Mechanics and Physics of Solids* **44**(7): 1179–1218. {*362, 363, 387, 447, 450*}

Avellaneda, M., Cherkaev, A. V., Lurie, K. A., and Milton, G. W. (1988). On the effective conductivity of polycrystals and a three-dimensional phase interchange inequality, *Journal of Applied Physics* **63**: 4989–5003. {*214, 267, 291, 292*}

Avellaneda, M., and Milton, G. W. (1989a). Bounds on the effective elastic tensor of composites based on two-point correlations, in D. Hui, and T. J. Kozik (eds), *Composite Material Technology*, ASME, pp. 89–93. {*381, 465, 466*}

Avellaneda, M., and Milton, G. W. (1989b). Optimal bounds on the effective bulk modulus of polycrystals, *SIAM Journal on Applied Mathematics* **49**(3): 824–837. {*382, 449, 466*}

Backus, G. E. (1962). Long-wave elastic anisotropy produced by horizontal layering, *Journal of Geophysical Research* **67**: 4427–4440. {*42, 180*}

Backus, G. E. (1970). A geometrical picture of anisotropic elastic tensors, *Reviews of Geophysics and Space Physics* **8**: 633–671. {*368*}

Bakhvalov, N. S., and Knyazev, A. (1994). Fictitious domain methods and computation of homogenized properties of composites with a periodic structure of essentially different components, in G. I. Marchuk (ed.), *Numerical Methods and Applications*, CRC Press, Boca Raton, pp. 221–276. {*47*}

Bakhvalov, N. S., and Panasenko, G. P. (1989). *Homogenization: Averaging Processes in Periodic Media*, Kluwer Academic Publishers Group, Dordrecht/Boston/London. {*xxiii, 45, 47*}

Balk, A., Cherkaev, A. V., and Slepyan, L. I. (1999). Dynamics of solids with non-monotone stress-strain relation, *Journal of the Mechanics and Physics of Solids* **submitted**. {*47*}

Ball, J. M. (1989). A version of the fundamental theorem for Young measures, *PDEs and continuum models of phase transitions (Nice, 1988)*, Vol. 344 of *Lecture Notes in Phys.*, Springer-Verlag, Berlin; New York, pp. 207–215. {*164, 319*}

Ball, J. M., Currie, J. C., and Oliver, P. J. (1981). Null-lagrangians, weak continuity, and variational problems of arbitrary order, *Journal of Functional Analysis* **41**: 4989–5003. {*150*}

Ball, J. M. (ed.) (1988). *Material instabilities in continuum mechanics: related mathematical problems: the proceedings of a symposium year on material instabilities in continuum mechanics*, Oxford University Press, Walton Street, Oxford OX2 6DP, UK. {*319*}

Ball, J. M., and James, R. (1987). Finite phase mixtures as minimizers of energy, *Archive for Rational Mechanics and Analysis* **100**: 13–52. {*319*}

Ball, J. M., and Murat, F. (1984). $W^{1,p}$-quasiconvexity and variational problems for multiple integrals, *Journal of Functional Analysis* **58**(3): 225–253. See errata (?). {*150, 158, 217*}

Ball, J. M., and Murat, F. (1986). Erratum: "$W^{1,p}$-quasiconvexity and variational problems for multiple integrals", *Journal of Functional Analysis* **66**(3): 439. {*497*}

Ball, J. M., and Murat, F. (1991). Remarks on rank-one convexity and quasiconvexity, *Ordinary and partial differential equations, Vol. III (Dundee, 1990)*, Longman Sci. Tech., Harlow, pp. 25–37. {*155, 158*}

Bandle, C. (1980). *Isoperimetric inequalities and applications*, Pitman (Advanced Publishing Program), Boston, Mass. {*477*}

Banichuk, N. V. (1977). Conditions of optimality in the problem of finding shapes of holes in elastic bodies, *Prikladnaia Matematika i Mekhanika (PMM)= Applied Mathematics and Mechanics* **41**(5): 920–925. {*410, 459*}

Banichuk, N. V., Bel'skiĭ, V. G., and Kobelev, V. V. (1984). Optimization in problems of elasticity with unknown boundaries, *Izvestiia Akademii Nauk SSSR. Mekhanika Tverdogo Tela* **19**(3): 46–52. {*410, 412*}

Beliaev, A. Y. A., and Kozlov, S. M. (1996). Darcy equation for random porous media, *Communications on Pure and Applied Mathematics (New York)* **49**(1): 1–34. {*47*}

Bendsøe, M. P. (1995). *Optimization of Structural Topology, Shape, and Material*, Springer-Verlag, Berlin; Heidelberg; London; etc. {*xxiii, 308, 399, 459*}

Bendsøe, M. P., Díaz, A. R., and Kikuchi, N. (1993). Topology and generalized layout optimization of elastic structures, *Topology design of structures (Sesimbra, 1992)*, Kluwer Academic Publishers Group, Dordrecht, pp. 159–205. {*189, 393, 464*}

Bendsøe, M. P., Díaz, A. R., Lipton, R., and Taylor, J. E. (1995). Optimal design of material properties and material distribution for multiple loading conditions, *International Journal for Numerical Methods in Engineering* **38**(7): 1149–1170. {*477*}

Bendsøe, M. P., Díaz, A. R., Lipton, R., and Taylor, J. E. (1996). Parametrization in laminate design for optimal compliance, *International Journal of Solids and Structures* **34**: 415–434. {*498*}

Bendsøe, M. P., Guedes, J. M., Haber, R. B., Pedersen, P., and Taylor, J. E. (1994). An analytical model to predict optimal material properties in the context of optimal structural design, *Journal of Applied Mechanics* **61**(4): 930–937. {*459, 464*}

Bendsøe, M. P., Guedes, J. M., Plaxton, S., and Taylor, J. E. (1996). Optimization of structure and material properties for solids composed of softening material, *International Journal of Solids and Structures* **33**(12): 1799–1813. {*477*}

Bendsøe, M. P., Hammer, V. B., Lipton, R., and Pedersen, P. (1995). Minimum compliance design of laminated plates, *Homogenization and applications to material sciences (Nice, 1995)*, Gakkōtosho, Tokyo, pp. 45–56. see also (?). {*393, 447, 477*}

Bendsøe, M. P., and Kikuchi, N. (1988). Generating optimal topologies in structural design using a homogenization method, *Computer Methods in Applied Mechanics and Engineering* **71**(2): 197–224. {*121, 393*}

Bendsøe, M. P., and Lipton, R. (1997). Extremal overall elastic response of polycrystalline materials, *Journal of the Mechanics and Physics of Solids* **45**(11-12): 1765–1780. {*447*}

Bendsøe, M. P., and Mota Soares, C. A. (eds) (1992). *Topology Design of Structures*, Vol. 227 of *NATO ASI series. Series E, Applied sciences*, Kluwer Academic Publishers Group, Dordrecht. Boston. London. {*xxiii, 447*}

Bensoussan, A., Boccardo, L., and Murat, F. (1992). *H* convergence for quasilinear elliptic equations with quadratic growth, *Applied Mathematics and Optimization* **26**(3): 253–272. {*64*}

Bensoussan, A., Lions, J. L., and Papanicolaou, G. (1978). *Asymptotic analysis of periodic structures*, North-Holland Publ. {*xxiii, xxv, 45, 47, 48, 60, 63*}

Benveniste, Y. (1994). Exact connections between polycrystal and crystal properties in two-dimensional polycrystalline aggregates, *Proceedings of the Royal Society of London. Series A, Mathematical and Physical Sciences* **447**: 1–22. {*72*}

Benveniste, Y. (1995). Correspondence relations among equivalent classes of heterogeneous piezoelectric solids under anti-plane mechanical and in-plane electrical fields, *Journal of the Mechanics and Physics of Solids* **43**(4): 553–571. {*72*}

Berdichevsky, V. (1983). *Variational principles of continuum mechanics*, Nauka, Moscow, Russia. In Russian. {*63*}

Berdichevsky, V. (1997). *Thermodynamics of chaos and order*, Longman, Harlow. {*47*}

Berdichevsky, V., Jikov, V., and Papanicolaou, G. (eds) (1999). *Homogenization*, World Scientific Publishing Co., Singapore; Philadelphia; River Edge, NJ. {*47, 63*}

Berdichevsky, V., Kunin, I., and Hussain, F. (1991). Negative temperature of vortex motion, *Physical Review A (Atomic, Molecular, and Optical Physics)* **43**(4): 2050–2051. {*47*}

Bergman, D. J. (1978). The dielectric constant of a composite material—a problem in classical physics, *Physics Reports* **43**(9): 377–407. {*294, 341*}

Bergman, D. J. (1980). Exactly solvable microscopic geometries and rigorous bounds for the complex dielectric constant of a two-component composite material, *Physical Review Letters* **44**: 1285–1287. {*341*}

Bergman, D. J. (1991). Effective medium approximation for nonlinear conductivity of a composite medium, *Composite media and homogenization theory (Trieste, 1990)*, Birkhäuser Boston, Boston, MA, pp. 67–79. {*63*}

Berlyand, L., and Golden, K. (1994). Exact result for the effective conductivity of a continuum percolation model, *Physical Review B (Solid State)* **50**: 2114–2117. {*47*}

Bhattacharya, K., Firoozye, N. B., James, R. D., and Kohn, R. V. (1994). Restrictions on microstructure, *Proceedings of the Royal Society of Edinburgh. Section A, Mathematical and Physical Sciences* **124**(5): 843–878. {*176, 319, 321*}

Bhattacharya, K., and Kohn, R. V. (1997). Elastic energy minimization and the recoverable strains of polycrystalline shape-memory materials, *Archive for Rational Mechanics and Analysis* **139**: 99–180. {*319*}

Bliss, G. A. (1946). *Lectures on the Calculus of Variations*, University of Chicago Press, Chicago, IL. {*10*}

Bourgeat, A., Kozlov, S. M., and Mikelić, A. (1995). Effective equations of two-phase flow in random media, *Calculus of Variations and Partial Differential Equations* **3**(3): 385–406. {*47, 63*}

Bratus', A. S. (1986). Multiple eigenvalues in problems of optimization of spectral properties of systems with a finite number of degrees of freedom, *Zhurnal Vychislitel'noi Matematiki i Matematicheskoi Fiziki* **26**(5): 645–654, 797. {*477*}

Bratus', A. S., and Seyranian, A. P. (1983). Bimodal solutions in eigenvalue optimization problems, *Prikladnaia Matematika i Mekhanika (PMM)= Applied Mathematics and Mechanics* **47**: 451–457. {*477*}

Bruggemann, D. A. G. (1930). *Elastizitätskonstanten von Kristallaggregaten*, PhD thesis, Universiteit te Verdedigen, Den Haag. {*183*}

Bruggemann, D. A. G. (1935). Berechnung verschiedener physikalischer Konstanten von heterogenen Substanzen. I. Dielectrizitátkonstanten und Leitfähigkeiten der Mischkörper aus isotropen Substanzen, *Annalen der Physik (1900)* **22**: 636–679. {*xxiii, 54, 183*}

Bruggemann, D. A. G. (1937). Berechnung verschiedener physikalischer Konstanten von heterogenen Substanzen. III. Die elastische Konstanten der Quasi-isotropen Mischkörper aus isotropen Substanzen, *Annalen der Physik (1900)* **29**: 160–178. {*54*}

Bruno, O. P. (1991). The effective conductivity of strongly heterogeneous composites, *Proceedings of the Royal Society of London. Series A, Mathematical and Physical Sciences* **433**(1888): 353–381. {*72, 304*}

Burns, T., and Cherkaev, A. V. (1997). Optimal distribution of multimaterial composites for torsional beams, *Structural Optimization* **13**(1): 14–23. {*81, 111, 114, 332*}

Buttazzo, G. (1989). *Semicontinuity, relaxation and integral representation in the calculus of variations*, Longman Scientific and Technical, Harlow, Essex, UK. {*145*}

Carathéodory, C. (1967). *Calculus of variations and partial differential equations of the first order. Part II: Calculus of variations*, Holden-Day Inc., San Francisco, Calif. Translated from the German by Robert B. Dean, Julius J. Brandstatter, translating editor. {*16*}

Carothers, S. D. (1912). Plane strain in a wedge, *Proceedings of the Royal Society of Edinburgh* **23**: 292. {*415*}

Cherepanov, G. P. (1974). Inverse problems of the plane theory of elasticity, *Journal of Applied Mathematics and Mechanics* **38**(6): 963–979. {*410, 412*}

Cherkaev, A. V. (1993). Stability of optimal structures of elastic composites, *Topology design of structures (Sesimbra, 1992),*, Vol. 227 of *NATO Adv. Sci. Inst. Ser. E Appl. Sci.*, Kluwer Academic Publishers Group, Dordrecht, pp. 547–558. {*461, 470*}

Cherkaev, A. V. (1994). Relaxation of problems of optimal structural design, *International Journal of Solids and Structures* **31**(16): 2251–2280. {*123, 202, 472, 475*}

Cherkaev, A. V. (1998). Variational approach to structural optimization, in C. T. Leondes (ed.), *Structural Dynamical Systems: Computational Technologies and Optimization*, Gordon and Breach Science Publ., pp. 199–237. {*472*}

Cherkaev, A. V. (1999). Necessary conditions technique in structural optimization, *International Journal of Solids and Structures* . To appear. {*237, 307, 318*}

Cherkaev, A. V., and Gibiansky, L. V. (1988). Optimal design of nonlinear-elastic and elastic-plastic torsioned bars, *Mechanics of Solids* **5**: 168–174. {*92, 103*}

Cherkaev, A. V., and Gibiansky, L. V. (1992). The exact coupled bounds for effective tensors of electrical and magnetic properties of two-component two-dimensional composites, *Proceedings of the Royal Society of Edinburgh. Section A, Mathematical and Physical Sciences* **122**(1-2): 93–125. First version (?) (in Russian). {*72, 193, 194, 198, 199, 265, 294, 297, 299, 303*}

Cherkaev, A. V., and Gibiansky, L. V. (1993). Coupled estimates for the bulk and shear moduli of a two-dimensional isotropic elastic composite, *Journal of the Mechanics and Physics of Solids* **41**(5): 937–980. {*294, 422, 425, 428, 436, 447*}

Cherkaev, A. V., and Gibiansky, L. V. (1994). Variational principles for complex conductivity, viscoelasticity, and similar problems in media with complex moduli, *Journal of Mathematical Physics* **35**(1): 127–145. {*341, 342, 351*}

Cherkaev, A. V., and Gibiansky, L. V. (1996). Extremal structures of multiphase heat conducting composites, *International Journal of Solids and Structures* **33**(18): 2609–2618. {*309, 313, 339*}

Cherkaev, A. V., Grabovsky, Y., Movchan, A. B., and Serkov, S. K. (1998). The cavity of the optimal shape under the shear stresses, *International Journal of Solids and Structures* **35**(33): 4391–4410. {*240, 410, 415*}

Cherkaev, A. V., Krog, L., and Kucuk, I. (1998). Stable optimal design of two-component elastic structures, *Control and Cybernetics* **27**(2): 265–282. {*465, 470*}

Cherkaev, A. V., Lurie, K. A., and Milton, G. W. (1992). Invariant properties of the stress in plane elasticity and equivalence classes of composites, *Proceedings of the Royal Society of London. Series A, Mathematical and Physical Sciences* **438**(1904): 519–529. {*219, 369*}

Cherkaev, A. V., and Miettinen, M. (1999). Necessary condition in the polycrystal *G*-closure problem. in preparation. {*338, 339*}

Cherkaev, A. V., and Palais, R. (1997). Optimal design of three-dimensional axisymmetric structures, *Structural Optimization* **13**(1): 10–23. {*383, 403*}

Cherkaev, A. V., and Palais, R. (1998). Optimal design of three-dimensional axisymmetric elastic structures, in C. T. Leondes (ed.), *Structural Dynamical Systems: Computational Technologies and Optimization*, Gordon and Breach Science Publ., pp. 237–267. {*403, 459*}

Cherkaev, A. V., and Robbins, T. C. (1999). Optimization of heat conducting structures, *Proceedings of the 3rd World Congress of Structural and Multi-disciplinary Optimization*, ISSMO, Buffalo, New York, p. 89. {*138*}

Cherkaev, A. V., and Slepyan, L. (1995). Waiting element structures and stability under extension, *International Journal of Damage Mechanics* **4**: 58–82. {*47*}

Cherkaeva, E. (1997). Optimal source control and resolution in nondestructive testing, *Structural Optimization* **13**(1): 12–16. {*482*}

Cherkaeva, E., and Cherkaev, A. V. (1995). Bounds for detectability of material damage by noisy electrical measurements, *Structural and multidisciplinary optimization*, Pergamon, New York, pp. 543–548. {*242, 477, 484*}

Cherkaeva, E., and Cherkaev, A. V. (1998). Optimal design for uncertain loading conditions, *Homogenization*, World Scientific Publishing Co., Singapore; Philadelphia; River Edge, NJ, p. in print. {*477*}

Cherkaeva, E., and Cherkaev, A. V. (1999). Structural optimization and biological "designs", in P. Pedersen, and M. Bendsøe (eds), *Synthesis in bio solid mechanics*, Solid mechanics and its applications, Kluwer Academic Publishers Group, Dordrecht. Boston. London, pp. 247–264. {*477*}

Cherkaeva, E., and Tripp, A. C. (1996). Inverse conductivity problem for inaccurate measurements, *Inverse Problems* **12**(6): 869–883. {*482*}

Chiheb, R., and Panasenko, G. P. (1998). Optimization of finite rod structures and L-convergence, *Journal of Dynamical and Control Systems* **4**(2): 273–304. {*64, 464*}

Christensen, R. M. (1971). *Theory of Viscoelasticity*, Academic Press, New York. {*351*}

Christensen, R. M. (1979). *Mechanics of Composite Materials*, Wiley-Interscience, New York. {*47, 54, 63, 423*}

Cioranescu, D., and Murat, F. (1982). Un terme étrange venu d'ailleurs. II, pp. 154–178, 425–426. Translated to English in (?). {*47*}

Cioranescu, D., and Murat, F. (1997). A strange term coming from nowhere, **31**: 45–93. Translation of the paper (Cioranescu and Murat, 1982). {*501*}

Clark, K. E., and Milton, G. W. (1994). Modelling the effective conductivity function of an arbitrary two-dimensional polycrystal using sequential laminates, *Proceedings of the Royal Society of Edinburgh. Section A, Mathematical and Physical Sciences* **124**(4): 757–783. {*291*}

Clark, K. E., and Milton, G. W. (1995). Optimal bounds correlating electric, magnetic and thermal properties of two-phase, two-dimensional composites, *Proceedings of the Royal Society of London. Series A, Mathematical and Physical Sciences* **448**(1933): 161–190. {*294*}

Clements, D. J., and Anderson, B. D. O. (1978). *Singular optimal control: the linear-quadratic problem*, Springer-Verlag, Berlin. Lecture Notes in Control and Information Sciences, Vol. 5. {*23*}

Courant, R., and Hilbert, D. (1962). *Methods of Mathematical Physics*, Interscience Publishers, New York. {*10, 36, 43, 104, 411*}

Cox, S. J., and Overton, M. L. (1992). On the optimal design of columns against buckling, *SIAM Journal on Mathematical Analysis* **23**(2): 287–325. {*477, 484*}

Dacorogna, B. (1982). *Weak continuity and weak lower semicontinuity of nonlinear functionals*, Springer-Verlag, Berlin. {*118, 120, 150, 158, 162, 221, 472*}

Dacorogna, B. (1989). *Direct methods in the calculus of variations*, Springer-Verlag, Berlin; Heidelberg; London; etc. {*10, 14, 145*}

Dacorogna, B., and Haeberly, J.-P. (1996). Remarks on a numerical study of convexity, quasiconvexity, and rank one convexity, *Variational methods for discontinuous structures (Como, 1994)*, Birkhäuser, Basel, pp. 143–154. {*155*}

Dal Maso, G. (1993). *An introduction to Γ-convergence*, Birkhäuser Boston Inc., Cambridge, MA. {*64, 118*}

Dal Maso, G., and Dell'Antonio, G. (eds) (1991). *Composite media and homogenization theory*, Birkhäuser Boston Inc., Cambridge, MA. {*47*}

De Giorgi, E. (1994a). New ideas in calculus of variations and geometric measure theory, *Motion by mean curvature and related topics (Trento, 1992)*, Walter de Gruyter, New York, USA and Berlin, Germany, pp. 63–69. {*xxiii*}

De Giorgi, E. (1994b). On the convergence of solutions of some evolution differential equations, *Set-Valued Analysis* **2**(1-2): 175–182. Set convergence in nonlinear analysis and optimization. {*xxiii*}

Dem'yanov, V. F., and Malozëmov, V. N. (1990). *Introduction to minimax*, Dover Publications, Inc., New York. Translated from the Russian by D. Louvish, Reprint of the 1974 edition. {*103, 487*}

Díaz, A. R., and Bendsøe, M. P. (1992). Shape optimization of structures for multiple loading situations using a homogenization method, *Structural Optimization* **4**: 17–22. {*477*}

Díaz, A. R., Kikuchi, N., Papalambros, P. Y., and Taylor, J. E. (1993). Design of an optimal grid for finite element methods, *Journal of Structural Mechanics* **11**(2): 215–230. {*393, 459*}

Díaz, A. R., and Lipton, R. (1997). Optimal material layout for 3D elastic structures, *Structural Optimization* **13**: 60–64. Preliminary version in (Lipton and Díaz, 1995). {*393, 403, 511*}

Díaz, A. R., Lipton, R., and Soto, C. A. (1994). A new formulation of the problem of optimum reinforcement of Reissner-Mindlin plates, *Computer Methods in Applied Mechanics and Engineering* . {*189*}

Díaz, A. R., and Sigmund, O. (1995). Checkerboard patterns in layout optimization, *Structural Optimization* **10**: 40–45. {*463, 464*}

Ding, Y. (1986). Shape optimization of structures: A literature survey, *Computers and Structures* **24**(6): 985–1004. {*464*}

Dolzmann, G., and Müller, S. (1995). The influence of surface energy on stress-free microstructures in shape memory alloys, *Meccanica (Milan)* **30**(5): 527–539. Microstructure and phase transitions in solids (Udine, 1994). {*145*}

Dundurs, J. (1989). Cavities vis-à-vis rigid inclusions and some related general results in plane elasticity, *Journal of Applied Mechanics* **56**(4): 786–790. {*369*}

Dundurs, J., and Jasiuk, I. (1997). Effective elastic moduli of composite materials: Reduced parameters dependence, *Applied Mechanics Reviews* **50**: S39–S43. {*369*}

Dvořák, J. (1994). A reliable numerical method for computing homogenized coefficients, *Technical Report MAT-Report 31*, Mathematical Institute, Danish Technical University, Lyngby, Denmark. {*81*}

Dykhne, A. M. (1970). Conductivity of a two-dimensional two-phase system, *Journal of Experimental and Theoretical Physics (JETP)* **7**: 110–116. {*302*}

Dykhne, A. M. (1971). Conductivity of a two-dimensional two-phase system, *Journal of Experimental and Theoretical Physics (JETP)* **32**: 63–65. {*290, 302*}

Ekeland, I., and Temam, R. (1976). *Convex analysis and variational problems*, North-Holland Publishing Co., Amsterdam, The Netherlands. Translated from the French, Studies in Mathematics and its Applications, Vol. 1. {*8, 28, 30, 44, 45, 145, 155*}

Eshelby, J. D. (1957). The determination of the elastic field of an ellipsoidal inclusion, and related problems, *Proceedings of the Royal Society of London. Series A, Mathematical and Physical Sciences* **241**: 376–396. {*240*}

Eshelby, J. D. (1961). Elastic inclusions and inhomogeneities, *Progress in Solid Mechanics, Vol. II*, North-Holland Publishing Co., Amsterdam, The Netherlands, pp. 87–140. {*240*}

Eshelby, J. D. (1975). The elastic energy-momentum tensor, *Journal of Elasticity* **5**(3-4): 321–335. Special issue dedicated to A. E. Green. {*240*}

Ewing, G. M. (1969). *Calculus of variations with applications*, W. W. Norton & Co. Inc., New York. {*xxv*}

Fannjiang, A., and Papanicolaou, G. (1994). Convection enhanced diffusion for periodic flows, *SIAM Journal on Applied Mathematics* **54**(2): 333–408. {*341*}

Fedorov, A. V., and Cherkaev, A. V. (1983). Choosing of optimal orientation of axes of elastic symmetry of orthotropic plate, *Mechanics of Solids* **3**: 135–142. {*447*}

Fenchel, W. (1949). On conjugate convex functions, *Canadian Journal of Mathematics* **1**: 73–77. {*28, 30*}

Firoozye, N. B. (1991). Optimal use of the translation method and relaxations of variational problems, *Communications on Pure and Applied Mathematics (New York)* **44**(6): 643–678. {*214*}

Firoozye, N. B., and Kohn, R. V. (1993). Geometric parameters and the relaxation of multiwell energies, in D. Kinderlehrer, R. James, M. Luskin, and J. L. Ericksen (eds), *Microstructure and phase transition*, Springer-Verlag, New York, pp. 85–109. {*307*}

Fonseca, I. (1987). Variational methods for elastic crystals, *Archive for Rational Mechanics and Analysis* **97**(3): 189–220. {*447*}

Fonseca, I. (1989). Phase transitions of elastic solid materials, *Archive for Rational Mechanics and Analysis* **107**(3): 195–223. {*xxiii, 319*}

Fonseca, I. (1992). Lower semicontinuity of surface energies, *Proceedings of the Royal Society of Edinburgh. Section A, Mathematical and Physical Sciences* **120**(1-2): 99–115. {*145*}

Fonseca, I., Kinderlehrer, D., and Pedregal, P. (1994). Energy functionals depending on elastic strain and chemical composition, *Calculus of Variations and Partial Differential Equations* **2**(3): 283–313. {*xxiii*}

Fox, C. (1987). *An introduction to the calculus of variations*, Dover Publications, Inc., New York. Reprint of the 1963 edition. {*10, 42*}

Francfort, G. A., and Milton, G. W. (1987). Optimal bounds for conduction in two-dimensional, multiphase, polycrystalline media, *Journal of Statistical Physics* **46**(1-2): 161–177. {*272, 275*}

Francfort, G. A., and Murat, F. (1986). Homogenization and optimal bounds in linear elasticity, *Archive for Rational Mechanics and Analysis* **94**(4): 307–334. {*185, 189, 191, 379, 383, 384, 491*}

Francfort, G. A., Murat, F., and Tartar, L. (1995). Fourth-order moments of nonnegative measures on S^2 and applications, *Archive for Rational Mechanics and Analysis* **131**(4): 305–333. {*384, 465*}

Francfort, G. A., and Tartar, L. (1991). Comportement effectif d'un mélange de matériaux élastiques isotropes ayant le m^eme module de cisaillement, *Comptes Rendus de l'Académie des Sciences, Paris* **312**(3): 301–307. {*420*}

Fu, Y., Klimkowski, K. J., Rodin, G. J., Berger, E., Browne, J. C., Singer, J. K., van de Geijn, R. A., and Vemaganti, K. S. (1998). A fast solution method for three-dimensional many-particle problems of linear elasticity, *International Journal for Numerical Methods in Engineering* **42**: 1215–1229. {*47*}

Fuchs, M. B., and Hakim, S. (1996). Improved multivariate reanalysis of structures based on the structural variation method, *Mechanics of Structures and Machines* **24**(1): 51–70. {*477*}

Gabasov, R., and Kirillova, F. M. (1973). *Singular optimal controls*, Nauka, Moscow, Russia. Monographs in Theoretical Foundations of Technical Cybernetics. {*23*}

Gamkrelidze, R. V. (1962). On sliding optimal states, *Doklady Akademii Nauk SSSR* **143**: 1243–1245. {*7, 14, 23*}

Gamkrelidze, R. V. (1985). Sliding modes in optimal control theory, *Trudy Matematicheskogo Instituta imeni V. A. Steklova = Proceedings of the Steklov Institute of Mathematics* **169**: 180–193, 255. Topology, ordinary differential equations, dynamical systems. {*14*}

Gelfand, I. M., and Fomin, S. V. (1963). *Calculus of variations*, Prentice-Hall, Upper Saddle River, NJ 07458. Revised English edition translated and edited by Richard A. Silverman. {*xxv, 10, 28, 123, 222, 347, 349, 350*}

Gibiansky, L. V., and Cherkaev, A. V. (1984). Design of composite plates of extremal rigidity, *Report 914*, Ioffe Physico-technical Institute, Acad of Sc, USSR, Leningrad, USSR. English translation in (?). {*81, 213, 379, 393, 395, 397, 405–407*}

Gibiansky, L. V., and Cherkaev, A. V. (1987). Microstructures of composites of extremal rigidity and exact estimates of the associated energy density, *Report 1115*, Ioffe Physico-Technical Institute, Acad of Sc, USSR, Leningrad, USSR. English translation in (?). {*185, 191, 214, 227, 379, 383, 387, 393, 403, 410, 455*}

Gibiansky, L. V., and Cherkaev, A. V. (1988). The set of tensor pairs of dielectric and magnetic permeabilities of two-phase composites, *Technical Report 1286*, Ioffe Physico-technical Institute, Acad of Sc, USSR, Leningrad, USSR. English translation in (Cherkaev and Gibiansky, 1992). {*500*}

Gibiansky, L. V., and Cherkaev, A. V. (1997a). Design of composite plates of extremal rigidity, in A. V. Cherkaev, and R. V. Kohn (eds), *Topics in the mathematical modelling of composite materials*, Vol. 31 of *Progress in nonlinear differential equations and their applications*, Birkhäuser Boston, Boston, MA, pp. 95–137. Translation. The original publication in (Gibiansky and Cherkaev, 1984). {*505*}

Gibiansky, L. V., and Cherkaev, A. V. (1997b). Microstructures of composites of extremal rigidity and exact bounds on the associated energy density, in A. V. Cherkaev, and R. V. Kohn (eds), *Topics in the mathematical modelling of composite materials*, Vol. 31 of *Progress in nonlinear differential equations and their applications*, Birkhäuser Boston, Boston, MA, pp. 273–317. Translation. The original publication is in (Gibiansky and Cherkaev, 1987). {*505*}

Gibiansky, L. V., Lurie, K. A., and Cherkaev, A. V. (1988). Optimal focusing of heat flow by heterogeneous heat conducting media (thermolens problem), *Zhurnal Tehniceskoj Fiziki* **58**(1): 67–74. {*130, 135*}

Gibiansky, L. V., and Milton, G. W. (1993). On the effective viscoelastic moduli of two-phase media: I. Rigorous bounds on the complex bulk modulus, *Proceedings of the Royal Society of London. Series A, Mathematical and Physical Sciences* **440**(1908): 163–188. {*341, 351, 353*}

Gibiansky, L. V., and Sigmund, O. (1998). Multiphase composites with extremal bulk modulus, *Technical Report 600*, DCAMM, DTU, Lyngby, Denmark. {*307, 336, 339*}

Gibiansky, L. V., and Torquato, S. (1993). Link between the conductivity and elastic moduli of composite materials, *Physical Review Letters* **71**(18): 2927–2930. {*294, 353, 436*}

Gibiansky, L. V., and Torquato, S. (1995a). Geometrical-parameter bounds on the effective moduli of composites, *Journal of the Mechanics and Physics of Solids* **43**(10): 1587–1613. {*436*}

Gibiansky, L. V., and Torquato, S. (1995b). Rigorous link between the conductivity and elastic moduli of fiber-reinforced composite materials, *Philosophical Transactions of the Royal Society of London Series A* **353**: 243–278. {*294, 353, 436*}

Gibiansky, L. V., and Torquato, S. (1996a). Bounds on the effective moduli of cracked materials, *Journal of the Mechanics and Physics of Solids* **44**(2): 233–242. {*353, 436*}

Gibiansky, L. V., and Torquato, S. (1996b). Connection between the conductivity and bulk modulus of isotropic composite materials, *Proceedings of the Royal Society of London. Series A, Mathematical and Physical Sciences* **452**(1945): 253–283. {*353*}

Golden, K. (1986). Bounds on the complex permittivity of a multicomponent material, *Journal of the Mechanics and Physics of Solids* **34**(4): 333–358. {*307, 341*}

Golden, K., and Papanicolaou, G. (1983). Bounds for effective parameters of heterogeneous media by analytic continuation, *Communications in Mathematical Physics* **90**(4): 473–491. {*341*}

Golden, K., and Papanicolaou, G. (1985). Bounds for effective parameters of multicomponent media by analytic continuation, *Journal of Statistical Physics* **40**(5-6): 655–667. {*307, 341*}

Goodman, J., Kohn, R. V., and Reyna, L. (1986). Numerical study of a relaxed variational problem from optimal design, *Computer Methods in Applied Mechanics and Engineering* **57**: 107–127. {*81, 90, 110*}

Grabovsky, Y. (1993). The G-closure of two well-ordered anisotropic conductors, *Proceedings of the Royal Society of Edinburgh* **123A**: 423–432. {*280*}

Grabovsky, Y. (1996). Bounds and extremal microstructures for two-component composites: A unified treatment based on the translation method, *Proceedings of the Royal Society of London. Series A, Mathematical and Physical Sciences* **452**(1947): 945–952. {*393*}

Grabovsky, Y. (1998). Exact relations for effective tensors of polycrystals. I. necessary conditions, *Archive for Rational Mechanics and Analysis* **143**(4): 309–329. {*72*}

Grabovsky, Y., and Kohn, R. V. (1995a). Microstructures minimizing the energy of a two phase elastic composite in two space dimensions. I: the confocal ellipse construction, *Journal of the Mechanics and Physics of Solids* **43**(6): 933–947. {*410, 412*}

Grabovsky, Y., and Kohn, R. V. (1995b). Microstructures minimizing the energy of a two phase elastic composite in two space dimensions. II: the Vigdergauz microstructure, *Journal of the Mechanics and Physics of Solids* **43**(6): 949–972. {*410, 412*}

Grabovsky, Y., and Milton, G. W. (1998). Exact relations for composites: towards a complete solution, *Proceedings of the International Congress of Mathematicians, (Berlin, 1998)*, Vol. III, pp. 623–632 (electronic). {*72, 270, 303*}

Grabovsky, Y., Milton, G. W., and Sage, D. S. (1999). Exact relations for effective tensors of polycrystals: Necessary conditions and sufficient conditions, *in preparation* . {*291*}

Grabovsky, Y., and Sage, D. S. (1998). Exact relations for effective tensors of polycrystals. II. Applications to elasticity and piezoelectricity, *Archive for Rational Mechanics and Analysis* **143**(4): 331–356. {*72*}

Greengard, L., and Helsing, J. (1998). On the numerical evaluation of elastostatic fields in locally isotropic two-dimensional composites, *Journal of the Mechanics and Physics of Solids* **46**(8): 1441–1462. {*47*}

Greengard, L., and Rokhlin, V. (1997). A new version of the fast multipole method for the laplace equation in three dimensions, *Acta numerica, 1997*, Cambridge University Press, Cambridge, UK, pp. 229–269. {*47*}

Guedes, J. M., and Taylor, J. E. (1996). On the prediction of material properties and topology for optimal continuum structures, *Structural Optimization* **14**: 193–199. {*464*}

Gurtin, M. E. (1972). The linear theory of elasticity, *Handbuch der Physik*, Vol. 6a/2, Springer-Verlag, New York, pp. 1–295. {*357*}

Haber, R. B., Jog, C., and Bendsøe, M. P. (1994). Variable-topology shape optimization with a constraint on perimeter, in B. Gilmore et al. (eds), *Advances in Design Automation*, Vol. DE 69-2, ASME, New York, pp. 261–272. {*464*}

Haber, R. B., Jog, C., and Bendsøe, M. P. (1996). A new approach to variable-topology shape design using a constraint on perimeter, *Structural Optimization* **11**: 1–12. {*393, 464*}

Haftka, R. T., and Gürdal, Z. (1992). *Elements of structural optimization*, third edn, Kluwer Academic Publishers Group, Dordrecht. {*xxiii, 459*}

510   References

Hardy, G. H., Littlewood, J. E., and Pólya, G. (1988). *Inequalities*, Cambridge University Press, Cambridge, UK. {*8, 9*}

Hashin, Z. (1965). On elastic behaviour of fibre reinforced materials of arbitrary transverse phase geometry, *Journal of the Mechanics and Physics of Solids* **13**: 119–134. {*424*}

Hashin, Z. (1970a). Theory of composite materials, in F. W. Wendt, U. Liebowitz, and N. Perrone (eds), *Mechanics of composite materials*, Pergamon, New York, pp. 257–282. {*423*}

Hashin, Z. (1970b). Theory of fiber reinforced materials, *Report Contract NAS1-8818*, University of Pennsylvania, Philadelphia, PA. {*63, 423*}

Hashin, Z. (1983). Analysis of composite materials: a survey, *Journal of Applied Mechanics* **50**: 481–505. {*63*}

Hashin, Z. (1992). Extremum principles for elastic heterogenous media with imperfect interfaces and their application to bounding of effective moduli, *Journal of the Mechanics and Physics of Solids* **40**(4): 767–781. {*63*}

Hashin, Z., and Shtrikman, S. (1962a). A variational approach to the theory of the effective magnetic permeability of multiphase materials, *Journal of Applied Physics* **35**: 3125–3131. {*xxiii, 54, 56, 67, 192, 251, 253, 278, 281*}

Hashin, Z., and Shtrikman, S. (1962b). A variational approach to the theory of the elastic behavior of polycrystals, *Journal of the Mechanics and Physics of Solids* **10**: 343–352. {*xxiii, 382, 384*}

Hashin, Z., and Shtrikman, S. (1963). A variational approach to the theory of the elastic behavior of multiphase materials, *Journal of the Mechanics and Physics of Solids* **11**: 127–140. {*307, 423, 491*}

Haslinger, J., and Dvořák, J. (1995). Optimum composite material design, *Mathematical Modelling and Numerical Analysis = Modelisation Mathématique et Analyse Numérique: M^2AN* . {*81*}

Haug, E. J., Choi, K. K., and Komkov, V. (1986). *Design Sensitivity Analysis of Structural System*, Mathematics in Science and Engineering. Volume 177, Academic Press, New York. {*459*}

Helsing, J., Milton, G. W., and Movchan, A. B. (1997). Duality relations, correspondences and numerical results for planar elastic composites, *Journal of the Mechanics and Physics of Solids* **45**(4): 565–590. {*47, 369*}

Hill, R. (1963). Elastic properties of reinforced solids: Some theoretical principles, *Journal of the Mechanics and Physics of Solids* **11**: 357–372. {*xxiii, 377*}

Hill, R. (1964). Theory of mechanical properties of fibre-strengthened materials. I. Elastic behaviour, *Journal of the Mechanics and Physics of Solids* **12**: 199–212. {*xxiii, 419–421*}

Hornung, U. (ed.) (1997). *Homogenization and porous media*, Springer-Verlag, New York. {*47*}

Ioffe, A. D., and Tihomirov, V. M. (1979). *Theory of extremal problems*, North-Holland Publishing Co., Amsterdam, The Netherlands. Translated from the Russian by Karol Makowski. {*7, 13, 23, 28*}

Jäger , W., Oleĭnik, O. A., and Schamajev, A. S. (1993). Averaging problem for Laplace equation in a partially perforated domain (in Russian), *Technical Report 333/4*, Dokladi Akademii Nauk (DAN) (Papers of the Academy of Sciences). {*xxiii*}

Jikov, V. V., Kozlov, S. M., and Oleĭnik, O. A. (1994). *Homogenization of differential operators and integral functionals*, Springer-Verlag, Berlin. Translated from the Russian by G. A. Yosifian [G. A. Iosif'yan]. {*xxiii, xxv, 45, 48–50, 60, 63, 118, 290*}

Jog, C., and Haber, R. B. (1996). Stability of finite element models for distributed-parameter optimization and topology design, *Journal for Computational Methods in Applied Mechanics and Engineering* **130**(8): 203–226. {*463*}

Jog, C., Haber, R. B., and Bendsøe, M. P. (1993). A displacement-based topology design method with self-adaptive layered materials, in M. P. Bendsøe, and C. A. Mota Soares (eds), *Topology Design of Structures, Sesimbra, Portugal*, NATO ASI, Kluwer Academic Publishers Group, Dordrecht. Boston. London, pp. 219–243. {*463, 464*}

Jog, C., Haber, R. B., and Bendsøe, M. P. (1994). Topology design with optimized self-adaptive materials, *International Journal for Numerical Methods in Engineering* **37**(8): 1323–1350. {*459*}

Kaganova, I., and Roitburd, A. (1987). Equilibrium shape of an inclusion in a solid, *Soviet Physics—Doklady* **32**: 925–927. {*xxiii*}

Karp, S. N., and Karal, F. C. (1962). The elastic-field behaviour in the neighbourhood of a crack of arbitrary angle, *Communications on Pure and Applied Mathematics (New York)* **15**: 413–421. {*415*}

Keller, J. B. (1964). A theorem on the conductivity of a composite medium, *Journal of Mathematical Physics* **5**: 548–549. {*xxiii, 290, 302*}

Khachaturyan, A. G. (1983). *Theory of structural transformation in solids*, John Wiley and Sons, New York; London; Sydney. {*xxiii, 319*}

Khruslov, E. Y. (1991). Homogenized models of composite media, *Composite media and homogenization theory (Trieste, 1990)*, Birkhäuser Boston Inc., Cambridge, MA, pp. 159–182. {*47*}

Khruslov, E. Y. (1995). Homogenized models of strongly inhomogeneous media, *Proceedings of the International Congress of Mathematicians, Vol. 1, 2 (Zürich, 1994)*, Birkhäuser Verlag, Basel, Switzerland, pp. 1270–1278. {*63*}

Kinderlehrer, D., James, R., Luskin, M., and Ericksen, J. L. (eds) (1993). *Microstructure and phase transition*, Springer-Verlag, New York. {*319*}

Kinderlehrer, D., and Pedregal, P. (1991). Characterizations of Young measures generated by gradients, *Archive for Rational Mechanics and Analysis* **115**(4): 329–365. {*164*}

Kinderlehrer, D., and Pedregal, P. (1994). Gradient Young measures generated by sequences in Sobolev spaces, *Journal of Geometric Analysis* **4**(1): 59–90. {*164*}

Klosowicz, B., and Lurie, K. A. (1971). On the optimal nonhomogeneity of a torsional elastic bar, *Archives of Mechanics = Archiwum Mechaniki Stosowanej* **24**(2): 239–249. {*97, 110*}

Klouček, P., and Luskin, M. (1994). Computational modeling of the martensitic transformation with surface energy, *Mathematical and Computer Modelling* **20**(10-11): 101–121. {*145*}

Kohn, R. V. (1991). The relaxation of a double-well energy, *Continuum Mechanics and Thermodynamics* **3**: 193–236. {*158*}

Kohn, R. V., and Lipton, R. (1988). Optimal bounds for the effective energy of a mixture of isotropic, incompressible, elastic materials, *Archive for Rational Mechanics and Analysis* **102**(4): 331–350. {*189, 393, 423*}

Kohn, R. V., and Milton, G. W. (1986). On bounding the effective conductivity of anisotropic composites, in J. Ericksen et al. (eds), *Homogenization and Effective Moduli of Materials and Media (Minneapolis, Minn., 1984/1985)*, Springer-Verlag, Berlin; Heidelberg; London; etc., pp. 97–125. {*267, 280*}

Kohn, R. V., and Müller, S. (1994). Surface energy and microstructure in coherent phase transitions, *Communications on Pure and Applied Mathematics (New York)* **47**: 405–435. {*145*}

Kohn, R. V., and Strang, G. (1982). Structural design, optimization, homogenization, and relaxation of variational problems, in R. Burridge, G. Papanicolaou, and S. Childress (eds), *Macroscopic Properties of Disordered Media*, Vol. 154 of *Lect. Notes Phys.*, Springer-Verlag, pp. ??–??. {*214, 217*}

Kohn, R. V., and Strang, G. (1983). Explicit relaxation of a variational problem in optimal design, *Bulletin of the American Mathematical Society* **9**: 211–214. {*214, 217*}

Kohn, R. V., and Strang, G. (1986a). Optimal design and relaxation of variational problems, *Communications on Pure and Applied Mathematics (New York)* **39**: 1–25 (part I), 139–182 (part II) and 353–357 (part III). {*81, 85, 90, 121, 155, 158, 165, 166, 214, 217*}

Kohn, R. V., and Strang, G. (1986b). Optimal design in elasticity and plasticity, *International Journal for Numerical Methods in Engineering* **22**(1): 183–188. {*121, 189, 423*}

Kozlov, S. M., and Piatnitski, A. L. (1996). Degeneration of effective diffusion in the presence of periodic potential, *Annales de l'Institut Henri Poincaré. Section B, Probabilités et Statistiques* **32**(5): 571–587. {*47*}

Krasnosel'skiĭ, M. A., and Rutickiĭ, J. B. (1961). *Convex functions and Orlicz spaces*, P. Noordhoff Ltd., Groningen. Translated from the first Russian edition by Leo F. Boron. {*8*}

Krog, L. A., and Olhoff, N. (1995). Topology optimization of plate and shell structures with multiple eigenfrequencies., in N. Olhoff, and G. I. N. Rozvany (eds), *Structural and Multidisciplinary Optimization*, Pergamon, Oxford, UK, 1995, Goslar, Germany, pp. 675–682. {*477*}

Kuttler, J. R. (1982). Bounds for Stekloff eigenvalues, *SIAM Journal on Numerical Analysis* **19**(1): 121–125. {*477*}

Landau, L. D., and Lifshitz, E. M. (1984). *Electrodynamics of continuous media*, Pergamon, New York. {*36, 341, 343, 345*}

Larsen, U. D., Sigmund, O., and Bouwstra, S. (1997). Design and fabrication of compliant mechanisms and material structures with negative Poisson's ratio, *Journal of Microelectromechanical Systems* **6**(2): 99–106. {*388*}

Lavrent'ev, M. A. (1989). *Variational methods for boundary value problems for systems of elliptic equations*, second edn, Dover Publications, Inc., New York. Translated from the Russian by J. R. M. Radok. {*10*}

Lavrov, N., Lurie, K. A., and Cherkaev, A. V. (1980). Non-homogeneous bar of extremal rigidity in torsion (in russian), *Izvestiia Akademii Nauk SSSR. Mekhanika Tverdogo Tela* **6**. {*81, 86, 89, 92, 110*}

Leitmann, G. (1981). *The calculus of variations and optimal control*, Plenum Press, New York; London. An introduction, Posebna Izdanja [Special Editions], 14. {*10*}

Levin, V. M. (1999). The elastic wave propagation in the piezoelectric fiber reinforced composites, in K. Markov, and E. Inan (eds), *Continuum models and discrete systems*, World Scientific Publishing Co. Inc., River Edge, NJ. {*47*}

Lipton, R. (1988). On the effective elasticity of a two-dimensional homogenised incompressible elastic composite, *Proceedings of the Royal Society of Edinburgh. Section A, Mathematical and Physical Sciences* **110**(1-2): 45–61. {*393, 423*}

Lipton, R. (1992). Bounds and perturbation series for incompressible elastic composites with transverse isotropic symmetry, *Journal of Elasticity* **27**(3): 193–225. {*189*}

Lipton, R. (1993). On implementation of partial relaxation for optimal structural compliance problems, in P. Pedersen (ed.), *Optimal Design with Advanced Materials. Lyngby, Denmark*, IUTAM, Elsevier, Amsterdam, The Netherlands, pp. 451–468. {*393*}

Lipton, R. (1994a). On optimal reinforcement of plates and choice of design parameters, *Control and Cybernetics* **23**(3): 481–493. {*383*}

Lipton, R. (1994b). Optimal bounds on effective elastic tensors for orthotropic composites, *Proceedings of the Royal Society of London. Series A, Mathematical and Physical Sciences* **444**(1921): 399–410. {*393*}

Lipton, R. (1998). The second Stekloff eigenvalue and energy dissipation inequalities for functionals with surface energy, *SIAM Journal on Mathematical Analysis* **29**(3): 673–680 (electronic). {*145*}

Lipton, R., and Díaz, A. R. (1995). Moment formulations for optimum layout in 3D elasticity., in N. Olhoff, and G. I. N. Rozvany (eds), *Structural and Multidisciplinary Optimization, Goslar, Germany*, Pergamon, New York, pp. 161–168. Journal version in (Díaz and Lipton, 1997). {*189, 502*}

Lurie, A. I. (1970a). *Theory of elasticity*, Nauka, Moscow, Russia. In Russian. {*357, 358*}

Lurie, A. I. (1990a). *Nonlinear theory of elasticity*, North-Holland Publishing Co., Amsterdam, The Netherlands. Translated from the Russian by K. A. Lurie. {*357*}

Lurie, K. A. (1963). The Mayer-Bolza problem for multiple integrals and optimization of the behavior of systems with distributed parameters, *Prikladnaia Matematika i Mekhanika (PMM)= Applied Mathematics and Mechanics* **27**(2): 842–853. In Russian. {*92, 122, 238*}

Lurie, K. A. (1967). The Mayer-Bolza problem for multiple integrals: Some optimum problems for elliptic differential equations arriving in magnetohydrodynamics, in G. Leitmann (ed.), *Topics in Optimization*, Academic Press, New York, pp. 147–193. {*92, 118, 122, 238*}

Lurie, K. A. (1970b). On the optimal distribution of the resistivity tensor of the working medium in the channel of a MHD generator, *Prikladnaia Matematika i Mekhanika (PMM)= Applied Mathematics and Mechanics* **34**(2): 67–81. {*122, 238*}

Lurie, K. A. (1975). *Optimal control in problems of mathematical physics*, Nauka, Moscow, Russia. In Russian. {*81, 92, 94, 238*}

Lurie, K. A. (1990b). The extension of optimisation problems containing controls in the coefficients, *Proceedings of the Royal Society of Edinburgh. Section A, Mathematical and Physical Sciences* **114**(1-2): 81–97. {*122, 133, 158*}

Lurie, K. A. (1993). *Applied Optimal Control Theory of Distributed Systems*, Plenum Press, New York; London. {*238*}

Lurie, K. A. (1994). Direct relaxation of optimal layout problems for plates, *Journal of Optimization Theory and Applications* **80**(1): 93–116. {*122, 133, 158*}

Lurie, K. A., and Cherkaev, A. V. (1978). Nonhomogeneous bar of extremal torsional rigidity, *Nonlinear problems in structural mechanics. Structural optimization*, Naukova Dumka, Kiev, Ukraine, pp. 64–68. {*81, 85, 89, 110, 121*}

Lurie, K. A., and Cherkaev, A. V. (1981a). G-closure of a set of anisotropic conducting media in the case of two dimensions, *Doklady Akademii Nauk SSSR* **259**(2): 328–331. In Russian. {*67, 270, 272, 290*}

Lurie, K. A., and Cherkaev, A. V. (1981b). G-closure of a set of anisotropically conducting media in the two-dimensional case, *Report 213*, DCAMM, Technical University of Denmark. Published later as (?). {*270, 272, 290*}

Lurie, K. A., and Cherkaev, A. V. (1981c). G-closure of some particular sets of admissible material characteristics for the problem of bending of thin elastic plates, *Report 214*, DCAMM, Technical University of Denmark. Published later as (?). {*67, 213, 369, 420–422, 447, 448*}

Lurie, K. A., and Cherkaev, A. V. (1982). Exact estimates of conductivity of mixtures composed of two isotropical media taken in prescribed proportion, *Technical Report Ref. 783*, Physical-Technical Institute, Acad. Sci. USSR. In Russian. Published in English as (?). {*189, 192, 213, 278*}

Lurie, K. A., and Cherkaev, A. V. (1984a). Exact estimates of conductivity of a binary mixture of isotropic compounds, *Report 894*, Ioffe Physico-technical Institute, Acad of Sc, USSR. In Russian. Published in English as (?). {*192, 193, 213, 282, 285*}

Lurie, K. A., and Cherkaev, A. V. (1984b). Exact estimates of conductivity of composites formed by two isotropically conducting media taken in prescribed proportion, *Proceedings of the Royal Society of Edinburgh. Section A, Mathematical and Physical Sciences* **99**(1-2): 71–87. Published earlier as (Lurie and Cherkaev, 1982) (in Russian). {*512*}

Lurie, K. A., and Cherkaev, A. V. (1984c). *G*-closure of a set of anisotropically conducting media in the two-dimensional case, *Journal of Optimization Theory and Applications* **42**(2): 283–304. First published as (Lurie and Cherkaev, 1981b). See errata (?). {*511*}

Lurie, K. A., and Cherkaev, A. V. (1984d). *G*-closure of some particular sets of admissible material characteristics for the problem of bending of thin elastic plates, *Journal of Optimization Theory and Applications* **42**(2): 305–316. Published first as (Lurie and Cherkaev, 1981c). See errata (?). {*512*}

Lurie, K. A., and Cherkaev, A. V. (1984e). G-closure of some particular sets of admissible material characteristics for the problem of bending of thin plates, *Journal of Optimization Theory and Applications* **42**: 305–316. {*362*}

Lurie, K. A., and Cherkaev, A. V. (1985). Optimization of properties of multicomponent isotropic composites, *Journal of Optimization Theory and Applications* **46**(4): 571–580. {*183, 187, 307, 312, 315, 316, 339*}

Lurie, K. A., and Cherkaev, A. V. (1986a). Effective characteristics of composite materials and the optimal design of structural elements. (russian), *Uspekhi Mekhaniki = Advances in Mechanics* **9**: 3–81. English translation in (?). {*67, 81, 121, 135, 214, 263, 268, 270, 272, 275*}

Lurie, K. A., and Cherkaev, A. V. (1986b). Exact estimates of the conductivity of a binary mixture of isotropic materials, *Proceedings of the Royal Society of Edinburgh. Section A, Mathematical and Physical Sciences* **104**(1-2): 21–38. Published earlier as (Lurie and Cherkaev, 1984a). {*512*}

Lurie, K. A., and Cherkaev, A. V. (1988). On a certain variational problem of phase equilibrium, in J. M. Ball (ed.), *Material instabilities in continuum mechanics (Edinburgh, 1985–1986)*, Oxford Sci. Publ., Oxford Univ. Press, New York, pp. 257–268. {*214, 319*}

Lurie, K. A., and Cherkaev, A. V. (1997). Effective characteristics of composite materials and the optimal design of structural elements [partial translation of MR 88e:73019], in A. V. Cherkaev, and R. V. Kohn (eds), *Topics in the mathematical modelling of composite materials*, Vol. 31 of *Progress in nonlinear differential equations and their applications*, Birkhäuser, Boston, MA, pp. 175–258. Translation. Originally published as (Lurie and Cherkaev, 1986). {*512*}

Lurie, K. A., Cherkaev, A. V., and Fedorov, A. V. (1982). Regularization of optimal design problems for bars and plates. I, II, *Journal of Optimization Theory and Applications* **37**(4): 499–522, 523–543. The first and the second parts were separately published as (Lurie et al., 1980b) and (Lurie et al., 1980a). {*92, 121, 183, 189, 219, 238, 362, 512*}

Lurie, K. A., Cherkaev, A. V., and Fedorov, A. V. (1984). On the existence of solutions to some problems of optimal design for bars and plates, *Journal of Optimization Theory and Applications* **42**(2): 247–281. {*121*}

Lurie, K. A., Fedorov, A. V., and Cherkaev, A. V. (1980a). On the existence of solutions of certain optimal design problems for bars and plates, *Report 668*, Ioffe Physico-technical Institute, Acad of Sc, USSR, Leningrad, USSR. In Russian. English translation in (Lurie et al., 1982), part II. {*81, 213, 512*}

Lurie, K. A., Fedorov, A. V., and Cherkaev, A. V. (1980b). Relaxation of optimal design problems for bars and plates and eliminating of contradictions in the necessary conditions of optimality, *Report 667*, Ioffe Physico-technical Institute, Acad of Sc, USSR, Leningrad, USSR. In Russian. English translation in (Lurie et al., 1982), part I. {*81, 94, 512*}

Marcellini, P. (1985). Approximation of quasiconvex functions, and lower semicontinuity of multiple integrals, *Manuscripta Mathematica* **51**(1-3): 1–28. {*158*}

Marino, A., and Spagnolo, S. (1969). Un tipo di approssimazione dell'operatore $\sum_1^n ij D_i(a_{ij}(x)D_j)$ con operatori $\sum_1^n j D_j(\beta(x)D_j)$, *Ann. Scuola Norm. Sup. Pisa (3)* **23**: 657–673. {*63*}

Markenscoff, X. (1997/98). On the shape of the Eshelby inclusions, *Journal of Elasticity* **49**(2): 163–166. {*410*}

Markenscoff, X., and Paukshto, M. (1998). The wedge paradox and a correspondence principle, *Proceedings of the Royal Society of London. Series A, Mathematical and Physical Sciences* **454**(1968): 147–154. {*415*}

Markov, K., and Inan, E. (eds) (1999). *Proceedings of the Ninth International Symposium on Continuum Models and Discrete Systems: 29 June–3 July 1998, Istanbul, Turkey*, World Scientific Publishing Co., Singapore; Philadelphia; River Edge, NJ. {*47, 63*}

Markov, K., and Preziosi, L. (eds) (1999). *Heterogeneous Media: Micromechanics Modelling Methods and Simulation*, Birkhäuser Verlag, Basel, Switzerland. {*47, 63*}

Matos, J. P. (1992). Young measures and the absence of fine microstructures in a class of phase transitions, *European Journal of Applied Mathematics* **3**(1): 31–54. {*164*}

Mendelson, K. S. (1975). A theorem on the conductivity of two-dimensional heterogeneous medium, *Journal of Applied Physics* **46**: 4740–4741. {*302*}

Mikhlin, S. G. (1964). *Variational methods in mathematical physics*, The Macmillan Co., New York. Translated by T. Boddington; editorial introduction by L. I. G. Chambers. A Pergamon Press Book. {*10*}

Milgrom, M. (1997). Some more exact results concerning multifield moduli of two-phase composites, *Journal of the Mechanics and Physics of Solids* **45**(3): 399–404. {*294, 303*}

Milgrom, M., and Shtrikman, S. (1989). Linear response of polycrystals to coupled fields: Exact relations among the coefficients, *Physical Review B (Solid State)* **40**(9): 5991–5994. {*72, 294, 303*}

Milton, G. W. (1980). Bounds on complex dielectric constant of a composite material, *Applied Physics Letters* **37**(3): 300–302. {*56, 286, 294, 308, 341*}

Milton, G. W. (1981a). Bounds on the complex permittivity of a two-component composite material, *Journal of Applied Physics* **52**: 5286–5293. {*183, 198, 200, 294, 341*}

Milton, G. W. (1981b). Bounds on the electromagnetic, elastic and other properties of two-component composites, *Physical Review Letters* **46**(8): 542–545. {*281, 294, 428*}

Milton, G. W. (1981c). Bounds on the transport and optical properties of a two-component composite material, *Journal of Applied Physics* **52**: 5294–5304. {*198, 200, 294, 302*}

Milton, G. W. (1981d). Concerning bounds on transport and mechanical properties of multicomponent composite materials, *Applied Physics* **A26**: 125–130. {*307, 312, 315, 316, 339*}

Milton, G. W. (1986). Modelling the properties of composites by laminates, in J. Ericksen et al. (eds), *Homogenization and Effective Moduli of Materials and Media (Minneapolis, Minn., 1984/1985)*, Springer-Verlag, Berlin; Heidelberg; London; etc., pp. 150–175. {*189, 275*}

Milton, G. W. (1987). Multicomponent composites, electrical networks and new types of continued fraction. I, *Communications in Mathematical Physics* **111**(2): 281–327. {*307, 339*}

Milton, G. W. (1990a). A brief review of the translation method for bounding effective elastic tensors of composites, *Continuum models and discrete systems, Vol. 1 (Dijon, 1989)*, Longman Sci. Tech., Harlow, pp. 60–74. {*214*}

Milton, G. W. (1990b). On characterizing the set of possible effective tensors of composites: the variational method and the translation method, *Communications on Pure and Applied Mathematics (New York)* **43**(1): 63–125. {*195, 214, 228, 234, 423, 424*}

Milton, G. W. (1991a). The field equation recursion method, *Composite media and homogenization theory (Trieste, 1990)*, Birkhäuser Boston, Boston, MA, pp. 223–245. {*193*}

Milton, G. W. (1991b). A proof that laminates generates all possible effective conductivity functions of two dimensional two-phase materials, *Advances in Multiphase Flow and Related Problems*, Vol. 5 of *Progress in Non-linear Differential Equations and Their Applications*, SIAM Press, Philadelphia, PA, pp. 136–146. {*189, 198*}

Milton, G. W. (1992). Composite materials with Poisson's ratios close to −1, *Journal of the Mechanics and Physics of Solids* **40**(5): 1105–1137. {*47, 201, 369, 387, 388, 475*}

Milton, G. W. (1994). A link between sets of tensors stable under lamination and quasiconvexity, *Communications on Pure and Applied Mathematics (New York)* **47**(7): 959–1003. {*183, 189, 268, 270*}

Milton, G. W. (2000). *Effective properties of composites*, Cambridge University Press, Cambridge, UK. in preparation. {*176, 219, 275*}

Milton, G. W., and Berryman, J. G. (1997). On the effective viscoelastic moduli of two-phase media. II. Rigorous bounds on the complex shear modulus in three dimensions, *Proceedings of the Royal Society of London. Series A, Mathematical and Physical Sciences* **453**(1964): 1849–1880. {*341, 351, 428*}

Milton, G. W., and Cherkaev, A. V. (1995). What elasticity tensors are realizable?, *Journal of Engineering Materials and Technology* **117**(4): 483–493. {*202, 204, 385, 388, 475*}

Milton, G. W., and Golden, K. (1985). Thermal conduction in composites, *Thermal Conductivity*, Vol. 18, Plenum Press, New York; London, pp. 571–582. {*193*}

Milton, G. W., and Golden, K. (1990). Representations for the conductivity functions of multicomponent composites, *Communications on Pure and Applied Mathematics (New York)* **43**(5): 647–671. {*195, 307, 341*}

Milton, G. W., and Kohn, R. V. (1988). Variational bounds on the effective moduli of anisotropic composites, *Journal of the Mechanics and Physics of Solids* **36**(6): 597–629. {*307, 314, 315, 336, 381*}

Milton, G. W., and Movchan, A. B. (1995). A correspondence between plane elasticity and the two-dimensional real and complex dielectric equations in anisotropic media, *Proceedings of the Royal Society of London. Series A, Mathematical and Physical Sciences* **450**(1939): 293–317. {*219, 369*}

Milton, G. W., and Movchan, A. B. (1998). Mapping certain planar elasticity problems to antiplane ones, *European Journal of Mechanics, A, Solids.* **17**(1): 1–11. {*219*}

Milton, G. W., and Nesi, V. (1999). Optimal G-closure bounds via stability under lamination, *Archive for Rational Mechanics and Analysis* . in print. {*307, 338*}

Milton, G. W., and Serkov, S. K. (1999). Bounding the current in nonlinear conducting composites, *Applied Mechanics Reviews* . submitted. {*63, 76*}

Morrey, C. B. (1952). Quasi-convexity and the lower semicontinuity of multiple integrals, *Pacific Journal of Mathematics* **2**: 25–53. {*158, 159, 166, 214*}

Morrey, C. B. (1966). *Multiple integrals in the calculus of variations*, Springer-Verlag, New York. Die Grundlehren der mathematischen Wissenschaften, Band 130. {*217*}

Movchan, A. B., and Movchan, N. V. (1995). *Mathematical modelling of solids with nonregular boundaries*, CRC Press, 2000 N.W. Corporate Blvd., Boca Raton, FL 33431-9868. {*240*}

Müller, S., and Fonseca, I. (1998). A-quasiconvexity, lower semicontinuity and Young measures, *Preprint*, Max-Planck-Institut für Mathematik in der Naturwissenshaften, Leipzig, Germany. {*158*}

Murat, F. (1972). Théorèmes de non-existence pour des problèmes de contrôle dans les coefficients, *Comptes Rendus Hebdomadaires des Séances de l'Académie des Sciences. Séries A et B* **274**: A395–A398. {*122*}

Murat, F. (1977). Contre-examples pour divers problèmes où le contrôle intervient dans les coefficients, *Annali di Matematica Pura ed Applicata. Series 4* pp. 49–68. {*81, 122, 214*}

Murat, F. (1981). Compacité par compensation: condition nécessaire et suffisante de continuité faible sous une hypothèse de rang constant, *Annali della Scuola Normale Superiore di Pisa, Classe di Scienze* 8(1): 69–102. {*133, 150, 220*}

Murat, F., and Tartar, L. (1985a). Calcul des variations et homogénéisation, **57**: 319–369. English translation as (Murat and Tartar, 1997). {*81, 158, 192, 214, 222, 278, 515*}

Murat, F., and Tartar, L. (1985b). Optimality conditions and homogenization, *Nonlinear variational problems (Isola d'Elba, 1983)*, Pitman Publishing Ltd., London, pp. 1–8. {*121, 158*}

Murat, F., and Tartar, L. (1997). Calculus of variations and homogenization, in A. V. Cherkaev, and R. V. Kohn (eds), *Topics in the mathematical modeling of composite materials*, Vol. 31 of *Progress in nonlinear differential equations and their applications*, Birkhäuser, Boston, pp. 139–173. Translation. The original publication in (Murat and Tartar, 1985a). {*108, 515*}

Muskhelishvili, N. I. (1943). *Some basic problems of the mathematical theory of elasticity*, Noordhoff, Groningen. {*412, 415*}

Muskhelishvili, N. I. (1953). *Some basic problems of the mathematical theory of elasticity*, P. Noordhoff Ltd., Groningen-Holland. {*370*}

Nemat-Nasser, S., and Hori, M. (1993). *Micromechanics: Overall properties of heterogeneous materials*, North-Holland Publishing Co., Amsterdam, The Netherlands. {*47, 63, 423*}

Nesi, V. (1995). Bounds on the effective conductivity of two-dimensional composites made of $n \geq 3$ isotropic phases in prescribed volume fraction: the weighted translation method, *Proceedings of the Royal Society of Edinburgh. Section A, Mathematical and Physical Sciences* **125**(6): 1219–1239. {*307, 317*}

Nesi, V., and Milton, G. W. (1991). Polycrystalline configurations that maximize electrical resistivity, *Journal of the Mechanics and Physics of Solids* **39**(4): 525–542. {*200, 292, 321, 323, 338*}

Norris, A. N. (1985). A differential scheme for the effective moduli of composites, *Mechanics of Materials: An International Journal* **4**: 1–16. {*183*}

Norris, A. N. (1999). Stress invariance and redundant moduli in three-dimensional elasticity, *Proceedings of the Royal Society of London. Series A, Mathematical and Physical Sciences* . In print. {*369*}

Oleĭnik, O. A., Yosifian, G. A., and Temam, R. (1995). Some nonlinear homogenization problems, *Applicable Analysis* **57**(1-2): 101–118. {*63*}

Olhoff, N. (1974). Optimal design of vibrating rectangular plates, *International Journal of Solids and Structures* **10**: 93–109. {*477*}

Olhoff, N., and Rasmussen, S. H. (1977). On bimodal optimum loads of clamped columns, *International Journal of Solids and Structures* **13**: 605–514. {*484*}

Olhoff, N., Scheel, J., and Rønholt, E. (1998). Topology optimization of three-dimensional structures using optimum microstructures, *Structural Optimization* **16**: 1–18. {*403*}

Olhoff, N., and Taylor, J. E. (1979). On optimal structural remodeling, *Journal of Optimization Theory and Applications* **27**(4): 571–582. {*xxiii, 402*}

Panasenko, G. P. (1994). Homogenization of lattice-like domains: L-convergence, *Report CNRS URA 740, 178*, Equipe d'Analyse Numérique, Lyon/Saint-Etienne. {*47*}

Papalambros, P. Y., and Wilde, D. J. (1988). *Principles of optimal design*, Cambridge University Press, Cambridge, UK. Modeling and computation. {*xxiii, 459*}

Parry, G. P. (1995). On the planar rank-one convexity condition, *Proceedings of the Royal Society of Edinburgh. Section A, Mathematical and Physical Sciences* **125**(2): 247–264. {*155*}

Parton, V. Z., and Perlin, P. I. (1984a). *Mathematical methods of the theory of elasticity. Vol. 1*, "Mir", Moscow. Translated from the Russian by Ram S. Wadhwa. {*357*}

Parton, V. Z., and Perlin, P. I. (1984b). *Mathematical methods of the theory of elasticity. Vol. 2*, "Mir", Moscow. Translated from the Russian by Ram S. Wadhwa. {*357*}

Pedersen, P. (1989). On optimal orientation of orthotropic materials, *Structural Optimization* **1**: 101–106. {*447*}

Pedersen, P. (1990). Bounds on elastic energy in solids of orthotropic materials, *Structural Optimization* **2**: 55–63. {*447*}

Pedersen, P. (1991). On thickness and orientational design with orthotropic materials, *Structural Optimization* **3**: 69–78. {*447*}

Pedersen, P., and Taylor, J. E. (1993). Optimal design based on power-law nonlinear elasticity, in P. Pedersen (ed.), *Optimal Design with Advanced Materials*, Elsevier, Amsterdam, The Netherlands, pp. 51–66. {*xxiii*}

Pedregal, P. (1997). *Parametrized measures and variational principles*, Birkhäuser Verlag, Basel, Switzerland. {*64, 118, 145, 155, 158, 164, 165*}

Pedregal, P., and Šverák, V. (1998). A note on quasiconvexity and rank-one convexity for 2 × 2 matrices, *Journal of Convex Analysis* **5**(1): 107–117. {*155, 217*}

Petukhov, L. V. (1989). Optimal elastic regions of maximum rigidity, *Prikladnaia Matematika i Mekhanika (PMM)= Applied Mathematics and Mechanics* **53**(1): 88–95. {*393*}

Petukhov, L. V. (1995). Necessary Weierstrass conditions for elliptic systems, *Prikladnaia Matematika i Mekhanika (PMM)= Applied Mathematics and Mechanics* **59**(5): 742–749. {*238*}

Pólya, G., and Szegö, G. (1951). *Isoperimetric Inequalities in Mathematical Physics*, Vol. 27 of *Annals of Mathematics Studies*, Princeton University Press, Princeton, NJ. {*10*}

Ponte Castañeda, P. (1992a). Bounds and estimates for the properties of nonlinear heterogeneous systems, *Philosophical Transactions of the Royal Society of London Series A* **340**(1659): 531–567. {*xxiii*}

Ponte Castañeda, P. (1992b). A new variational principle and its application to nonlinear heterogeneous systems, *SIAM Journal on Applied Mathematics* **52**(5): 1321–1341. {*xxiii*}

Ponte Castañeda, P. (1996). Exact second-order estimates for the effective mechanical properties of nonlinear composite materials, *Journal of the Mechanics and Physics of Solids* **44**(6): 827–862. {*xxiii, 63*}

Ponte Castañeda, P. (1997). Nonlinear composite materials: effective constitutive behavior and microstructure evolution, *Continuum micromechanics*, Springer-Verlag, Vienna, pp. 131–195. {*xxiii, 63*}

Ponte Castañeda, P., and Willis, J. R. (1988). On the overall properties of nonlinearly viscous composites, *Proceedings of the Royal Society of London. Series A, Mathematical and Physical Sciences* **416**(1850): 217–244. {*xxiii, 63*}

Pontryagin, L. S., Boltyanskii, V. G., Gamkrelidze, R. V., and Mishchenko, E. F. (1964). *The mathematical theory of optimal processes*, A Pergamon Press Book. The Macmillan Co., New York. Translated by D. E. Brown. {*92, 238*}

Prager, W. (1968). Optimality criteria in structural design, *Proceedings of the National Academy of Sciences of the United States of America* **61**(3): 794–796. {*411, 412*}

Prager, W. (1974). *Introduction to structural optimization*, Springer-Verlag, Vienna. Course held at the Department of Mechanics of Solids, International Centre for Mechanical Sciences, Udine, October, 1974, International Centre for Mechanical Sciences, Courses and Lectures, No. 212. {*459*}

Raĭtum, U. Ė. (1978). Extension of extremal problems connected with linear elliptic equations, *Doklady Akademii Nauk SSSR* **243**(2): 281–283. {*67, 72, 122*}

Raĭtum, U. Ė. (1979). On optimal control problems for linear elliptic equations, *Doklady Akademii Nauk SSSR* **244**(4): 828–830. {*121*}

Raĭtum, U. Ė. (1989). *Optimal control problems for elliptic equations*, "Zinatne", Riga. In Russian. {*76*}

Raĭtum, U. Ė. (1999). On the local representation of G-closure, *Technical Report 206*, Department of mathematics, University of Jyväskylä, Jyväskylä, Finland. {*64*}

Reshetnyak, Y. G. (1967). General theorems on semi-continuity and quasiconvexity of functionals, *Sibirskii Matematicheskii Zhurnal* **8**: 801–816. {*158, 217*}

Reuss, A. (1929). Berechnung der Fliessgrenze von Mischkristallen auf Grund Plastizitätsbedingung für Einkristalle, *Zeitschrift für Angewandte Mathematik und Mechanik* **9**(1): 49–58. {*62*}

Rockafellar, R. (1966). Extension of Fenchel's duality theorem for convex functions, *Duke Mathematical Journal* **33**: 81–89. {*30*}

Rockafellar, R. (1967). Duality and stability in extremum problems involving convex functions, *Pacific Journal of Mathematics* **21**: 167–187. {*28, 350*}

Rockafellar, R. (1997). *Convex analysis*, Princeton University Press, Princeton, NJ. Reprint of the 1970 original, Princeton Paperbacks. {*8, 16, 17, 28, 31, 103, 350*}

Rodrigues, H. C., Soto, C. A., and Taylor, J. E. (1999). A design model to predict optimal two-material composite structure, *Structural Optimization* **17**: 186–198. {*459, 464*}

Rosakis, P., and Simpson, H. C. (1994/95). On the relation between polyconvexity and rank-one convexity in nonlinear elasticity, *Journal of Elasticity* **37**(2): 113–137. {*155*}

Roubíček, T. (1997). *Relaxation in optimization theory and variational calculus*, Walter de Gruyter & Co., Berlin. {*145*}

Rozonoèr, L. I. (1959). L. S. Pontryagin maximum principle in the theory of optimum systems. I, II, III, *Automatic and Remote Control* **20**: 1288–1302, 1405–1421, 1517–1532. {*92, 238*}

Rozvany, G. I. N. (1989). *Structural design via optimality criteria*, Kluwer Academic Publishers Group, Dordrecht. Boston. London. The Prager approach to structural optimization. {*xxiii, 459*}

Rozvany, G. I. N., Bendsøe, M. P., and Kirsch, U. (1995). Layout optimization of structures, *Applied Mechanics Reviews* **48**(2): 41–119. {*447, 459, 464*}

Rozvany, G. I. N., Olhoff, N., Bendsøe, M. P. et al. (1987). Least weight design of perforated elastic plates, *International Journal of Solids and Structures* **23**: 521–536 (part I) and 537–550 (part II). {*464*}

Rykhlevskiĭ, Y. (1988). Symmetry of tensor functions and a spectral theorem, *Uspekhi Mekhaniki = Advances in Mechanics* **11**(3): 77–125. {*368*}

Ryzhik, L., Papanicolaou, G., and Keller, J. (1996). Transport equations for elastic and other waves in random media, *Wave Motion* **24**(4): 327–370. {*47*}

Sánchez-Palencia, E. (1980). *Nonhomogeneous media and vibration theory*, Springer-Verlag, Berlin. {*xxiii, 45*}

Sawicz, R., and Golden, K. (1995). Bounds on the complex permittivity of matrix-particle composites, *Journal of Applied Physics* **78**: 7240–7246. {*341*}

Schulgasser, K. (1976). Relationship between single-crystal and polycrystal electrical conductivity, *Journal of Applied Physics* **47**: 1880–1886. {*xxiii, 183, 292, 339*}

Schulgasser, K. (1977). Bounds on the conductivity of statistically isotropic polycrystals, *Journal of Physics C: Solid State Physics* **10**: 407–417. {*183, 292, 312*}

Seyranian, A. P. (1987). Multiple eigenvalues in optimization problems, *Prikladnaia Matematika i Mekhanika (PMM)= Applied Mathematics and Mechanics* **51**: 272–275. {*484*}

Seyranian, A. P., Lund, E., and Olhoff, N. (1994). Multiple eigenvalues in structural optimization problems, *Structural Optimization* **8**(4): 207–227. {*477, 484*}

Shilov, G. E. (1996). *Elementary functional analysis*, Dover Publications, Inc., New York. Revised English edition translated from the Russian and edited by Richard A. Silverman, Corrected reprint of the 1974 English translation. {*21, 38*}

Sigmund, O. (1994). Materials with prescribed constitutive parameters: an inverse homogenization problem, *International Journal of Solids and Structures* **31**(17): 2313–2329. {*476*}

Sigmund, O. (1996). Some inverse problems in topology design of materials and mechanisms, *IUTAM Symposium on Optimization of Mechanical Systems (Stuttgart, 1995)*, Kluwer Academic Publishers Group, Dordrecht. Boston. London, pp. 277–284. {*388, 476*}

Sigmund, O. (1997). On the design of compliant mechanisms using topology optimization, *Mechanics of Structures and Machines* **25**(4): 495–526. {*476*}

Sigmund, O. (1998). A new class of extremal composites, *Report 599*, DCAMM, Technical University of Denmark. {*212, 428*}

Sigmund, O., and Petersson, J. (1998). Numerical instabilities in topology optimization: A survey on procedures dealing with checkerboards, mesh-dependencies and local minima, *Structural Optimization* **16**: 68–75. A review. {*463, 464*}

Sigmund, O., and Torquato, S. (1997). Design of materials with extreme thermal expansion using a three-phase topology optimization method, *Journal of the Mechanics and Physics of Solids* **45**(6): 1037–1067. {*47, 388, 476*}

Šilhavý, M. (1997). *The mechanics and thermodynamics of continuous media*, Springer-Verlag, Berlin. {*357*}

Šilhavý, M. (1998). Polar decomposition of rank 1 perturbations in two dimensions, *Mathematics and Mechanics of Solids : MMS* **3**(1): 107–112. {*155*}

Smyshlyaev, V. P., and Willis, J. R. (1999). On the relaxation of a three-well energy, *R. Soc. Lond. Proc. Ser. A Math. Phys. Eng. Sci.* **455**(1983): 779–814. {*307*}

Sokolnikoff, I. S. (1983). *Mathematical theory of elasticity*, second edn, Robert E. Krieger Publishing Co. Inc., Melbourne, Fla. {*110, 357*}

Strang, G. (1976). *Linear algebra and its applications*, Academic Press [Harcourt Brace Jovanovich Publishers], New York. {*303*}

Strang, G. (1986). *Introduction to applied mathematics*, Wellesley-Cambridge Press, Wellesley, MA. {*27*}

Strang, G., and Kohn, R. V. (1988). Optimal design of a two-way conductor, in J. Moreau et al. (eds), *Topics in Nonsmooth Mechanics*, Birkhäuser, pp. 143–155. {*189, 214*}

Suzuki, K., and Kikuchi, N. (1991). A homogenization method for shape and topology optimization, *Computer Methods in Applied Mechanics and Engineering* **93**: 291–318. {*393*}

Šverák, V. (1992a). New examples of quasiconvex functions, *Archive for Rational Mechanics and Analysis* **119**(4): 293–300. {*158, 217*}

Šverák, V. (1992b). Rank-one convexity does not imply quasiconvexity, *Proceedings of the Royal Society of Edinburgh. Section A, Mathematical and Physical Sciences* **120**(1-2): 185–189. {*155, 176*}

Šverák, V. (1993). On optimal shape design, *Journal de Mathématiques Pures et Appliquées* **72**(6): 537–551. {*464*}

Talbot, D. R. S., and Willis, J. R. (1985). Variational principles for inhomogeneous nonlinear media, *IMA Journal of Applied Mathematics* **35**(1): 39–54. {*63*}

Talbot, D. R. S., and Willis, J. R. (1992). Some simple explicit bounds for the overall behaviour of nonlinear composites, *International Journal of Solids and Structures* **29**(14-15): 1981–1987. {*xxiii, 63*}

Talbot, D. R. S., and Willis, J. R. (1995). Upper and lower bounds for the overall properties of a nonlinear elastic composite, *Anisotropy, inhomogeneity and nonlinearity in solid mechanics (Nottingham, 1994)*, Kluwer Academic Publishers Group, Dordrecht. Boston. London, pp. 409–414. {*63*}

Talbot, D. R. S., and Willis, J. R. (1996). Three-point bounds for the overall properties of a nonlinear composite dielectric, *IMA Journal of Applied Mathematics* **57**(1): 41–52. {*xxiii*}

Talbot, D. R. S., and Willis, J. R. (1997). Bounds of third order for the overall response of nonlinear composites, *Journal of the Mechanics and Physics of Solids* **45**(1): 87–111. {*xxiii, 63*}

Talbot, D. R. S., Willis, J. R., and Nesi, V. (1995). On improving the Hashin-Shtrikman bounds for the effective properties of three-phase composite media, *IMA Journal of Applied Mathematics* **54**(1): 97–107. {*xxiii, 63, 317*}

Tartar, L. (1975). Problèmes de contrôle des coefficients dans des équations aux dérivées partielles, **107**: 420–426. {*67, 72, 122, 214*}

Tartar, L. (1979a). Compensated compactness and applications to partial differential equations, *Nonlinear analysis and mechanics: Heriot-Watt Symposium, Vol. IV*, Pitman Publishing Ltd., London, pp. 136–212. {*133, 220*}

Tartar, L. (1979b). Estimation de coefficients homogénéisés, *Computing methods in applied sciences and engineering (Proc. Third Internat. Sympos., Versailles, 1977), I,*, Vol. 704 of *Lecture Notes in Mathematics*, Springer-Verlag, Berlin; Heidelberg; London; etc., pp. 364–373. English translation published in (Tartar, 1997). {*214*}

Tartar, L. (1983). The compensated compactness method applied to systems of conservation laws, *Systems of nonlinear partial differential equations (Oxford, 1982)*, Reidel, Dordrecht, pp. 263–285. {*222*}

Tartar, L. (1985). Estimation fines des coefficients homogénéisés, in P. Kree (ed.), *E. De Giorgi colloquium (Paris, 1983)*, Pitman Publishing Ltd., London, pp. 168–187. {*214, 227, 278, 282, 286*}

Tartar, L. (1986). Oscillations in nonlinear partial differential equations: compensated compactness and homogenization, *Nonlinear systems of partial differential equations in applied mathematics, Part 1 (Santa Fe, NM, 1984)*, American Mathematical Society, Providence, RI, pp. 243–266. {*150*}

Tartar, L. (1987). The appearance of oscillations in optimization problems, *Nonclassical continuum mechanics (Durham, 1986)*, Cambridge University Press, Cambridge, UK, pp. 129–150. {*121*}

Tartar, L. (1989). Nonlocal effects induced by homogenization, *Partial differential equations and the calculus of variations, Vol. II*, Birkhäuser Boston, Boston, MA, pp. 925–938. {*222*}

Tartar, L. (1990). *H*-measures, a new approach for studying homogenisation, oscillations and concentration effects in partial differential equations, *Proceedings of the Royal Society of Edinburgh. Section A, Mathematical and Physical Sciences* **115**(3-4): 193–230. {*64*}

Tartar, L. (1992). On mathematical tools for studying partial differential equations of continuum physics: *H*-measures and Young measures, *Developments in partial differential equations and applications to mathematical physics (Ferrara, 1991)*, Plenum, New York, pp. 201–217. {*164*}

Tartar, L. (1993). Some remarks on separately convex functions, in D. Kinderlehrer, R. James, M. Luskin, and J. L. Ericksen (eds), *Microstructure and phase transition*, Springer-Verlag, New York, pp. 191–204. {*155, 321*}

Tartar, L. (1994). Remarks on optimal design problems, in G. Buttazzo, G. Bouchitte, and P. Suquet (eds), *Calculus of variations, homogenization and continuum mechanics*, Vol. 18 of *Series on advances in mathematics for applied sciences*, World Scientific Publishing Co., Singapore; Philadelphia; River Edge, NJ, pp. 279–296. {*121*}

Tartar, L. (1995). Beyond Young measures, *Meccanica (Milan)* **30**(5): 505–526. Microstructure and phase transitions in solids (Udine, 1994). {*164*}

Tartar, L. (1997). Estimation of homogenized coefficients, in A. V. Cherkaev, and R. V. Kohn (eds), *Topics in the mathematical modeling of composite materials*, Vol. 31 of *Progress in nonlinear differential equations and their applications*, Birkhäuser, Boston, pp. 9–20. {*222, 521*}

Taylor, J. E. (1993). A global extremum principle for the analysis of solids composed of softening material, *International Journal of Solids and Structures* **30**(15): 2057–2069. {*xxiii*}

Taylor, J. E. (1996). Surface motion due to crystalline surface energy gradient flows, *Elliptic and parabolic methods in geometry (Minneapolis, MN, 1994)*, A. K. Peters, Ltd., Wellesley, MA, pp. 145–162. {*145*}

Taylor, J. E. (1998). An energy model for the optimal design of linear continuum structures, *Structural Optimization* **16**: 116–127. {*368, 464*}

Taylor, J. E. (1999). Optimal modification and evolution of elastic continuum structures, in P. Pedersen, and M. P. Bendsøe (eds), *Synthesis in bio solid mechanics*, Solid mechanics and its applications, Kluwer Academic Publishers Group, Dordrecht. Boston. London, pp. 115–128. {*xxiii*}

Telega, J. J. (1990). Variational methods and convex analysis in contact problems and homogenization, *IFTR Reports* **38**: 119–134. {*47*}

Telega, J. J. (1995). Epi - limit on HB and homogenization of heterogeneous plastic plates, *Nonlinear Analysis, Theory, Methods and Applications* **25**: 499–529. {*63*}

Timoshenko, S. P. (1970). *Theory of elasticity*, McGraw-Hill, New York. {*4, 357*}

Torquato, S. (1999). Diffusion-controlled reaction and flows in porous media, in K. Markov, and L. Preziosi (eds), *Heterogeneous Media: Micromechanics Modelling Methods and Simulation*, Birkhäuser Verlag, Basel, Switzerland. {*47, 63*}

Torquato, S., and Rubinstein, J. (1991). Improved bounds on the effective conductivity of high-contrast suspensions, *Journal of Applied Physics* **69**: 7118–7125. {*304*}

Tortorelli, D. A., and Wang, Z. (1993). A systematic approach to shape sensitivity analysis, *International Journal of Solids and Structures* **30**(9): 1181–1212. {*459*}

Truesdell, C. A. (1991). *A first course in rational continuum mechanics. Vol. 1*, second edn, Academic Press, New York. General concepts. {*357*}

Truesdell, C. A., and Toupin, R. (1960). The classical field theories, *Handbuch der Physik, Bd. III/1*, Springer-Verlag, Berlin, pp. 226–793; appendix, pp. 794–858. With an appendix on tensor fields by J. L. Ericksen. {*358, 373*}

Truesdell, C., and Noll, W. (1965). The non-linear field theories of mechanics, *Handbuch der Physik, Band III/3*, Springer-Verlag, Berlin, pp. 1–602. {*357*}

Truskinovsky, L., and Zanzotto, G. (1996). Ericksen's bar revisited: energy wiggles, *Journal of the Mechanics and Physics of Solids* **44**(8): 1371–1408. {*319*}

van Keulen, F., Polynkine, A. A., and Toropov, V. V. (1997). Shape optimization with adaptive mesh refinement: Target error selection strategies, *Engineering Optimization* **28**: 95–125. {*459*}

Vigdergauz, S. B. (1986). Effective elastic parameters of a plate with a regular system of equal-strength holes, *Izvestiia Akademii Nauk SSSR. Mekhanika Tverdogo Tela* **21**(2): 165–169. {*410*}

Vigdergauz, S. B. (1988). Stressed state of an elastic plane with constant-stress holes, *Izvestiia Akademii Nauk SSSR. Mekhanika Tverdogo Tela* **23**(3): 96–99. {*281, 410, 412, 413*}

Vigdergauz, S. B. (1989). Regular structures with extremal elastic properties, *Izvestiia Akademii Nauk SSSR. Mekhanika Tverdogo Tela* **24**(3): 57–63. {*308, 410, 412*}

Vigdergauz, S. B. (1994a). Three-dimensional grained composites of extreme thermal properties, *Journal of the Mechanics and Physics of Solids* **42**(5): 729–740. {*410, 412*}

Vigdergauz, S. B. (1994b). Two-dimensional grained composites of extreme rigidity, *Journal of Applied Mechanics* **61**(2): 390–394. {*410*}

Vigdergauz, S. B. (1996). Rhombic lattice of equi-stress inclusions in an elastic plate, *Quarterly Journal of Mechanics and Applied Mathematics* **49**(4): 565–580. {*410*}

Vigdergauz, S. B., and Cherkaev, A. V. (1986). A hole in a plate, optimal for its biaxial extension-compression, *Prikladnaia Matematika i Mekhanika (PMM)= Applied Mathematics and Mechanics* **50**(3): 524–528. {*410, 415*}

Voigt, W. (1928). *Lehrbuch der Kristallphysik*, Teubner, Stuttgart; Leipzig. {*xxiii, 62*}

von Neumann, J., and Morgenstern, O. (1980). *Theory of games and economic behavior*, third edn, Princeton University Press, Princeton, NJ. {*106*}

Walpole, L. (1966). On bounds for the overall elastic moduli of inhomogeneous systems I and II, *Journal of the Mechanics and Physics of Solids* **14**: 151–162, 289–301. {*425*}

Warga, J. (1972). *Optimal control of differential and functional equations*, Academic Press, New York. {*14*}

Weinstock, R. (1974). *Calculus of variations*, Dover Publications, Inc., New York. With applications to physics and engineering, Reprint of the 1952 edition. {*xxv, 10*}

Wiener, W. (1912). Die Theorie des Mischkörpers für das Feld des stationaären Störmung. Erste Abhandlung die Mittelswertsätze für Kraft, Polarization und Energie, *Abhandlungen der Mathematisch Physischen Klasse der Königlischen-Sachsischen Gesellschaft der Wissenschaften* **32**: 509–604. {*xxiii, 62, 67*}

Willis, J. R. (1981). Variational and related methods for the overall properties of composite materials, in C.-S. Yih (ed.), *Advances in Applied Mechanics*, Vol. 21, Academic Press, New York, pp. 2–78. {*xxiii*}

Young, L. C. (1942a). Generalized surfaces in the calculus of variations, *Annals of Mathematics (2)* **43**: 84–103. {*14, 164*}

Young, L. C. (1942b). Generalized surfaces in the calculus of variations. II, *Annals of Mathematics (2)* **43**: 530–544. {*14*}

Young, L. C. (1969). *Lectures on the calculus of variations and optimal control theory*, W. B. Saunders Co., Philadelphia. Foreword by Wendell H. Fleming. {*7, 14*}

Zhang, K. (1997). On non-negative quasiconvex functions with unbounded zero sets, *Proceedings of the Royal Society of Edinburgh. Section A, Mathematical and Physical Sciences* **127**(2): 411–422. {*217*}

Zhikov, V. V. (1986). Estimates for the trace of an averaged matrix, *Matematicheskie Zametki* **40**(2): 226–237, 287. {*214, 280, 314*}

Zhikov, V. V. (1988). Estimates for the trace of an averaged tensor, *Doklady Akademii Nauk SSSR* **299**(4): 796–800. {*381*}

Zhikov, V. V. (1991a). Estimates for an averaged matrix and an averaged tensor, *Russian Mathematical Surveys* **46**(3(279)): 49–109, 239. {*xxiii, 214*}

Zhikov, V. V. (1991b). Estimates for the homogenized matrix and the homogenized tensor, *Russian Mathematical Surveys* **46**(3): 65–136. {*307*}

Zhikov, V. V. (1993). Lavrent'ev phenomenon and homogenization for some variational problems, *Comptes Rendus des Séances de l'Académie des Sciences. Série I. Mathématique* **316**(5): 435–439. {*13, 162*}

Zhikov, V. V. (1995). On the averaging of nonlinear variational problems in punctured domains, *Doklady Akademii Nauk SSSR* **345**(2): 156–160. {*63*}

Zhikov, V. V. (1996). Connectedness and averaging. Examples of fractal conductivity, *Matematicheskii Sbornik / Izdavaemyi Moskovskim Matematicheskim Obshchestvom = Recueil Mathématique de la Société Mathématique de Moscou* **187**(8): 3–40. {*47*}

Zohdi, T. I., Oden, J. T., and Rodin, G. J. (1996). Hierarchical modeling of heterogeneous bodies, *Computer Methods in Applied Mechanics and Engineering* **138**(1-4): 273–298. {*47*}

Author/Editor Index

NOTE: Primary authors are given in **bold** type, and are cross-referenced to secondary authors. Secondary authors are in roman type, and are cross-referenced to primary authors.

Subject Index

Applied Mathematical Sciences

(continued from page ii)

(continued on next page)

Applied Mathematical Sciences

(continued from previous page)